MICROBIAL BIOTECHNOLOGY

MICROBIAL BIOTECHNOLOGY

Progress and Trends

Edited by
Farshad Darvishi Harzevili
Hongzhang Chen

CRC Press is an imprint of the
Taylor & Francis Group, an **informa** business

CRC Press
Taylor & Francis Group
6000 Broken Sound Parkway NW, Suite 300
Boca Raton, FL 33487-2742

First issued in paperback 2017

ISBN-13: 978-1-4822-4520-2 (hbk)
ISBN-13: 978-1-138-74869-9 (pbk)

Visit the Taylor & Francis Web site at
http://www.taylorandfrancis.com

and the CRC Press Web site at
http://www.crcpress.com

Contents

Preface

Microbial biotechnology is a technology based on microbiology. Microbial biotechnology is the use of microorganisms and their derivatives to make or modify specific products or processes. Microbial biotechnology is closely related to applied and industrial microbiology. Microbial biotechnology is sometimes considered synonymous with modern industrial microbiology.

The microorganisms used in industrial processes are natural, laboratory-selected mutants, or genetically engineered strains to obtain an economically valuable product or activity on a commercial and large scale. Natural and mixed microbial strains were used in ancient or traditional industrial microbiology, whereas pure or mutant microbial strains are used in classic industrial microbiology. The use of genetically engineered microbial strains began with modern industrial microbiology or microbial biotechnology.

Microbial biotechnology is an interdisciplinary field, and successful development in this field requires major contributions in a wide range of disciplines, particularly microbiology, biochemistry, genetics, molecular biology, chemistry, biochemical engineering, bioprocess engineering, and so on.

Recently, new methods of metabolic engineering, industrial systems biology, bioinformatics, and X-omics science such as genomics, metagenomics, transcriptomics, proteomics, metabolomics, fluxomics, and even nanobiotechnology, have been used to find and modify microorganisms with industrial capacity and their valuable products.

In general, the production of products in industrial microbiology and microbial biotechnology are typically investigated under upstream processes, fermentation processes, and downstream processes.

Microbial Biotechnology: Progress and Trends covers recent developments in some fields of microbial biotechnology. Chapter 1 reviews microbial biotechnology from its historical roots to its different processes. Chapters 2 through 5 discuss some of the new developments in upstream processes. Chapter 6 considers solid-state fermentation as an interesting field in fermentation processes. Chapters 7 through 12 argue about recent developments in the production of valuable microbial products such as biofuels, organic acids, amino acids, probiotics, healthcare products, and edible biomass. Chapters 13 through 15 discuss important microbial activities such as biofertilizer, biocontrol, biodegradation, and bioremediation.

The book is written in simple and clear text, and we also used many figures and tables to make the book easier to understand. Furthermore, case studies are included at the end of some chapters.

Overall, this book will serve as a suitable reference for students, scientists, and researchers at universities, industries, corporations, and government agencies interested in biotechnology, applied microbiology, bioprocess/fermentation technology, healthcare/

pharmaceutical products, food innovations/food processing, plant agriculture/crop improvement, energy and environment management, and all disciplines related to microbial biotechnology.

Farshad Darvishi Harzevili
Harzevil, Gilan, Iran

Hongzhang Chen
Beijing, China

Acknowledgments

I thank Hongzhang Chen as coeditor, and the experienced authors for their sound and enlightening contributions. I am extremely grateful to Michael Slaughter (acquiring editor) for his continued interest, critical evaluation, constructive criticism, and support.

On behalf of the authors, I would like to thank Michele Smith (editorial assistant), Marsha Pronin (project coordinator), Rachael Panthier (production editor), Scott Shamblin (cover designer) and their respective teams at CRC Press/Taylor & Francis Group for their valuable efforts to develop our manuscript into a high-quality book.

Farshad Darvishi Harzevili
Editor-in-Chief

Editors

Dr. Farshad Darvishi Harzevili earned a BSc in biology at the University of Guilan, Iran. He earned his MSc and PhD in industrial microbiology and microbial biotechnology from the University of Isfahan, Iran. He is currently a faculty member and head of the microbial biotechnology and bioprocess engineering (MBBE) group at the University of Maragheh, Iran. His main interest is in the biotechnological and environmental applications of yeasts, especially the use of agro-industrial wastes and renewable low-cost substrates in the production of biotechnologically valuable products such as microbial enzymes, organic acids, single-cell oils, biofuels, and so forth. He is also interested in the expression of heterologous proteins, metabolic engineering and synthetic biology of yeasts.

Dr. Hongzhang Chen earned his master's degree in microbiology from Shandong University in 1991, and his PhD in biochemical engineering from the Institute of Process Engineering, Beijing, China in 1998. He is currently a member of the Chinese Academy of Sciences and the vice director of the State Key Laboratory of Biochemical Engineering. His research is focused on ecological biochemical engineering and cellulose biotechnology. Currently, his research explores new types of solid-state fermentation techniques, devices and processes, and clean fractionation using steam explosion techniques for efficient pretreatment in cellulose conversion to useful biorenewable materials.

Contributors

Muna M. Abbas
Department of Biological Sciences
Faculty of Sciences
University of Jordan
Amman, Jordan

Siba Prasad Adhikary
Department of Biotechnology
Fakir Mohan University
Odisha, India

Temitope Banjo
Institute of Human Resources and Development
Federal University of Agriculture
Ogun State, Nigeria

R. Baskaran
Department of Ocean Studies and Marine Biology
Pondicherry University
Port Blair, India

Hongzhang Chen
National Key Laboratory of Biochemical Engineering
Institute of Process Engineering
Chinese Academy of Sciences
Beijing, China

Farshad Darvishi Harzevili
Division of Microbiology
Department of Biology
University of Maragheh
Maragheh, Iran

Patrick Fickers
Biotechnology and Bioprocess
Université Libre de Bruxelles
Brussels, Belgium

S. Ganesamoorthy
Department of Biotechnology
Himachal Pradesh University
Shimla, India

and

Synthetic Chemistry
Syngene International Ltd.
Bangalore, India

Armando Hernández García
Department of Microbiology
Swedish University of Agricultural Sciences
Uppsala, Sweden

Rikita Gupta
School of Biotechnology
University of Jammu
Jammu, India

Pankaj Kumar Jain
Biotechnology Laboratory
Department of Biological Science FASC
MITS University
Rajasthan, India

Sarafadeen Olateju Kareem
Department of Microbiology
Federal University of Agriculture
Ogun State, Nigeria

Guanhua Li
National Key Laboratory of
Biochemical Engineering
Institute of Process Engineering
Chinese Academy of Sciences
Beijing, China

Catherine Madzak
INRA
UMR1319 Micalis
Domaine de Vilvert
Jouy-en-Josas, France

Adel M. Mahasneh
Department of Biological Sciences
Faculty of Sciences
University of Jordan
Amman, Jordan

Saad Bin Zafar Mahmood
Aga Khan University Hospital
Stadium Road
Karachi, Pakistan

Zafar Alam Mahmood
Colorcon Limited
Kent, United Kingdom

P.M. Mohan
Department of Chemistry
National Institute of Technology
Tiruchirappalli, India

and

Synthetic Chemistry
Syngene International Ltd.
Bangalore, India

Poonam Singh Nigam
Faculty of Life and Health Sciences
University of Ulster
Coleraine, United Kingdom

Ashok Kumar Nadda
Department of Biotechnology
Himachal Institute of Life Sciences
Paonta Sahib, India

and

En-vision Enviro Engineers Pvt. Ltd.
Surat, Gujarat, India

Jayanti Kumari Sahu
Department of Botany
Dhenkanal Autonomous College
Odisha, India

Santosh Kumar Sethi
Department of Biotechnology
Utkal University
Odisha, India

Sakshi Sharma
School of Biotechnology
University of Jammu
Jammu, India

Andriy Sibirny
Department of Molecular Genetics and
Biotechnology
Institute of Cell Biology
National Academy of Sciences of Ukraine
Lviv, Ukraine

Anoop Singh
Department of Scientific and Industrial
Research
Ministry of Science and Technology
Government of India
New Delhi, India

Jyoti Vakhlu
University of Jammu
Jammu, India

chapter one

Microbial biotechnology

An introduction

Farshad Darvishi Harzevili

Contents

1.1 Introduction

There are many definitions for biotechnology. Simply, biotechnology is a technology based on biology, or the use of living systems and organisms to make or develop useful products. The most comprehensive definition was given by the United Nations Convention on Biological Diversity at a meeting held in Rio de Janeiro, Brazil in 1992. In this convention, biotechnology was defined as "any technological application that uses biological systems, living organisms, or derivatives thereof, to make or modify products or processes for specific use." This definition was signed and accepted by 168 countries up to now.

The biological processes of microorganisms have been used to make and preserve useful food products for more than 6000 years. Microbial biotechnology or industrial microbiology is the use of microorganisms to obtain an economically valuable product or activity at a commercial or large scale. The microorganisms used in industrial processes are natural, laboratory-selected mutant or genetically engineered strains. Economically valuable products such as alcohols, solvents, organic acids, amino acids, enzymes, fermented dairy products, food additives, vitamins, antibiotics, recombinant proteins and hormones, biopolymers, fertilizers, and biopesticides are produced by microorganisms that are used in chemical, food, pharmaceutical, agricultural, and other industries. Biodegradation and biotransformation of complex compounds, domestic and industrial wastewater treatment, biomining, and enhanced oil recovery are examples of microbially valuable activities. According to the UN Convention on Biological Diversity, microbial biotechnology can be defined as any technological application that uses microbiological systems, microbial organisms, or derivatives thereof, to make or modify products or processes for specific use.

1.2 A brief history

Industrial microbiology has a long history and its roots can be traced back to the ancient times of human life, when microbiology had not yet been accepted and developed as a science. Approximately 7000 years BC, the Sumerians and Babylonians used yeast to convert sugar to alcohol. By 4000 years BC, the Egyptians used leaven containing yeast to improve bread quality. Moreover, the ancients knew how to use bacteria and molds for vinegar and cheese production.

The Chinese used molds as antibiotics for the treatment of purulent wounds approximately 500 years BC. This first period, from several thousand years to 150 years ago, is known as the ancient or traditional industrial microbiological period, in which mixed and impure cultures of microorganisms were used in nonsterile conditions to make products (El-Mansi et al. 2012; Soetaert and Vandamme 2010).

Approximately 150 years ago, Louis Pasteur proved the microbial source of fermentation and established industrial microbiology as a science based on scientific principles. During World War I, Chaim Weizmann used *Clostridium* bacterium for the production of acetone and butanol. Then *Aspergillus* mold was used to produce citric acid.

Following Alexander Fleming's discovery of the antibiotic properties of *penicillium* mold, Florey and Chain were able to prepare a pure form of penicillin during the Second World War. In the 1940s, Waksman discovered several aminoglycoside antibiotics such as streptomycin and neomycin. In the late 1950s and 1960s, microorganisms were used to produce amino acids and single-celled proteins, respectively. This second period, 150 years to 40 years ago, is known as the classic industrial microbiological period, in which pure cultures of microorganisms were used in sterile conditions for the manufacture of products. Furthermore, the microbial strains were improved by classic genetic methods such as protoplasm fusion and mutagenesis with physical and chemical mutagens (Glazer and Nikaido 2007).

The third period, which began in the 1970s and continues to the present, is known as the modern industrial microbiology or microbial biotechnology period. The prominent features of this period are its use of recombinant DNA or genetic engineering methods for the improvement of industrial strains and the production of recombinant proteins (Figure 1.1).

1.3 Nature of industrial microbiology and microbial biotechnology

In some sources, the term *biotechnology* was incorrectly substituted for genetic engineering or modification. This mistake originated in the United States where, several years ago, new genetic methods were considered as awful and demonic procedures. Therefore, the term biotechnology was used instead of genetic engineering and producing transgenic organisms to reduce worry and diversion of public opinion. Later, the term was used by the media and politicians, and thus it entered legislation and government documents.

Humans have been using genetic modification with the selective breeding of plants and animals for more productivity over tens of thousands of years. For more than 50 years, classic methods such as protoplasm fusion and mutagenesis have been used for the genetic modification of organisms. Genetic engineering and its equivalent terms, genetic modification or genetic manipulation, are advanced molecular biology techniques that have been used since the 1970s (Smith 2009).

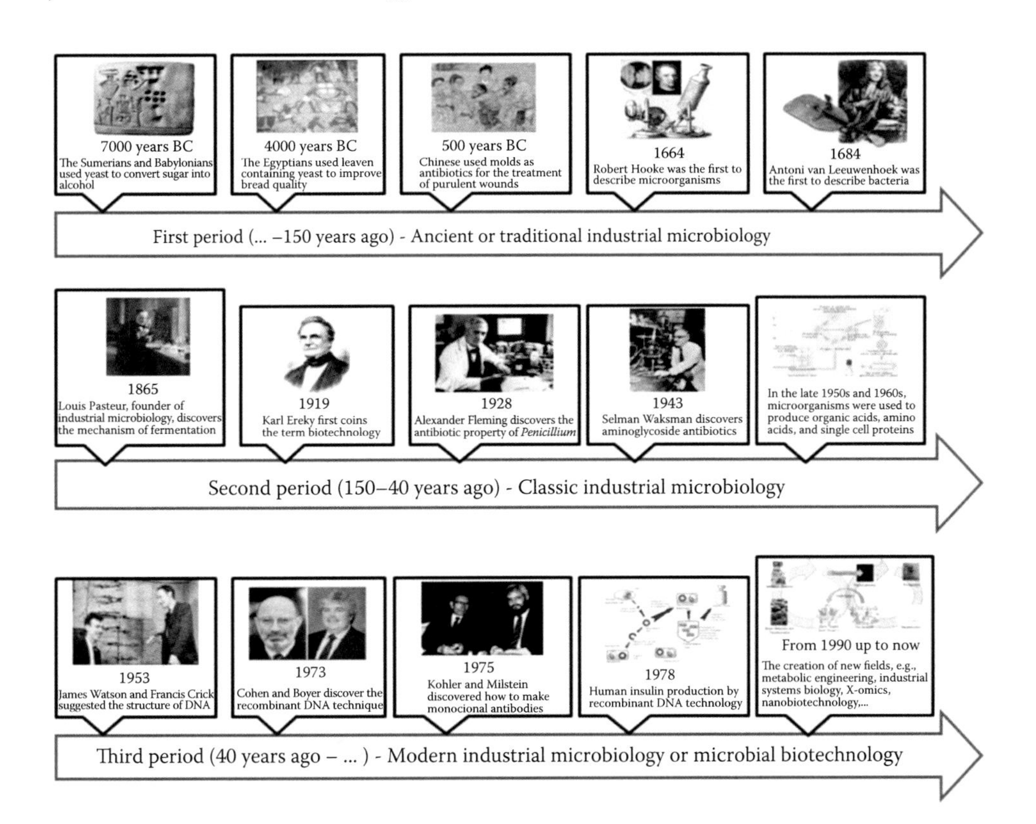

Figure 1.1 Timeline of industrial microbiology and microbial biotechnology.

The term biotechnology was first coined by Karl Ereky in 1917, when molecular genetics and genetic engineering had not yet been discovered. By 1919, in his book entitled *Biotechnology of Meat, Fat and Milk Production in an Agricultural Large-scale Farm*, Ereky described biotechnology as a technology based on converting raw materials into useful products.

Afterward, many definitions were proposed for biotechnology. The application of biological systems, living organisms, or derivatives thereof, to make or modify products is the most comprehensive definition for biotechnology. Biotechnology is not only a science or a set of methods but it is also the interdisciplinary science that encompasses microbiology, plant and animal science, biochemistry, cellular and molecular biology, genetic modification, and engineering fields with biological perspectives such as mechanics, electronics, information technology, robotics, and so on.

The European Federation of Biotechnology considers biotechnology in two categories—"traditional or old" and "new or modern" biotechnology. Several thousand-years–old traditional methods are used to produce beverages, foods, and dairy in traditional or old biotechnology, which is the equivalent of traditional industrial microbiology and classic industrial microbiology. New methods of genetic engineering, which were used from the 1970s to the early 1980s, began the development and evolution of traditional biotechnology to new or modern biotechnology, which is the equivalent of modern industrial microbiology or microbial biotechnology (Smith 2009).

Today, the third wave of biotechnology, known as industrial biotechnology or white biotechnology, is expanding. It has made considerable progress in comparison with the second wave, namely, red biotechnology or medical biotechnology and the first wave, namely, green biotechnology or agricultural biotechnology.

Industrial biotechnology uses biological systems, especially microorganisms, in industrial fermentation processes to produce large quantities of pure materials and energy including alcohols, organic acids, amino acids, vitamins, solvents, antibiotics, biopolymers, biopesticides, enzymes, alkaloids, steroids, and others. Industrial biotechnology is founded on biological catalysts and fermentation technology, and it is associated with advances in molecular genetics, protein engineering, and metabolic engineering of microorganisms and cells.

Recently, new methods of metabolic engineering, industrial systems biology, bioinformatics, X-omics such as genomics, metagenomics, transcriptomics, proteomics, metabolomics, fluxomics, and even nanobiotechnology, have been used to find and modify microorganisms with industrial capacity and their valuable products (Soetaert and Vandamme 2010).

1.4 *Review of main processes in microbial biotechnology*

In the past, many diverse products were derived from natural sources or synthesized through chemical processes; nowadays, some of these products are commercially produced through microbial fermentation and biological conversion processes. The benefits of using microorganisms in such processes includes the ease of mass production, the high growth rate, and the use of cheap substrates which, in many cases, are considered waste products of some industries. In general, the products of industrial microbiology and microbial biotechnology are typically investigated under upstream processes, fermentation processes, and downstream processes (Stanbury et al. 1995).

1.4.1 *Upstream processes in microbial biotechnology*

The upstream processes consider the isolation and screening of microorganisms, strain improvement to produce cheap and abundant amounts of the desired product, industrial strain preservation, development of inoculum, substrate selection, optimization and development of appropriate media, and industrial sterilization of media (Figure 1.2).

Microorganisms (as producing agents) and raw materials (as substrates of fermentation) are important in the upstream processes. Strain improvement is critical in the development of most fermentation industries because it increases production efficiency and reduces production costs. In many cases, strain improvement is performed by conventional mutagenesis and recombinant DNA techniques. Then, a suitable strain for the industrial fermentation process is selected through mutant and recombinant strains.

The choice of medium depends on the scale of fermentation. In the small laboratory scale, mostly pure chemicals are used in media composition. This is not possible for large-scale fermentation due to the high costs. Hence, low-cost complex substrates are mainly used in large and commercial-scale fermentations. Most of these compounds are obtained from plant and animal residues, as well as from other industrial wastes that have variable compositions. The effects of fluctuation must be considered between each group of substrates. These effects on performance and product recovery are examined in the small scale with each group of substrate (Waites et al. 2009).

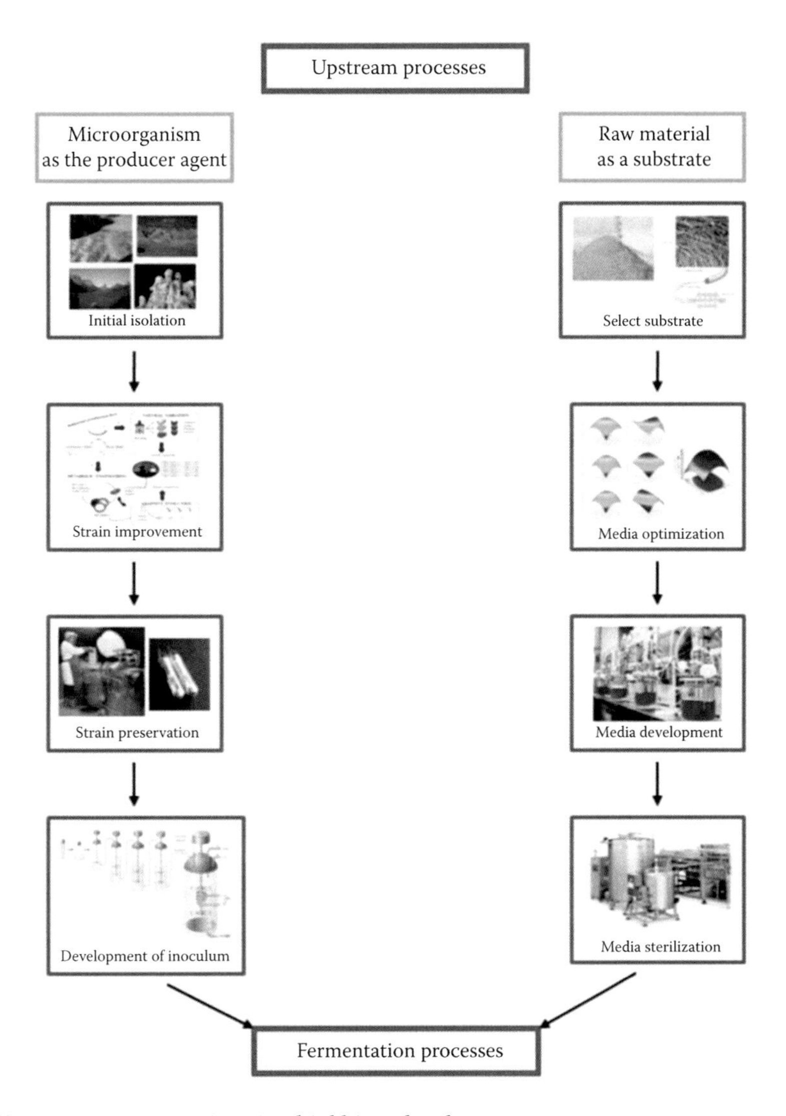

Figure 1.2 Upstream processes in microbial biotechnology.

1.4.2 *Fermentation processes in microbial biotechnology*

The root word of fermentation is derived from the Latin verb *fervere*, meaning to boil, which describes the boiling appearance of the action of yeast on extracts of fruit or grain during fermentation. Fermentation, from the viewpoint of an industrial microbiologist, is the production of a product by the mass culture of microorganisms under aerobic or anaerobic conditions. Fermentation is usually carried out in a fermentor or bioreactor set, whose main objective is to provide a suitable environment for organisms to produce biomass and metabolites. Fermentation system performance depends on many factors, but the main physicochemical properties that need to be controlled are temperature, pH, oxygen transfer, agitation, and foam level.

Fermentation is performed in simple to complex fermentors consisting of a tank equipped with (or without) an agitator controlled by an integrated computer system.

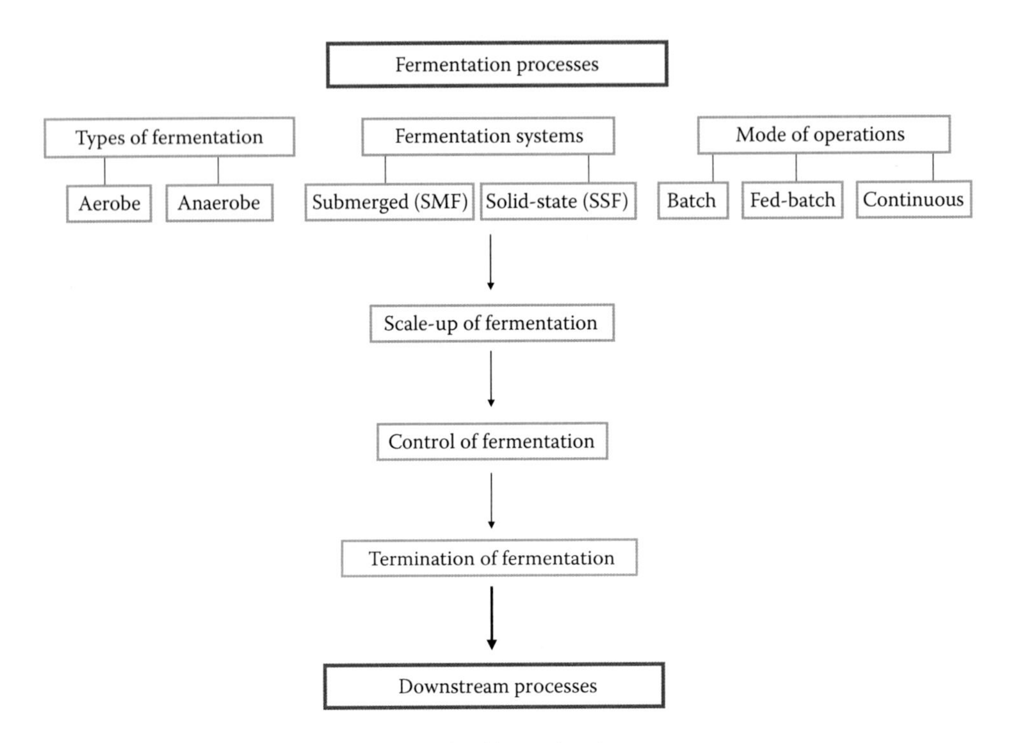

Figure 1.3 Fermentation processes in microbial biotechnology.

Industrial fermentation, depending on the type of end product, is operated by batch, semi-continuous, or continuous models (Stanbury et al. 1995).

The type of fermentation, fermentation system, mode of operation, fermentation scale-up, characteristics of the fermentor, physicochemical control conditions, addition of inoculum and additives to sterile medium, maintenance of sterile conditions during fermentation, and choice of a suitable time to stop fermentation are generally investigated in the fermentation processes (Figure 1.3).

1.4.3 Downstream processes in microbial biotechnology

Downstream processes involve all operations after fermentation, the main aim of which is to increase efficiency and ensure product recovery with the desired purity and biological activity. The primary process includes cell separation and cell wall disruption, followed by isolation and purification of the product from the cell extract or culture medium.

Biomass with insoluble bodies are separated by centrifugation or filtering from the liquid part of the medium. When the goal is the isolation of intracellular material, cells are disrupted using chemical, physical, and biological methods. The amount of desired product in the fermented medium is generally low, therefore a concentration process is required.

Afterward, purification methods such as different types of chromatography, dialysis, reverse osmosis, distillation, and solvent extraction are used depending on the type and desired purity level of the final product. In the final stages, evaporation, the addition of chemicals, or drying are used to make the powdered or crystal form of the product.

The downstream processes should use effective and rapid methods for the isolation and purification of the product. This is very important when impure products are

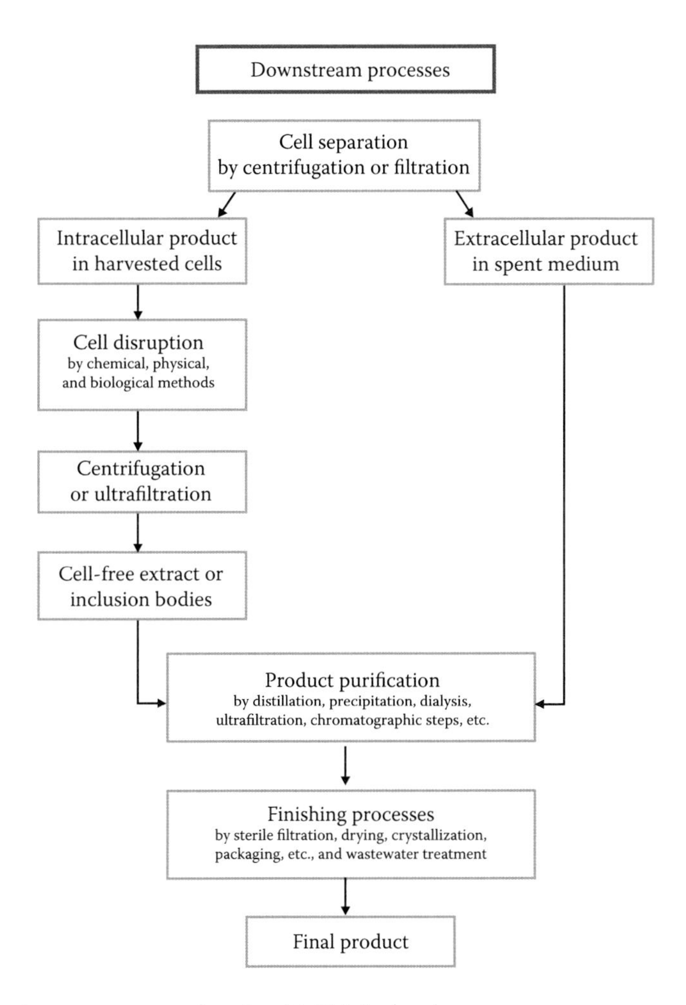

Figure 1.4 Downstream processes in microbial biotechnology.

unstable and affected by environmental conditions. In the case of some products, particularly enzymes, it is critical to maintain their biological activity during the downstream processes. Finally, any wastes produced during the fermentation process should be eliminated using cheap and safe methods (Waites et al. 2009).

In general, biomass separation, cell disruption to release intracellular metabolites, concentration, purification, quantity and quality control, packaging and storage of products, and sterilization and wastewater treatment (especially for industrial and recombinant microorganisms) are investigated in the downstream processes (Figure 1.4).

1.5 Conclusion

Microbial biotechnology is the use of microorganisms and their derivatives to make or modify specific products or processes. Microbial biotechnology is an interdisciplinary field, and successful development in this field requires major contributions from a wide range of other disciplines, especially microbiology, biochemistry, genetics, molecular biology, chemistry, biochemical engineering, bioprocess engineering, and so on.

Typical microbial biotechnology processes involve upstream processes, fermentation processes, and downstream processes. Recent findings and new methods from other disciplines are used in the microbial biotechnology processes. For example, in the upstream processes, new methods in metabolic engineering, industrial systems biology, bioinformatics, omics science, and even nanobiotechnology are used to find and modify microorganisms with industrial capacity and their valuable products.

References

El-Mansi, E.M.T., Bryce, C.F.A., Dahhou, B. et al. *Fermentation Microbiology and Biotechnology*, 3rd ed. Boca Raton: CRC Press/Taylor & Francis Group, 2012.

Glazer, A.N., and Nikaido, H. *Microbial Biotechnology: Fundamentals of Applied Microbiology*, 2nd ed. New York: Cambridge University Press, 2007.

Smith, J.E. *Biotechnology*, 5th ed. New York: Cambridge University Press, 2009.

Soetaert, W., and Vandamme, E.J. *Industrial Biotechnology*. Weinheim: WILEY-VCH, 2010.

Stanbury, P.F., Whitaker, A., and Hall, S.J. *Principles of Fermentation Technology*, 2nd ed. Oxford: Pergamon Press, 1995.

Waites, M.J., Morgan, N.L., Rockey, J.S. et al. *Industrial Microbiology: An Introduction*. Oxford: Blackwell Science, 2009.

chapter two

Screening of microbial metabolites and bioactive components

R. Baskaran, P.M. Mohan, S. Ganesamoorthy, and Ashok Kumar Nadda

Contents

2.1 Introduction

For many decades, synthetic chemicals have been used as drugs in the treatment of many life-threatening diseases. The pharmaceutical industry has synthesized more than 3 million new chemicals in their effort to produce novel drugs. Despite their success in developing drugs to treat or cure many diseases, the treatment of certain diseases such as cancer, AIDS, heart disease, and diabetes has not been addressed completely due to the complexity of these diseases. Over the centuries, people have been living in close association with the environment and relying on its flora and fauna as a source of food and medicine. As a result, many societies have their own rich plant pharmacopeias. In developing countries, due to economic concerns, nearly 80% of the population still depends on the use of plant extracts as a source of medicine. Natural products are playing an important role in the health care system. In 1816, the isolation of the analgesic morphine from the opium poppy, *Papaver somniferum*, led to the development of many highly effective pain relievers (Benyhe 1994).

The penicillin from the filamentous fungus *Penicillium notatum* was discovered by Fleming in 1929. It had a great effect on the investigation of nature as a source of new bioactive agents (Bennett and Chung 2001). For the past 50 years, antibiotics have transfigured medicine by providing cures for formerly life-threatening diseases. However, strains of bacteria and fungi have recently emerged as virtually unresponsive to antibiotics. Such multidrug resistance arising through antibiotic misuse is recognized as a life-threatening problem worldwide. The condition is exacerbated by the fact that no novel class of antibiotics have been discovered in the last 20 years. Although many preexisting antibiotics have been customized to yield new derivatives, microorganisms have the capability to mutate their known resistance mechanism (Knowles 1997; Levy 1998).

Resistance to antibiotics is a phenomenon in which a microorganism is no longer affected by an antimicrobial compound to which it was previously sensitive to. It is the capacity of certain microorganisms such as bacteria, viruses, and parasites to neutralize the effect of the medicine. The resistance can come from the mutation of the microorganisms or from the acquisition of resistance genes. The infections caused by resistant microorganisms do not respond to ordinary treatment, which result in prolonged illness and the risk of death. A high percentage of the infections contracted in hospitals are caused by very resistant bacteria, such as methicillin-resistant *Staphylococcus aureus* (MRSA), as well as *Enterococcus faecium* and several microorganisms (that are) Gram-negative resistant to vancomycin. New mechanisms indicate that the ability of antimicrobial drugs to act against some bacteria has been completely cancelled. The World Health Organization (WHO) is focusing their efforts on standardizing—through alertness, technical support, generation of knowledge, alliances, anticipation, and control of certain illnesses such as tuberculosis, malaria, human immunodeficiency virus (HIV), proper illnesses of infancy, sexually transmitted diseases, and hospital infections—the quality, supply, and rational use of essential medicines. On the other hand, the struggle against antimicrobial-resistant compounds was one of the topics of the 2011 World Health Day. On this occasion, WHO sought to incorporate the growing issue of resistance to antimicrobial compounds by alerting the political bodies so that governments could initiate proactive steps to solve the problem (WHO 2011).

The importance of finding and using secondary metabolites can be justified in two ways: (1) to know the natural substances that could be beneficial for humans, and (2) to identify the organisms that produce these substances because these may be the only carriers of useful compounds to combat pathogenic microbes (Abbott 1925; Altschul et al. 1990). The screening of new bioactive compounds necessarily involves a vast study of microorganisms, their microbial habitat, and their life cycles. It is more convenient to design cultivation techniques for the different microbial genera present in a particular environment and to identify the organisms that produce bioactive secondary metabolites that are of interest. It will also help to sketch a plan of use and preservation for those species. These procedures represent a potential source of new drug development; especially those obtained from bacteria because bacteria have their own cultivation characteristics. Principally, it has attracted much attention on a global scale from a large number of investigators in search of new natural products with anticancer and antibiotic activity.

2.2 Bioactive compounds from microorganisms

The scientific investigations in the field of natural products have been emphasized particularly because of the following facts:

1. To find biologically active compounds. Research plants or organisms are approved for bioactivity-driven isolation and identification of secondary metabolites if initial extracts (of higher plants and broths) showed useful biological activity
2. To screen transgenic organisms for a single metabolite of commercial importance. This is a natural extension of the first objective, with biotechnology skills playing a major role.
3. To reclassify a confused phylogeny. Secondary metabolites from each species of a genus are elaborately isolated, characterized, and compared with one another (Kopcke et al. 2002).

Microorganisms, including certain bacteria, fungi, and algae, which will produce secondary metabolites with some degree of bioactivity, either against another microorganism or acting against certain physiological states of a diseased body, are known as bioactive substances. New bioactive compounds may be prepared using chemical synthesis or it may be isolated from nature. The newly emerging field of combinatorial chemistry, besides the capacity to produce new compounds, has its own challenges in the synthesis of useful leads as new drugs. Berdy (2005) states that natural resources will provide structurally and mechanistically new molecules serving as useful direct drugs or lead compounds with better chances than any chemical approaches. At present, pharmaceutical manufacturers have synthesized several million (according to some estimations, approximately three to four million) new organic chemical structures, but only a negligible portion of them (no more than one in several ten thousand, ~0.001%) have become accepted as drugs.

From the tens of thousands of known microbial metabolites, approximately 150 to 160 (~0.2% to ~0.3%) compounds have been utilized commercially as therapeutic products. A comparison of a library of natural products (including bioactive microbial metabolites) with random libraries of synthetic and combinatorial chemistry shows that screening microbial product libraries is more efficient and economic than the discovery of new bioactive compounds. The inherent molecular diversity of natural products outweighs any chemical-combinatorial libraries (Feher and Schmidt 2003). Perhaps only a combination of the natural product libraries with these chemical libraries will provide more promising chances for additional target-based screenings to discover more active bioactive and useful derivatives of the natural products. Natural products usually result in new mechanisms of action and have a higher probability of exhibiting novel therapeutic activities compared with synthetic or combinatorial compounds.

The reason for the higher efficiency of natural products in contrast with random chemicals is that natural products are somewhat prescreened by nature and have already evolved specific biological interactions. Furthermore, the products of microbes have many similarities with the metabolites of other living systems, for example, mammalian systems. For a compound to have a higher chance of exhibiting its drug-like actions, it must be absorbable and metabolizable. Therefore, it would require minimal modification to develop as an effective, orally active, and easily marketable product. The overestimated capability of rational drug design or the structure–activity relationship approaches was thought of several years ago as a replacement for natural product research.

It has to be emphasized that combinatorial chemistry has a complementary role only and not a replacing role. It may assist in the development of favorable derivatives of lead compounds originally derived by screening from microbial or other natural sources. It will only help modify the structurally and mechanistically new, inherently active natural compounds with respect to absorption; transport and uptake make them effective *in vivo*. It is unquestionable that microorganisms represent the biggest opportunity for the development of useful compounds, which may serve not only as direct drugs or pesticides/herbicides but also as lead compounds and templates for rational drug design. It is an important point that microbes from higher plant and marine animal sources give better chances for satisfactory scale-up (Berdy 2005). Further, organic chemists could not have imagined the unprecedented architectures of these natural products. It is a strong belief that, in most respects, nature is the first; the chemists of today are still able to only copy nature (Berdy 2005). The approximate number of known natural products derived from the main types of plants, animals, and microorganisms are summarized in Table 2.1.

Table 2.1 A Data of Approximate Number of Known Useful Products from Natural Resources

Source	No. of compounds	Bioactives	Antibiotics
Natural products	>1 million	200,000–250,000	25,000–30,000
Plant kingdom	600,000–700,000	150,000–200,000	~25,000
Microbes	>50,000	22,000–23,000	~17,000 (bacteria, 2900; actinomycetales, 8700; fungi, 4900)
Algae, lichen	3000–5000	1500–2000	~1000
Higher plants	500,000–600,000	~100,000	10,000–12,000
Animal kingdom	300,000–400,000	50,000–100,000	~5000
Protozoa	Several hundred	100–200	~50
Invertebrates	~100,000	–	~500
Marine animals	20,000–25,000	7000–8000	3000–4000
Insects/worms/etc.	8000–10,000	800–1000	150–200
Vertebrates (mammals, fishes, amphibians, etc.)	200,000–250,000	50,000–70,000	~1000

Source: Berdy, J., *The Journal of Antibiotics* 58(1): 1–26, 2005.

2.2.1 *Bioactive compounds from bacteria*

The ocean is called the "mother of life" and occupies more than 70% of the Earth's surface. Due to its depth, it encompasses approximately three hundred times the habitable volume of the terrestrial habitats on Earth. In the past, the oceans had been considered as a rich source of extremely potent compounds, which represented a considerable number of drug candidates (Gochfeld et al. 2003; Haefner 2003). Although very little is known about the microbial diversity of marine sediments, similar to terrestrial soil, marine sediments also contain considerable amounts of readily available organic matter and carbon, which is present in complex form (i.e., cellulose and chitin). However, culture-independent methods have demonstrated that marine sediments contain a wide range of unique microorganisms that are not present in terrestrial environments (Ravenchlang et al. 1999; Stach et al. 2003). Although macroorganisms from the oceans have proven to be good sources of novel bioactive metabolites, large-scale production of these bioactive metabolites has been difficult (Bernan et al. 1997). Marine microorganisms such as bacteria and fungi have been reported to produce antibacterial, antifungal, antiviral, and antitumor substances (Rosenfeld and Zobell 1947; Bernan et al. 1997). The biodiversity of marine microbes and the versatility of their bioactive metabolites have not been fully explored. Due to the complex nature of the oceans, marine bacteria have developed sophisticated physiological and biochemical systems with which they uniquely adapt to extreme habitats and various unfavorable marine environmental conditions. They live in a biologically competitive environment with unique conditions of pH, temperature, pressure, oxygen, light, nutrients, and salinity, as well as being especially rich in chlorine and bromine elements.

Microbes can sense, adapt, and respond to their environment quickly and can compete for defense and survival by the generation of unique secondary metabolites. Even though these compounds are produced in response to stress, many have shown value in biotechnological and pharmaceutical applications (Wenzel and Müller 2005). The marine microbial metabolites exhibit unique biological activities compared with terrestrial bacteria (Blunt et al. 2004; Berdy 2005). Microorganisms that are present in different environments continue to provide pharmacologically important secondary metabolites, which are commercially explored as drugs and other products. The importance of marine sources for the discovery of novel natural products with a pharmaceutical potential has been proved during the last decade and has been highlighted in various reviewed articles (Faulkner 2000; Blunt et al. 2003, 2004; Hentschel et al. 2006).

Knowledge of the distributions of various bacterial groups in complex marine environments is essential for searching and developing a new chemical resource. However, studies on the distributions of marine bacteria in the marine habitat are limited. The actinobacteria or filamentous bacteria, which have been the single most important source of exciting metabolites from soil bacteria, are also members of the Gram-positive family of bacteria.

In past decades, many studies were carried out to understand the diversity of true marine bacteria in their natural marine environment (Jensen and Fenical 1996). Marine bacteria are generally involved in the mineralization of organic matter, nutrient cycling, and energy transfer in aquatic environments (Azam and Worden 2004). However, their potential to synthesize novel chemical compounds with antimicrobial properties was first recognized by Rosenfeld and Zobell (1947) and Grein and Meyers (1958). The bactericidal property of seawater was observed during the same period of study, and it was subsequently realized that this was due to the production of antibiotics by planktonic algae and marine bacteria, which were found abundantly distributed in the seawater (Baam et al. 1966; Baslow 1969).

Currently, the number of life-threatening infections caused by various microorganisms is rapidly increasing as the number of microorganisms that are found to be highly resistant to available antibiotics as well as other drugs has became a serious problem. Many strains of bacteria are resistant to one or more of the 100 antibiotics that are now in use, and these antibiotic-resistant bacteria also spreading globally. Microorganisms, particularly bacteria have tremendously influenced the development of medical science. More than 10,000 of these are biologically active and approximately 8000 are recognized as antibiotics and antitumor agents (Berdy 1989; Stierle et al. 1993; Tomasz 1995). Nearly a hundred microbial products have now found clinical applications as antibiotics, antitumor agents, and agrochemicals (Demain 1983; Clark 1996; Cragg et al. 1997). Despite these early observations, relatively little attention was directed toward the study of natural products from marine bacteria. This was due to the difficulties encountered during the isolation and cultivation of marine bacteria. Only a small percentage of the viable bacterial cells in marine samples ultimately grow under standard culture conditions. Not many studies about the diversity and distribution of marine bacteria have been conducted and only a small percentage has been explored. Hence, only few marine bacteria have been the subject of comprehensive chemical analysis and most of their structurally unique metabolites have been discovered through fermentation studies (Clark 1996; Cragg et al. 1997). This was mainly to establish significant pharmacological applications of the novel metabolites to replace failed antibiotics (Umezawa 1972, 1982).

Actinobacteria are Gram-positive bacteria that have characteristically high (>55%) G + C content (Goodfellow and Williams 1983). The name "actinomycetes" was derived from the Greek *aktis* (a ray beam) and *mykes* (fungus). This terminology was used for these beam-like organisms from the initial observations based on their morphology. The actinobacteria were originally considered to be an intermediate between bacteria and fungi but are now recognized as prokaryotes (Das et al. 2008). The majority of actinobacteria are free-living, saprophytic bacteria found widely distributed in soil, water, and colonizing plants. In the soil, they are involved in the decomposition of organic matter and in the mineralization cycle through the production of extracellular enzymes, such as cellulases, chitinases, and lignin peroxides.

Genera of actinobacteria such as *Streptomyces*, *Sacchropolyspora*, and *Amycolatopsis* are adapted to survive in highly erratic and competitive soil environments. They are not only equipped with a wide array of enzymes for exploiting nutrients, but also to produce a broad range of bioactive metabolites of industrial and medical importance, for example, compounds with antibiotic activity for fungal and bacterial inhibition. Actinobacteria are recognized as a source of novel antibiotic and anticancer agents with unusual structure and properties (Jensen et al. 2005) and are also known to produce chemically diverse compounds with a wide range of biological activities (Bredholt et al. 2008). Therefore, actinobacteria hold a prominent position due to their diversity and proven ability to produce new compounds.

Streptomyces is the largest genus of actinobacteria belonging to the family Streptomycetaceae (Kampfer 2006). Approximately 500 species of *Streptomyces* have been described thus far (Euzeby 2008). It is predominantly found in soil and decaying vegetation. Most streptomycetes produce spores and are noted for their distinct "earthy" odor, which results from the production of a volatile metabolite, geosmin. They produce more than two-thirds of the clinically useful antibiotics (e.g., neomycin and chloramphenicol) of natural origin (Kieser et al. 2000). Members of the actinobacteria group are primarily recognized as organisms of academic curiosity because they are potential degraders and organisms of antibiotic producers. Actinobacteria are the main sources of clinically important antibiotics, most of which are too complex to be synthesized using combinatorial chemistry. Actinobacteria represent a ubiquitous group of microorganisms from various natural ecosystems around the world. In addition to its widespread distribution in terrestrial and aquatic habitats, it is also abound in extreme environments including marine ecosystems. It is well known that members of the genus *Streptomyces* produce numerous antibiotics and other classes of biologically active secondary metabolites. Approximately two-thirds of the known antibiotics are produced with actinobacteria and among them nearly 80% are made from members of the genus *Streptomyces*, with other genera trailing numerically. If we include secondary metabolites with biological activities other than antimicrobial systems, actinobacteria are still at the forefront, with more than 60% (again *Streptomyces* sp. account for 80% of these). Many reports on the effects of antagonistic actinobacteria against pathogenic bacteria and fungi are given in Table 2.2.

In recent years, marine microorganisms have gained important status in the pharmaceutical industry as they are potential sources of microbial products exhibiting antimicrobial, antiviral, antitumor, anticoagulant, and cardioactive properties. Although bacteriologists have long been fascinated with the identification of novel actinobacteria and novel compounds from diverse environments, the marine environment is a relatively recent source. Some of the industrially important strains of actinobacteria such as *Salinospora* have become a source of novel secondary metabolites. These molecules with potent biological activity and these marine actinobacteria represent an important future resource for small-molecule drug discovery. Compounds isolated from marine actinobacteria have a broad spectrum of biological activities such as antibiotic, antifungal, cytotoxic, neurotoxic, antimitotic, antiviral, and antineoplastic activities (Newman and Cragg 2007). Recently, new targets have been added to the general screening of actinobacteria producing bioactive compounds for AIDS, immunosuppression, anti-inflammation, Alzheimer disease, aging processes, and other tropical diseases. However, these new discoveries on marine actinobacteria are facing some unexpected problems. For many of the marine

Table 2.2 Antagonistic Activities of Bioactive Compounds Producing Actinobacteria Against Bacteria and Fungi

Organism(s)	Activity against	Reference(s)
Actinobacteria	*Fusarium oxysporum F. cubense*, and *Penicillium graminicolum*	Meredith 1946
Actinobacteria	*Mycobacterium tuberculosis*	Johnstone 1947
Streptomyces spp.	*Trichophyton* sp., *Fusarium* sp., *Penicillium* sp., and certain bacteria	Leben et al. 1952
Actinobacteria	Yeast	Takahashi et al. 1993
Actinobacteria	*S. aureus* and *E. coli*	Krasilnikov et al. 1953
Streptomyces sp.	Soil fungi and bacteria	Lockwood 1959; Zeeck et al. 1987
Streptomyces, *Micromonospora*, and actinobacteria	Gram-positive and Gram-negative bacteria, yeast, and filamentous fungi	Pisano et al. 1989, 1992
Actinobacteria	*S. aureus* and *B. subtilis*	Choi and Park 1993
Microbispora spp.	*S. aureus* and *A. niger*	Hayakawa et al. 1995
S. aureus	*H. oryzae*, *Curvularia lunata*, and *T. mentagrophytes*	Chakrabarty and Chandra 1979
S. hydroscopicus	*T. mentagrophytes* and *C. albicans*	Gurusiddaiah et al. 1979
S. violaceoniger	*Macrophomia phaseolina* and *C. albicans*	Hussain and El-Gammal 1980
S. globus	*Alternaria solani*, *A. niger*, *C. pallescens*, *T. rubrum*, *T. mentagrophytes*, *C. albicans*, and *Phytophthora* sp.	Nair et al. 1994
S. arabicus	*A. brassicae*	Sharma et al. 1985
S. anandii	Bacteria	Christopher et al. 1987
S. roseiscleroticus (sultriecin) and *S. hygroscopicus* (yatakemycin)	*C. albicans*, *Cryptococcus neoformans*, *A. fumigatus*, *F. moniliforme*, *T. mentagrophytes*, *Blastomyces dermatitides*, and *Petriellidium boydii*	Ohkuma et al. 1992; Satoh et al. 1993; Zheng et al. 2000; Datta et al. 2001
S. auermitilis (Avermectins)	Antiparasitic—nematodes and arthropods	Ikeda and Omura 1993
Streptomyces sp. (nonpolyene)	*C. albicans*, *C. tropicalis*, *B. cinerea*, *A. fumigatus*, *F. solani*, *F. oxysporum*, *P. irregulare*, *Pseudomonas* sp., *Salmonella* sp. and *Legionella* sp.	Hacene et al. 1994; Yon et al. 1995
S. violaceusniger (new macrolide)	*C. neoformans*, *C. albicans*, *C. tropicalis*, *C. parapsiolis*, *C. glabrata*, *A. fumigatus*, *A. flavus*, *T. mentagrophytes*, *T. rubrum*, and *M. canis*	Kook and Kim 1995; Woo and Kamei 2003
Streptomyces sp.	*Aeromonus hydrophila*, *A. sobria*, *Edwardsiella tarda*, *Vibrio alginolyticus*, and *V. harveyi*	Patil et al. 2001a,b
S. bottropensis and *S. griseoruber*	*Agrobacterium tumefaciensis* and *C. albicans*	Saadoun and Al-Momani 1998, 2000
Actinomadura spp.	*Bacillus subtilis*, *S. aureus*, *M. luteus*, *C. albicans*, and fungi	Srivibool 2000

(continued)

Table 2.2 (Continued) Antagonistic Activities of Bioactive Compounds Producing Actinobacteria Against Bacteria and Fungi

Organism(s)	Activity against	References(s)
Streptomyces sp.	Bacteria and fungi	Ellaiah et al. 2002
Streptomyces sp.	*B. subtilis, S. aureus, M. luteus, A. fumigatus, A. niger, M. hiemalis, P. roqueforti, P. rariotii, C. albicans, Cryptococcus humicolus, S. cerevisiae, P. cinnamoni, Pestalotiopsis sydowiana, B. cinerea,* and *Sclerotinia homoeocarpa*	Frandberg et al. 2000; Moncheva et al. 2002; Narayana et al. 2004
S. antibioticus and *S. rimosus*	Gram-positive and Gram-negative bacteria and yeasts	Sahin and Ugar 2003
Actinobacteria	Gram-positive and Gram-negative	Kokare et al. 2004a,b
Streptomycetes	Phytopathogenic and human pathogenic fungi	Oskay et al. 2004
Streptomyces spp.	Rhizotonia solani	Cao et al. 2004
Marine actinobacteria	*Salmonella* sp., *E. coli, Klebsiella* sp., *R. solani, Pyricularia oryzae, H. oryzae,* and *Colletotrichum falcatum*	Kathiresan et al. 2005; Dhanasekaran et al. 2005a,b
S. albovinaceus	*B. subtilis, B. pumilus, S. aureus, E. coli,* and *P. aeroginosa*	Ellaiah et al. 2005

actinobacteria, the taxonomy of the strain is poorly defined, so that binomial identification becomes cumbersome (Findlay et al. 1986).

In contrast with marine invertebrates, marine actinobacteria seem to be a promising source as a producer of drugs, for example, thiocoraline, a new anticancer drug produced by a marine *Micromonospora* strain (Romero et al. 1997). Also, the marine actinobacteria represent an important future resource for small-molecule drug discovery. Although the positive effects of actinomycete secondary metabolites on human health is clear, an intensive research by the pharmaceutical industry has exhausted the supply of compounds that can be discovered from this group of bacteria. This perception has been a driving force behind the recent shift away from natural products as a resource for the discovery of small-molecule therapeutics toward other drug discovery platforms, including high-throughput combinatorial synthesis and rational drug design. The potential of secondary metabolites, derived from marine microorganisms to inhibit *Plasmodium* growth, and the effect of proteasome inhibitors such as salinosporamide A on *in vitro* and *in vivo* parasite development by a secondary metabolite of marine bacterium with significant antimalarial activity was studied (Prudhomme et al. 2008). Salinosporamide A, which is derived from a marine actinobacteria, showed inhibitory activity against parasite development *in vitro* (*Plasmodium falciparum*) and *in vivo* (*Plasmodium yoelli*) in which the 20S proteasome is the molecular target. This facilitates the discovery of an increased specificity of *Plasmodium* proteasome inhibitors in the future from some other sources from similar habitats (Prudhomme et al. 2008). The antiviral drug cyclomarin was also produced by a marine streptomycetes strain (Rosenberg et al. 2000). In both cases, fermentation of the microorganisms enabled the production of the amounts required for clinical applications.

The actinobacteria were first recognized by Gasperni (1890) as potential destroyers of fungi and bacteria. The actinobacteria antagonistic of *Pythium* from sugarcane was reported by Tims (1932). Waksman (1937) made a detailed survey of actinobacteria possessing antagonistic effects on the activity of other microorganisms in their studies on decomposition. The distribution of antagonistic activities of actinobacteria against fungi was studied by Alexopolus (1941). Carvajal (1946) observed that only a few strains of *Streptomyces griseus* isolated from various sources were found to produce streptomycin, and these varied greatly in their ability to do so. Emerson et al. (1946) found that actinobacteria and molds isolated from the soil produced inhibitory substances.

Systematic screening of soil actinobacteria from different parts of India had already been carried out for the production of antibiotic compounds (Banerjee et al. 1954; Rangaswami et al. 1967; Agarwal et al. 1968). Nine percent of the marine actinobacteria isolated required seawater for normal growth and to produce strong antibiotic compounds for antibacterial and antifungal activities (Bredholt et al. 2008). Rizk et al. (2007) reported that 188 *Streptomyces* species, isolated from Egyptian soils, were screened for their antagonistic activity against 19 fungal and bacterial species. Substantial antagonistic activities were attained by five *Streptomyces* isolates against *Candida albicans*. Actinobacteria are capable of producing a large number of medically useful antibiotics (Nolan and Cross 1998). Due to the presence of a large genome and thousands of different genes that encode for a large number of proteins, actinobacteria are capable of utilizing a vast number of substrates. They occur in a multiplicity of natural and man-made environments (Goodfellow and Williams 1983) with different morphological, cultural, biochemical, and physiological characters. Thus, to survive in extreme natural environments, they are capable of producing various secondary metabolites and new antibiotics (Lechevalier and Lechevalier 1967). Nithya and Ponmurugan (2008) reported that strains such as KN-2 and KL-3 showed the production of secondary metabolites for antibacterial and antifungal activities. Muriru et al. (2008) reported that the concentration of antibiotic metabolites by partial purification and freeze-drying resulted in, respectively, 21.5% and 20.1% enhanced antimicrobial affinity. Pugazhvendan and coworkers (2010) reported 10 antimicrobial actinobacteria from the Tamil Nadu coastal area.

Ethyl acetate extract from the *Streptomyces* sp. isolate B865 delivered the trioxaccerains A to C (2a–2c) and additionally these new derivatives were designated as trioxacarcins D to F (2d–2F). All trioxacarcins showed high antibacterial activity, whereas some showed high antitumor and antimalarial activity. The structure of the new antibiotics were derived from mass, one-dimensional and two-dimensional nuclear magnetic resonance spectra, and confirmed by comparison. The absolute configuration of the trioxacaricins was deduced from the x-ray analysis of gutingimycin (2g) and from known *Streptomyces* of the L-trioacarcinoses A and B (Markey et al. 2004). Some useful bioactive compounds of actinobacteria are given in Table 2.3.

The difficulty and high cost of isolating novel structures of antimicrobial agents with new modes of action led to the phase of decline in this field of research. However, in recent decades, the chemistry of natural marine products has emerged as a mature field after years of intensive research. As a result, studies on secondary metabolites from microorganisms are a rapidly growing field. Typically, primary metabolites are found across all species within broad phylogenetic groupings, and are produced using the same pathway or using similar pathways in all these species. However, microorganisms have received very little attention in the field of drug discovery. This is mainly due to the noncultivability of most bacteria (Hugenholtz et al. 1996). Several studies showed that marine bacteria are capable of producing unusual bioactive compounds that are not isolated from terrestrial sources (Fenical et al. 1993, 1997).

Many unusual compounds such as thermostable proteases, lipases, esterases, starch, and xylan-degrading enzymes have been isolated mainly from marine bacteria and hyperthermophilic archaea (Bertoldo 2002). Unusual Gram-positive bacteria from deep sea sediments, which have produced a series of new natural products, Macrolectin A to F of C24

Table 2.3 Bioactive Compounds of Actinobacteria

Antibiotic	Production	Application(s)
Actinomycin D	*Streptomyces* spp.	Antitumor
Antimycin A	*Streptomyces* spp.	Telocidal
Avermectin	*S. avermitilis*	Antiparasitic
Bambermycin	*S. bambergiensis*	Growth promotant
Bialaphos	*S. hygroscopicus*	Herbicidal
Bleomycin	*S. verticillus*	Antitumor
Candicidin	*S. griseus*	Antifungal
Cephamycin C	*Nocardia lactamdurans*	Antibacterial
Chloramphenicol	*S. venezuelae*	Antibacterial
Chlortetracycline	*S. aureofaciens*	Antibacterial
Cycloserine	*S. orchidaceus*	Antibacterial
Daptomycin	*S. roseosporus*	Antibacterial
Daunorubicin	*S. peucetius*	Antitumor
Doxorubicin (adriamycin)	*S. peucetius var. caesius*	Antitumor
Erythromycin	*Sac. erythraea*	Antibacterial
FK506 (tacrolimus)	*S. hygroscopicus*	Immunosuppressant
Fosfomycin	*Streptomyces* spp.	Antibacterial
Gentamicin	*Micromonospora* spp.	Antibacterial
Hygromycin B	*S. hygroscopicus*	Antihelminthic
Kanamycin	*S. kanamyceticus*	Antibacterial
Lincomycin	*S. lincolnensis*	Antibacterial
Milbemycin	*S. hygroscopicus*	Antiparasitic
Mithramycin	*S. argillaceus*	Antitumor
Mitomycin C	*S. caespitosus*	Antitumor
Natamycin	*S. nataensis*	Antifungal
Neomycin	*S. fradiae*	Antibacterial
Nikkomycin	*S. tendae*	Antifungal; insecticidal
Nocardicin	*Nocardia uniformis*	Antibacterial
Nosiheptide	*S. actuosus*	Growth promotant
Novobiocin	*S. niverus*	Antibacterial
Nystatin	*S. noursei*	Antifungal
Oleandomycin	*S. antibioticus*	Antibacterial
Oxytetracycline	*S. rimosus*	Antibacterial
Phleomycin	*S. verticillus*	Antitumor
Polyoxins	*S. cacaoi var. asoensis*	Antifungal (Plant protection)
Pristinamycin	*S. pristinaespiralis*	Antibacterial
Puromycin	*S. alboniger*	Research
Rapamycin	*S. hygroscopicus*	Immunosuppressant
Ristocetin	*Nocardia lurida*	Antibacterial
Spectinomycin	*S. spectabilis*	Antibacterial
Spinosyns	*Sac. spinosa*	Insecticidal
Spiramycin	*S. ambofaciens*	Antibacterial
Streptogramins	*S. graminofacients*	Antibacterial

(*continued*)

Table 2.3 (Continued) Bioactive Compounds of actinobacteria

Antibiotic	Production	Application(s)
Streptomycin	*S. griseus*	Antibacterial
Teichoplanin	*Actinoplanes teichomyceticus*	Antibacterial
Tetracycline	*S. aureofaciens*	Antibacterial
Thienamycin	*S. cattleya*	Antibacterial
Thiostrepton	*S. azureus*	Growth promotant
Tobramycin	*S. tenebrarius*	Antibacterial
Tylosin	*S. fradiae*	Growth promotant
Validamycin	*S. hygroscopicus*	Plant protectant
Vancomycin	*Amycolatopsis orientalis*	Antibacterial
Virginiamycin	*S. virginiae*	Growth promotant

Source: Kieser, T. et al., *Practical Streptomyces Genetics*, 2nd ed. Norwich, England: John Innes Foundation, 2000 (ISBN 0-7084-0623-8).

linear acetogen origin, have been isolated (Gustafson et al. 1989). The major metabolite, Macrolectin A, inhibits B16-F10 murine melanoma cells in *in vitro* assays, showing significant inhibition of mammalian herpes simplex virus (HSV type I and II) and protecting T lymphocytes against HIV replication (Carte 1996). In another study, a microbial metabolite from *Alteromonas* spp. with anti-HIV potential as reverse transcriptase inhibitor has been reported.

The first marine bacterial metabolite to be reported was the highly brominated pyrrole antibiotic. This was isolated by Burkholder and coworkers (1966) through fermentation studies of a Gram-negative bacterium obtained from the surface of the Caribbean sea grass Thalasia. The molecule showed impressive *in vitro* antibiotic properties against Gram-positive bacteria, with minimum inhibitory concentrations (MICs) ranging from 0.0063 to 0.2 pg/mL. The compound synthesized by this bacterium was known to be pentabromopseudiline and its antitumor properties were also reported. The Faulkner group in California isolated a purple-pigmented bacterium, which also proved to be a potent antibiotic. Chemical analysis showed that the *Alteromonas* sp. produces several antimicrobial compounds such as pyrrole, tetrabromopyrrole, hexabromo-2,2′-bipyrrole, and several simple phenolics including 4-hydroxybenzaldehyde, *n*-propyl-4-hydroxybenzoate. Tetrabromopyrrole showed moderate antimicrobial activity *in vitro* against *S. aureus, Escherichia coli, Pseudomonas aeruginosa,* and *C. albicans.* It was even more active against several groups of marine bacteria and showed autotoxicity against the producing *Chromobacterium* sp.

Pseudomonas sp. was reported to produce 6-bromoindolecarboxaldehyde, its analogue, and a mixture of 2-*n*-pentyl- and 2-*n*-heptylquinolinol (Wratten et al. 1977). The 2-heptylquinolinol is a known antibiotic produced by strains of *P. aeruginosa*. The most potent of these simple antibacterial agents was 2-*n*-pentylquinolinol, which showed its greatest activity against *S. aureus.* Another marine bacteria, *Alteromonas rubra,* was isolated and identified by researchers at the Roche Research Institute in Australia. This bacterium provided a novel new target for the isolation of unique natural products. Under a fermentation study, the bacterium produced a series of C16 aromatic acids, which are acetogenins of fatty acid synthetic origin. *P. magnesiorubra* isolated from the surfaces of the tropical marine green algae, *Caulerpa peltata,* also provided a novel metabolite. Two antibiotic pigments, the magnesidins were isolated as 1-L mixtures of methylene homologues (4- and 6-methylene groups), and identified as minor metabolites of this marine bacterium (Gandhi et al. 1976). These unique pigments were considered to be oxidation products of prodigiosin, a common tripyrrolic pigment produced by marine as well as terrestrial bacteria. The seawater-derived bacteria studied thus far have been taxonomically very limited and seemingly driven by screening processes for new antibiotics.

The marine sediments are a more nutrient-rich microhabitat that varies greatly in organic content. Such a microhabitat provides a diversity of bacterial flora; usually not found in more nutrient-limited habitats. The presence of free surfaces in these microhabitats is known to stimulate and support enhanced bacterial colonization as well as the growth of chemically prolific marine bacteria. The antimicrobial agent producers present in marine sediments are generally Gram-positive bacteria, such as those from the genus *Bacillus* and the actinobacteria, and Gram-negative bacteria mainly *Pseudomonas* spp. and *Alteromonas* spp. The actinobacteria group comprises the major sources of antimicrobial compounds. Approximately 10% to 33% of the total bacterial community present in soil belongs to *Streptomyces* sp. and *Nocardia* sp., which are the most abundant actinobacteria in soil (Osborne et al. 2000; Pandey et al. 2002). These microbes exhibit a vast metabolic versatility. Hence, they have many physiological cycles that produce intermediate molecules such as enzymes or secondary metabolites with antibacterial, antifungal, and antiviral capabilities.

A slow-growing, Gram-positive bacterium C-237 was isolated from deep-sea sediment samples obtained from the coast of California. Fermentation studies in a salt-based medium yielded a series of novel cytotoxic and antiviral macrolides, the Macrolactins A to F (Gustafson et al. 1989). Under standard fermentation conditions at atmospheric pressure, this bacterium produced six macrolides and two open-chain hydroxy acids in varying amounts. Macrolactin A was produced as the major metabolite (4.8 mg/L) in most of the fermentations. The majority of the biological properties were due to Macrolactin A, which showed modest antibacterial activity, but was active against B16-FlO murine melanoma *in vitro* with IC_{50} values of 3.5 μg/mL. More importantly, Macrolactin A inhibited several viruses including HSV (IC_{50} = 5.0 μg/mL) and HIV (IC_{50} = 10 μg/mL). Another group of Gram-positive bacteria, from the genus *Bacillus* sp. present in soil, also have the ability to produce antimicrobial agents with clinical and agricultural significance. *Bacillus thuringiensis* secrete certain toxins with insecticidal properties. This chemical has found wide application and is now used in the biocontrol of insects. In the case of *B. thuringiensis* serovar Israelensis (*Bti*), it produces a toxin that is used for the biocontrol of black flies and mosquitoes. Among the Gram-negative bacteria, *Pseudomonas* sp., *Alteromonas* sp., and *Vibrios* sp. are known to produce antimicrobial substances. Bioactive molecules produced by *Pseudomonas* sp. have found wide application mainly in the agriculture sector.

Another group of Gram-positive bacteria present in soil and responsible for the production of antimicrobial agents with clinical and agricultural importance is the genus *Bacillus*. This genus is characterized by being Gram-positive, and having spore-forming rods. It has been demonstrated that these microbes produce antimicrobial agents in various stages of their growth curve. For example, *Bacillus subtilis* 168 can produce nonribosomal oligopeptides with antifungal and antimicrobial properties such as surfactins, inturinics, and bacilysin (Oskay 2004). The antibiotics pyoluteorin (Plt), pyrrolnitrin (Prn), phenazine-1-carboxylic acid (PCA), and 2,4-diacetylphloroglucinol (Phl) are currently a major focus of research in biological control. Strains from *Pseudomonas* sp. have been isolated from soils that exhibited suppressive effects against plant diseases such as Take-all of wheat, Black root rot of tobacco, *Fusarium* wilt of tomato, and damping off of tomatoes caused by *Rhizoctonia solani*.

Bacteria produce a variety of antimicrobial substances that are able to kill or inhibit other microorganisms. Bacterial antibiotics can be subdivided into two types on the basis of their chemical nature: (i) nonpeptide antibiotics and (ii) peptide antibiotics. One group comprises nonribosomally synthesized peptides produced on large enzymatic complexes such as surfactin (Carrillo et al. 2003). The second group comprises ribosomally synthesized peptides such as bacteriocins. Bacteriocins produced by Gram-positive bacteria exhibit a number of characteristics that make them attractive for both the food industry and for biomedical applications (Bower et al. 2001; Chen and Hoover 2003). Several bacteriocins or bacteriocin-like substances (BLIS) produced by the genus *Bacillus* have been reported such as a bacteriocin of *B. brevis*, subtilin by *B. subtilis*, thruicin 7 by *B. thuringiensis*, and many others. The newly characterized bacteriocins include the lantibiotics ericin A (2986 Da) and ericin S (3342 Da) produced by *B. subtilis*; cerein 7A (3940 Da) and cerein 7B (4893 Da) from *Bacillus cereus* and the subclass IIa peptide SRCAM 1580 (3486 Da) produced by *Bacillus circulans* (Klein et al. 1992; Oscariz et al. 1999; Cherif et al. 2001; Hyung et al. 2001; Stein et al. 2002; Svetoch et al. 2005; Oskouie et al. 2006).

Brammavidhya and Usharani (2013) reported that a low-molecular weight peptide from sponge associated with *B. cereus* was inhibitory to human pathogens and food-spoilage bacteria, such as *C. albicans*, *Listeria monocytogenes*, *P. aeruginosa*, *B. cereus*, *B. subtilis*, and *Lactobacillus bulgaris* and *Proteus vulgaris*.

2.2.2 *Secondary metabolites of fungi*

In the last decade, marine fungi have become increasingly important sources of new bioactive natural products (Konig et al. 2006; Laatsch 2006; Ramaswamy et al. 2006; Shimizu and Li 2006). The biodiversity of fungi is enormous. Out of the estimated 1.5 million species of fungi recorded worldwide, approximately 4000 secondary metabolites of fungal origin are recognized to possess biological activities, the vast majority coming from the species of *Penicillium*, *Aspergillus*, *Acremonium*, and *Fusarium* (Hawksworth 1991; Dreyfuss and Chapela 1994; Onifade 2007).

The initial report of a bioactive natural product from a marine-derived fungus dates back to the 1940s, when the fungus *Acremonium chrysogenum* was isolated. This fungus was isolated from a sewage outlet in the Mediterranean Sea close to the island of Sardinia (Proksch et al. 2008). The fungus was the source of Cephalosporin C, the parent compound of modern cephalosporin antibiotics, which are indispensible for the treatment of numerous bacterial infections (Abraham and Loder 1972). Primarily, progress with rigorous evaluation of marine fungal metabolites was slow. This situation changed dramatically in the 1990s, when there was a sharp increase in interest in marine microbial metabolites, which continues until today. Up to the year 2002, 272 new natural products had been isolated from marine-derived fungi (Bugni and Ireland 2004). A dramatic increase in the number of elucidated marine fungal structures began afterward, illustrated by the fact that between 2002 and 2004, 240 additional compounds were described (Ebel 2006).

Marine-derived fungi as a new source of bioactive metabolites is now commonly accepted, there is still much debate on the nature of fungi that are isolated from various marine substrates, such as driftwood, algae, or invertebrates (Proksch et al. 2008). Because numerous, if not most, of these fungi belong to genera already well known from the terrestrial environment, such as *Aspergillus*, *Penicillium*, *Cladosporium*, *Phoma*, and *Fusarium*, the true marine origin of these fungal strains is frequently doubted (Holler et al. 2000; Kohlmeyer and Volkmann-Kohlmeyer 2003). It is possible that several marine-derived fungi thus far investigated originated from terrestrial habitats (e.g., soil) from which they were washed to sea and survived (as spores) until they were recovered by a marine chemist looking for new compounds from the sea. On the other hand, in the last few years, more and more evidence has accumulated indicating an adaptation of these "ubiquitous" fungi to the marine environment (Geiser et al. 1998a,b; Alker et al. 2001; Duncan et al. 2002; Zuccaro et al. 2004).

Interestingly, sponges continue to be one of the most important sources for the isolation of metabolite-producing marine-derived fungi, even though the presence of fungal mycelia growing in sponges has not yet been proven. In contrast to sponge-derived natural products, such as jaspaklinolide L, theopederin C, or mycalamide A, which bear obvious structural resemblances to known bacterial metabolites and are assumed to originate from bacterial sponge symbionts, there are no examples of known fungal metabolites that have been isolated from sponges. For the time being, the true nature of sponge–fungal associations remains unclear and far more research needs to be devoted to this field (Holler et al. 2000; Jensen and Fenical 2002; Proksch et al. 2002; Bugni and Ireland 2004; Butzke and Piel 2006; Ebel 2006).

Proksch et al. (2008) reported that fungi isolated from sponge produced 19 different metabolites, including three new natural products, that were isolated and structurally identified. *Aspergillus ustus* yielded two sesquiterpenes, a drimane derivative, and deoxyuvidin as well as a sesterterpene ophiobolin H. The drimane derivative had an ED_{50} value against L5178Y cells of 1.9 mg/mL *in vitro*. *Penicillium* sp. yielded the largest number of metabolites. Viridicatin, viridicatol, cyclopenin, and cyclopenol suppressed larval growth of the polyphagous pest insect *Spodoptera littoralis* when incorporated into an artificial diet at an arbitrarily chosen concentration of 237 ppm (Proksch et al. 2008).

Screening of *Aspergillus gorakhpurensis* for the production of bioactive secondary metabolites results in the production of 4-(*N*-methyl-*N*-phenyl amino) butan-2-one and itaconic acid. Biological evaluation of the two compounds against test microorganisms showed strong inhibitory activity of 4-(*N*-methyl-*N*-phenyl amino) butan-2-one toward bacteria, fungi, and in the *Spodoptera litura* larvicidal bioassay. An investigation of the secondary metabolite from *Diaporthe helianthi* yielded the phenolic compound 2(-4 hydroxyphenyl)-ethanol (Tyrosol). Its antimicrobial reaction was tested and the ensuing antagonistic effects on the human pathogenic bacteria *Enterococcus hirae*, *E. coli*, *Micrococcus luteus*, *Salmonella typhi*, *S. aureus*, phytopathogenic *Xanthomonas campestris* pv. *phaseoli* and phytopathogenic fungi were demonstrated. Results show that bioactive compounds and Tyrosol produced by *D. helianthi* have a biotechnological potential (Specian et al. 2012). Recently, bioactive compounds such as an antihypertensive angiotensin I converting enzyme inhibitor into an antiangiogenic compound, antidementia β-secretase inhibitor and ginsenoside-Rg3 were produced and characterized to form *Saccharomyces cerevisiae* (Jeong et al. 2006; Lee and Lee 2007; Kim et al. 2009).

Previously, researchers have reported many secondary metabolites from marine fungi. These include *Cephalosporin* sp. and *Stachybotrys* sp. which were isolated from seawater and these were reported to produce Cephalosporins and Stachybotrins A and B, respectively. Some other fungal genera of marine origin such as *Aspergillus* sp., *Penicillin* sp., *Penicillium* sp., and *Aspergillus* sp. were reported to produce Gliotoxin, Epolactaene,

Acetophthalidin, and Aspergillamides A and B, respectively. Sponge-associated fungi such as *Trichoderma harzianum, Microascus longirostris, Microsphaeropsis olivacea, Exophiala pisciphila, Aspergillus niger, Gymnascella dankaliensis, T. harzianum, Paecilomyces* cf. *javanica, Trichoderma longibranchiatum, Microsphaeropsis* sp., *Conoithyrium* sp., and *Drechslera hawaiiensis* produced industrially important secondary metabolites such as Trichoharzin, Cathestatins A–C, Crebroside, Exophilin A, Asperazine, Gymnastatins A–E, Trichodenones A–C, Harzialactones A–B, *R*-mavalonnlactones, Deoxynortrichoharzin and Speciferones A and B, Speciferol A.

2.2.3 *Bioactive compounds from cyanobacteria*

Cyanobacteria or blue-green algae are a charming group of primitive phototrophic prokaryotic organisms whose long evolutionary histories date back to the Proterozoic era. These organisms, endowed with tremendous genome plasticity, are distributed in all possible biotypes of the world. These organisms have tremendous potential in environmental management as soil conditioners. Due to their occurrence in diverse habitats, these organisms make excellent subject material for investigation by ecologists, physiologists, biochemists, pharmacists, and molecular biologists. Accordingly, looking for cyanobacteria with antimicrobial activity has gained importance in recent years. Biologically active substances were proved to be extracted by cyanobacteria (Borowitzka 1995; Kreitlow et al. 1999; Mundt et al. 2001; Volk and Furkert 2006). Various strains of cyanobacteria are known to produce intracellular and extracellular metabolites with diverse biological activities such as antialgal, antibacterial, antifungal, and antiviral activity (Moore et al. 1989; Jaki et al. 1999; Ghasemi et al. 2003, 2007; Isnansetyo et al. 2003; Kundim et al. 2003; Soltani et al. 2005).

Antibiotic resistance in bacteria is one of the emerging health-related problems in the world nowadays. Plants, among them algae, are valuable natural sources that are effective against infectious agents. Extensive efforts for the identification of bioactive compounds derived from natural resources have been made worldwide to develop safe, nontoxic, and efficient antimicrobial agents of valuable practice in pharmacology. Algae of marine and terrestrial origins have been the best choice among natural resources within the aquaculture and agriculture fields. Screening bioactivity of algal crude extracts is mandatory in biomedical practice, where antibacterial, antifungal, antiviral, and even more antialgal activity have been assessed to these metabolites. Emergence concerns have been raised to establish the structural and functional properties of the bioactive compounds described in algal crude extracts; thus far, more than 2400 bioactive metabolites have been isolated and identified from a diverse group of algal communities (Faulkner 2001; Hellio et al. 2002; Tang et al. 2002; Serkedjieva 2004; Tuney et al. 2006).

In a recent study, some biotechnologically important compounds from the crude extracts of some cynobacteria were reported (Al-Wathnani et al. 2012). These include *Phormidium autumnale* (1-hexyl-2-nitrocyclohexane; 91.7%), *Chlorella vulgaris* (3-methyl-2-butanol; 90.8%), *Spirulina platensis* (2-hexyl-1-nitrocyclohexane; 92.1%), *Nostoc nostoc* (octadecanal and aldehyde; 86.8%), *Dunaliella salina* (3-methyl-2-(2-oxopropyl) and furan; 90%), *Tolypothrix distorta* (boronic acid, ethyl, dimethyl ester; 83.9%), and last, *Microcystis aeruginosa*, which was reported to produce approximately 92% (*S*)-(+)-1-cyclohexylethylamine. Al-Wathnani et al. (2012) reviewed the potential compounds from the algae listed in Table 2.4.

Table 2.4 Cyanobacterial Metabolites and the Species Responsible for Synthesis

Organism(s)	Bioactive component(s)	Target organism
M. aeruginosa	Lipid	Algaecide
M. aeruginosa	Micropeptin 478-A Micropeptin 478-B Microginin 299-A,B Kawaguchipeptin B	Plasma inhibitor Leucin aminopeptidase Inhibitor Bactericide
M. aeruginosa	Aqueous extract	Antiviral (influenza A)
Microcystis viridis	Aeruginosin 102-A Aeruginosin 102-B	Thrombin inhibitor
Microcystis viridis	Micropeptin 103	Chymotrypsin inhibitor
M. aeruginosa (NIES-88)	Cyclic undecapeptide	*S. aureus*
Chlorella stigmatophora	N/R	*S. aureus; B. subtilis*
Synechococcus leopoliensis	N/R	*B. subtilis*
Laurencia chondrioides	Sesquiterpine	*S. aureus; B. subtilis, M. flavus*
Cystoseira tamariscifolia	Meroditerpenoid	*A. tumafasciens, E. coli*
Fucus vesiculosus	Polyhydroxylated Fucophlorethol	*S. aureus, E. coli, P. aeruginosa*
Scenedesmus sp.	N/R	*Alternaria* sp.
Scytonema julianum	Acetyl-sphingomyelin	Causes platelet aggregation
Scytonema varium	Scytovirin; Cyanovirin N	Anti-HIV protein
Haloferax alexandrinus	Canthaxanthin	Carotenoid
Microcystis PCC 7806	Cyanopeptolin 963A	Chymotrypsin inhibitor
Aphanizomenon flos-aquae	Phycocyanin	Antioxidant
M. aeruginosa	Zeaxanthin	Carotenoid
Oscillatoria redekei HUB 051	Unsaturated fatty acids	Antibacterial
Fischerella sp. CENA 19	Fischerellin A; 12-epi-hapalindole	Algaecide
Microcystis sp.	N/R	Antiviral (influenza A)
M. aeruginosa	Microviridin	Serum protease inhibitor
Schizotrix sp.	Schizotrin A	Antimicrobial

Source: Al-Wathnani, H. et al., *Journal of Medicinal Plants Research* 6(18), 3425–3433, 2012.

2.3 *Case study: Fatty acid analysis of antifungal active* Streptomyces parvulus *DOSMB D105 isolated from mangrove sediments of Andaman sea coast*

The term mangrove, in a broader sense, refers to the highly adapted plants found in tropical intertidal forest communities or ecosystem. The term "mangrove" may have been derived from a combination of the Malay word *manggi-manggi*, for a type of mangrove tree (Avicenna) and the Arabic *el gurm* for the same, as *mang-gurm*. As an ecosystem, mangroves are one of the most productive wetlands on earth. The mangrove ecosystem does not have as many species of plants and animals as other ecosystems due to the harsh and constantly changing environment (Lee et al. 2005).

The mangrove ecosystem is one of the largest ecosystems, and one that is widely distributed throughout the world. This also includes forests that occupy several million hectares of coastal area worldwide; distributed in more than 112 countries and territories comprising a total area of approximately 181,000 km^2 in more than one-fourth of the world's coastlines (Spalding et al. 1997; Alongi 2002). Mangroves are specialized ecosystems developed along estuarine seacoast and river mouths in tropical and subtropical regions of the world, mainly in the intertidal zone. Hence, the ecosystem and its biological components are under the influence of both marine and freshwater conditions. It has developed a set of physiological adaptations to overcome problems of anoxia, salinity, and frequent tidal inundations. This leads to the assemblage of a wide variety of plant and animal species with special adaptations suited to that ecosystem. The carbon fixed in mangroves is highly important in the coastal food webs, the litter from mangroves, and the subsequent formation of detritus and its tidal export have a profound effect on promoting the rich biodiversity. Mangroves are ecological habitats for a variety of microorganisms. They also play a valuable role in foreshore production, reducing erosion from cyclones and lessening the effects of storm surges.

Mangrove mud is rich in nutrients. Bacteria, fungi, and algae thrive on debris such as fallen mangrove leaves that are washed down by the river. These in turn will feed creatures that are present higher in the food chain. The roots themselves are covered by a vast range of sponges, anemones, marine algae, seaweed, oysters, and barnacles. These form the rich mangrove community. Mangrove microorganisms are an untapped source of metabolites and products with novel properties. Bacteria perform varied activities in the mangrove ecosystem such as photosynthesis, nitrogen fixation, methanogenesis, production of antibiotics and enzymes of specific activities (arylsulfatase, L-glutaminase, chitinase, L-asparginase, cellulase, protease, phosphatase), and others.

Chhiaki and coworkers (2007) stated that microorganisms that are present in the mangrove environment produce naturally and traditionally biologically active compounds, namely, bioactive substances. These natural organic compounds produced by microorganisms are important in screening the target for a variety of bioactive substances. Biologically active compounds produced by microorganisms in the mangrove may be a source for producing new bioactive compounds (Kokare et al. 2004a,b).

Bacteriological studies in the mangrove ecosystems are quantitative and a serious effort is necessary to identify the bacteria (Kathiresan et al. 1990). Most of the marine bacteria belong to the Gram-negative group, and the Gram-positive bacteria make up less than 10% of the total bacterial population (Zobell and Upham 1944). Now, evidence indicates that Gram-positive bacteria do occur at a higher percentage in sediments. Marine environments are a largely untapped source for the isolation of new microorganisms with the potential to produce active secondary metabolites (Baskaran et al. 2011).

It is now accepted that actinobacteria can be indigenous to the marine environment. It is promising to yield many unusual actinobacteria that have great potential as producers of novel antibiotics and other compounds. These actinobacteria are important in the field of pharmaceutical industries and also agriculture. Because of the discovery of novel antibiotic and nonantibiotic lead molecules through microbial secondary metabolite screening, it is becoming increasingly important.

Members of the marine actinobacteria are poorly understood and few reports are available regarding the existence of actinobacteria from mangroves (Lakshmanaperumalsamy 1978; Ratnakala and Chandtika 1993; Vikineshwari et al. 1997; Sivakumar 2001). Searching for novel actinobacteria constitutes an essential component in natural product-based drug discovery.

The marine sediment of India, especially in the Andaman and Nicobar Islands, have potentially enormous microbial biodiversity. Surprisingly, they have not been extensively explored for the registration of novel actinobacteria and its bioactive compounds. Keeping these in mind and recognizing the significance of marine actinobacteria as a source of novel bioactive compounds, the present study was made in an effort to isolate and screen for antifungally active *Streptomycetes* from the mangrove sediments collected from Andaman Islands, India.

The analysis of cellular fatty acid profiles in *Streptomyces* spp. has been well studied (Bowers et al. 1996; Seong et al. 2001; Magervey et al. 2004). Rapidly growing pathogenic aerobic actinobacteria was separated into two large groups based on the presence of a major amount of branched chain or of saturated or unsaturated straight-chain fatty acids (McNabb et al. 1997).

2.3.1 *Isolation of* Streptomyces *spp.*

The sediment samples were air-dried and ground aseptically with a mortar and pestle. The powdered samples were mixed thoroughly and passed through a 2-mm sieve to remove gravel and debris. The sample was kept at 70°C for 15 min (Hayakawa and Nonomura 1987; Hayakawa et al. 1991; Seong et al. 2001). Then, 10-fold serial dilutions of the sediment samples were made using sterile 50% seawater. The culture medium was prepared by using 50% seawater and contained the antibiotics cycloheximide and nalidixic acid (80 and 75 µg/mL, respectively; HiMedia, Mumbai) to prevent other bacterial and fungal growth (Baskaran et al. 2011). Approximately 0.1 mL of the serially diluted samples were spread over Kuster's agar containing glycerol (10 mL), casein (0.3 g), KNO_3 (2 g), K_2HPO_4 (2 g), NaCl (2 g), $MgSO_4·7H_2O$ (0.05 g), $CaCO_3$ (0.02 g), $FeSO_4$ (0.01 g), agar (18 g), seawater (500 mL), and distilled water (500 mL at pH 7 ± 2).

The plates were incubated at 28°C ± 2°C. After 15 days, the actinobacteria colonies grown on Petri plates were counted and recorded. The actinobacterial colony was purified on yeast extract–malt extract agar medium ISP2 (International *Streptomyces* Project, as described by Shirling and Gottlieb 1966) by streak plate technique and were maintained in ISP2 agar slants for further investigation. The isolate was identified based on the colony morphology, aerial mycelia, substrate mycelia, and microscopic examination, which was subsequently used for further study.

2.3.2 *Primary screening for antifungal activity*

The Streptomycetes was screened using a cross streak method (Lemos et al. 1985) for antagonistic activity. The strain was streaked on potato dextrose agar medium and incubated at 28°C ± 2°C. After observing a good ribbon-like growth of the *Streptomyces* on the Petri plates, the fungal pathogens such as *A. niger, Aspergillus flavus, Aspergillus fumigates, Penicillium* sp. *Fusarium* sp., and *Candida magnolia* were streaked at right angles to the original streak of *Streptomyces* and incubated at 28°C ± 2°C. The inhibition zone was measured after 48 to 72 h of incubation.

2.3.3 In vitro *assay for antifungal activity*

Streptomyces was inoculated into a 500 mL conical flask containing 200 mL of production medium containing dextrose (20 g), soya bean (20 g), soluble starch (5 g), peptone (5 g), $(NH_4)_2SO_4$ (2.5 g), $MgSO_4·7H_2O$ (0.25 g), K_2HPO_4 (0.024 g), NaCl (4 g), $CaCO_3$ (2 g), seawater (500 mL), and distilled water (500 mL, pH 7 ± 2) and incubated at 28°C ± 2°C for 7 days in rotary shaker at 250 rpm. The fermented broth was centrifuged at 10,000 rpm at 4°C for 20 min and the supernatant was filtered using a 0.45 µm membrane filter (Millipore).

An equal volume of ethyl acetate, ethanol, methanol, chloroform, and hexane was added separately to the cell-free culture filtrates and shaken for 12 h. The solvent extracts were evaporated in aseptic conditions and the crude powder was collected as the antimicrobial compound. This powder was mixed with dimethyl sulfoxide (DMSO) to evaluate its activity against pathogens (Sambamurthy and Ellaiah 1974). One yeast-like fungus and five filamentous fungi were inoculated into Sabouraud's dextrose broth and incubated for 24 to 48 h. The test organisms were swabbed over on the surface of Sabouraud's dextrose media. After solidification, wells were made using sterile cork borer. Then, the extracts were added separately in the wells in triplicate, and incubated for 48 to 72 h. After incubation, the zone of inhibition around the well was measured as millimeters in diameter. The maximum zone of inhibition obtained from ethyl acetate extract of *Streptomyces parvulus* DOSMB-D105 against *A. niger* was 25.66 mm, followed by *C. magnolia* (20.66 mm), *A. flavus* (20.33 mm), *A. fumigatus* (19.33 mm), *Penicilium* sp. (18.66 mm), and *Fusarium* sp. (17.33 mm; Figure 2.1).

Out of five different solvents used to extract the antimicrobial compounds of streptomycetes, only ethyl acetate extract possesses remarkable activity against all the pathogens tested. Whereas the methanol, chloroform, ethanol, and hexane extract exhibited moderate to minimum activity. They did not show inhibition to all the pathogens tested. Methanol extract of *S. parvulus* DOSMB-D105 showed moderate activity against 11 pathogens tested followed by chloroform, hexane, and ethanol (Table 2.5).

Ethyl acetate extract from both *Streptomyces* spp. showed maximum activity against all the pathogens when compared with standard antibiotic (erythromycin). The present

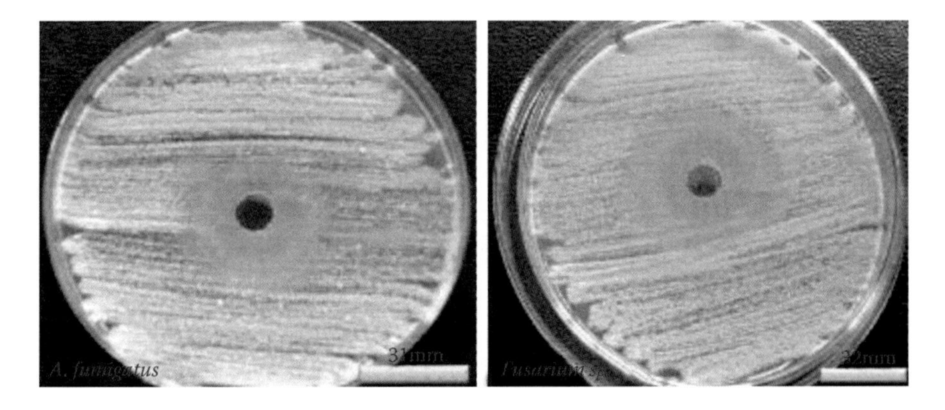

Figure 2.1 Antifungal activities of ethyl acetate extract of *Streptomyces parvulus* DOSMB-D105.

Table 2.5 Antifungal Activity of *S. parvulus* DOSMB.D105 Extracted by Different Solvents

S. No	Pathogens	Methanol	Chloroform	Ethanol	Ethyl acetate	Hexane	Control DMSO
1	*A. niger*	23	–	14.66	25.66	18.66	3.33
2	*A. flavus*	18.66	13.33	–	20.33	16	–
3	*A. fumigatus*	16.33	–	10.66	19.33	14.66	3.66
4	*Penicillium* sp.	15.66	13	14	18.66	13.33	–
5	*Fusarium* sp.	14.33	11	11.66	17.33	–	3
6	*C. magnolia*	17.66	12.66	12.66	20.66	14.33	–

results suggest that the *Streptomyces* sp. DOSMB-D105 has potential antibiotic activity against human pathogens when compared with standard antibiotics. Therefore, the ethyl acetate extract of *Streptomyces* spp. were forwarded for GC-MS analysis for the detailed investigation of antimicrobial compounds.

Similarly, Vijayakumar and coworkers (2011) have reported on the antimicrobial activity of different solvent extracts of *Streptomyces* sp. (VPTSA18) against two species of fungi. The ethyl acetate extract was highly active against *C. albicans* (17 mm), the other solvent extracts had moderate to minimum inhibitory effects against the test pathogens. This also confirmed the efficacy of the extracted compounds from the present study.

Researchers evaluated the antimicrobial efficacies of the actinobacteria using separate solvents including *n*-butanol, chloroform, ethyl acetate, and methanol; ethyl acetate, methanol, chloroform, and alcohol; and petroleum ether, *n*-butanol, and ethyl acetate (Sahin and Ugur 2003; Thangadurai et al. 2004; Ilic et al. 2005; Taechowisan et al. 2005; Remya and Vijayakumar 2008; Vimal et al. 2009; Usha et al. 2010), and they have reported that the ethyl acetate solvent extract had promising activity against most pathogens.

Similar types of studies have also been reported by many researchers (Saadoun and Al-Momani 2000; Balagurunathan and Subramanian 2001; Bordoloi et al. 2001; Narayana et al. 2005). The above reports confirm the present approach of the ethyl acetate solvent for the extraction of antimicrobial compounds from marine actinobacteria.

2.3.4 *Analysis of fatty acids using the FAME method*

Relationships among the *Streptomyces* isolates was evaluated based on the analysis of cellular fatty acids (Kroppenstedt 1985). *Streptomyces* isolates were grown on starch casein agar for 7 to 10 days at 28°C ± 2°C. Spores were scraped from the surface of the medium and inoculated into 20 mL of starch casein broth in clean conical flask containing approximately 5.0 g of 3 mm diameter glass beads. The culture was incubated on a rotary shaker at 225 rpm at 28°C ± 2°C for 72 h. Mycelia were harvested by centrifugation (10 min at 4000 rpm on a benchtop centrifuge) and the growth medium was decanted. Approximately 200 to 220 mg wet weight of mycelia was placed in a screw cap test tube and the samples were processed. In this process, cellular fatty acids were saponified, and methylated to form fatty acid methyl esters (FAME). FAME was then extracted from the aqueous phase into an organic phase, washed, and stored in the freezer until analysis. FAME was separated using a Hewlett-Packard 5890, a gas chromatograph fitted with column DEGC flame ionization detector. Chromatography output of FAME peak was analyzed, named, measured, and expressed in milligrams of the total fatty acid content. The fatty acid profile was determined by gas chromatography. The fatty acid retention times, responses, area/heights, response factors, equivalent chain lengths, peak names, and percentage shown by gas chromatography for both *Streptomyces* sp. is depicted in Figure 2.2 and Table 2.6.

Streptomyces sp. DOSMB-A105 had 26 different types of fatty acids in their cell. Among them, seven were saturated fatty acids and 19 were unsaturated fatty acids (Table 2.7). The strain DOSMB-D105 showed similar indices of 0.538 for *Streptomyces lavendulae*.

Analysis of the fatty acid profile to characterize taxa at the generic and suprgeneric level have certain limitations. However, quantitative analysis of fatty acid profiles provided useful taxonomic information at the species and, in some cases, at the subspecies level. Differences in fatty acid composition influenced membrane functions, particularly membrane permeability, which consequently could favor the synthesis of antimicrobial compounds (Gesheva and Rachev 1998, 2000; Kinked et al. 1998).

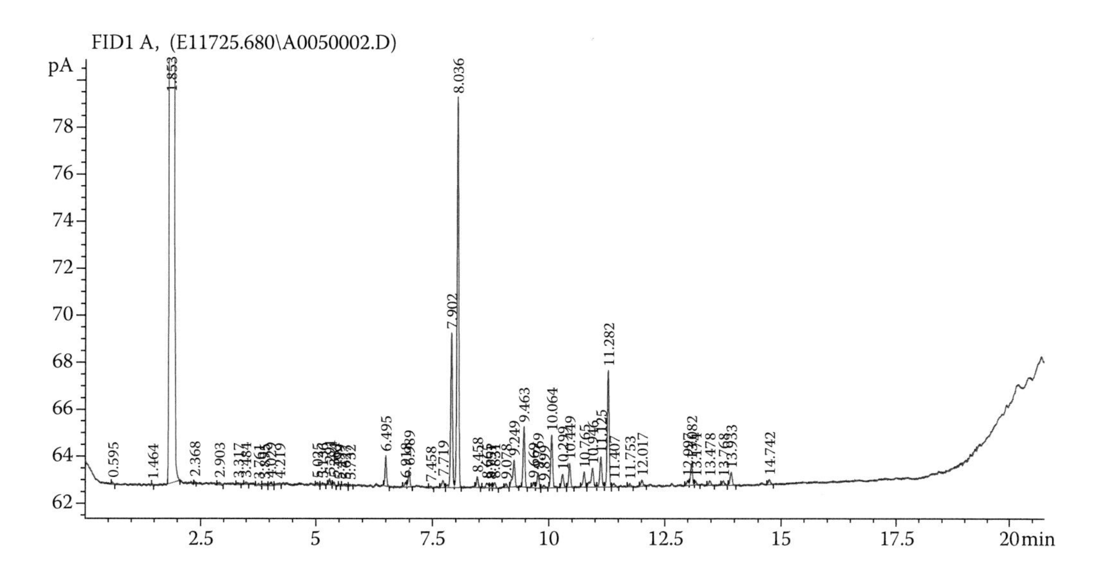

Figure 2.2 GC spectrum of potential strain DOSMB-D105 for fatty acid analysis.

Table 2.6 GC data of Fatty Acids Analysis of *Streptomyces* sp. DOSMB-D105

RT	Response	Ar/Ht	RFact	ECL	Peak name	%	Comment 1	Comment 2
0.595	408	0.027	–	4.227		–	< min rt	
1.464	145	0.015	–	6.168		–	< min rt	
1.853	4.07E+8	0.029	–	7.036	SOLVENT PEAK	–	< min rt	
2.368	357	0.022	–	8.186		–	< min rt	
2.903	574	0.041	1.264	9.381	8:0 3OH	0.31	ECL deviates −0.004	
3.761	328	0.032	1.126	10.930	Sum In Feature 3	0.16	ECL deviates 0.002	unknown 10.928
5.291	1072	0.030	1.024	12.613	13:0 ISO	0.47	ECL deviates 0.001	Reference −0.003
5.384	595	0.032	1.020	12.701	13:0 ANTEISO	0.26	ECL deviates 0.000	Reference −0.004
5.499	200	0.027	–	12.809		–		
5.627	634	0.053	1.010	12.928	13:1 AT 12-13	0.27	ECL deviates −0.003	
6.495	6106	0.038	0.984	13.616	14:0 ISO	2.57	ECL deviates −0.002	Reference −0.005

(continued)

Table 2.6 (Continued) GC Data of Fatty Acids Analysis of *Streptomyces* sp. DOSMB-D105

RT	Response	Ar/Ht	RFact	ECL	Peak name	%	Comment 1	Comment 2
6.989	3712	0.040	0.971	14.000	14:0	1.54	ECL deviates 0.000	Reference −0.003
7.458	188	0.022	–	14.319		–		
7.719	2034	0.045	0.958	14.497	unknown 14.503	0.83	ECL deviates −0.006	
7.902	35,447	0.042	0.955	14.622	15:0 ISO	14.50	ECL deviates 0.001	Reference −0.002
8.036	87,321	0.042	0.953	14.713	15:0 ANTEISO	35.64	ECL deviates 0.002	Reference −0.001
8.458	2402	0.043	0.947	15.000	15:0	0.97	ECL deviates 0.000	Reference −0.003
9.078	814	0.034	0.940	15.386	16:1 ISO E	0.33	ECL deviates 0.000	
9.249	7860	0.051	0.939	15.492	Sum In Feature 3	3.16	ECL deviates 0.004	14:0 3OH/16:1 ISO I
9.463	13,928	0.043	0.937	15.625	16:0 ISO	5.59	ECL deviates −0.001	Reference −0.004
9.669	1381	0.039	0.935	15.753	16:1 A	0.55	ECL deviates −0.004	
9.769	3872	0.043	0.934	15.815	16:1 CIS 9	1.55	ECL deviates −0.002	
9.899	1032	0.056	–	15.896		–		
10.064	13,131	0.046	0.932	15.999	16:0	5.24	ECL deviates −0.001	Reference −0.005
10.299	3518	0.048	0.931	16.138	15:0 ISO 3OH	1.40	ECL deviates 0.003	
10.449	5680	0.041	–	16.227		–		
10.765	4093	0.047	0.928	16.415	16:0 9? METHYL	1.63	ECL deviates −0.001	
10.946	6860	0.061	0.927	16.523	17:1 ANTEISO C	2.72	ECL deviates −0.002	

(continued)

Table 2.6 (Continued) GC Data of Fatty Acids Analysis of *Streptomyces* sp. DOSMB-D105

RT	Response	Ar/Ht	RFact	ECL	Peak name	%	Comment 1	Comment 2
11.125	8073	0.050	0.926	16.629	17:0 ISO	3.20	ECL deviates 0.000	Reference −0.005
11.282	28,577	0.046	0.926	16.722	17:0 ANTEISO	11.33	ECL deviates 0.000	Reference −0.005
11.407	862	0.039	0.925	16.796	17:1 CIS 9	0.34	ECL deviates 0.004	
	732	0.042	0.924	17.001	17:0	0.29	ECL deviates 0.001	Reference −0.004
12.017	1579	0.047	0.923	17.154	16:0 ISO 3OH	0.62	ECL deviates 0.009	
12.997	1256	0.046	0.921	17.721	Sum In Feature 6	0.50	ECL deviates 0.001	18:2 CIS 9,12/18:0a
13.082	5733	0.050	0.921	17.771	18:1 CIS 9	2.26	ECL deviates 0.002	
13.174	773	0.032	0.921	17.823	Sum In Feature 7	0.30	ECL deviates 0.001	18:1 CIS 11/t 9/t 6
13.478	1200	0.039	0.921	18.000	18:0	0.47	ECL deviates 0.000	Reference −0.007
13.768	1123	0.039	0.921	18.168	17:0 ISO 3OH	0.44	ECL deviates 0.004	
13.933	3562	0.051	–	18.263		–		
14.742	1377	0.043	0.920	18.732	19:0 ANTEISO	0.54	ECL deviates 0.003	
–	8189	–	–	–	Summed Feature 3	3.32	12:0 ALDE?	unknown 10.928
–	–	–	–	–		–	16:1 ISO I/14:0 3OH	14:0 3OH/16:1 ISO I
–	1256	–	–	–	Summed Feature 6	0.50	18:2 CIS 9,12/18:0a	18:0 ANTEISO/ 18:2 c
–	773	–	–	–	Summed Feature 7	0.30	18:1 CIS 11/t 9/t 6	18:1 TRANS 9/ t6/c11

Note: ECL deviation, 0.003; reference ECL shift, 0.004; number of reference peaks, 13; total response, 267,477; total named, 246,967; percent named, 92.33%; total amount, 233,575. Matches: library, ACTIN1 3.80; similar index, 0.538; and entry name, *Streptomyces lavendulae*.

Table 2.7 Presence of Saturated and Unsaturated Fatty Acid in *S. parvulus* DOSMB-D105

Peak name/Fatty acid name	Systemic name	DOSMB-D105	Saturated/Unsaturated
13:0 ISO	Isotridecanioc acid	+	Unsaturated
13:0 ANTEISO	Anteisotridecanioc acid	+	Unsaturated
14:0 ISO	Isotetradecanioc acid	+	Unsaturated
15:0 ISO	Isopentadecanioc acid	+	Unsaturated
15:0 ANTEISO	Anteisopentadecanioc acid	+	Unsaturated
16:0 ISO	Isopalmitic acid	+	Unsaturated
16:0	Palmitic acid	+	Saturated
15:0 ISO 3OH	Trihydroxydecanioc acid	+	Unsaturated
17:1 ANTEISO C	Anteisoheptadecanoic acid	+	Unsaturated
17:0 ISO	Isoheptadecanoic acid	+	Unsaturated
17:0 ANTEISO	Anteisoheptadecanoic acid	+	Unsaturated
16:0 ISO 3OH	Isotrihydroxyshexadecanoic acid	+	Unsaturated
18:1 CIS 9	Oleic acid	+	Unsaturated
18:0	Stearic acid	+	Saturated
17:0 ISO 3OH	Isotrihydroxyheptadecanoic acid	+	Unsaturated
19:0 ANTEISO	Anteisononadecylic acid	+	Unsaturated
08:03 OH	Trihydroxyoctanoic acid	+	Unsaturated
13:1 AT 12-13	Tridecenoic acid	+	Saturated
15:0	Pentadecylic acid	+	Saturated
16:1 ISO E	Isopalmitoleic acid	+	Unsaturated
16:1 A	Palmitoleic acid	+	Unsaturated
16:0 9 METHYL	9-Methyltetradecanoate	+	Saturated
14:0	Myristic acid	+	Saturated
17:1 CIS 9	*cis*-9-Heptadecenoic acid	+	Unsaturated
17:0	Margaric acid	+	Saturated
16:1 CIS 9	*cis*-9-Hexadecenoic	+	Unsaturated

2.4 Conclusion

New antibiotic compounds with novel mechanisms of action have become essential in response to the recent paradigm of increasingly resistant strains of bacteria. Secondary metabolites as natural and bioactive compounds from fungus, cyanobacteria, and bacteria such as actinobacteria serve as lead compounds for the development of pharmaceutical drugs. These drugs are widely used to fight bacterial, viral, and fungal infections, as well as cancer and immune system disorders.

References

Abbott, W.S. A method for computing the effectiveness of an insecticide. *J. Econ. Entomol.* (1925): ***Vol. 18***: pp. 265–267.

Abraham, E.P., and Loder, P.B. In: *Cephalosporins and Penicillins; Chemistry and Biology*, Flynn, E.H. ed. Academic Press, New York (1972): pp. 1–26.

Agarwal, A.K., Mehrotra, S., Mondal, H., Trivedi, P.D., Sultana, S., Arif, A.J., Jaffri, B.J., Chandra, B., and Srivastava, O.P. Isolation of actinobacteria from different soil types of Uttar Pradesh and their antagonistic properties. *Indian J. Microbiol.* (1968): ***Vol. 8***: pp. 225–235.

Ahern, M., Verschueren, S., and van Sinderen, D. Isolation and characterization of a novel bacteriocin produced by *Bacillus thuringiensis* strain B439. *FEMS Microbiol. Lett.* (2003): ***Vol. 220***: pp. 127–131.

Alexopolus, C.J. Studies on antibiosis between bacteria and fungi II. Species of actinobacteria inhibiting the growth of *Colletotrichum gloesporioides* Penz. In culture. *Ohio. J. Sci.* (1941): ***Vol. 41***: pp. 425–430.

Alker, A.P., Smith, G.W., and Kim, K. Characterisation of *Aspergillus sydowii* (Thom *et* Church), a fungal pathogen of Caribbean sea fan corals. *Hydrobiologia.* (2001): ***Vol. 460***: pp. 105–111.

Alongi, D.M. Present state and future of the world's mangrove forests. *Environ. Conserv.* (2002): ***Vol. 29***: pp. 331–349.

Altschul, S.F., Gish, W., Miller, W., Myers, E.W., and Lipman, D.J. Basic local alignment search tool. *J. Mol. Biol.* (1990): ***Vol. 215***: pp. 403–411.

Al-Wathnani, H., Ara, I., Tahmaz, R.R., Al-Dayel, T.H., and Bakir, M.A. Bioactivity of natural compounds isolated from cyanobacteria and green algae against human pathogenic bacteria and yeast. *J. Med. Plants Res.* (2012): ***Vol. 6(18)***: pp. 3425–3433.

Azam, F., and Worden, A.Z. Microbes, molecules and marine ecosystems. *Science.* (2004): ***Vol. 303***: pp. 1622–1624.

Baam, R.B., Gandhi, N.M., and Freitas, Y.M. Antibiotic activity of marine microorganisms: The antibacterial spectrum. *Helgol. Wiss. Meeresunters.* (1966): ***Vol. 13***: pp. 188–191.

Balagurunathan, R., and Subramanian, A. Antagonistic *Streptomyces* from marine sediments. *Adv. Biosci.* (2001): ***Vol. 20***: pp. 71–76.

Banerjee, A.K., Mukherjee, S.K., and Nandi, P. Production of antifungal substances by *Streptomyces* spp. isolated from Indian soil. *Sci. Cult.* (1954): ***Vol. 20***: pp. 141–143.

Baskaran, R., Vijayakumar, R., and Mohan, P.M. Enrichment method for the isolation of bioactive actinobacteria from mangrove sediments of Andaman Island, India. *Malaysian J. Microbiol.* (2011): ***Vol. 7***: pp. 22–28.

Baslow, M.H. *A Study of Twins and Other Biological Active Substances as Marine Origin: Marine Pharmacology*. The Williams and Wilkins Co., Baltimore, MD (1969): p. 286.

Bechard, J., Eastwell, K.C., Sholberg, P.L., Mazza, G., and Skura, B. Isolation and partial chemical characterization of an antimicrobial peptide produced by a strain of Bacillus subtilis. *J. Agric. Food Chem.* (1998): ***Vol. 46***: pp. 5355–5361.

Bennett, J.W., and Chung, K.T. Alexander Fleming and the discovery of penicillin. *Adv. Appl. Micro.* (2001): ***Vol. 49***: pp. 163–184.

Benyhe, S. Morphine: New aspects in the study of an ancient compound. *Life Sci.* (1994): ***Vol. 55***: pp. 969–979.

Berdy, J. The discovery of new bioactive microbial metabolites: Screening and identification. In: Bushell, M.E., and Grafe, U. eds. Bioactive metabolites from microorganisms. *Progress in Industrial Microbiology.* (1989): ***Vol. 27***: pp. 3–25.

Berdy, J. Bioactive microbial metabolites: A personal view. *J. Antibiot.* (2005): ***Vol. 58(1)***: pp. 1–26.

Bernan, V.S., Greenstein, M., and Maiese, W.M. Marine microorganisms as a source of new natural products. *Adv. Appl. Microbiol.* (1997): ***Vol. 43***: pp. 57–90.

Bertoldo, C., and Antranikian, G. Starch-hydrolyzing enzymes from thermophilic archea and bacteria. *Curr. Opin. Chem. Biol.* (2002): ***Vol. 2***: pp. 151–160.

Bizani, D., and Brandelli, A. Characterization of a bacteriocin produced by a newly isolated *Bacillus* sp. strain 8 A. *J. Appl. Microbiol.* (2002): ***Vol. 93***: pp. 512–519.

Blunt, W., Copp, B.R., Muro, M.H., Northcote, P.T., and Prinsep, M.R. Marine natural products. *Nat. Prod. Rep.* (2003): ***Vol. 20***: pp. 48.192.

Blunt, W., Copp, B.R., Muro, M.H., Northcote, P.T., and Prinsep, M.R. Marine natural products. *Nat. Prod. Rep.* (2004): ***Vol. 21***: pp. 1–49.

Bordoloi, G.N., Kumarim, B., Guha, A., Bordoloi, M., Yadav, R.N., Roy, M.K., and Bora, T.C. Isolation and structure elucidation of a new antifungal and antibacterial antibiotic produced by *Streptomyces* sp. 201. *Biosci. Biotechnol. Biochem.* (2001): ***Vol. 65***: pp. 1856–1858.

Borowitzka, M.A. Microalgae as sources of pharmaceuticals and other biologically active compounds. *J. Appl. Phycol.* (1995): ***Vol. 7***: pp. 3–15.

Bower, C.K., Bothwell, M.K., and McGuire, J. Lantibiotics as surface active agents for biomedical applications. *Colloids Surf. B Biointerfaces.* (2001): ***Vol. 22***: pp. 259–265.

Bowers, J.H., Kinkel, L.L., and Jones, R.K. Influence of disease suppressive strains of *Streptomyces* in soil as determined by the analysis of cellular fatty acids. *Can. J. Microbiol.* (1996): ***Vol. 42***: pp. 27–37.

Brammavidhya, S., and Usharani. Isolation and purification of low molecular weight peptide from marine *B. cereus* and its antimicrobial activity. *Int. J. Res. Mar. Sci.* (2013): ***Vol. 2(1)***: pp. 1–5.

Bredholt, H., Fjaervik, E., Jhon, G., Sergey, B., and Zotechev. Actinobacteria from sediments in the Trondheim Fjord. Norway, Diversity and biological activity. *Mar. Drugs.* (2008): ***Vol. 6***: pp. 12–24.

Bugni, T.S., and Ireland, C.M. Marine-derived fungi: A chemically and biologically diverse group of microorganisms. *Nat. Prod. Rep.* (2004): ***Vol. 21***: pp. 143–163.

Burkholder, P.R., Pfister, R.M., and Leitz, F.H. Production of a pyrrole antibiotic by a marine bacterium. *Appl. Environ. Microbiol.* (1966): ***Vol. 14***: pp. 649–653.

Busi, S., Peddikotla, P., Suryanarayana, Upadyayula, M., and Yenamandra, V. Isolation and biological evaluation of two bioactive metabolites from *Aspergillus gorakhpurensis. Rec. Nat. Prod.* (2009): ***Vol. 3(3)***: pp. 161–164.

Butzke, D., and Piel, J. Genomic and metagenomic strategies to identify biosynthetic gene clusters in uncultivated symbionts of marine invertebrates. In: *Frontiers in Marine Biotechnology*, Proksch, P., and Müller, W.E.G. eds. Horizon Bioscience, Norfolk, NE (2006): pp. 327–356.

Cao, I., Qiu, Z., You, J., Tan, H., and Zhou, S. Isolation and characterization of endophytic *Streptomyces* strains from surface sterilized tomato (*Lycopersicon esculentum*) roots. *Lett. Appl. Microbiol.* (2004): ***Vol. 39***: pp. 425–430.

Carrillo, C., Teruel, J.A., Aranda, F.J., and Ortiz, A. Molecular mechanism of membrane permeabilization by the peptide antibiotic surfactin. *Biochim. Biophys. Acta.* (2003): ***Vol. 1611***: pp. 91–97.

Carte, B.K. Biomedical potential of marine natural products. *Bioscience.* (1996): ***Vol. 46***: pp. 271–286.

Carvajal, F. Studies on the structure of *Streptomyces griseus. Mycologia.* (1946): ***Vol. 38***: pp. 587–595.

Chakrabarty, S., and Chandra, A.L. Antifungal activity of A-7, a new tetraene antibiotic. *Indian J. Exp. Biol.* (1979): ***Vol. 17***: pp. 313–315.

Chen, H., and Hoover, D.G. Bacteriocins and their food applications. *Comp. Rev. Food. Sci. Food Saf.* (2003): ***Vol. 2***: pp. 82–100.

Cherif, A., Hassen, H., Japua, S., and Boudabous, A. Thurin 7: A novel bacteriocin produced by *Bacillus thuringensis* BMG 1.7, a new strain isolated from soil. *Lett. Appl. Microbiol.* (2001): ***Vol. 32***: pp. 243–247.

Chhiaki, I., Noako, K., Masazumi, K., Takeshi, and Naoko, H. Isolation and characterization of antibacterial substances produced by marine actinobacteria in the presence of seawater. *Actinomcetologica.* (2007): ***Vol. 21***: pp. 27–31.

Choi, J.D., and Park, U.Y. Identification of the marine microorganisms producing bioactives. 1. Isolation and cultural conditions of the marine actinobacteria no. 101 producing antimicrobial compounds. *Bull. Korean Fish. Soc.* (1993): ***Vol. 26***: pp. 305–311.

Christopher, F.M.M., Gandhi, N., Chatterjee, S., and Ganguli, B.N. Swalpamycin, a new macrolide antibiotic. *J. Antibiot.* (1987): ***Vol. 11***: pp. 1361–1367.

Clark, A.M. Natural products as a resource for new drugs. *Pharm. Res.* (1996): ***Vol. 13(8)***: pp. 1133–1141.

Cragg, G.M., Newman, D.J., and Snader, K.M. Natural products in drug discovery and development. *J. Nat. Prod.* (1997): ***Vol. 60***: pp. 52–60.

Das, S., Lyla, P.S., and Khan, S.A. Characterization and identification of marine actinomycetes existing systems, complexities and future directions. *Natl. Acad. Sci. Lett.* (2008): ***Vol. 31***: pp. 149–160.

Datta, I., Banerjee, M., Mukherjee, S.K., and Majumdar, S.K. JU-2, a novel phosphorous-containing antifungal antibiotic from *Streptomyces kanamyceticus* M8. *Indian J. Exp. Biol.* (2001): ***Vol. 39***: pp. 604–606.

Demain, A.L. New applications of microbial products. *Science.* (1983): ***Vol. 219***: pp. 709–714.

Dhanasekaran, D., Sivamani, P., Panneerselvam, A., Thajuddin, N., Rajakumar, G., and Selvamani, S. Biological control of Tomato seedling damping off with *Stretomyces* sp. *Pl. Pathol. J.* (2005a): ***Vol. 4***: pp. 91–95.

Dhanasekaran, D., Rajakumar, G., Sivamani, P., Selvamani, S., Panneerselvam, A., and Thajuddin, N. Screening of saltpan actinobacteria for antibacterial agents. *Intern. J. Microbiol.* (2005b): ***Vol. 1***: pp. 1–6.

Dreyfuss, M.M., and Chapela, I.H. Potential of fungi in the discovery of novel, low molecular weight pharmaceuticals. In: *The Discovery of Natural Products with Therapeutic Potential*, Gullo, V.P. ed. Butterworth-Heinemann, Boston (1994): pp. 49–80.

Duncan, R.A., Sullivan, J.R., Alderman, S.C., Spatafora, J.W., and White, J.J.F. *Claviceps purpurea* var. *spartinae* var. nov.: An ergot adapted to the aquatic environment. *Mycotaxonomy.* (2002): ***Vol. 81***: pp. 11–25.

Ebel, R. Secondary metabolites from marine-derived fungi. In: *Frontiers in Marine Biotechnology*, Proksch, P., and Müller, W.E.G. eds. Horizon Bioscience, Norfolk, NE (2006): pp. 73–144.

Ellaiah, P., Adhinarayana, K., Naveen Babu, K., Thaer, A., Srinivasulu, B., and Prabhakar, T. Bioactive actinomycetes from marine sediments off Bay of Bengal near Machilipatinam. *Geobios.* (2002): ***Vol. 29***: pp. 97–100.

Ellaiah, P., Adhinarayana, K., Saisha, V., and Vasu, P. An oligoglyeosidic antibiotic from a newly isolated *Streptomyces albovinaceus. Ind. J. Microbiol.* (2005): ***Vol. 45***: pp. 33–36.

Emerson, R.L., Whiffen, A.J., Bohonos, N., and De Boe, C. Studies on the production of antibiotics by actinomycetes and molds. *J. Bacteriol.* (1946): ***Vol. 52***: pp. 357–365.

Euzeby, J.P. Genus *Streptomyces*. List of Prokaryotic names with staining in Nomenclature. (2008). Available at http://www.bacterio.cict.fr/S/Streptomycesa.html.

Faulkner, D.J. Marine natural products. *Nat. Prod. Rep.* (2000): ***Vol. 17***: pp. 7–55.

Faulkner, D.J. Marine natural products. *Nat. Prod. Rep.* (2001): ***Vol. 18***: pp. 1–4.

Feher, M., and Schmidt, J.M. Property distributions: Differences between drugs, natural products, and molecules from combinatorial chemistry. *J. Chem. Inf. Comput. Sci.* (2003): ***Vol. 43***: pp. 218–227.

Fenical, W. Chemical studies of marine developing a new resources. *Chem. Rev.* (1993): ***Vol. 93***: pp. 1673–1683.

Fenical, W. New pharmaceuticals from marine organisms. *Trends Biotechnol.* (1997): ***Vol. 15***: pp. 339–341.

Findlay, R.H., Fell, J.W., Colemanm, N.K., and Vestal, J.R. Biochemical indicators of the role of fungi and thraustochytrids in mangrove detrital systems. In: *The Biology of Marine Fungi*, Moss, S.T. ed. Cambridge University Press, Cambridge (1986): pp. 91–104.

Frandberg, E., Peterson, C., Lundgren, L.N., and Schnurer, J. *Streptomyces halstedii* K122 produces the antifungal compounds bafilomycin B1 and C1. *Can. J. Microbiol.* (2000): ***Vol. 46***: pp. 753–758.

Gandhi, N.M., Patell, J.R., Gandhi, J., De Souza, N.J., and Kohl, H. Prodigiosin metabolites of a marine Pseudomonas species. *Marine Biol.* (1976): ***Vol. 34(3)***: pp. 223–227.

Gasperni, G. Researches morphologiques et biologiques sur un microorganism de atmosphere, le *Streptothrix foersterii* Cohn. *Ann. Microgr.* (1890): ***Vol. 2***: pp. 449–474.

Geiser, D.M., Frisvad, J.C., and Taylor, J.W. Evolutionary relationships in Aspergillus section Fumigati inferred from partial-tubulin and hydrophobin DNA sequences. *Mycologia.* (1998a): ***Vol. 90***: pp. 831–845.

Geiser, D.M., Taylor, J.W., Ritchie, K.B., and Smith, W.G. Cause of sea fan death in the West Indies. *Nature.* (1998b): ***Vol. 394***: pp. 137–138.

Gesheva, V., and Rachev, R. Fatty acid profile of a phosphate deregulated and antibiotic hyperproducer mutant of *Streptomyces hygroscopicus* population. *Actinobacteria.* (1998): ***Vol. 9***: pp. 49–56.

Gesheva, V., and Rachev, R. Fatty acid composition of *Streptomyces hygroscopicus. Actinobacteria.* (2000): ***Vol. 9***: pp. 49–56.

Ghasemi, Y., Moradian, A., Mohagheghzadeh, A., Shokravi, S., and Morowvat, M.H. Antifungal and antibacterial activity of the microalgae collected from paddy fields of Iran: Characterization of antimicrobial activity of *Chroococcus disperses. J. Biol. Sci.* (2007): ***Vol. 7***: pp. 904–910.

Ghasemi, Y., Yazdi, M.T., Shokravi, S., Soltani, N., and Zarrini, G. Antifungal and antibacterial activity of paddy-fields cyanobacteria from the north of Iran. *J. Sci. Islamic Repub. Iran.* (2003): ***Vol. 14***: pp. 203–209.

Gochfeld, D.J., El Sayed, K.A., Yousat, M., Wilkins, S.P., Zjawiony, J.K., Schinazi, R.F., Schlueter, W.S., Pharnish, P.M., and Hamann, M.T. Marine natural products as lead anti-HIV agents. *Mini Rev. Med. Chem.* (2003): ***Vol. 3***: pp. 401–424.

Goodfellow, M., and Williams, S.T. Ecology of actinobacteria. *Ann. Rev. Microbiol.* (1983): ***Vol. 37***: pp. 189–216.

Grein, A., and Meyers, S.P. Growth characteristics and antibiotic production of actinomycetes isolated from littoral sediments and materials suspended in seawater. *J. Bacteriol.* (1958): ***Vol. 76***: pp. 457–468.
Gurusiddaiah, S., Winward, L.D., Burger, D., and Graham, S.O. Pantomycin, a new antimicrobial antibiotic. *Mycologia.* (1979): ***Vol. 71***: pp. 103–118.
Gustafson, K., Roman, M., and Fenical, W. The macrolactins, a novel class of antiviral and cytotoxic macrolides from a deep-sea marine bacterium. *J. Am. Chem.* Soc. (1989): ***Vol. 111***: pp. 7519–7524.
Hacene, H., Sabaou, N., Bounaga, N., and Lefebrvre, G. Screening for non-polyenic antifungal antibiotics produced by rare actinomycetales. *Microbios.* (1994): ***Vol. 79***: pp. 81–85.
Haefner, B. Drugs from the deep: Marine natural products as drug candidates. *Drug Discov. Today.* (2003): ***Vol. 8***: pp. 536–544.
Hawksworth, D.L. The fungal dimension of biodiversity: Magnitude, significance and conservation. *Mycol. Res.* (1991): ***Vol. 95***: pp. 641–655.
Hayakawa, M., Ishizawa, K., Yamazaki, T., and Nonomura, H. Distribution of antibiotic-producing *Microbispora* strains in soils with different pHs. *Actinobacteria.* (1995): ***Vol. 6***: pp. 12–17.
Hayakawa, M., and Nonomura, H. Efficacy of artificial humic acid is a selective nutrient in HV agar used for the isolation of actinobacteria. *J. Ferm. Tech.* (1987): ***Vol. 65***: pp. 609–616.
Hayakawa, M., Sadaka, T., Kayiura, T., and Nonomura, H. New methods for the highly selective isolation *Micromonospora* and *Microbispora. J. Tech.* (1991): ***Vol. 72***: pp. 320–326.
Hellio, C., Berge, J.P., Beaupoil, C., Le Gal, Y., and Bourgougnon, N. Screening of marine algal extracts for anti-settlement activities microalgae and macroalgae. *Biofouling.* (2002): ***Vol. 18***: pp. 205–215.
Hentschel, U., Usher, K.M., and Taylor, M.W. Marine sponges as mictobial fermenters. *FEMS Microbial. Ecol.* (2006): ***Vol. 55***: pp. 167–177.
Holler, U., Wright, A.D., Matthee, G.F., König, G.M., Draeger, S., Aust, H.J., and Schulz, B. Fungi from marine sponges: Diversity, biological activity and secondary metabolites. *Mycol. Soc. Phytochem.* (2000): ***Vol. 104***: pp. 1354–1365.
Hugenholtz, P., and Pace, N.R. Identifying microbial diversity in natural environment a molecular phylogenetic approach. *Trends Biotechnol.* (1996): ***Vol. 14***: pp. 190–197.
Hussain, A.M., and El-Gammal, A. An antibiotic produced by *Streptomyces violaceoniger. Egypt. J. Bot.* (1980): ***Vol. 23***: pp. 187–190.
Hyung, M.J., Kwang-Soo, K., Jong-Hyun, P., Myung-Woo, B., Young-Bae, K., and Han-Joon, H. Bacteriocin with a broad antimicrobial spectrum, produced by *Bacillus* sp. isolated from Kimchi. *J. Microbiol. Biotechnol.* (2001): ***Vol. 11(4)***: pp. 577–584.
Ikeda, H., and Omura, S. Genetic aspects of the selective production of useful components in the avermectin producer *Streptomyces avermitilis. Actinomycetologica.* (1993): ***Vol. 7***: pp. 133–144.
Ilic, S.B., Kontantinovic, S.S., and Todorovic, Z.B. UV/VIS analysis and antimicrobial activity of *Streptomyces* isolates. *Facta Universitatis, Med. Biol.* (2005): ***Vol. 12***: pp. 44–46.
Isnansetyo, A., Cui, L., Hiramatsu, K., and Kamei, Y. Antibacterial activity of 2,4-diacetylphloroglucinol produced by *Pseudomonas* sp. AMSN isolated from a marine algae, against vancomycin-resistant *Staphylococcus aureus. Int. J. Antimicrob. Agents.* (2003): ***Vol. 22***: pp. 545–547.
Jaki, B., Orjala, J., and Sticher, O. A novel extracellular diterpenoid with antibacterial activity from the cyanobacterium *Nostoc commune* EAWAG 122b. *J. Nat. Prod.* (1999): ***Vol. 62***: pp. 502–503.
Jensen, P.R., and Fenical, W. Marine bacterial biodiversity as a resources for novel microbial products. *J. Ind. Microbiol.* (1996): ***Vol. 17***: pp. 346–451.
Jensen, P.R., and Fenical, W. Secondary metabolites from marine fungi. In: *Fungi in Marine Environments*, Hyde, K.D. ed. Fungal Diversity Research Series 7. Fungal Diversity Press, Hong Kong (2002): pp. 293–315.
Jensen, P., Gontang, R., Mafnas, E., Mincer, T.T., and Fenical, W. Culturable marine actinobacteria diversity from tropical Pacific Ocean sediments. *Appl. Environ. Microbiol.* (2005): ***Vol. 95***: pp. 1039–1048.
Jeong, S.C., Lee, D.H., and Lee, J.S. Production and characterization of an anti-angiogenic agent from *Saccharomyces cerevisiae* K-7. *J. Microbiol. Biotechnol.* (2006): ***Vol. 16***: pp. 1904–1911.
Johnstone, D.B. Soil actinobacteria of Bikini Atoll with special reference to their antagonistic properties. *Soil Sci.* (1947): ***Vol. 64***: pp. 453–458.
Kampfer, P. The family Streptomycetaceae, Part I: taxonomy. In: *The Prokaryotes: A Handbook on the Biology of Bacteria*, Dworkin, M. et al. eds. Springer, Berlin (2006): pp. 538–604.

Kathiresan, K.S., Balagurunathan, R., and Masilamani Selvam, M. Fungicidal activity of marine actinobacteria against phytopathogenic fungi. *Indian J. Botechnol.* (2005): ***Vol. 4***: pp. 271–276.
Kathiresan, K., Ravikumar, S., Ravichandran, D., and Sakaravathy, K. Auxin-phenol-induced rooting in a mangrove, *Rhizophora apiculata* Blume. *Curr. Sci.* (1990): ***Vol. 59***: pp. 430–432.
Kieser, T., Bibb, M.J., Buttner, M.J., Chapter, K.F., and Hopwood, D.A. *Practical Streptomyces Genetics* (2nd ed.). John Innes Foundation, Norwich, England (2000). ISBN 0-7084-0623-8.
Kim, N.M., Lee, S.K., Cho, H.H., So, S.H., Jang, D.P., Han, S.T., and Lee, J.S. Production of ginsenoside-Rg3 enriched yeast biomass using ginseng steaming effluent. *J. Ginseng Res.* (2009): ***Vol. 33***: pp. 183–188.
Kinked, L.L., Boers, J.H., Shimizu, K., Neeno-Eclwall, E.C., and Schottel, J.S. Quantitative relationships among Thaxtomin A production, potato scap severity and fatty acid composition in *Streptomyces. Can. J. Microbiol.* (1998): ***Vol. 44***: pp. 768–776.
Klein, C., Kaletta, C., Schnell, N., and Entian, K. Analysis of genes involved in biosynthesis of the lantibiotic subtilin. *Appl. Environ. Microbiol.* (1992): ***Vol. 5***: pp. 132–142.
Knowles, D.J.C. New strategies for—drug design. *Trends Microbial.* (1997): ***Vol. 5(10)***: pp. 379–383.
Kohlmeyer, J., and Volkmann-Kohlmeyer, B. Fungi from coral reefs: A commentary. *Mycol. Res.* (2003): ***Vol. 107***: pp. 386–387.
Kokare, C.R., Mahadik, K.R., Kadam, S.S., and Chopade, B.A. Isolation, characterization and antimicrobial activity of marine halophilic *Actinopolyspora* species AH1 from west coast of India. *Curr. Sci.* (2004a): ***Vol. 86***: pp. 593–597.
Kokare, C.R., Mahadik, K.R., Kadam, S.S., and Chopade, B.A. Isolation of bioactive marine actinobacteria from marine sediments isolated from Goa and Maharashtra coastlines (West coast of India). *Indian J. Mar. Sci.* (2004b): ***Vol. 33***: pp. 248–256.
Konig, G.M., Kehraus, S., Seibert, S.F., Abdel-Lateff, A., and Muller, D. Natural products from marine organisms and their associated microbes. *ChemBioNews.* (2006): ***Vol. 7***: pp. 229–238.
Kook, H.B., and Kim, B.S. In vivo efficacy and in vitro activity of tubercidin, an antibiotic nucleoside, for control of *Phytophthora capsici* blight in *Capsicum annuum. Pest. Sci.* (1995): ***Vol. 44***: pp. 255–260.
Kopcke, B., Weber, R.W.S., and Anke, H. Galiellalactone and its biogenetic precursors as chemotaonomic markers of the Sarccosmataceae (Ascomycota). *Ohytochem.* (2002): ***Vol. 60***: pp. 709–714.
Krasilnikov, N.A., Koreniako, A.I., and Artamonava, O.I. The distribution of actinomycete—antagonists in soils. *Microbiology.* (1953): ***Vol. 22***: pp. 3–10.
Kreitlow, S., Mundt, S., and Linequist, U. Cyanobacteria a potential source of new biologically active substances. *J. Biotechnol.* (1999): ***Vol. 70***: pp. 61–63.
Kroppenstedt, R.M. Fatty acid and menaquinone analysis of actinobacteria and related organisms. In: *Chemical Methods in Biochemical Systematics*, Goodfollow and Minnikin eds. Society for Applied Bacteriology, Technical Series No. 20. Academic Press, London (1985): pp. 173–179.
Kundim, B.A., Itou, Y., Sakagami, Y., Fudou, R., Iizuka, T., Yamanaka, S., and Ojika, M. New haliangicin isomers, potent antifungal metabolites produced by a marine myxobacterium. *J. Antibiot.* (2003): ***Vol. 56***: pp. 630–638.
Laatsch, H. Marine bacterial metabolites. In: *Frontiers in Marine Biotechnology*, Proksch, P., and Muller, W.E.G. eds. Horizon Bioscience, Norfolk, NE (2006): pp. 225–288.
Lakshmanaperumalsamy, P. Studies on actinobacteria with special reference to antagonistic streptomycetes from sediments of Porto Novo coastal one. PhD Thesis, Annamalai University (1978).
Leben, C., Stessel, G.J., and Keitt, G.W. Helixin, and antibiotic active against fungi and bacteria. *Mycologia.* (1952): ***Vol. 44***: pp. 159–169.
Lechevalier, H.A., and Lechevalier, M.P. Biology of actinobacteria. *Annu. Rev. Microbiol.* (1967): ***Vol. 21***: pp. 71–100.
Lee, D.H., and Lee, J.S. Characterization of a new antidementia β-secretase inhibitory peptide from *Saccharomyces cerevisiase. Enzyme Microb. Technol.* (2007): ***Vol. 42***: pp. 83–88.
Lee, S.Y., Nakajima, I., Ihara, F., Kinoshita, H., and Nihira, T. Cultivation of entomophathogenic fungi for the search of antibacterial compounds. *Mycopathologia.* (2005): ***Vol. 160***: pp. 321–325.
Lemos, M.L., Toranzo, A.E., and Barja, J.L. Antibiotic activity of epiphytic bacteria isolated from intertidal seaweeds. *J. Microb. Ecol.* (1985): ***Vol. 11***: pp. 149–163.
Levy, S.P. The challenge of antibiotic resistance. *Sci. Am.* (1998): ***Vol. 278(3)***: pp. 32–39.

Lockwood, J.L. *Streptomyces* spp. as a case of natural fungi toxicity in soils. *Phytopathology.* (1959): ***Vol. 49***: pp. 327–331.
Magervey, N.A., Keller, J.M., Bernan, V., Dworkin, M., and Sherman, D.H. Isolation and characterization of novel marine-derived actinomycete taxa rich in bioactive metabolite. *Appl. Environ. Microbiol.* (2004): ***Vol. 70***: pp. 7520–7529.
Markey, R.P., Helmke, E., Kayser, O., Fiebig, H.H., and Mater, A. Anti-cancer and antibacterial Trioxacarcins with high anti malaria activity from A marine Streptomycete and their absolute Stereochemistry. *J. Antibiot.* (2004): ***Vol. 57***: pp. 771–779.
McNabb, A., Shuttleworth, R., Behme, R., and Colby, W.D. Fatty acid characterization of rapidly growing pathogenic aerobic actinobacteria as a means of identification. *J. Clin. Microbiol.* (1997): ***Vol. 35***: pp. 1361–1368.
Meredith, C.H. Soil actinobacteria applied to banana plants in the field. *Phytopathology.* (1946): ***Vol. 36***: pp. 983–987.
Moncheva, P., Tishkov, S., Dimitrova, N., Chipeva, V., Nikolova, S.A., and Bogatzevska, N. Characteristics of soil actinobacteria from Antartica. *J. Cul. Coll.* (2002): ***Vol. 3***: pp. 3–14.
Moore, R.E., Cheuk, C., Yang, X.G., and Patterson, G.M.L. Hapalindoles, antibacterial and antimycotic alkaloids from the cyanophyte *Hapalosiphon fontinalis. J. Org. Chem.* (1989): ***Vol. 52***: pp. 1036–1043.
Mundt, S., Kreitlow, S., Nowotny, A., and Effmert, U. Biochemical and pharmacological investigation of selected cyanobacteria. *Int. J. Hyg. Environ. Health.* (2001): ***Vol. 203***: pp. 327–334.
Muriru, W.M., Mutitu, E.W., and Mukunya, D.M. Identification of related actinobacteria. Isolates and characterization of their antibiotic metabolites. *J. Biol. Sci.* (2008): ***Vol. 8***: pp. 1021–1026.
Nair, M.G., Amitabh, C., Thorogod, D.L., and Chandra, A. Gopalamicin, an antifungal macrodiolide produced by soil actinobacteria. *J. Agric. Food Chem.* (1994): ***Vol. 42***: pp. 2308–2310.
Narayana, K.J.P., Ravikiran, D., and Vijayalakshmi, M. Production of antibiotics from *Streptomyces* species isolated from virgin soil. *Indian J. Microbiol.* (2004): ***Vol. 44***: pp. 147–148.
Narayana, K.J.P., Ravikiran, D., and Vijayalakshmi, M. Screening of *Streptomyces* species from cultivated soil for broad-spectrum antimicrobial compounds. *Asian J. Microbiol. Biotechnol. Environ. Sci.* (2005): ***Vol. 7***: pp. 121–124.
Newman, D.J., and Cragg, M.G. Natural products as sources of new drugs over the last 25 years. *J. Nat. Prod.* (2007): ***Vol. 70***: pp. 461–477.
Nithya, B., and Ponmurugan, P. Plasmid DNA of antibiotic producing strains of *Streptomyces sonnanesis* isolated from deferent states in southern India. *Biotechnology.* (2008): ***Vol. 7***: pp. 487–492.
Nolan, R.D., and Cross, T. Isolation and screening of actinobacteria. In: *Actinobacteria in Biotechnology*, Goodfellow, M., Williams, S.T., and Mordarski, M. eds. Academic Press, London (1998): pp. 1–32.
Ohkuma, H., Naruse, N., Nishiyama, Y., Tsuno, T., Hoshino, Y., Sawada, Y., Konishi, M., and Oki, T. Sultriecin, a new antifungal and antitumor antibiotic from *Streptomyces roseiscleroticus.* Production, isolation, structure and biological activity. *J. Antibiot.* (1992): ***Vol. 45***: pp. 1239–1249.
Onifade, A.K. Research trends: Bioactive metabolites of fungal origin. *Res. J. Biol. Sci.* (2007): ***Vol. 2(I)***: pp. 81–84.
Osborn, A.M., Moore, E.R.B., and Timmis, K.N. An evaluation of terminal-restriction fragment length polymorphisms (TRFLP) analysis for the study of microbial community structure and dynamics. *Environ. Microbiol.* (2000): ***Vol. 2***: pp. 39–50.
Oscariz, J.C., Lasa, I., and Pisabarro, A.G. Detection and characterization of cerein 7, a new bacteriocin produced by *Bacillus cereus* with a broad spectrum of activity. *FEMS Microbiol. Lett.* (1999): ***Vol. 178***: pp. 337–341.
Oskay, M., Tamer, A.U., and Azeri, C. Antibacterial activity of some actinobacteria isolated from farming soils of Turky. *Afr. J. Biotech.* (2004): ***Vol. 3***: pp. 441–446.
Oskouie, S.F.G., Tabandeh, F., Yakhchali, B., and Eftekhar, F. Enhancement of alkaline protease production by *Bacillus clausii* using Taguchi experimental design. *Afr. J. Biotechnol.* (2006): ***Vol. 6***: pp. 2559–2564.
Pandey, B., Ghimirel, P., Prasad, V., Thomas, M., Chan, Y., and Ozanick, S. Studies of the antimicrobial activity of the actinomycetes isolkated from the Khumby region of Nepal. Technical Report aquatic ecosystem and management society (AEHMS). (2002). Available at http://www.aehms.org.

Patil, R., Jeyasekaran, G., Shanmugam, S.A., and Shakila, R.J. Control of bacterial pathogens, associated with fish diseases, by antagonistic marine actinobacteria isolated from marine sediments. *Indian J. Mar. Sci.* (2001a): ***Vol. 30***: pp. 264–267.
Patil, R., Jeyasekran, G., and Shanmugam, S.A. Occurrence and activity of marine actinobacteria against shrimp bacterial pathogens. *Appl. Fish. Aqua.* (2001b): ***Vol. 1***: pp. 79–81.
Pisano, M.A., Sommer, M.J., and Brancaccio, L. Isolation of bioactive actinomycetes from marine sediments using Rifamcin. *Appl. Microbiol. Biotechnol.* (1989): ***Vol. 31***: pp. 609–612.
Pisano, M.A., Sommer, M.J., and Taras, L. Bioactivity of chitinolytic actinomycetes of marine origin. *Appl. Microbiol. Biotechnol.* (1992): ***Vol. 36***: pp. 553–555.
Proksch, P., Ebel, R., Edrada, R., Riebe, F., Liu, H., Diesel, A., Bayer, M. et al. Sponge-associated fungi and their bioactive compounds, the *Suberites* case. *Bot. Mar.* (2008): ***Vol. 51***: pp. 209–218.
Proksch, P., Edrada, R.A., and Ebel, R. Drugs from the seas–current status and microbiological implications. *Appl. Microbiol. Biotech.* (2002): ***Vol. 59***: pp. 125–134.
Prudhomme, J., McDaniel, E., Ponts, N., Bertani, S.P., Fenical, W., Jensen, P., and Le Roch, K. Marine actinobacteria: A new source of compounds against the human malaria parasite. *Plos One.* (2008): ***Vol. 3***: p. e2335.
Pugazhvendan, S.R., Kumaran, S., Alagappan, K.M., and Prasad, G. Inhibition of fish bacteriology pathogens by antagonistic marine actinobacteria. *Euro. J. App. Sci.* (2010): ***Vol. 2***: pp. 41–43.
Ramaswamy, A.V., Flatt, P.M., Edwards, D.J., Simmons, T.L., Han, B., and Gerwick, W.H. The secondary metabolites and biosynthetic gene clusters of marine cyanobacteria. Applications in biotechnology. In: *Frontiers in Marine Biotechnology*, Proksch, P., and Müller, W.E.G. eds. Horizon Bioscience, Norfolk, NE (2006): pp. 175–224.
Rangaswami, G., Oblisami, G., and Swaminathan, R. *Antagonistic Actinobacteria in Soils of South India.* Univer. Agri. Sci. Bangalore and USDA P.L., Visveswarapurum (1967): 480 (FG In129).
Ratnakala, R., and Chandtika, V. Effect of different media for isolation, growth and maintenance of actinobacteria from mangrove sediments. *Indian J. Mar. Sci.* (1993): ***Vol. 22***: pp. 297–299.
Ravenchlang, K., Salam, K., Pernthater, J., and Amann, R. High bacterial diversity in permanently cold marine sediments. *Appl. Environ. Microbial.* (1999): ***Vol. 65***: pp. 3982–3989.
Remya, M., and Vijayakumar, R. Isolation and characterization of marine antagonistic actinobacteria from west coast of India. *Med. Biol.* (2008): ***Vol. 15***: pp. 13–19.
Rizk, M., Rahman, T.A., and Metwally, H. Screening of antagonistic species against some pathogenic microorganisms. *J. Biol. Sci.* (2007): ***Vol. 7***: pp. 1418–1423.
Romero, F., Espliego, F., Perez Baz, J., Garcia de Quesada, T., Gravalos, D., De La Calle, F., and Fernadez-Puentes, J.L. Thiocoraline, a new depsipeptide with antitumor activity produced by a marine *Micromonospora.* I. Taxonomy, fermentation, isolation, and biological activities. *J. Antibiot (Tokyo).* (1997): ***Vol. 50***: pp. 734–737.
Rosenberg, J.M., Lin, Y.M., Lu, Y., and Miller, M.J. Studies and syntheses of siderophores, microbial iron chelators, and analogs as potential drug delivery agents. *Curr. Med. Chem.* (2000): ***Vol. 7***: pp. 159–197.
Rosenfeld, W.D., and ZoBell, C.E. Antibiotic production by marine microorganisms. *J. Bacteriol.* (1947): ***Vol. 54***: pp. 393–398.
Saadoun, I., and Al-Momani, F. Frequency of grey series streptomycetes in Jordan soils. *Actinobacteria.* (1998): ***Vol. 9***: pp. 61–65.
Saadoun, I., and Al-Momani, F. Activity of North Jordan streptomycete isolates against *Candida albicans. World J. Microbiol. Biotech.* (2000): ***Vol. 16***: pp. 139–142.
Sahin, N., and Ugur, A. Investigation of the antimicrobial activity of some isolates. *Turk. J. Biol.* (2003): ***Vol. 27***: pp. 79–84.
Sambamurthy, K., and Ellaiah, P. A new Streptomycete producing Neomycin (B and C) complex *S. marinensis* (Part–I). *Hind. Antibiot. Bull.* (1974): ***Vol. 17***: pp. 24–28.
Satoh, A., Murakami, T., Takebe, H., Imai, S., and Satoa, H. Industrial development of bialaphos, a herbicide from the metabolites of *Streptomyces hygroscopicus* (SF1293). *Actinomycetologica.* (1993): ***Vol. 7***: pp. 128–132.
Seong, C.H., Choi, J.H., and Baik, K.S. An improved selective isolation of rare actinobacteria from forest soil. *J. Microbiol.* (2001): ***Vol. 17***: pp. 23–39.
Serkedjieva, J. Antiviral activity of the red marine alga *Ceramium rubrum. Phytother. Res.* (2004): ***Vol. 18***: pp. 480–483.

Sharma, A.K., Gupta, J.S., and Singh, S.P. Effect of temperature on the antifungal activity of *Streptomyces arabicus* against *Alternaria brassicae*. *Geobios*. (1985): ***Vol. 12***: pp. 168–169.
Shimizu, Y., and Li, B. Microalgae as a source of bioactive molecules: Special problems and methodology. In: *Frontiers in Marine Biotechnology*, Proksch, P., and Müller, W.E.G. eds. Horizon Bioscience, Norfolk, NE (2006): pp. 145–176.
Shirling, E.B., and Gottlieb, D. Methods for characterization of *Streptomyces* species. *Int. J. Syst. Bact.* (1966): ***Vol. 16***: pp. 313–340.
Sivakumar, K. Actinobacteria of an Indian mangrove (Pitchavaram) environment: An inventory. PhD Thesis, Annamalai Univeristy, Tamil Nadu, India. (2001): p. 91.
Soltani, N., Khavari-Nejad, R.A., Tabatabaei Yazdi, M., Shokravi, S.H., and Fernandez-Valiente, E. Screening of soil cyanobacteria for antibacterial antifungal activity. *Pharm. Biol.* (2005): ***Vol. 43(5)***: pp. 455–459.
Spalding, M., Blasco, F., and Hield, C. *World Mangrove Atlas*. The International Society for Mangrove Ecosystem, Okinawa, Japan (1997): 178 pp.
Specian, V., Sarragiotto, M.H., Pamphile, J.A., and Clemente, E. Chemical characterization of bioactive compounds from the endophytic fungus *Diaporthe helianthi* isolated from *Luehea divaricata*. *Braz. J. Microbiol.* (2012): pp. 1174–1182.
Srivibool, R. Antimicrobial activity of *Actinomadura* isolates from tropical Island soils. *Actinobacteria.* (2000): ***Vol. 10***: pp. 10–12.
Stach, J.E.M., Maldonado, L.A., Masson, D.G., Ward, A.C., Goodfellow, M., and Bull, A.T. Statistical approaches for estimating actinobacterial diversity in marine sediments. *Appl. Environ. Microbiol.* (2003): ***Vol. 69***: pp. 6189–6200.
Stein, T., Borchert, S., Conrad, B., Feesche, J., Hofemeister, B., Hofemeister, J., and Entian, K.D. Two different lantibiotic-like peptides originate from the ericin gene cluster of *Bacillus subtilis* A1/3. *J. Bacteriol.* (2002): ***Vol. 184***: pp. 1703–1711.
Stierle, A., Strobel, G.A., and Stierle, D. Taxol and taxane production by *Taxomyces andreanae*. *Science.* (1993): ***Vol. 260***: pp. 214–216.
Svetoch, E.A., Stern, N.J., Eruslanov, B.V., Kovalev, Y.N., Volodina, L.I., Perelygin, V.V., Mitsevich, E.V. et al. Isolation of *Bacillus circulans* and *Paenibacillus polymyxa* strains inhibitory to *Campylobacter jejuni* and characterization of associated bacteriocins. *J. Food Prot.* (2005): ***Vol. 68***: pp. 11–17.
Taechowisan, T., Lu, C., Shen, Y., and Lumyong, S. 4-Arylcomarins from endophytic *Streptomyces aureofaciens* CMUAc130 and their antifungal activity. *Ann. Microbiol.* (2005): ***Vol. 55***: pp. 63–66.
Takahashi, S., Miyaoka, H., Tanaka, K., Enokita, R., and Okazaki, T. Milbemycins alpha11, alpha12, alpha13, alpha14 and alpha15: A new family of milbemycins from *Streptomyces hygroscopicus* sub sp. *aureolacrimosus*. *J. Antibiot.* (1993): ***Vol. 46***: pp. 1364–1371.
Tang, H.F., Yang-Hua, Y., Yao, X.S., Xu, Q.Z., Zhang, S.Y., and Lin, H.W. Bioactive steroids from the brown alga *Sargassum carpophyllum*. *J. Asian Nat. Prod. Res.* (2002): ***Vol. 4***: pp. 95–101.
Thangaduri, D., Murthy, K.S.R., Prased, P.J.N., and Pullaiah, T. Antimicrobial screening of Dcalepishamiltonii Wight and Arn (Asclepiadaceae) root extracts against food-related microorganisms. *J. Food Safety*. (2004): ***Vol. 24***: pp. 239–245.
Tims, F.C. An actinomycete antagonistic to a *Pythium* root parasite of sugarcane. *Phytopathology.* (1932): pp. 22–27.
Tomasz, M. Mitomycin C: Small, fast and deadly (but very selective). *Chem. Biol.* (1995): ***Vol. 2***: pp. 575–579.
Tuney, I., Cadirci, B.H., Unal, D., and Sukatar, A. Antimicrobial activities of the extracts of marine algae from the coast of Urla (Izmir, Turkey). *Turk. J. Biol.* (2006): ***Vol. 30***: pp. 171–175.
Umezawa, H. *Enzyme Inhibitors of Microbial Origin*. University Park Press, Baltimore (1972): p. 50.
Umezawa, H. Low-molecular-weight inhibitors of microbial origin. *Annu. Rev. Microbiol.* (1982): ***Vol. 36***: pp. 75–99.
Usha, R., Ananthavalli, P., Venil, G.K., and Palaniswamy, M. Antimicrobial and antiangiogenesis activity of *Streptomyces parvulus* KUAP106 from mangrove soil. *Eur. J. Biol. Sci.* (2010): ***Vol. 2***: pp. 77–83.
Vijayakumar, R., Panneerselvam, K., Muthukumar, C., Thajuddin, N., Panneerselvam, A., and Saravanamuthu, R. Optimization of antimicrobial production by a marine actinomycete *Streptomyces afghaniensis* VPTS3-1 isolated from Palk Strait, East Coast of India. *Indian. J. Microbiol.* (2011): pp. 011–0138.

Vikineshwari, S., Nsadaraj, P., Wong, W.H., and Balabaskeran, S. Actinobacteria from tropical mangrove ecosystem. Antifungal activity of selected strains. *Asian Pacific J. Mol. Biol. Biotechnol.* (1997): ***Vol. 5***: pp. 81–86.

Vimal, V., Mercy, R.B., and Kannabiran, K. Antimicrobial activity of marine actinomycete, *Nocardiopsis* sp. VITSVK5 (FJ973467). *Asian J. Med. Sci.* (2009): ***Vol. 1***: pp. 57–63.

Volk, R.B., and Furkert, F.H. Antialgal, antibacterial and antifungal activity of two metabolites produced and excreted by cyanobacteria during growth. *Microbiol. Res.* (2006): ***Vol. 161***: pp. 180–186.

Waksman, S.A. Associative and antagonistic effects of microorganisms I. Historical review of antagonistic relationships. *Soil Sci.* (1937): ***Vol. 43***: pp. 51–68.

Wenzel, S.C., and Muller, R. Recent developments towards the heterologous expression of complex bacterial natural product biosynthetic pathways. *Curr. Opin. Biotechnol.* (2005): ***Vol. 16***: pp. 594–606.

Woo, J.H., and Kamei, Y. Antifungal mechanism of an anti-pythium protein (SAP) from the marine bacterium *Streptomyces* sp. strain AP77 is specific for *P. porphyrae,* a causative agent of red rot disease in *Porhyra* spp. *Appl. Microbiol. Biotechnol.* (2003): ***Vol. 62***: pp. 407–413.

World Health Organization. World health statistics. WHO Library Cataloguing-in-Publication Data (2011). ISBN 978 92 4 156419 9. Available at http://www.who.int/whosis/whostat/EN_WHS2011_Full.pdf.

Wratten, S.J., Wolfe, M.S., Andersen, R.J., and Faulkner, D.J. Antibiotic metabolites from a marine Pseudomonas. *Antimicrob. Agents Chemother.* (1977): ***Vol. 11***: pp. 411–414.

Yon, C., Suh, J.W., Chang, J.H., Lim, Y., Lee, C.H., Lee, Y.S., and Lee, Y.W. Al702, a novel anti-legionella antibiotic produced by *Streptomyces* sp. *J. Antibiot.* (1995): ***Vol. 48***: pp. 773–779.

Zeeck, A., Schroeder, K., Frobel, K., Grote, R., and Thiericke, R. The structure of manumycin I. characterization. Structure elucidation and biological activity. *J. Antibiot.* (1987): ***Vol. 40***: pp. 1530–1540.

Zheng, Z., Zeng, W., Huang, Y., Yang, Z., Li, J., Ai, H., and Su, W. Detection of antitumor and antimicrobial activities in marine organisms. Associated Actinobacteria isolated from the Taiwan Strait, China. *FEMS Microbial. Lett.* (2000): ***Vol. 188***: pp. 87–91.

Zobell, C.E., and Upham, H.C. A list of marine bacteria including sixty new species. *Bull. Seripps Inst. Oceanogr.* (1944): ***Vol. 5***: pp. 239–292.

Zuccaro, A., Summerbell, R.C., Gams, W., Schroers, H.J., and Mitchell, J.I. A new *Acremonium* species associated with *Fucus* spp., and its affinity with a phylogenetically distinct marine *Emericellopsis* clade. *Stud. Mycol.* (2004): ***Vol. 50***: pp. 283–297.

chapter three

Metagenomics as advanced screening methods for novel microbial metabolites

Sakshi Sharma and Jyoti Vakhlu

Contents

3.1 Introduction

Metagenomics is the study of metagenomes, that is, microbial genetic material recovered directly from environmental samples, and is broadly referred to as environmental or community genomics. Metagenomics is used as a means of systematically investigating, classifying, and manipulating the entire microbial genetic material isolated from environmental samples (Zeyaullah et al. 2009). Metagenomics combines the power of genomics, bioinformatics, and systems biology. The term *metagenomics* was first used by Jo Handelsman and coworkers in 1998 and they defined it as "cloning and functional analysis of collective microbial genomes" (Handelsman et al. 1998). Recently, the definition has been modified by Kevin Chen and Lior Pachter at University of California, Berkeley as "the application of modern genomics techniques to the study of communities of microbial organisms directly in their natural environments, bypassing the need for isolation and lab cultivation of individual species" (Chen and Pachter 2005).

The emergence of metagenomics, as a field to study microbial diversity, was a consequence of the discovery of the fact that the microbial diversity in any environment is much greater than previously realized and accessed by cultivation. This unharvested diversity provides an unlimited source for novel microbial metabolites and some of them may find a place in the market. Considering that more than 99% of microorganisms in most environments are not amenable to cultivation, very little is known about their genomes, genes, and encoded enzymatic activities (Ferrer et al. 2005).

Thus, many of these microbes remain unnoticed and the potential of these microbes remains hidden, which limits the spectrum of search for novel metabolites. The metagenomic technologies developed are used to complement or replace culture-based approaches and bypass some of their inherent and established limitations. The isolation and analysis of environmental DNA has enabled the screening of microbial diversity, allowing access to their genomes, enabling the identification of protein coding sequences, and even the reconstruction of biochemical pathways and genomes, providing insights into the properties and functions of these organisms (Kennedy et al. 2008).

This rapidly growing research area provides new insights into microbial life and access to novel biomolecules. The construction and analysis of metagenomic libraries is thus a powerful approach to harvest and archive environmental genetic resources. It provides a way to identify which organisms are present, what they do, and how their genetic information can be beneficial to mankind. It can also be complemented with metatranscriptomic or metaproteomic approaches to describe expressed activities (Gilbert et al. 2008; Wilmes and Bond 2006). Metagenomics is also a powerful tool for generating novel hypotheses related to microbial functions; the remarkable discoveries of proteorhodopsin-based photoheterotrophy attest to this fact (Beja et al. 2000).

One of the main application areas of metagenomics is in the mining of metagenomes for genes encoding novel metabolites. Due to the complexity of most metagenomes, new sensitive and efficient high-throughput screening techniques that allow for fast and reliable identification of genes encoding beneficial novel metabolites from complex metagenomes have been invented. Screens of metagenomic DNA have been based either on nucleotide sequence as a sequence-driven approach or on metabolic activity as a function-driven approach (Hall 2007; Kennedy et al. 2008; Li et al. 2009; Simon et al. 2009; Uchiyama and Watanabe 2007; Yun and Ryu 2005).

The key factor on which the success of the metagenomic process depends is the extraction of metagenomic DNA from an environmental sample (Purohit and Singh 2008). Once an appreciable amount of eDNA has been obtained, further analysis is based on two approaches: function-based screening and sequence-based screening. Both function-based and sequence-based screening have individual advantages and disadvantages, and they have been applied successfully for the discovery of novel metabolites from the metagenome.

3.2 *Metagenomes as source of novel metabolites*

The total number of microbial cells on Earth is estimated to be 10^{30} (Turnbaugh and Gordon 2008). Prokaryotes represent the largest proportion of individual organisms, comprising 10^{6} to 10^{8} separate genospecies (Sleator et al. 2008). They are the most ubiquitous organisms on Earth, and are represented in all habitats including soil, sediment, marine and terrestrial subsurfaces, animals, and plant tissues, which represent an enormous reservoir of novel valuable molecules for health or industry. Microbes harbor a far greater genetic and metabolic diversity than plants and animals. Microbial diversity is a precious source for modern biotechnology because it has the potential to provide innovative and sustainable solutions to a broad range of important and hard-to-solve problems of modern society. Because it is estimated that only 0.1% of the bacterial and 5% of the total fungal species has been reported, there exists a huge potential of finding new and important metabolites.

Moreover, among the microbes already studied, only a few have been explored with respect to metabolite production. Bioactive metabolites may be extracted from microbes isolated from diverse habitats. A crude or untreated extract from microbes typically contains novel and structurally diverse metabolites. The effort to search for these biologically active metabolites is known as *bioprospecting*. The probability of finding novel products from microbes depends on the number of strains being explored and the level of diversity between the explored strains, their uniqueness, and their ability to produce metabolites. These factors must be seriously thought of before screening metagenomes for bioactive metabolites. Because a large number of microbial products have already been discovered, exploiting microbes for novel metabolites is a challenging endeavor. However, microorganisms possess a remarkable ability to synthesize novel metabolites and variants of known metabolites under different conditions and environments. This ability of microbes results in a vast number of structurally original and potent bioactive compounds that are otherwise difficult or impossible to obtain through chemical methods. Biological and geographical diversity is responsible for the diverse nature of metabolites produced by microbes isolated from different habitats.

3.3 *Prospecting from metagenomes*

The techniques for the recovery of novel biomolecules from environmental samples comprise the selection of the environmental samples, extraction of metagenomic DNA, and cloning and construction of metagenomic libraries. This is followed by screening the libraries using various methods described later in the chapter. Figure 3.1 shows the overall process involved in the search of novel metabolites from microbial metagenomes.

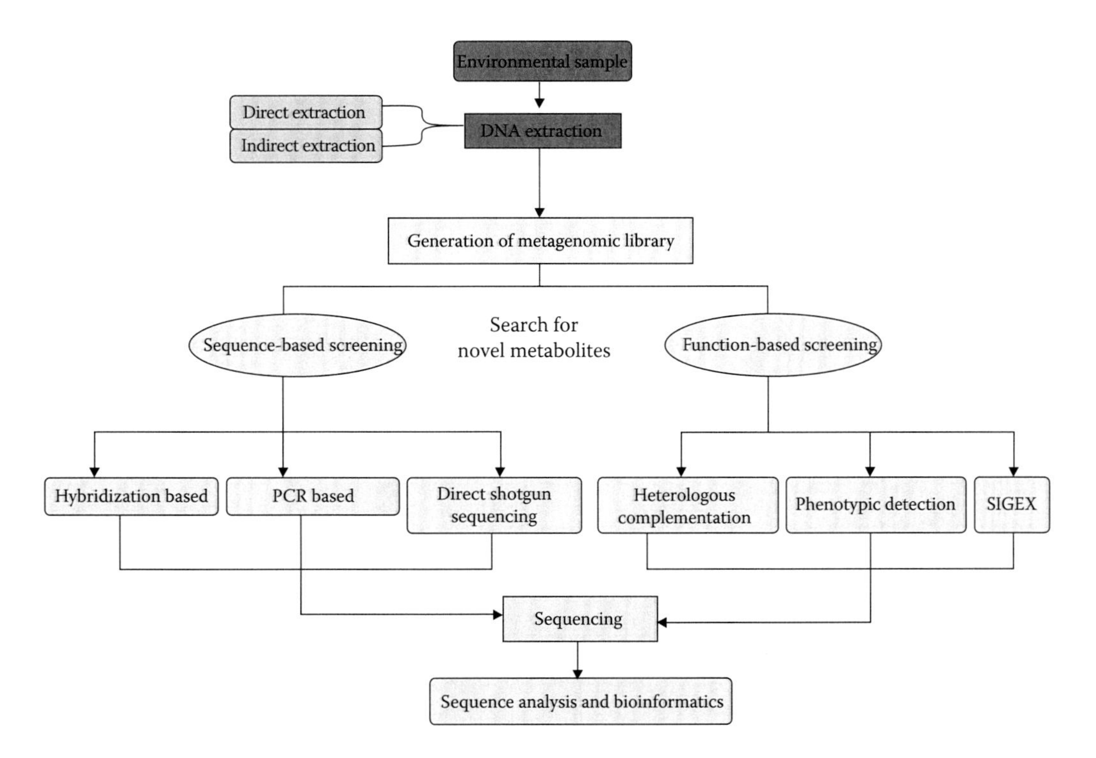

Figure 3.1 Different approaches for metagenome mining for new metabolites.

3.3.1 Sampling

Because there is hardly any place on Earth where microbes are not present, metagenomic sampling can be done from both abiotic as well as biotic habitats. Abiotic habitats include air, soil, or water including both marine and freshwater sources, whereas biotic habitats consist of animals, insects, plants, and others. Researchers are constantly looking forward to new unexplored environments that are likely to provide new insights into genetic and metabolic diversity. A step forward in this direction is the exploitation of extreme environments such as hot springs, geographical poles, very dry deserts, volcanoes, deep ocean trenches, cold habitats, and so on for the detection of extremophiles producing novel metabolites (Schiraldi and De Rosa 2002). It is now well established that microbes not only thrive in these previously considered hostile environments but are also metabolically active, producing various metabolites (Herbert 1992). To survive in such conditions, microorganisms have developed unique defenses and this has frequently led them to produce novel molecules. This capability had made them a mine of natural and novel metabolites (Simon and Daniel 2011).

Metagenomic studies fall into three main categories in terms of their selection of environments (Steele et al. 2008).

i. The first category selects highly diverse environmental samples such as soil and marine waters. The sample is either directly subjected to DNA extraction or the sample is first enriched with microbes producing the target metabolite.
ii. The second category selects environments naturally enriched for the target biocatalyst such as the search for xylanases in insect gut (Brennan et al. 2004).
iii. The third category targets extreme environments for metabolites active under extreme conditions.

These metabolites may be active under extreme conditions, but the variation may be quite limited such as alkalophilic enzymes, which function only over a narrow alkaline pH range, because the extremophiles produce metabolites that are functional only in those particular environmental conditions. Highly genomically diverse environments, which are naturally less extreme and more heterogeneous in terms of being subjected to greater fluctuations on temperature, pH or salinity, may provide biocatalysts that show a greater level of stability and activity over a wider range of conditions (Steele et al. 2008).

3.3.2 Metagenomic DNA extraction

Microbial communities from any habitat are composed of a mixture of microbes having different cell wall characteristics and thus having variable susceptibility to lysis. This implies that some special methods are required for the lysis of diverse bacterial cells in a sample to extract their genomes simultaneously. The isolation of appreciable amounts of quality environmental DNA is one of the key factors responsible for the success of a metagenomic process (Kennedy et al. 2007; Sharma et al. 2007; Voget et al. 2003). The metagenomic DNA extracted should be representative of all cells present in the sample and should be manipulable at the molecular level. For successful isolation of DNA from diverse environmental samples, different isolation protocols have been developed and some of these protocols are commercially available as DNA extraction kits (Rondon et al. 2000; Purohit and Singh 2008). The methods vary with respect to purity and quantity vis-a-vis degradation and inhibitors present in the isolated DNA. All protocols rely on the basic principle of using enzymes or chemicals and hot detergent treatments to disrupt the cells (Rondon et al. 2000). Lysozyme and sodium dodecyl sulfate (SDS) are most commonly used to lyse the cells. Several protocols also recommend the application of mechanical forces generated by bead beating, freeze-thawing, and sonication methods to disrupt the rigid cell structure (Kennedy et al. 2007; Voget et al. 2003). Metagenomic DNA extraction from one site with more than one isolation protocol is recommended for the maximum representation of the microbial community (Delmont et al. 2011). The metagenomic extraction protocols can be categorized broadly into two types: direct extraction and indirect extraction (Gabor et al. 2003; Purohit and Singh 2008).

3.3.2.1 Direct DNA extraction

Direct DNA extraction is an *in situ* extraction, in which the cells are lysed within the environmental sample matrix. The DNA is then separated from the sample matrix and cell debris. This method provides a less biased extraction of microbial DNA with high nucleic acid yield, and is preferred when large quantities of nucleic acids are required. However, the resulting nucleic acid extracts are commonly sheared and contaminated. Furthermore, extracts often contain unknown amounts of extracellular or eukaryotic DNA (or both). Currently, two types of lysis methods are used either alone or in combination. These are the soft and harsh lysis methods (Purohit and Singh 2008).

The soft lysis method is based on the disruption of microorganism solely by enzymatic and chemical means. Many protocols based on enzymatic lysis have been developed. Lysozyme treatment is one of the most common (Niemi et al. 2001; Rochelle et al. 1992; Tebbe and Vahjen 1993). Another enzyme, achromopeptidase, was used to improve the lysis of the recalcitrant Gram-positive bacteria *Frankia*, whereas proteinase K has been used to digest contaminating proteins (Rochelle et al. 1992; Simonet et al. 1984; Zhou et al. 1996). Proteinase K is a broad-spectrum protease digesting most of the proteins. Chemical methods are extensively used either alone or in combination with physical methods. The most common chemical used is a detergent SDS, which dissolves the hydrophobic material of cell membranes. Detergents have often been used in combination with heat treatment and with chelating agents such as EDTA. The addition of cetyltrimethylammonium bromide or cetrimonium bromide (CTAB) and polyvinylpolypyrrolidone (PVPP) has also been reported. CTAB forms insoluble complexes with denatured proteins, polysaccharides, and cell debris. PVPP is used to remove polyphenols (Desai and Madamwar 2007; Verma and Satyanarayana 2011).

On the other hand, the harsh lysis approach involves mechanical cell disruption by bead beating, sonication, freeze-thawing, and grinding. Bead beating is an easy method based on the ballistic disintegration of the cells (Leff et al. 1995; Siddhapura et al. 2010). Tiny beads made of glass, ceramic, or steel are mixed with the sample already suspended in aqueous media, and subjected to agitation by stirring or shaking. The efficiency of cell disruption depends on the time of agitation and bead size. Sonication uses high-frequency ultrasound ranging from 20 to 50 kHz for disrupting the cells. The efficiency of sonication depends on the energy input (Siddhapura et al. 2010). Even under optimized conditions, harsh treatment may sometimes result in shearing of high molecular-weight DNA, low yields, and small fragment sizes.

3.3.2.2 Indirect DNA extraction

Indirect DNA extraction involves the separation of microbial cells from the sample matrix prior to cell lysis. Although time-consuming, indirect extraction methods are preferred for targeting prokaryotic communities when high DNA purity is required for inhibitor-sensitive methods, and when the recovery of high molecular-weight DNA is necessary (Robe et al. 2003). Indirect DNA extraction methods yielded 10-fold to 100-fold lower amounts of DNA compared with direct procedures (Steffan et al. 1988), but the bacterial diversity of DNA recovered by indirect means was distinctly higher (Gabor et al. 2003; Leff et al. 1995). The indirect DNA extraction protocols involve blending cation-exchange methods or other techniques such as the use of super paramagnetic silica–magnetite nanoparticles for the isolation and purification of DNA from the samples (Gabor et al. 2003; Gupta and Vakhlu 2012; Jacobsen et al. 1992).

Most DNA extraction protocols require standardization because the nature of different environmental samples varies with respect to their matrices, organic and inorganic compounds, and biotic factors. Environmental samples are heterogeneous and contain different substances, which may sometimes interfere with the extraction of quality DNA. Thus, it is quite obvious that the extraction procedures would have to be case specific. For example, soil is a complex matrix containing many substances such as humic acids, which can be coextracted during DNA isolation. It is thus essential to modify the extraction protocol in such a way so as to remove these acids from the extracted DNA before it can be further processed. Optimizing the DNA extraction protocols also reduces the bias caused by unequal lysis of different members of the microbial community. Improved DNA extraction techniques could help ensure a metagenomic library that efficiently represents the entire community's genome without inhibitory substances.

Inhibitor-free metagenomic DNA of high quality and high yield is essential for the success of metagenomics. Various inhibitory substances extracted along with the metagenome are a major problem because these compounds hinder the molecular biological manipulation such as DNA digestion, ligation, and polymerase activity and hence the DNA needs to be purified of such inhibitors. Some examples of such inhibitors include the humic acid and polyphenols present in soil samples. Samples from polluted environments consist of humic and fulvic acids, along with a large number of other organic pollutants, metal ions, and chemical impurities as inhibitors (Fortin et al. 2004; Hinoue et al. 2004; Sharma et al. 2007; Verma and Satyanarayana 2011). The phenolic groups in humic acids denature biological molecules by bonding to amides or are oxidized to form a quinone, which covalently bonds to DNA (Young et al. 1993). In most studies, cell lysis is followed by the purification of DNA using organic solvent extraction. This is done either by using phenol–chloroform or phenol–chloroform–isoamyl alcohol as solvents. This preliminary purification step is followed by the precipitation of DNA using ethanol, isopropanol, or polyethylene glycol (Steffan et al. 1988). These basic purification steps have been found to be insufficient to remove contaminants from soil metagenome. Many other purification methods have been tried by various workers to remove humic acids. These include gel elution, gel filtration (using Sephadex G150; Sepharose 2B, 4B, and 6B; and Biogel P100 and P200), ion exchange (using Q-Sepharose), and activated charcoal and amberlite resins, which are efficient removers of metal ions, organics, and humic acids (Harry et al. 2000; Jackson et al. 1997; More et al. 1994; Rochelle et al. 1992; Sharma et al. 2007; Verma and Satyanarayana 2011). Using all these methods for removing contaminants is time-consuming and laborious, hence new techniques need to be developed for obtaining inhibitor-free DNA.

One solution to this problem is the use of multiple displacement amplification (MDA) for whole genome amplification from single cells, which will allow the study of the entire biochemical potential of single uncultured microbes from complex microbial communities (Stepanauskas and Sieracki 2007; Woyke et al. 2009). In this approach, microbial cells are first sorted using fluorescence-activated cell sorting (FACS) and then collected. Then, MDA is used to amplify the entire genome from the sorted cells. The amplified genomic

DNA can then be analyzed to access the full metabolic potential of the microorganism. This approach has great potential for the discovery of novel enzymes, as it affords easier access to rare microbiota.

3.3.3 Screening metagenomes for novel metabolites

Once the manipulable metagenomic DNA is isolated, it can be analyzed for microbial community dynamics, construction of novel microbial genomes, and the isolation of novel genes and gene products for screening novel metabolites.

The focus of the present chapter is on the use of metagenomics as a tool for the isolation of novel metabolites. Hence, we restrict ourselves only to novel metabolite screening using metagenomics. For metabolite screening, cloning and expression of metagenomic DNA is compulsory in contrast to microbial diversity analysis or genome reconstruction. The extracted metagenomes can be screened for the presence of genes encoding novel metabolites using two different approaches: *sequence-based screening* and *function-based screening* (Hall 2007; Kennedy et al. 2008; Li et al. 2009; Simon et al. 2009; Uchiyama and Watanabe 2007; Yun and Ryu 2005).

Although sequence-based metagenomic approaches rely on comparing sequence data, which is obtained with sequences deposited in databases, functional metagenomics focuses on screening DNA library clones directly for a phenotype, that is, the genes are recognizable by their function rather than by their sequence. The power of such an approach is that it does not require the genes of interest to be recognized by already existing sequences. This ensures the identification of entirely new classes of genes for both known and indeed novel functions. Moreover, the results from a functional metagenomics approach are unambiguous. However, for the approach to be successful, specific and sensitive assays should be available and heterologous gene expression is still a bottleneck.

Sequence-based metagenomic approaches depend on both cloning-dependent as well as cloning-independent methods, whereas function-based approach is cloning-dependent. Cloning-based screening of metagenomes requires the generation of libraries. DNA and cDNA as well as specific amplicons can be used as inserts to generate libraries. These libraries are known as metagenomic, metatranscriptomic, and amplicon libraries, respectively. Table 3.1 shows some examples of the products derived from metagenomic libraries along with the screening methods.

Once the DNA has been isolated from the sample and purified, the next step is the construction of DNA libraries. Successful generation of DNA libraries require suitable cloning vectors and host strains and the selection of host–vector system depends primarily on the metabolite to be isolated. The most common approach used includes the generation of small insert libraries in a standard sequencing vector and in *Escherichia coli* as a host strain (Henne et al. 1999). The vector system is decided, depending on the desirable insert size, which in turn is based on the metabolite to be screened, DNA quality, targeted genes, and screening strategy. The insert size in small insert libraries is less than 10 Kb, and hence plasmid is mostly used as a vector. These small insert libraries suffer from the limitation of not being able to allow the detection of large gene clusters and operons. To overcome this limitation, many large insert libraries have been used by researchers. These libraries include cosmid DNA libraries with insert sizes ranging from 25 to 35 kb or bacterial artificial chromosome (BAC) libraries with insert sizes up to 200 kb and fosmids with inserts of 40 kb of foreign DNA (Beja et al. 2000; Entcheva et al. 2001; Rondon et al. 2000).

Metagenomic libraries constructed using DNA extracted directly from naturally occurring bacterial populations are now used extensively to screen for clones that have the

Table 3.1 Recent Examples for Metagenome-Derived Biocatalysts and the Screening Strategy Used

Sampling site	Target	Vector used	No. of clones screened	Screening technique	Reference
Soil from a meadow, sugar beet field, and river valley (Germany)	Na+/H+ antiporters	Plasmids	1,480,000	Heterologous complementation	Majernik et al. 2001
Crude oil–contaminated groundwater microbial flora (Japan)	Aromatic hydrocarbon catabolic operon fragments	Plasmids	152,000	SIGEX	Uchiyama et al. 2005
Collection of soil samples (United States and Costa Rica)	Oxidative coupling enzyme (OxyC)	Cosmids	10,000,000	Sequence-based	Banik and Brady 2008
Soil from a botanical garden (Germany)	Blue light photoreceptor	Cosmids	2500	Sequence-based	Pathak et al. 2009
Groundwater	Phenol degradation	Plasmids	152,000	SIGEX	Uchiyama et al. 2005
Copper waste–exposed sediment	Copper P-type ATPase	–	–	PCR homology	Iglesia et al. 2010
Soil metagenome	*N*-acyl-homoserine lactones	Phagemid	7392	Functional	Bijtenhoorn et al. 2011
Antarctic soil	Esterase	Fosmids	10,000	Functional	Hu et al. 2012
Mountain soil	Thioesterase	Plasmid	10,000	Functional	Sudan and Vakhlu 2013
Woodland soil	Cellulose	–	–	PCR homology	Cucurachi et al. 2013

genetic capacity to produce new biocatalysts as well as small molecule products (Healy et al. 1995). Screening of metagenomic libraries unraveled many important biological products that had previously gone unnoticed. Banik and Brady explored 10,000,000 unique cosmid clones and successfully recovered glycopeptide-encoding gene clusters from soil metagenomic libraries (Banik and Brady 2008). In 2005, two separate groups reported the cloning and heterologous expression of biosynthetic gene clusters for the cyanobactins patellamide from metagenomic libraries of uncultured cyanobacterial symbionts associated with marine *Didemnidae* sponges (Long et al. 2005; Schmidt et al. 2005). Cyanobactins display interesting cytotoxic activities. Heath and coworkers screened 100,000 fosmid clones from a metagenomic library of Antarctic desert soil and identified a novel alkaliphillic esterase active clone (Hu et al. 2012).

Pathak and coworkers developed a metagenomic library of soil from a botanical garden in Germany using cosmid vectors and screened 2500 clones to detect the presence of a blue light photoreceptor (Pathak et al. 2009). Avneet and coworkers isolated a novel thioesterase gene from the metagenome of a mountain peak in the Himalayas (Sudan and Vakhlu 2013). All these studies suggested that exploring metagenomic libraries have led to the discovery of many novel products and that these libraries can still further be explored for the same.

3.3.4 Function-based screening

Unculturable bacteria are predicted to be a significant reservoir of novel molecules. One means to access the biosynthetic potential contained within the genomes of uncultured bacteria is functional metagenomics (Riesenfeld et al. 2004b). Function-based screening of novel genes encoding novel metabolites is based on the metabolic activities of metagenomic library containing clones. This method can be thought of as the only approach that has the potential to identify novel classes of genes encoding known or novel functions because it does not depend on the sequence information or sequence similarity to known genes (Daniel 2005; Ferrer et al. 2009; Riesenfeld et al. 2004a).

The success of function-based screening, that is, the probability of identifying a certain gene depends on various parameters. These parameters include the host, vector system, size of the target gene, its abundance in the source metagenome, the assay method, and the efficiency of heterologous gene expression in a surrogate host (Uchiyama and Miyazaki 2009).

E. coli is still the most preferred host for the cloning and expression of any metagenome-derived genes. Although quite a large number of genes derived from Enterobacteriaceae will be readily expressed in the most common *E. coli* host, many genes from more distantly related organisms may not be expressed. This is because the promoter regions of these genes are not recognized by the *E. coli* transcriptional machinery or are expressed at low levels due to differences in codon usage (Simon et al. 2009).

Even where transcription and translation of foreign genes results in efficient protein expression, additional problems can arise when proteins need to be posttranslationally modified or exported for activity. It has been observed that significant differences exist in expression modes between different taxonomic groups of prokaryotes and only 40% of the enzymatic activities may be detected by random cloning in *E. coli* (Gabor et al. 2004). Thus, additional hosts, such as *Streptomyces* spp., *Thermus thermophilus, Sulfolobus solfataricus,* and diverse Proteobacteria have been used to expand the range of detectable activities in metagenomic screens (Albers et al. 2006; Angelov et al. 2009; Craig et al. 2010; Wang et al. 2000).

Although functional metagenomic-based approaches have proven to be successful for the discovery of novel enzymes, it is quite likely that many novel enzymes from rare microbes in complex communities are poorly represented in metagenomic libraries. In addition, function-driven screening often requires the analysis of more clones than sequence-based screening for the recovery of a few positive clones. The major advantage of a function-based screening approach is that only full-length genes and functional gene products are detected (Simon et al. 2009).

The following three different types of function-driven approaches have been used for the screening of metagenomic libraries. All the three methods have their own benefits and limitations.

3.3.4.1 Heterologous complementation of host strains or mutants

It is a simple and a fast screening technique in which clones expressing the desired function are identified. The technique is based on the principle of heterologous complementation of host strains or mutants of host strains that require targeted genes for growth under selective conditions. Only recombinant clones harboring the targeted gene and producing the corresponding gene product in an active form are able to grow. The method is highly selective as no false-positives occur because the clones that do not contain the target genes are not able to express. One recent example is the identification of DNA polymerase-encoding genes from metagenomic libraries derived from microbial communities present in glacier ice. The host used in the study was an *E. coli* mutant carrying a cold-sensitive lethal mutation in the 5′-3′ exonuclease domain of the DNA polymerase I. At a growth temperature of 20°C, only recombinant *E. coli* strains complemented by a gene conferring DNA polymerase-activity were able to grow (Simon et al. 2009).

3.3.4.2 *Phenotypical detection of the desired activity*

This method uses chemical dyes or chromophore-containing derivatives of the enzyme substrates in the growth medium. The target gene or the metabolite can be identified by visualizing the change on the growth medium associated with them. Hence, the specific metabolic capabilities of the clones can be registered. A recent example of such an activity-driven screen targeted genes is encoding bacterial β-D-glucuronidases, which are part of the human intestinal microbiome. These enzymes have putatively beneficial effects on human health. A metagenomic library comprising 4600 clones derived from bacterial DNA extracted from pools of feces was screened using an *E. coli* strain that is deficient in β-D-glucuronidase activity. Nineteen positive clones were detected, with one that exhibited strong β-D-glucuronidase activity after cloning of the corresponding gene into an expression vector (Gloux et al. 2011).

3.3.4.3 *Substrate-induced gene expression*

The technique was first proposed by Uchiyama and coworkers in 2005, and is based on the fact that catabolic genes require some environmental stimulus for their expression. In many cases, expression is usually controlled by regulatory elements situated close to the genes to be expressed.

Substrate-induced gene expression (SIGEX) screens the clones harboring the desired catabolic genes that are expressed in the presence of substrates but are not expressed in the absence of substrates.

The SIGEX procedure is composed of the following four steps:

i. Construction of a metagenomic library in liquid culture. The metagenome to be analyzed is restriction digested. The metagenome fragments are then cloned upstream of *gfp* gene in the operon-trap p18GFP vector, thereby placing green fluorescent protein (GFP) expression under the control of promoters in the metagenomic DNA. A library is subsequently constructed in a liquid culture by transforming suitable host, for example, *E. coli*.
ii. Removal of clones containing self-ligation plasmids and those expressing GFP constitutively. All clones that fluoresce in the absence of the substrate are discarded by FACS after cultivation and induction with isopropyl-β-D-thiogalactopyranoside (IPTG). This procedure greatly improves the efficiency for selecting positive clones because false-positive clones can almost completely be eliminated.
iii. Selection of clones expressing GFP in the presence of a target substrate. Cells selected in step ii are grown in liquid cultures in the presence of a target substrate and subjected to FACS to select for fluorescent cells.

iv. Colony isolation of the sorted cells on agar plates. A sorted-cell fraction is spread on agar plates to isolate positive clones as colonies. Isolated colonies are then grown under the same conditions as step iii and subjected to FACS to check if they really are positive. Positive clones are then used for sequence analyses and expression of enzymatic activities.

SIGEX has many advantages in metagenomic library screening. The most important advantage is that it saves time, labor, and expenses because it is semiautomated. Moreover, it can be considered as a high-throughput approach because it is aided by FACS, which allows the cloning of many different genes in a comparatively shorter time. SIGEX can also detect catabolic genes for which other screening methods such as colorimetry or on-plate screening have not been developed. In addition, SIGEX does not require modified toxic and costly substrates for the detection of catabolic genes. SIGEX enable us to access genes that are otherwise difficult to obtain (Uchiyama and Watanabe 2007; Yun and Ryu 2005).

Despite the great advantages that make SIGEX a highly attractive technique for screening metagenomic libraries to identify novel metabolites, the technique has many limitations as well. In this approach, the catabolic genes are expressed only when the transcriptional regulator encoded in the SIGEX fragment interacts effectively with the transcriptional factors in host cells. The affinity of interaction may be low when the SIGEX fragment is derived from an organism distantly related to the host organism. This implies a possibility that SIGEX selection may bias itself toward genes in organisms closely related to the host. Even if the regulatory genes and catabolic genes interact efficiently, the success is still not sure because they are not always proximate to each other. It has also been observed that the expression of catabolic genes is not always induced by their substrates, which also limits the success rate of SIGEX. Substrates that do not migrate to the cytoplasm cannot be used with SIGEX. Genes for catabolic enzymes obtained using SIGEX may sometimes be partial when they are situated at the end of a cloned genome fragment (Uchiyama and Watanabe 2007).

Finally, the gate setting in FACS and the media conditions containing the inducer are critical for discriminating false-positive and false-negative results. Therefore, when SIGEX is applied, these drawbacks should be considered carefully. Despite the limitations of SIGEX, which can only detect genes that are correctly induced in a heterologous host, the method is extremely high-throughput because a large number of clones can be screened in relatively short timescales (Yun and Ryu 2005). SIGEX can be a very powerful tool for the isolation of novel metabolites only when the conditions are appropriate, that is, compatible substrate and target genes are selected and the gate settings of FACS are optimized.

3.3.5 Sequence-based screening

In sequence-based screening, the isolation of novel proteins is based on homology searches. It is an efficient method of isolating metabolites with different levels of similarity from previously identified genes. Generally, it is based on the conserved DNA sequences of target genes. It provides information on the distribution of functions in a community, linkage of traits, genomic organization, and horizontal gene transfer. Approaches typically involve either the sequencing of random clones to accumulate vast stores of sequence information or the identification of clones based on methods that detect a particular sequence. With both of these approaches, phylogenetic markers are sought on the clone of interest to link cloned sequences with the probable origin of the DNA. The discovery of novel metabolites or genes encoding novel metabolites by sequence-based screening can be carried out either by exploring metagenomic libraries or by directly analyzing the community DNA as such.

Cloning-based approaches require the generation of metagenomic libraries as discussed previously. The screening of these libraries is based on a common procedure of designing oligonucleotide probes or degenerate primers, which are derived from conserved regions of already-known genes or protein families (Schmidt et al. 1991). The library is then screened for the genes of interest using these probes or primers in combination with polymerase chain reaction. Subsequently, the gene products are subcloned, expressed, and characterized. Alternatively, the metagenomic libraries can be directly sequenced by direct shotgun sequencing method.

The sequence-based screening of metagenomic libraries generally experience two major limitations. First, primer/probe design is purely dependent on known sequences of enzymes and the reference sequence is of the proteins that have mostly been purified and cloned from easily cultured laboratory isolates of microorganisms. It is, therefore, possible that the designed primers and probes are biased toward genes that are similar to known genes and thus eliminate less homologous genes encoding novel enzymes. The second major drawback is that PCR amplification results in only partial fragments of target genes. Hence, the recovery of whole genes necessary for their activities may require additional steps.

In contrast to cloning-based screening, cloning independent approach escapes the most tedious step of generating metagenomic libraries. Instead, cloning independent strategies directly exploit the advantages of sequencing methods. The total DNA extracted from an environment is subjected to sequencing. The metagenomic DNA is either sequenced as such or it is first subjected to 16S-based PCR amplification, the amplicons purified, and then sequenced. The sequences obtained are then subsequently analyzed to detect the presence of any genes, functions, or novel traits and metabolites. Currently, two main sequencing strategies are used in metagenomics: shotgun sequencing and metagenomics analysis using next generation sequencing (NGS) technology.

Advances in high-throughput sequencing technology and lower cost sequencing technologies have made random shotgun sequencing of environmental DNA economically feasible. In this method, the metagenomic DNA is broken up randomly into numerous small segments using different restriction enzymes. These fragmented segments are then sequenced using a chain termination method to obtain reads. Multiple overlapping reads for the target DNA are obtained by performing several rounds of this fragmentation and sequencing. The overlapping ends of different reads are then assembled into a continuous sequence with the help of various computer programs, and therefore the whole metagenome can be analyzed. This approach generates vast amount of data (Carola and Daniel 2009), reveals millions of novel genes, and can be used to deduce metabolic pathways from uncultured bacteria. However, the technique is costly and requires considerable efforts. Moreover, the assembled sequence is a result of DNA from different organisms and hence may contain chimeric regions. There has been a great demand for a low-cost technique for sequencing, which led to the development and popularization of high-throughput sequencing methods. These techniques produce thousands or millions of sequences simultaneously (Church 2006; Hall 2007). In ultrahigh-throughput sequencing, as many as 500,000 sequencing-by-synthesis operations may be run in parallel (Bosch and Grody 2008).

NGS is based on the same principle as that of capillary electrophoresis-based Sanger sequencing. In both these methods, a small fragment of DNA is resynthesized from a template strand and the base sequence is identified depending on the signals emitted. NGS differs from traditional Sanger's method in that the sample can consist of a population of DNA molecules that do not require clonal purification and the whole process can be completed within hours, regardless of genome size. Whereas Sanger's sequencing method requires library preparation in most cases. Moreover, the sample must contain a single template that requires purification from single bacterial and yeast colonies or phage plaques (Ronaghi et al. 1996). These advances enable rapid sequencing of large stretches of DNA fragments in a short period. Developed instruments, which produce hundreds of gigabases of data in a single sequencing run, are like a boon adding to the benefits of these advances. NGS methods are high-throughput methods involving single-molecule real-time sequencing, ion semiconductors, pyrosequencing, sequencing by synthesis, and sequencing by ligation.

Sequence-based strategies have led to the successful identification of a large number of genes encoding novel enzymes. Some of them include dimethylsulfoniopropionate-degrading enzymes, dioxygenases, nitrite reductases, [Fe-Fe]-hydrogenases, and chitinases (Bartossek et al. 2010; Hjort et al. 2010; Lee et al. 2011; Varaljay et al. 2010; Zaprasis et al. 2010). The genomic sequencing of *Streptomyces avermitilis* was done using a shotgun sequencing technique and revealed the presence of more than 20 biosynthetic gene clusters for metabolites; before sequencing, it was known to produce three to four secondary metabolites only (Omura et al. 2001; Schmidt et al. 2010).

3.3.6 Metagenomic data analysis

Analyzing metagenomic data is not easy as the data is much more complex than what has previously been seen in genomics. Metagenomic sequence data has lower sequence redundancy, lower sequence quality, short read lengths, increased polymorphisms, and relative abundance. Moreover, the size of the data increases rapidly. The scientific community has already seen the size of these data sets quickly move from mega base pairs (Mbps) to giga base pairs (Gbps) and now terabases, which require significant computational resources and expertise as well.

Sufficient computational technologies are available and the raw data can be subjected to various different approaches. The results depend on the analysis path taken and the tools used. Each approach has its own strengths and limitations. The bioinformatic tools available for metagenomic analysis are still evolving.

The metagenomic data obtained is subjected to a series of bioinformatic tools to generate valuable information (Figure 3.2). The data generated by metagenomics experiments

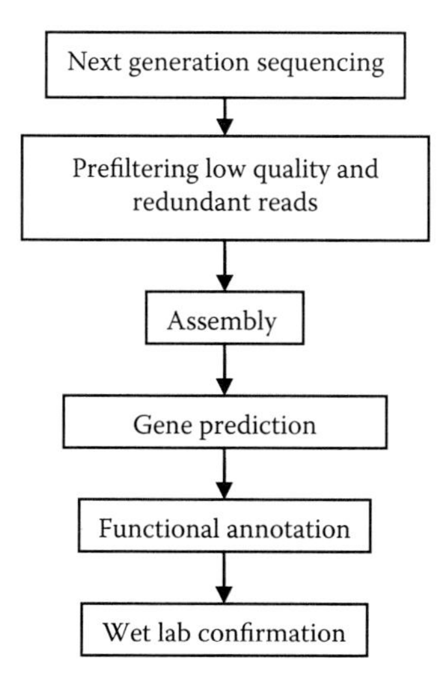

Figure 3.2 Analysis of the metagenomic data.

are both enormous and inherently noisy, containing fragmented data representing as many as 10,000 species (Wooley et al. 2010). The first step in metagenomic data analysis requires certain prefiltering steps, which include the removal of redundant, low-quality sequences and sequences of probable eukaryotic origin (Balzer et al. 2013). The tools available for prefiltering include *Eu-Detect* and *DeConseq* (Schmeider and Edwards 2011).

Prefiltering is followed by data assembly, which, in case of metagenomics, is difficult and unreliable because the data is highly redundant and is error prone due to short reads (Handelsman et al. 2007). Many assembly programs are available such as *Phrap* or *Celera* assembler and *Velvet* assembler.

Assembly of the metadata is followed by gene prediction. Metagenomics uses two approaches for predicting the gene. The first approach is to identify genes based on homology with genes that are already available in sequence databases. This is usually done using simple BLAST searches. MEGAN4 is usually used to implement such approach (Huson et al. 2011). The second approach is an *ab initio* method that uses intrinsic features of the sequence to predict coding regions based on gene training sets from related organisms. This is the approach taken by programs such as *GeneMark* and *GLIMMER* (Wenhan et al. 2010).

Once the gene involved in the metabolite production has been predicted, the next step is the functional annotation of the gene. All the possible products that can be produced by a particular gene are identified using bioinformatic tools. If the gene does not correspond to any of the previously available products, it can be a novel metabolite. All these results can then be confirmed by wet laboratory experiments such as gene expression. A few bioinformatics programs have been established for analyzing metagenome sequences, gene prediction, and annotation. The European Union-funded "MetaFunctions" project covers the development of "metagenomes Mapserver." It is a data-mining system that correlates genetic patterns in genomes and metagenomes with contextual environmental data. Nevertheless, more innovative and sophisticated bioinformatics tools must be devised to assure continued valuable progress in the field of metagenomics.

3.3.7 *Future prospects*

Metagenomics is one of the fastest growing research areas and has proven effective for isolating novel metabolites from the environment. The field has made great progress in its scale and scopes, which have expanded since the concept was first introduced. Initially, metagenomics was applied only to the microbial communities of temperate environments, but with the advances made in DNA isolation methods, cloning strategies, and screening techniques, microbial communities of extreme environments have also been analyzed for novel metabolites. However, the whole metagenomic process still needs to be improved. Development of standard protocols for isolating sufficiently purified metagenomic DNA from environmental samples is required. Screening strategies need to be revised because each strategy has its own limitations. Conventional screening methods are time-consuming and laborious and hence it remains necessary to develop more effective and economic strategies. SIGEX is a good alternative but better high-throughput screening methods need to be developed. In addition, we need to understand the utility and limitations of each screening method to facilitate the selection of the appropriate one for a desired gene or set of genes. Despite all these shortcomings, metagenomics is still an important and indispensable tool for the identification of novel biomolecules.

References

Albers, S.V., Jonuscheit, M., Dinkelaker, S., Urich, T., Kletzin, A., Tampe, R., Driessen, A.J.M. and Schleper, C. 2006. Production of recombinant and tagged proteins in the hyperthermophilic archaeon *Sulfolobus solfataricus*. *Applied and Environmental Microbiology, 72*: 102–111.

Angelov, A., Mientus, M., Liebl, S. and Liebl, W. 2009. A two-host fosmid system for functional screening of (meta)genomic libraries from extreme thermophiles. *System Applied Microbiology, 32*: 177–185.

Balzer, S., Malde, K., Grohme, M.A. and Jonassen, I. 2013. Filtering duplicate reads from 454 pyrosequencing data. *Bioinformatics, 29 (7)*: 830–836.

Banik, J.J. and Brady, S.F. 2008. Cloning and characterization of new glycopeptide gene clusters found in an environmental DNA megalibrary. *Proceedings of the National Academy of Sciences of the United States of America, 105 (45)*: 17273–17277.

Bartossek, R., Nicol, G.W., Lanzen, A., Klenk, H.P. and Schleper, C. 2010. Homologues of nitrite reductases in ammonia-oxidizing archaea: Diversity and genomic context. *Environmental Microbiology, 12 (4)*: 1075–1088.

Beja, O., Aravind, L., Koonin, E.V., Suzuki, M.T., Hadd, A., Nguyen, L.P., Jovanovich, S.B. et al. 2000. Bacterial rhodopsin: Evidence for a new type of phototrophy in the sea. *Science, 289 (5486)*: 1902–1906.

Bijtenhoorn, P., Schipper, C., Hornung, C., Quitschau, M., Grond, S., Weiland, N. and Streit, W.R. 2011. BpiB05, a novel metagenome-derived hydrolase acting on *N*-acyl-homoserine lactones. *Journal of Biotechnology, 155 (1)*: 86–94.

Bosch, J.R. and Grody, W.W. 2008. Keeping up with the next generation: Massively parallel sequencing in clinical diagnostics. *The Journal of Molecular Diagnostics, 10 (6)*: 484–492.

Brennan, Y., Callen, W.N., Christoffersen, L., Dupree, P., Goubet, F., Healey, S., Hernández, M. et al. 2004. Unusual microbial xylanases from insect guts. *Applied and Environmental Microbiology, 70 (4)*: 3609–3617.

Carola, S. and Daniel, R. 2009. Achievements and new knowledge unraveled by metagenomic approaches. *Applied Microbiology and Biotechnology, 85 (2)*: 265–276.

Chen, K. and Pachter, L. 2005. Bioinformatics for whole-genome shotgun sequencing of microbial communities. *PLoS Computational Biology, 1 (2)*: e24.

Church, G.M. 2006. Genomes for all. *Scientific American, 294 (1)*: 46–54.

Craig, J.W., Chang, F.Y., Kim, J.H., Obiajulu, S.C. and Brady, S.F. 2010. Expanding small-molecule functional metagenomics through parallel screening of broad-host-range cosmid environmental DNA libraries in diverse proteobacteria. *Applied and Environmental Microbiology, 76 (5)*: 1633–1641.

Cucurachi, M., Busconi, M., Marudelli, M., Soffritti, G. and Fogher, C. 2013. Direct amplification of new cellulase genes from woodland soil purified DNA. *Molecular Biology Reports, 40 (7)*: 4317–4325.

Daniel, R. 2005. The metagenomics of soil. *Nature Reviews Microbiology, 3 (6)*: 470–478.

Delmont, T.O., Robe, P., Cecillon, S., Clark, I.M., Constancias, F., Simonet, P., Hirsch, P.R. and Vogel, T.M. 2011. Accessing the soil metagenome for studies of microbial diversity. *Applied and Environmental Microbiology, 77 (4)*: 1315–1324.

Desai, C. and Madamwar, D. 2007. Extraction of inhibitor-free metagenomic DNA from polluted sediments, compatible with molecular diversity analysis using adsorption and ion-exchange treatments. *Bioresource Technology, 98 (4)*: 761–768.

Entcheva, P., Liebl, W., Johann, A., Hartsch, T. and Streit, W.R. 2001. Direct cloning from enrichment cultures, a reliable strategy for isolation of complete operons and genes from microbial consortia. *Applied and Environmental Microbiology, 67 (1)*: 89–99.

Ferrer, M., Martínez-Abarca, F. and Golyshin, P.N. 2005. Mining genomes and "metagenomes" for novel catalysts. *Current Opinion in Biotechnology, 16 (6)*: 588–593.

Ferrer, M., Beloqui, A., Timmis, K.M. and Golyshin, P.N. 2009. Metagenomics for mining new genetic resources of microbial communities. *Journal of Molecular Microbiology and Biotechnology, 16 (1–2)*: 109–123.

Fortin, N., Beaumier, D., Lee, K. and Greer, C.W. 2004. Soil washing improves the recovery of total community DNA from polluted and high organic content sediments. *Journal of Microbiological Methods, 56 (2)*: 181–191.

Gabor, E.M., Alkema, W.B. and Janssen, D.B. 2004. Quantifying the accessibility of the metagenome by random expression cloning techniques. *Environmental Microbiology, 6 (9)*: 879–886.

Gabor, E.M., Vries, E.J. and Janssen, D.B. 2003. Efficient recovery of environmental DNA for expression cloning by indirect extraction methods. *FEMS Microbiology Ecology, 44 (2)*: 153–163.

Gilbert, J.A., Field, D., Huang, Y., Edwards, R., Li, W., Gilna, P. and Joint, I. 2008. Detection of large numbers of novel sequences in the metatranscriptomes of complex marine microbial communities. *PLoS One, 3 (8)*: e3042.

Gloux, K., Berteau, O., Béguet, F., Leclerc, M. and Doré, J. 2011. A metagenomic β-glucuronidase uncovers a core adaptive function of the human intestinal microbiome. *Proceedings of the National Academy of Sciences, 108 (Supplement 1)*: 4539–4546.

Gupta, P. and Vakhlu, J. 2012. Metagenomics: Techniques, applications and challenges. In: *Omics: Applications in Biomedical, Agricultural, and Environmental Sciences*. Editors: Debmalya Barh, Vasudeo Zambare, Vasco Azevedo. Chapter No. *23*: 572, CRC Press, Boca Raton, Florida, USA.

Hall, N. 2007. Advanced sequencing technologies and their wider impact in microbiology. *Journal of Experimental Biology, 210 (9)*: 1518–1525.

Handelsman, J., Rondon, M.R., Brady, S.F., Clardy, J. and Goodman, R.M. 1998. Molecular biological access to the chemistry of unknown soil microbes: A new frontier for natural products. *Chemistry and Biology, 5 (10)*: R245–R249.

Handelsman, J., Tiedje, J., Alvarez-Cohen, L., Ashburner, M., Cann, I.K.O., Delong, E.F. and Doolittle, W.F. 2007. *The New Science of Metagenomics: Revealing the Secrets of Our Microbial Planet*. National Academy of Sciences, Washington, DC.

Harry, M., Gambier, B. and Sillam, E.G. 2000. Soil conservation for DNA preservation for bacterial molecular studies. *European Journal of Soil Biology, 36 (1)*: 51–55.

Healy, F.G., Ray, R.M., Aldrich, H.C., Wilkie, A.C., Ingram, L.O. and Shanmugam, K.T. 1995. Direct isolation of functional genes encoding cellulases from the microbial consortia in a thermophilic, anaerobic digester maintained on lignocellulose. *Applied Microbiology and Biotechnology, 43 (4)*: 667–674.

Henne, A., Daniel, R., Schmitz, R.A. and Gottschalk, G. 1999. Construction of environmental DNA libraries in *Escherichia coli* and screening for the presence of genes conferring utilization of 4-hydroxybutyrate. *Applied and Environmental Microbiology, 65 (9)*: 3901–3907.

Herbert, R.A. 1992. A perspective on the biotechnological potential of extremophiles. *Trends in Biotechnology, 10*: 395–402.

Hinoue, M., Fukuda, K., Wan, Y., Yamauchi, K., Ogawa, H. and Taniguchi, H. 2004. An effective method for extracting DNA from contaminated soil due to industrial waste. *Journal of University of Occupational and Environmental Health, 26 (1)*: 13–21.

Hjort, K., Bergstrom, M., Adesina, M.F., Jansson, J.K., Smalla, K. and Sjoling, S. 2010. Chitinase genes revealed and compared in bacterial isolates, DNA extracts and a metagenomic library from a phytopathogen-suppressive soil. *FEMS Microbiology Ecology, 71 (2)*: 197–207.

Hu, X.P., Heath, C., Taylor, M.P., Tuffin, M. and Cowan, D. 2012. A novel, extremely alkaliphilic and cold-active esterase from Antarctic desert soil. *Extremophiles, 16 (1)*: 79–86.

Huson, D.H., Mitra, S., Weber, N., Ruscheweyh, N. and Schuster, S.C. 2011. Integrative analysis of environmental sequences using MEGAN4. *Genome Research, 21 (9)*: 1552–1560.

Iglesia, D.L., Valenzuela-Heredia, R.D., Pavissich, J.P., Freyhoffer, S., Andrade, S., Correa, J.K. and Gonzalez, B. 2010. Novel polymerase chain reaction primers for the specific detection of bacterial copper P-type ATPases gene sequences in environmental isolates and metagenomic DNA. *Letters in Applied Microbiology, 50 (6)*: 552–562.

Jackson, C.R., Harper, J.P., Willoughby, D., Roden, E.E. and Churchill, P.F. 1997. A simple, efficient method for the separation of humic substances and DNA from environmental samples. *Applied and Environmental Microbiology, 63 (12)*: 4993–4995.

Jacobsen, C.S., Carsten, S. and Rasmussen, O.F. 1992. Development and application of a new method to extract bacterial DNA from soil based on separation of bacteria from soil with cation-exchange resin. *Applied and Environmental Microbiology, 58 (8)*: 2458–2462.

Kennedy, J., Marchesi, J.R. and Dobson, A.D. 2007. Metagenomic approach to exploit the biotechnological potential of the microbial consortia of marine sponges. *Applied Microbiology and Biotechnology, 75 (1)*: 11–20.

Kennedy, J., Marchesi, J.R. and Dobson, A.D. 2008. Marine metagenomics: Strategies for the discovery of novel enzymes with biotechnological applications from marine environments. *Microbial Cell Factories, 7 (1)*: 27.

Lee, T.K., Lee, J., Sul, W.J., Iwai, S., Chai, B., Tiedje, J.M. and Park, J. 2011. Novel biphenyl-oxidizing bacteria and dioxygenase genes from a Korean tidal mudflat. *Applied and Environmental Microbiology, 77 (11)*: 3888–3891.

Leff, L.G., Dana, J.R., McArthur, J.V. and Shimkets, L.J. 1995. Comparison of methods of DNA extraction from stream sediments. *Applied and Environmental Microbiology, 61 (3)*: 1141–1143.

Li, L.L., McCorkle, S.R., Monchy, S., Taghavi, S. and van der Lelie, D. 2009. Bioprospecting metagenomes: Glycosyl hydrolases for converting biomass. *Biotechnology Biofuels, 2 (10).*

Long, P.F., Dunlap, W.C., Battershill, C.N. and Jaspars, M. 2005. Shotgun cloning and heterologous expression of the patellamide gene cluster as a strategy to achieving sustained metabolite production. *ChemBioChem, 6 (10)*: 1760–1765.

Majernik, A., Gerhard, G. and Rolf, D. 2001. Screening of environmental DNA libraries for the presence of genes conferring Na+ (Li+)/H+ antiporter activity on *Escherichia coli*: Characterization of the recovered genes and the corresponding gene products. *Journal of Bacteriology, 183 (22)*: 6645–6653.

More, M.I., James, B.H., Margarida, C.S., William, C.G. and Eugene, L.M. 1994. Quantitative cell lysis of indigenous microorganisms and rapid extraction of microbial DNA from sediment. *Applied and Environmental Microbiology, 60 (5)*: 1572–1580.

Niemi, R.M., Heiskanen, I., Wallenius, K. and Lindstrom, K. 2001. Extraction and purification of DNA in rhizosphere soil samples for PCR-DGGE analysis of bacterial consortia. *Journal of Microbiological Methods, 45 (3)*: 155–165.

Omura, S., Ikeda, H., Ishikawa, J., Hanamoto, A., Takahashi, C., Shinose, M. and Hattori, M. 2001. Genome sequence of an industrial microorganism *Streptomyces avermitilis*: Deducing the ability of producing secondary metabolites. *Proceedings of the National Academy of Sciences, 98 (21)*: 12215–12220.

Pathak, G.P., Ehrenreich, A., Losi, A., Streit, W.R. and Gartner, W. 2009. Novel blue light-sensitive proteins from a metagenomic approach. *Environmental Microbiology, 11 (9)*: 2388–2399.

Purohit, M.K. and Singh, S.P. 2008. Assessment of various methods for extraction of metagenomic DNA from saline habitats of coastal Gujarat (India) to explore molecular diversity. *Letters in Applied Microbiology, 49 (3)*: 338–344.

Riesenfeld, C.S., Goodman, R.M. and Handelsman, J. 2004a. Uncultured soil bacteria are a reservoir of new antibiotic resistance genes. *Environmental Microbiology, 6 (9)*: 981–989.

Riesenfeld, C.S., Schloss, P.D. and Handelsman, J. 2004b. Metagenomics: Genomic analysis of microbial communities. *Annual Review of Genetics, 38*: 525–552.

Robe, P., Nalin, R., Capellano, C., Vogel, T.M. and Simonet, P. 2003. Extraction of DNA from soil. *European Journal of Soil Biology, 39 (4)*: 183–190.

Rochelle, P.A., Fry, J.C., Parkes, R.J. and Weightman, A.J. 1992. DNA extraction for 16S rRNA gene analysis to determine genetic diversity in deep sediment communities. *FEMS Microbiology Letters, 100 (1)*: 59–65.

Ronaghi, M., Karamohamed, S., Pettersson, B., Uhlen, M. and Nyren, P. 1996. Real-time DNA sequencing using detection of pyrophosphate release. *Analytical Biochemistry, 242 (1)*: 84–89.

Rondon, M., August, P., Bettermann, A., Brady, S., Grossman, T., Liles, M., Loiacona, K. and Lynch, B. 2000. Cloning the soil metagenome: A strategy for accessing the genetic and functional diversity of uncultured microorganisms. *Applied and Environmental Microbiology, 66 (6)*: 2541–2547.

Schiraldi, C. and De Rosa, M. 2002. The production of biocatalysts and biomolecules from extremophiles. *TRENDS in Biotechnology, 20 (12)*: 515–521.

Schmeider, R. and Edwards, E. 2011. Fast identification and removal of sequence contamination from genomic and metagenomic datasets. *PLoS One, 6 (3)*: e17288.

Schmidt, E.W., Nelson, J.T., Rasko, D., Sudek, S., Eisen, J., Haygood, M.G. and Ravel, J. 2005. Patellamide A and C biosynthesis by a microcin-like pathway in Prochloron didemni, the cyanobacterial symbiont of Lissoclinumpatella. *Proceedings of National Academy of Sciences of the United States of America, 102 (20)*: 7315–7320.

Schmidt, O., Drake, H.L. and Horn, M.A. 2010. Hitherto unknown (Fe-Fe)-hydrogenase gene diversity in anaerobes and anoxic enrichments from a moderately acidic fen. *Applied and Environmental Microbiology, 76 (6)*: 2027–2031.

Schmidt, T.M., DeLong, E.F. and Pace, N.R. 1991. Analysis of a marine picoplankton community by 16S rRNA gene cloning and sequencing. *Journal of Bacteriology, 173 (14)*: 4371–4378.

Sharma, P., Capalash, N. and Kaur, J. 2007. An improved method for single step purification of metagenomic DNA. *Molecular Biotechnology, 36 (1)*: 61–63.

Siddhapura, P.K., Vanparia, S., Purohit, M.K. and Singh, S.P. 2010. Comparative studies on the extraction of metagenomic DNA from the saline habitats of Coastal Gujarat and Sambhar Lake, Rajasthan (India) in prospect of molecular diversity and search for novel biocatalysts. *International Journal of Biological Macromolecules, 47 (3)*: 375–379.

Simon, C. and Daniel, R. 2011. Metagenomic analyses: Past and future trends. *Applied and Environmental Microbiology, 77 (4)*: 1153–1161.

Simon, C., Herath, J., Rockstroh, S. and Daniel, R. 2009. Rapid identification of genes encoding DNA polymerases by function-based screening of metagenomic libraries derived from glacial ice. *Applied and Environmental Microbiology, 75 (9)*: 2964–2968.

Simonet, P., Capellano, A., Navarro, E., Bardin, R. and Moiroud, A. 1984. An improved method for lysis of Frankia with acharomopeptidase allows detection of new plasmids. *Canadian Journal of Microbiology, 30 (10)*: 1292–1295.

Sleator, R.D., Shortall, C. and Hill, C. 2008. Metagenomics. *Letters in Applied Microbiology, 47 (5)*: 361–366.

Steele, H.L., Jaeger, K.E., Daniel, R. and Streit, W.R. 2008. Advances in recovery of novel biocatalysts from metagenomes. *Journal of Molecular Microbiology and Biotechnology, 16 (1–2)*: 25–37.

Steffan, R.J., Goksøyr, J., Bej, A.K. and Atlas, R.M. 1988. Recovery of DNA from soils and sediments. *Applied and Environmental Microbiology, 54 (12)*: 2908–2915.

Stepanauskas, R. and Sieracki, M.E. 2007. Matching phylogeny and metabolism in the uncultured marine bacteria, one cell at a time. *Proceedings of the National Academy of Sciences, 104 (21)*: 9052–9057.

Sudan, A.K. and Vakhlu, J. 2013. Isolation of a thioesterase gene from the metagenome of a mountain peak, Apharwat, in the northwestern Himalayas. *3 Biotech, 3 (1)*: 19–27.

Tebbe, C.C. and Vahjen, W. 1993. Interference of humic acids and DNA extracted directly from soil in detection and transformation of recombinant DNA from bacteria and yeast. *Applied and Environmental Microbiology, 59 (8)*: 2657–2665.

Turnbaugh, P.J. and Gordon, J.L. 2008. An invitation to the marriage of metagenomics and metabolomics. *Cell, 134 (5)*: 708–713.

Uchiyama, A. and Miyazaki, A. 2009. Functional metagenomics for enzyme discovery: Challenges to efficient screening. *Current Opinion in Biotechnology, 20 (6)*: 616–622.

Uchiyama, T. and Watanabe, K. 2007. The SIGEX scheme: High throughput screening of environmental metagenomes for the isolation of novel catabolic genes. *Biotechnology and Genetic Engineering Reviews, 24 (1)*: 107–116.

Uchiyama, T., Abe, T., Ikemura, T. and Watanabe, K. 2005. Substrate-induced gene-expression screening of environmental metagenome libraries for isolation of catabolic genes. *Nature Biotechnology, 23 (1)*: 88–93.

Varaljay, V.A., Howard, E.C., Sun, S. and Moran, M.A. 2010. Deep sequencing of a dimethylsulfoniopropionate-degrading gene (dmdA) by using PCR primer pairs designed on the basis of marine metagenomic data. *Applied and Environmental Microbiology, 76 (2)*: 609–617.

Verma, D. and Satyanarayana, T. 2011. An improved protocol for DNA extraction from alkaline soil and sediment samples for constructing metagenomic libraries. *Applied Biochemistry and Biotechnology, 165 (2)*: 454–464.

Voget, S., Leggewie, C., Uesbeck, A., Raasch, C., Jaeger, K.E. and Streit, W.R. 2003. Prospecting for novel biocatalyst in a soil metagenome. *Applied and Environmental Microbiology, 69 (10)*: 6235–6242.

Wang, G.Y.S., Graziani, E., Waters, B., Pan, W., Li, X., McDermott, J., Meurer, G. et al. 2000. Novel natural products from soil DNA libraries in a streptomycete host. *Organic Letters, 2 (16)*: 2401–2404.

Wenhan, Z., Alex, L. and Mark, B. 2010. Ab initio gene identification in metagenomic sequences. *Nucleic Acids Research, 38 (12)*: 132.

Wilmes, P. and Bond, P.L. 2006. Metaproteomics: Studying functional gene expression in microbial ecosystems. *Trends in Microbiology, 14 (2)*: 92–97.

Wooley, J.C., Godzik, A. and Friedberg, I. 2010. A primer on metagenomics. *PLoS Computational Biology, 6 (2)*: e1000667.

Woyke, T., Xie, G., Copeland, A., Gonzalez, J.M., Han, C., Kiss, H., Saw, J.H. et al. 2009. Assembling the marine metagenome, one cell at a time. *PLoS One, 4 (4)*: e5299.

Young, C.C., Burghoff, R.L., Keim, J.G., Berbero, V.M., Lute, J.R. and Hinton, S.M. 1993. Polyvinylpyrrolidone–agarose gel electrophoresis purification of polymerase chain reaction amplifiable DNA from soils. *Applied and Environmental Microbiology, 59 (6)*: 1972–1974.

Yun, J. and Ryu, S. 2005. Screening for novel enzymes from metagenome and SIGEX, as a way to improve it. *Microbial Cell Factories, 4 (1)*: 8.

Zaprasis, A., Liu, Y.J., Liu, S.J., Drake, H.L. and Horn, M.A. 2010. Abundance of novel and diverse tfdA-like genes, encoding putative phenoxyalkanoic acid herbicide-degrading dioxygenases, in soil. *Applied Environmental Microbiology, 76 (1)*: 119–128.

Zeyaullah, M., Kamli, M.R., Islam, B., Atif, M., Benkhayal, F.A., Nehal, M., Rizvi, M.A. and Arif, A. 2009. Metagenomics—An advanced approach for non-cultivable microorganisms. *Biotechnology and Molecular Biology Reviews 4 (3)*: 49–54.

Zhou, J., Bruns, M.A. and Tiedje, J.M. 1996. DNA recovery from soils of diverse composition. *Applied and Environmental Microbiology, 62 (2)*: 316–322.

chapter four

Genetic engineering of nonconventional yeasts for the production of valuable compounds

Andriy Sibirny, Catherine Madzak, and Patrick Fickers

Contents

4.1 *Introduction*

Yeasts are suitable host organisms for the production of heterologous proteins and other valuable compounds because they combine ease of genetic manipulation with rapid growth on inexpensive medium at high cell densities, and have the ability, in contrast with bacteria, to perform complex posttranslational modifications. Yeasts also offer the possibility of secreting proteins, thus avoiding toxicity from intracellularly accumulated materials. Protein secretion also simplifies the subsequent purification process. Heterologous protein production tools were first developed in *Saccharomyces cerevisiae* in the 1980s. However, plasmid instability, low production yield, and hyperglycosylation phenomena have limited the development of profitable industrial processes (Buckholz et al. 1991; Romanos 1995). In addition, *S. cerevisiae* produces proteins with *N*-linked glycosylation patterns ending with an α-1,3-linked mannose motif that is known to be allergenic (Çelik et al. 2012). All of these led to the development of alternative expression systems in other yeasts including *Pichia pastoris, Hansenula polymorpha, Kluyveromyces lactis, Schizosaccharomyces pombe, Arxula adeninivorans,* and *Yarrowia lipolytica.* In this chapter, the production of recombinant proteins and other valuable compounds using the nonconventional yeasts *P. pastoris, Y. lipolytica,* and *H. polymorpha* is discussed.

4.2 Pichia pastoris

4.2.1 *Historical background and main characteristics*

Only a limited number of yeast species are able to use methanol as a primary carbon source (Hazeu et al. 1972). In that regard, the Phillips Petroleum Company (PPC) was the first to develop media and fermentation protocols for growing *P. pastoris* in continuous cultures at high cell densities to produce single-cell proteins (SCP) for feedstock from methanol (Cereghino et al. 2000). Unfortunately, the oil crisis of the 1970s rendered this process unattractive due to the high cost of methanol. In the following decade, PPC and the Salk Institute Biotechnology/Industrial Associates Inc. (SIBIA, La Jolla, CA) developed tools for heterologous protein production in *P. pastoris*. As a result, recombinant expression vectors, methods for transformation, selectable markers, and fermentation processes were developed to exploit the productive potential of this yeast (Cregg et al. 1993; Rosenfeld 1999). Expression systems developed for *P. pastoris* were patented by Research Corporation Technologies (Tucson, AZ) in 1993 and are available as a kit from Invitrogen Corporation (Carlsbad, CA). Actually, more than 1000 proteins have been cloned and expressed in *P. pastoris* (Cregg et al. 2000; Damasceno et al. 2012). *P. pastoris* has also been selected by several protein production platforms for genomics programs (Yokoyama 2003).

The *P. pastoris* genome is organized into four chromosomes with a total estimated size of 9.7 Mbp by pulsed-field gel electrophoresis (Ohi et al. 1998) and 9.43 Mbp after genome sequencing of strain GS115 (De Schutter et al. 2009). Genome sequence analysis led to the identification of 5.313 protein-coding genes with a GC content of 41.6%. Protein encoding genes represent 80% of the genome sequence. A secretion signal peptide was predicted in

9% of these genes. Overall, the codon usage is similar to that of *S. cerevisiae.* The *S. cerevisiae* orthologue gene *MNN1*, encoding the α-1,3-mannosyltransferase, which is responsible for the incorporation of the immunogenic terminal α-1,3-mannosyl glycotopes, was not detected in GS115 genome. The genome sequence analysis allowed the identification of vacuolar and secreted proteases, which will ease the development of protease-deficient strains (De Schutter et al. 2009).

P. pastoris prefers a respiratory growth mode and thus does not produce fermentation by-products such as acetic acid or ethanol (Cereghino et al. 2002). This enables cultures to reach high cell densities (up to 200 g/L dry weight; Heyland et al. 2010). Another advantage of *P. pastoris* is its capacity to secrete large amounts of heterologous proteins into the culture medium. Because *P. pastoris* secretes its native (i.e., nonrecombinant) proteins at low levels, secretion provides an effective method of separating heterologous proteins from the bulk of intracellular host proteins and other cellular materials (Romanos 1995).

Methylotrophic yeasts such as *P. pastoris* are able to metabolize methanol as their sole carbon source for energy through a specific metabolic pathway involving several unique enzymes. The initial steps of this pathway take place in the peroxisomes where methanol is converted into formaldehyde by two specific alcohol dehydrogenases, which is in turn converted into dihydroxyacetone and glyceraldehyde-3-phosphate by a dihydroxyacetone synthase. In the peroxisome, a catalase converts the oxygen peroxide generated in the first steps of methanol oxidation into oxygen and water. The subsequent steps of methanol assimilation are localized in the cytosol. Extensive reviews on methanol metabolism in *P. pastoris* could be found in the works of Cereghino et al. (2000) and Gellissen (2000).

4.2.2 Genetic and molecular tools

All *P. pastoris* expression strains are derived from the wild-type strain NRRL-Y 11430 (Northern Regional Research Laboratories, Peoria, IL). Auxothrophic mutants could carry one auxotrophy such as GS115 (*his4*), GS190 (*arg4*), or JC254 (*ura3*); two auxotrophies such as GS200 (*arg4 his4*) or JC227 (*ade1 arg4*); three auxothrophies such as JC300 (*ade1 arg4 his4*); and even four auxotrophies such as JC308 (*ade1 arg4 his4 ura3*; Cereghino et al. 2000). Protease-deficient strains are also available such as SMD1163 (*his4 pep4 prb1*), SMD1165 (*his4 prb1*), and SMD1168 (*his4 pep4*). In addition, there are three types of *P. pastoris* host strains available that differ with regard to their ability to metabolize methanol. These are the wild-type strain (Mut$^+$) and those resulting from *AOX1* deletion (Mut^S, methanol utilization slow) or from the double *AOX1-AOX2* deletion (Mut$^-$, methanol utilization minus). The Mut$^+$ phenotype is mainly used for heterologous protein production, although the Mut^S phenotype has been used in some cases. Molecular genetic manipulations of *P. pastoris* such as DNA-mediated transformation, gene targeting, gene replacement, and cloning by functional complementation are similar to those described for other yeasts such as *S. cerevisiae* (Cereghino and Cregg 2000). *P. pastoris* can be transformed by electroporation (10^5 transformant/µg DNA) or by using the classic lithium chloride (10^2 transformant/µg DNA) or polyethylene glycol (PEG_{1000}, 10^3 transformant/µg DNA) methods (Cregg et al. 1985, 1998). *P. pastoris* has a propensity for homologous recombinations between native and exogenous DNAs. However, gene replacement occurs at a lower frequency compared with *S. cerevisiae* and requires longer flanking sequences to direct integration (Cregg et al. 1998).

The success of *P. pastoris* as a platform for the production of heterologous proteins is related to the strong and tightly regulated promoter of the *AOX1* gene encoding alcohol oxidase I. As stated above, alcohol oxidases are the first enzyme of the methanol catabolism pathway. In *P. pastoris*, there are two alcohol oxidase genes, *AOX1* and *AOX2*, with

AOX1 representing 90% of the total alcohol oxidase activity in the cell (Cos et al. 2006). In appropriate culture conditions (i.e., methanol feeding at a growth-limiting rate), the Aox1 level can represent more than 30% of the total soluble proteins (Cregg et al. 2000). Therefore, *AOX1* promoter (p*AOX1*) is mainly used to drive and control the synthesis of heterologous proteins. p*AOX1* is repressed in the presence of glucose, glycerol, and ethanol and is strongly induced in the presence of methanol. Other carbon sources such as sorbitol have no significant repressive effect on p*AOX1* induction (Çalık et al. 2010; Niu et al. 2013). Indeed, sorbitol does not repress cell growth or heterologous protein production at less than 50 g/L (Çelik et al. 2009).

Despite p*AOX1*-based expression presenting several advantages, it also has some drawbacks. First, methanol is a fire hazard and therefore its storage in large quantities is undesirable. Second, as a derivative of petroleum sources, it is unsuitable for use in the production processes of certain foods and additives. Third, its monitoring during the production process is often difficult due to the unreliability of on-line probes and the time-consuming methods of off-line measurements (Macauley-Patrick et al. 2005). Fourth, methanol accumulation in the culture broth could result in a decrease of cell viability and alter heterologous protein stability.

To circumvent the use of methanol or at least to reduce its utilization, efforts have been made to develop novel alternative promoters to p*AOX1*. Several inducible or constitutive promoters have been developed. For instance, the promoter of the gluthatione-dependent formaldehyde dehydrogenase (FLD) was used successively to drive the production of heterologous proteins. *FLD1* is a key enzyme of methanol metabolism and certain alkylated amines such as methylamine as nitrogen source (Zwart et al. 1980). Its promoter is strongly and independently induced by either methanol/ammonium sulfate or methylamine/glucose mixtures. Furthermore, induction levels obtained with either methanol or methylamine are comparable with those obtained with p*AOX1* in the presence of methanol (Shen et al. 1998).

Waterham and coworkers developed an expression vector based on the promoter of glyceraldehyde 3-phosphate dehydrogenase gene (p*GAP*). Promoter studies, using β-lactamase as a reporter, showed that p*GAP* is constitutively active. However, its strength varies depending on the carbon source used for cell growth. Expression levels in glucose media are significantly higher than those obtained with p*AOX1* in the presence of methanol. p*GAP* has been successfully used for the synthesis of heterologous proteins from bacterial, yeast, insect, and mammalian origins (Waterham et al. 1997). In addition, Sears et al. (1998) developed a moderate promoter from the *YPT1* gene that encodes a small GTPase involved in secretion. Expression levels from *YPT1* promoter are 10-fold to 100-fold lower than those from the *GAP* promoter. This promoter is particularly well suited for expressing genes that would be toxic for the cells when overexpressed. A similar approach for moderate expression level was based on the promoter from the *PEX8* gene encoding a peroxisomal matrix protein that is essential for peroxisome biogenesis (Liu et al. 1995). The use of the *AOX2* gene promoter (full or truncated version) has also been reported in the literature (Mochizuki et al. 2001). Despite its expression level being much lower than p*AOX1*, Kobayashi et al. (2000b) demonstrated that the addition of small amounts of oleic acid (0.01%) led to a twofold increased protein production. This could be associated with the presence of an oleic acid-responsive element within the p*AOX2*. Menendez and coworkers (2003) cloned the *ICL1* gene encoding an isocitrate lyase and developed expression vectors based on its promoter (p*ICL1*). The authors reported that high levels of heterologous gene expression could be obtained in the presence of low glucose concentrations or in the presence of ethanol as the sole carbon source. However, higher expression levels were obtained in the absence of glucose than after induction with ethanol. An extensive review of the promoters used for heterologous protein production in *P. pastoris* and their regulation has been published recently (Vogl et al. 2013).

Although molecular genetic techniques are well developed for *P. pastoris*, few selectable markers have been described in the literature. These are biosynthetic markers from histidine biosynthetic pathways, namely, *HIS1, HIS2, HIS4, HIS5,* and *HIS6*; arginine biosynthetic pathways, namely, *ARG1, ARG2, ARG3,* and *ARG4*; or uracil biosynthetic pathways, namely, *URA3* and *URA5* (Cereghino et al. 2001; Cregg et al. 1985; Nett et al. 2003, 2005). Others markers such as *ADE1* (PR-amidoimidazolsuccinocarboxamide synthase; Cereghino et al. 2001), *MET2* (homoserine-*O*-transacetylase; Thor et al. 2005), and *FLD1* (formaldehyde dehydrogenase, see below; Sunga et al. 2004) have been described. Besides biosynthetic markers, other selectable markers, based on bacterial antibiotic resistance genes, could be used in *P. pastoris* (Romanos et al. 1992). These are the *Sh ble* gene (Zeo^R), from the bacterium *Streptoalloteichus hindustanu*, which confers resistance to the bleomycin-like drug zeocin (Drocourt et al. 1990; Higgins et al. 1998), the blasticidin S-deaminase gene from *Aspergillus terreus*, which confers resistance to blasticidin (Kimura et al. 1994), and the kanamycin resistance gene (Kan^R), which confers resistance to the eukaryotic antibiotic G418 (Scorer et al. 1994). The use of these markers allows multicopy integrations of an expression vector to be obtained when the selection of transformants is performed at high antibiotic concentrations (Scorer et al. 1994). A disadvantage of these multicopy integrants is the integration of bacterial antibiotic resistance genes within the *P. pastoris* genome. Indeed, this could represent a potential DNA hazard such as the dissemination of these resistance genes into the environment. To circumvent this drawback, an amplification marker based on the *FLD1* gene was developed (Sunga et al. 2004). As stated previously, *P. pastoris* can metabolize methanol as its sole carbon source and certain alkylated amines such as methylamine and choline as sole nitrogen sources. Formaldehyde is a common toxic intermediate in the methanol and methylamine pathways that could be further oxidized to formate by formaldehyde dehydrogenase Fld1 and to carbon dioxide by formate dehydrogenase (Sunga et al. 2004). Thus, this oxidative pathway may help the cell in protecting itself from the toxic effects of an excess of formaldehyde (Sibirny et al. 1990). Sunga and coworkers (2004) isolated a *P. pastoris fld1* mutant with increased sensitivity to formaldehyde and demonstrated that the level of resistance to formaldehyde was proportional to the number of *FLD1* gene copies present in that disrupted strain. Thus, the utilization of this selectable marker in combination with this *fld1* mutant strain allows multicopy integrations of expression vectors without any potential biohazard problem.

Several methods for unmarked genetic modifications in *P. pastoris* have been developed. One uses either *URA3* or *URA5* as a counterselectable marker (Nett et al. 2003). However, uracil auxothrophic strains grow very slowly even in medium supplemented with uracil, which renders their utilization difficult and time-consuming (Cereghino et al. 2001). To overcome this limitation, alternative counterselectable markers that can be used in any *P. pastoris* strain have been developed. One of these is the T-*urf*13 gene from the mitochondrial genome of male sterile maize used in combination with the insecticide Methomyl (Soderholm et al. 2001). Indeed, the introduction of T-*urf*13 gene renders the strain sensitive to the insecticide. The problem with this method is that the toxicity of T-*urf*13 gene may cause conditional lethality with certain gene deletions (Nett et al. 2003). Another method is based on the *Escherichia coli* toxin gene *mazF*, which is used as a counterselectable marker (Yang et al. 2009a). The main disadvantage of this method is that gene disruptions require the construction of plasmids and the performance of time-consuming selectable marker rescue, which leaves undesired repeat sequences in the *P. pastoris* genome.

However, Pan et al. (2011) have developed a rapid method for unmarked gene deletion and marker rescue based on the Cre/mutated *lox* system and a zeocin resistance gene. The disruption cassette "5′ target region-*lox71*-Cre-ZeoR-*lox66*-3′ target region" is obtained by PCR, rendering plasmid construction unnecessary. The Cre gene expression is regulated by pAOX1 and thus is expressed in the presence of methanol. After gene disruption, transient expression of the Cre recombinase allows *lox71* and *lox66* to recombine with excision of the Cre-ZeoR cassette. The resulting hybrid *lox* site (known as *lox72*) in the genome displays a strongly reduced binding affinity for the Cre recombinase. This method could be used sequentially to disrupt genes without the introduction of any marker in the cell genome and without the possibility of genome rearrangement.

P. pastoris can produce heterologous proteins intracellularly or extracellularly. Protein secretion allows several steps of purification such as cell lysis and subsequent clarifications to remove cell debris to be avoided. The *S. cerevisiae* α-factor prepro-peptide or the *P. pastoris* acid phosphatase (*PHO1*) signals are the most widely used secretion signals (Cereghino et al. 2000). However, other secretion signals have also been used successfully. These are signal peptides from *Rhizopus oryzae* α-amylase (Li et al. 2011), *Trichoderma reesei* class 2 hydrophobins (Kottmeier et al. 2011), human serum albumin (Xiong et al. 2008), K28 yeast virus toxin (Eiden-Plach et al. 2004), *Pichia acacia* killer toxin (Crawford et al. 2002), *Phaseolus vulgaris* phytohemagglutinin (Raemaekers et al. 1999), and the *P. pastoris* endogenous signal peptides *PIR1* (Inan 2010), *SCW*, *DSE*, and *EXG* (Liang et al. 2013). The efficiency of these signal peptides is protein-dependent and must be investigated experimentally to isolate the most suited one. In some cases, the native signal sequence of the heterologous protein is used.

In contrast with bacteria, yeasts are able to produce glycosylated proteins. However, glycosylation patterns in yeasts are different from those of higher eukaryotes such as mammalian cells. *P. pastoris* is capable of adding *O*-linked and *N*-linked carbohydrate moieties to secreted proteins (Cregg et al. 2000). However, *O*-oligosaccharides in *P. pastoris* are composed of mannose residues compared with mammals in which *O*-oligosacharides are made up of mainly sialic acid, galactose, and *N*-acetylgalactosamine (Daly et al. 2005). Moreover, and in contrast with most eukaryotes that add *O*-linked saccharides onto hydroxyl groups of serine and threonine, *P. pastoris* has no preferred amino acids for *O*-glycosylation (Cregg et al. 2000). Glycosylation of proteins in *P. pastoris* could modify the functions and characteristics of the heterologous proteins produced, which could become an issue for some applications (Eckart et al. 1996). Moreover, the absence of sialic acid in glycoproteins leads to their rapid clearance from the bloodstream, which can also be a disadvantage in some applications (Khandekar et al. 2001). In the last decade, there have been great advances in humanizing the *N*-glycosylation pathway in *P. pastoris*. Vervecken and coworkers used a strategy that consists of the disruption of the *OCH1* gene coding for α-1,6-mannosyltransferase Och1p, which is responsible for hypermannolysation in yeasts (Jacobs et al. 2008; Vervecken et al. 2004). In a second step, they stepwise introduced heterologous glycosylation enzymes into the endoplasmic reticulum or Golgi complex. These enzymes are an α-1,2-mannosidase and two chimeric glycosyltransferases, namely, Ker2-GnTI and Ker2-β-1,4-galactosyltransferase. For that purpose, they developed specific vectors known as GlycoSwitch vectors, which allow the conversion of any wild-type *P. pastoris* strain into strains that modify their glycoproteins with Gal2GlcNAc2Man3GlcNAc2N-glycans (Jacobs et al. 2008). Other groups also succeeded in humanizing glycosylation pathways in *P. pastoris*. Hamilton and coworkers (2006) constructed strains capable of secreting terminally sialylated complex and biantennary glycoproteins. After knocking out four genes to eliminate yeast-specific glycosylation, they introduced 14 heterologous genes, allowing them to replicate the sequential steps of the human glycosylation pathway. In addition, Li et al. (2006) were able to optimize the production of humanized IgGs in a glycoengineered *P. pastoris* strain that presented antibody-mediated effector functions.

4.2.3 Bioprocess operations and optimization

Bioprocess parameters such as medium composition, pH, temperature, aeration rate, induction, and feeding strategies are of utmost importance because it directly affects the production yield. These parameters vary according to the strain used or the heterologous protein produced. However, there are some guidelines that allow high productivity yields to be attained. The most commonly used medium for heterologous protein production in *P. pastoris* is the basal salt medium (BSM), which was developed by Invitrogen Co. (Invitrogen 2000). Despite this medium being considered as the standard, it presents some negative points including unbalanced composition, precipitate formation, ionic strength, and others. Therefore, alternative media were developed (d'Anjou et al. 2000; Stratton et al. 1998). All defined media are used in combination with the trace salt solution PTM1 from Invitrogen Co. (Invitrogen 2000). Ammonium hydroxide is usually used as the nitrogen source and also to maintain pH at the setup value. Various combinations of glycerol and methanol are used as carbon and energy sources. Most of the heterologous proteins are produced under the control of the *AOX1* promoter in Mut^+ or Mut^S *P. pastoris* strains according to a protocol proposed by Invitrogen Co., which is mainly derived from the works of Brierley et al. (1990). This includes a glycerol batch phase (GBP), a glycerol-fed batch phase (GFBP), a transition phase (TP), and finally a methanol induction phase (MIP).

The objective of the GBP is to accumulate cells in the least amount of time possible. Glycerol is used as a carbon source because it is nonfermentable and allows higher maximum specific growth rate compared with methanol (0.18 vs. 0.14 h^{-1}; Cos et al. 2006). In GBP, glycerol concentration is limited to 40 g/L because higher values could inhibit cell growth (Invitrogen 2000). When glycerol from GBP is exhausted, as indicated by a sharp increase of the dissolved oxygen, the GFBP phase starts. Glycerol is added at a constant rate usually equal to the maximum glycerol-specific consumption rate (0.0688 g $(gh)^{-1}$; Cos et al. 2006). The primary objective of this phase is to further increase cell concentration without repressing growth. At the end of this phase, methanol is added at an increasing stepwise mode to acclimate cells and to derepress the *AOX1* promoter (TP phase). Finally, in the MIP, methanol is added in fed-batch mode at a specific feeding rate that varies according the *P. pastoris* strain genotype and the heterologous protein produced. Indeed, it has been shown that the specific growth rate on methanol influences heterologous protein production (Çelik et al. 2009; Kobayashi et al. 2000a; Zhang et al. 2005).

Methanol is a high-degree reductant with high heat of combustion (Niu et al. 2013). This leads to one major challenge during a large-scale MIP. Because heat production is almost linearly correlated with oxygen consumption, the challenge is how to reduce oxygen consumption without affecting protein productivity. One possible answer is to use a cosubstrate such as glycerol (Jungo et al. 2007a; Zalai et al. 2012; Zhang et al. 2003), glucose (Jorda et al. 2012; Paulová et al. 2012), or sorbitol (Çelik et al. 2009; Gao et al. 2012; Inan et al. 2001; Ramón et al. 2007; Wang et al. 2010; Zhu et al. 2011) to partially replace or at least complement methanol during the protein production phase. The most promising cosubstrate is sorbitol because it is a low-degree reductant and a nonrepressing carbon source for p*AXO1* (Inan et al. 2001). The benefits of mixed feeding of sorbitol and methanol have been widely characterized with the aims of reducing oxygen consumption without loss of protein productivity. For instance, Jungo et al. (2007b) developed a fed-batch strategy based on a methanol/sorbitol mixture (43:57, C-mol C-mol). However, very few studies have been performed to quantitatively analyze the cellular physiology during cofeeding (Çelik et al. 2010; Jungo et al. 2007b). Most of them focused only on the final product (i.e., heterologous protein) without any insights into p*AOX1* regulation. Niu et al. (2013) reported on the quantitative characterization of *P. pastoris* cell metabolism, with special emphasis on the quantification of p*AOX1* induction during a methanol/sorbitol cofeeding process using the LacZ reporter gene, transient continuous culture, and metabolic flux analysis (MFA). Their results demonstrated that cell-specific oxygen consumption (qO_2) could be reduced by decreasing the methanol fraction in the feeding media. More interestingly, optimal pAOX1 induction was achieved and maintained in the range of 0.45 to 0.75 C-mol/C-mol of the methanol fraction. In addition, the qO_2 was reduced by 30% at most in those conditions. Based on a simplified metabolic network, an MFA was performed to quantify intracellular metabolic flux distribution. Cofeeding of sorbitol can reduce the oxidation flux in the peroxisome leading to less oxygen consumption and heat production. At the same time, sorbitol in the mixture produces energy through the TCA cycle and provides carbon source for biomass synthesis. According to their MFA model, 61% of the sorbitol goes through the TCA cycle whereas the rest is used for biosynthesis at 0.5 C-mol/C-mol methanol.

Most of the processes of heterologous protein production by *P. pastoris* are performed at 30°C, which is the optimal growth temperature. However, Dragosits and coworkers demonstrated that a 10°C decrease led to a threefold increased heterologous protein yield together with a reduced flux through the TCA cycle, reduced levels of protein involved in oxidative stress response, and a lower cellular level of molecular chaperones. All these indicate that folding stress is generally decreased at lower cultivation temperatures, enabling more efficient heterologous protein secretion in *P. pastoris* host cells (Dragosits et al. 2009).

The pH of the culture medium also plays an important role because many parameters such as growth rate, enzyme stability, and proteolytic degradation are pH dependent. Although pH values in the range of 3.5 to 5.5 were reported to have a limited effect on cell growth of *P. pastoris* Mut^+ strain (Inan et al. 1999), different optimal pH values have been reported for the production of heterologous proteins (Macauley-Patrick et al. 2005).

4.2.4 Production of recombinant proteins for biopharmaceutical and medical uses

Antibodies are glycoproteins with high affinities to specific molecules. They are considered as a group of biopharmaceuticals with potent applications for the treatment of cancer as well as inflammatory and infectious diseases (Jeong et al. 2011). *P. pastoris* is the host cell of choice for antibody or antibody fragment production. It allows high protein titers to be obtained, free of bacterial endotoxins or potential viral contaminations associated with mammalian cell culture (Barnard et al. 2010). During the 1990s, only antibody fragments, single-chain IgG fragments (scFv), or fusion ScFv were produced in *P. pastoris* (Goncalves et al. 2013). The first success of full-length IgG production was published in 1999 (Ogunjimi et al. 1999). Despite this, IgG was biologically active, it was produced in *P. pastoris* strain SMD1168 (*pep4 his4*), which was unable to achieve human-like glycosylations. The glycan moiety of IgG is an important feature not only for its stability and solubility but also for its efficient binding to the target receptor. With the advances in strain glycoengineering such as GlycoFi (Beck et al. 2010) or GlycoSwitch (Jacobs et al. 2008), high titers of full-length antibodies with humanized glycan chains have been reported (Li et al. 2006; Potgieter et al. 2009; Ye et al. 2011).

Other proteins of medical interest have been produced in *P. pastoris* at various yields. These are insulin precursor (1.5 g/L; Wang et al. 2001), cytokines such as interleukin 2 (4 g/L; Cregg et al. 1993), human serum albumin (6 g/L; Cregg et al. 1993), tumor necrosis factor (10 g/L; Sreekrishna et al. 1989), tetanus toxin fragment C (12 g/L; Clare et al. 1991), coagulation inhibitors such as hirudin variant HV2 (1.5 g/L; Rosenfeld et al. 1996), ovine pregnancy recognition hormone interferon τ (0.28 g/L; Van Heeke et al. 1996), cytomegalovirus antigenic protein pp52 (0.1 g/L; Battista et al. 1996), $5HT_{5A}$ serotonin receptor (22 pmol/mg membrane protein; Weiß et al. 1995), and human μ-opioid receptor (400 fmol/mg protein; Talmont et al. 1996).

4.2.5 Production of recombinant proteins for industrial uses

P. pastoris has been utilized for the production of lytic enzymes widely used in the pulp and paper, textiles, detergent, agro-food, and chemical industries. Some examples are listed below:

- Xylanases such as *Aspergillus niger* endo-β-1,4-xylanase (based on p*AOX1* and *S. cerevisiae* invertase signal peptide, 60 mg/L; Berrin et al. 2000), thermostable *Thermomyces lanuginosus* IOC-4145 β-1,4-xylanase (based on p*AOX1* and *S. cerevisiae* α-factor signal peptide, 148 mg/L; Damaso et al. 2003), or yeast-like fungus *Aureobasidium pullulans* acidophilic endo-1,4-β-xylanase (based on p*AOX1* and native signal peptide, 178 mg/L; Tanaka et al. 2004).
- Cellulases such as acidic and thermostable carboxymethyl cellulase from the yeast *Cryptococcus* sp. S-2 (based on p*AOX1* and *HAS* signal peptide; Thongekkaew et al. 2008) or the edible straw mushroom *Volvariella volvacea* endo-β-1,4-glucanase EG1 (based on p*AOX1* and *S. cerevisiae* α-factor signal peptide, 65 mg/L; Ding et al. 2002).
- Proteases such as *Mucor pusillus* aspartic protease (based on p*AOX1* and *S. cerevisiae* invertase signal peptide; Montesino et al. 1999), *Aspergillus oryzae* alkaline protease (based on p*AOX1* and *S. cerevisiae* α-factor signal peptide, 513 mg/L; Guo et al. 2008), or *Thermomonospora fusca* YX thermostable serine protease (based on p*GAP* and *S. cerevisiae* α-factor signal peptide, 135 units/L; Kim et al. 2005).
- Lipases such as *Y. lipolytica* Lip2p (based on p*AOX1* and *S. cerevisiae* α-factor signal peptide, 630 mg/L; Yu et al. 2007), *Burkholderia cepacia* lipase (based on p*GAP* and *S. cerevisiae* α-factor signal peptide, 184 units/L; Jia et al. 2010), or *Candida antarctica* lipase B (based on p*GAP* and *S. cerevisiae* α-factor signal peptide, 44 mg/L; Larsen et al. 2008).
- Amylases such as mouse salivary α-amylase (based on p*AOX1* and pGkl killer protein signal peptide, 0.24 mg/L; Kato et al. 2001), alkalophilic *Bacillus* α-amylase (based on p*AOX1* and *PHO1* phosphatase signal peptide, 50 mg/L; Tull et al. 2001), or *Thermobifida fusca* α-amylase (based on p*GAP* and *S. cerevisiae* α-factor signal peptide, 510 units/L; Yang et al. 2010).
- Phosphatases such as human placental alkaline phosphatase (based on p*AOX1* and *PHO1* phosphatase signal peptide, 2 mg/L; Heimo et al. 1998), soybean root nodule phosphatase (based on p*AOX1* and *S. cerevisiae* α-factor signal peptide, 10 mg/L; Penheiter et al. 1998), or *Drosophila* type 1 serine/threonine phosphatase (based on p*AOX1* and *S. cerevisiae* α-factor signal peptide, 37 mg/L; Szöőr et al. 2001).
- Pectinase such as *Fusarium solani* pectate lyase (based on p*AOX1* and *PHO1* phosphatase signal peptide; Guo et al. 1995) or *Bacillus subtilis* pectate lyase based on p*AOX1* and *S. cerevisiae* α-factor signal peptide, 100 units/L; Zhuge et al. 2008).

- Phytases such as *A. niger* PhyA (based on p*AOX1* and *S. cerevisiae* α-factor signal peptide, 63 units/mL; Han et al. 1999), *Peniophora lycii* 6′-phytase (based on p*AOX1* and synthetic MF4I signal peptide, 10.540 units/mL; Xiong et al. 2006), or *A. niger* SK-57 phytase (based on p*AOX1* and *S. cerevisiae* α-factor signal peptide, 865 units/mL; Xiong et al. 2005).
- Laccases such as *Pycnoporus sanguineus* laccase (based on p*AOX1* and native signal peptide, 28 mg/L; Lu et al. 2009), or *Trametes* sp. laccase B (based on p*AOX1* and *S. cerevisiae* α-factor signal peptide, 32 units/mL; Li et al. 2007).

P. pastoris has also been used for the production of other compounds such as antifreeze proteins (Loewen et al. 1997), human gelatin (Williams et al. 2008), or human cathepsin K protease inhibitor (Linnevers et al. 1997).

4.3 Yarrowia lipolytica

4.3.1 *Main characteristics and industrial applications*

Y. lipolytica is an oleaginous yeast from the Saccharomycetes/Hemiascomycetes class, which exhibits remarkable lipolytic and proteolytic activities. This yeast appears ubiquitously in natural environments (soils, mycorrhizae, marine, and hypersaline waters) as well as in food, particularly in fermented dairy and meat products (e.g., cheeses and dry sausages), as reviewed in the article by Groenewald et al. (2014). Because *Y. lipolytica* is able to efficiently degrade hydrocarbons, especially alkanes, it is also frequently isolated from oil-polluted environments, hence its potential for bioremediation (reviewed in an article by Bankar et al. 2009). *Y. lipolytica* is a biosafety class 1 microorganism (Groenewald et al. 2014), which means that it is considered as "safe-to-use": several applications of this yeast have been given a "generally regarded as safe" (GRAS) status by the U.S. Food and Drug Administration. The complete sequence and annotation of *Y. lipolytica* genome by the Genolevures consortium (CNRS, France; Dujon et al. 2004) has been made available on the Genolevures website (http://www.genolevures.org/). This data, together with the recent design of the first genome-scale metabolic model (Loira et al. 2012), provided new tools for the study and valorization of this yeast, which has recently been the subject of volumes 24 and 25 from Springer's Microbiology Monographs series (Barth 2013a,b). For more than 50 years, *Y. lipolytica* has called the attention of scientists and manufacturers due to its potential for industrial applications. Oriented at first toward producing biomass or metabolites with commercial value from wild-type or traditionally improved strains, industrial use of this yeast could benefit from several decades of progress in genetic engineering, which would allow for modifications in its metabolism or its use as a host for heterologous protein production. In the 1950s, industrial use of *Y. lipolytica* was pioneered by the British Petroleum Company (UK), which exploited the alkane-degradating properties of this yeast to produce SCP for livestock feeding (Toprina G). This application was discontinued in 1978 as a consequence of the oil crisis (Groenewald et al. 2014). In the 1970s, Pfizer Inc. (US) developed industrial citric acid production from *Y. lipolytica*, a technology acquired in 1990 by Archer Daniels Midland Company (US), which is presumed to be still exploiting it nowadays (Groenewald et al. 2014). These large-scale industrial applications for producing SCP or citric acid have driven extensive safety and efficacy studies (GRAS status) and produced a large amount of data on *Y. lipolytica* cultivation. More recent applications of wild-type or traditionally improved *Y. lipolytica* strains include the production of erythritol for use as a food additive (Baolingbao Biology Co., China), the use of biomass as fodder yeast (Skotan SA, Poland), and the production of a starter for the bioremediation of lipid-rich wastewaters, composed of freeze-dried cells and extracellular lipase (Artechno, Belgium). The Polish company Skotan SA is also developing prebiotic and probiotic applications for the food industry (Groenewald et al. 2014). Because genetic and molecular tools have been developed, *Y. lipolytica* has emerged as an efficient heterologous production host for pharmaceutical or industrial proteins and enzymes (reviewed in the works of Madzak et al. 2004, 2013). Since the development by Pfizer Inc. (US) of a transformation method for *Y. lipolytica* (Davidow et al. 1985), the scientific literature has reported the successful expression of more than 120 proteins in this yeast, originating from approximately 75 species of various origins (reviewed in the works of Madzak et al. 2013).

The success of this research domain is largely due to the high secretion capacity of *Y. lipolytica* and to the predominance of the cotranslational secretion pathway in this yeast (Boisramé et al. 1998), a characteristic shared with higher eukaryotes (e.g., mammals), which optimizes the folding of complex heterologous proteins. The relatively recent development of different surface display systems now allows *Y. lipolytica* to be used as an arming yeast, further increasing its potential as a microbial cell factory (Yue et al. 2008; Yang et al. 2009b). However, most applications proposed in the scientific literature remain only prospective, their development rarely passing the exploratory stage. Indeed, there are currently only a few commercial or industrial applications for genetically modified (GM) strains of *Y. lipolytica*, which are described in Table 4.1 (data from Groenewald et al. 2014, from the companies' websites or personal communications). A commercial kit for heterologous expression/secretion is currently on the market (Yeastern Biotech Co., Taiwan) and *Y. lipolytica* has been used as a protein manufacturing platform (Protéus, France; Oxyrane, UK); both use technologies developed at INRA (French National Institute for Agricultural Research). Two categories of valuable compounds are presently produced using GM *Y. lipolytica*: carotenoids (Microbia, acquired by DSM in 2010) and eicosapentaenoic acid (EPA)-rich products (DuPont). DuPont has recently disclosed all information about *Y. lipolytica* metabolic engineering for the production of EPA-rich single-cell oils (SCO; Xue et al. 2013; see Section 4.3.3.3). However, if their EPA-rich *Y. lipolytica* biomass is currently used as an ω-3 supplement in fish feed (harmoniously raised salmon, Verlasso), the marketing of EPA-rich SCO for use as a dietary supplement (New Harvest) has been recently discontinued (personal communication from DuPont). New Harvest was only advertised as a vegetarian alternative to fish-based ω-3 oils; therefore, one can only wonder if public perception of its actual GMO-derived nature had something to do with DuPont's decision to stop its commercialization. In addition, two projects using recombinant enzymes, produced in engineered *Y. lipolytica* strains, for enzyme replacement therapies (ERTs), are now nearly on the marketing stage: human lysosomal enzymes, for the treatment of lysosomal storage diseases, for example, Pompe disease (Oxyrane, Belgium; Tiels et al. 2012; see Section 4.3.2.3) and overexpressed homologous *Y. lipolytica* lipase (Pignède et al. 2000) for the treatment of exocrine pancreatic insufficiency (Mayoly Spindler, France).

4.3.2 Genetic and molecular tools

The development of tools for engineering *Y. lipolytica* dates back nearly 30 years. This yeast was at first distinguished for its high secretion yield of numerous proteins, especially of alkaline extracellular protease (AEP), encoded by *YlXPR2* gene. Consequently, this gene provided the first (and the most frequently used) elements necessary for heterologous

Table 4.1 Commercial and Industrial Applications of GM Strains of *Y. lipolytica*

Years of activity	Company	Country	Product or use	Intellectual property, tools, safety, and efficacy data
Since 2004	Yeastern Biotech	Taiwan	YLEX Kit for expression of heterologous proteins in *Y. lipolytica*	INRA License, expression (pYLEX1) and secretion (pYLSC1) vectors + Po1g recipient strain
Since 2005	Protéus/PCAS Group	France	Use of *Y. lipolytica* as expression host in protein manufacturing platform	INRA License
2005–2007	Oxyrane	South Africa	Enantiomerically pure pharmaceuticals	INRA License, use of *Y. lipolytica* as expression platform for enantioselective epoxide hydrolases
Since 2007	Oxyrane	UK/ Belgium/ US	Idem + human lysosomal enzymes for ERTs of lysosomal storage diseases	INRA License, engineered *Y. lipolytica* strains for efficient targeting of therapeutic enzymes to lysosomes (high levels of mannose-6-phosphate); start of clinical trials for a treatment targeting Pompe disease
Since 2005	Microbia/DSM	US/ Netherlands	Carotenoids for use in food and feed	GRAS self-affirmation for β-carotene
Since 2006	Mayoly Spindler + Protea Biosciences (since 2011)	France/US	Recombinant *Y. lipolytica* lipase for ERT	INRA License, overproduction of homologous LIP2p, clinical phase II trials for chronic pancreatitis
2010–2013	DuPont Applied Biosciences	US	EPA-rich biomass for use in fish feed, EPA-rich SCO for use as dietary supplement	SCO safety and efficacy studies, GRAS status (GRN 355), marketing of New Harvest SCO discontinued in 2013

production in *Y. lipolytica*, namely, a promoter, a secretion signal (optional), and a terminator. The main genetic elements described in the literature for use in *Y. lipolytica* expression systems are listed in Table 4.2.

4.3.2.1 Promoters

Despite its strength and historical importance, the *XPR2* promoter (p*XPR2*) was shown to have a complex regulation (Ogrydziak et al. 1977), which hindered its industrial use,

Table 4.2 Genetic Elements Used in the Design of *Y. lipolytica* Expression System

Type of element (most frequently used in boldtype)	Characteristics (reference) (when unspecified: Barth et al. 1996)
Selection markers	
LEU2*, *URA3, *LYS5*, *ADE1*	Auxotrophy complementation
ura3d4	Idem, defective allele (Le Dall et al. 1994)
E. coli Phleo^R, **hph (Hyg^R)**	Antibiotic resistance: dominant
S. cerevisiae SUC2	Sugar utilization: dominant (Nicaud et al. 1989)
Promoters (source gene)	
p*LEU2* (β-isopropylmalate dehydrogenase)	Inducible (leucine precursor)
p*XPR2* (alkaline extracellular protease)	Inducible (peptones), complex regulation
p*POX2*, p*POT1* (acyl-CoA oxidase 2,3-oxo-acyl-CoA thiolase)	Inducible (fatty acids and derivatives, alkanes)
p*ICL1* (isocitrate lyase)	Inducible (fatty acids and derivatives, alkanes, ethanol, and acetate)
p*POX1*, p*POX5* (acyl-CoA oxidases 1 and 5)	Weakly inducible (alkanes)
p*G3P* (glycerol-3-phosphate dehydrogenase)	Inducible (glycerol)
p*MTP* (metallothioneins 1 and 2)	Bidirectional, inducible (metallic salts)
p*FBA1* and p*FBA1*$_{IN}$ [+native intron] (fructose 1,6-biphosphate aldolase)	Inducible (glucose), very strong especially when *FBA1* intron is included (Hong et al. 2012)
p*TEF1*, p*RPS7* (translation elongation factor-1α, ribosomal protein S7)	Constitutive (Müller et al. 1998)
hp4d (hybrid promoter derived from p*XPR2*)	Growth phase-dependent, based on four iterations of p*XPR2* UAS1B sequence (Madzak et al. 2000)
Hybrid promoters constructed on the "hpnd" model (derived from p*XPR2*)	Very strong, tunable, based on *n* iterations of UAS1B with *n* as high as 32 (Blazeck et al. 2011)
Secretion signals (processing enzyme for maturation)	
***XPR2* prepro**	157 aa, KR cleavage site (Xpr6p endoproteinase)
***XPR2* pre**	13 aa, LA cleavage site (signal peptidase complex)
***LIP2* prepro**	32 aa, KR cleavage site (Xpr6p endoproteinase)
LIP2 pre + dipeptides track	21 aa, XA or XP cleavage site (diamino-peptidase)
GPI anchor signals for surface display	
C terminal of YlCWP1p	110 aa (Yue et al. 2008)
C terminal of *S. cerevisiae* FLO1p	1097 aa (Yang et al. 2009b)
C terminal of YlCWP3p or YlCWP6p	121 and 139 aa, respectively (Yuzbasheva et al. 2011)
Terminators	
XPR2t, LIP2t, *PHO5*t	430, 150, and 320 bp, respectively, of noncoding 3′
minimal XPR2t	100 bp of noncoding 3′ + flanking restriction sites (obtained by PCR; Swennen et al. 2002)

Note: aa, amino acids; A, alanine; K, lysine; L, leucine; P, proline; R, arginine; X, any aa.

prompting the search for new promoters. A functional dissection of p*XPR2* showed that one of its upstream activating sequences, UAS1B, was only weakly affected by environmental conditions (Blanchin-Roland et al. 1994; Madzak et al. 1999). This UAS1B element was used to design the recombinant hp4d promoter, composed of four tandem copies of UAS1B upstream from a TATA box (minimal p*LEU2*; Madzak et al. 2000). Patented as a strong promoter whatever the pH and composition of culture medium, hp4d is not however a constitutive promoter because it has retained some unidentified elements that confer a growth phase-dependent expression (Madzak et al. 1996, 2000). Namely, hp4d-driven expression is limited during cell growth and increases at the beginning of the stationary phase, naturally ensuring a partial dissociation of growth and expression phases. This very peculiar characteristic is particularly interesting for heterologous protein production because it maximizes productivity and alleviates possible toxicity problems. Consequently, hp4d is by far the promoter most frequently used for such applications (>40% of cases reported in the scientific literature, as reviewed in the works of Madzak et al. 2013). Two strong constitutive promoters have been isolated, for the purpose of expression cloning, from *Y. lipolytica* TEF1 and RPS7 genes (Müller et al. 1998). Although p*TEF1* has also been used for heterologous production per se, these fully constitutive promoters are not recommended for such use because early expression of heterologous genes could impair cell growth. The use of inducible promoters responding to an easily controllable factor would ease industrial applications. Several laboratories have isolated such inducible promoters, among which the most interesting seem to be p*ICL1*, p*POT1*, and p*POX2*, isolated from key genes from fatty acid metabolisms, which are particularly well adapted for use in engineering *Y. lipolytica* hydrophobic substrate pathways (Juretzek et al. 2000). More recently, several promoters have been isolated from glycolytic genes and described in patents on metabolic engineering of *Y. lipolytica* (reviewed in the works of Madzak et al. 2013). The most interesting of these glucose-induced promoters seems to be p*FBA1*, not only because its activity is particularly high but also because heterologous production can be further increased (by a fivefold factor) when retaining the upstream region of the *FBA1* gene, carrying its 102 bp intron, in the recombinant construct (Hong et al. 2012). As demonstrated in the case of p$FBA1_{IN}$, the presence of the intron was able to simultaneously enhance promoter activity and increase mRNA stability, a phenomenon previously reported in various eukaryotes as intron-mediated enhancement (Hong et al. 2012). Blazeck and coworkers (2011) have recently performed an important improvement in the design of strong and tunable *Y. lipolytica* promoters based on the concept of the recombinant hp4d promoter. These authors constructed a large set of recombinant promoters, in which various copy numbers (from 1 to 32) of UAS1B were inserted upstream of either a minimal p*LEU2*, or different fragments of p*TEF1*. The strongest of these "hpnd-like" promoters exhibited unprecedented efficiency, eightfold higher than that of any known natural *Y. lipolytica* promoter (Blazeck et al. 2011). These strong tunable promoters constitute promising tools, but the genetic stability of those with a high number of tandem repeats may reveal an inability to comply with the very strict standards required for industrial processes. Specifically, Good Manufacturing Practice (GMP) guidelines (http://www.gmp-compliance.org/eca_link_navigator.html) state that producing strains should be stable over hundreds of generations, when a "hp12d-like" promoter was shown to exhibit recombinations in 17 out of 20 progeny isolates after only 36 generations (Blazeck et al. 2011). When inserting either eight or 16 UAS1B upstream from different fragments of p*TEF1* (from minimal to larger than full-length), the authors observed enhancing effects for all p*TEF1* variants. Thus, this work suggests that endogenous promoters are enhancer-limited and that this limitation can be alleviated through the addition of new UASs. Moreover, transcription factor availability does not seem to be a limiting factor, even for the strongest promoters constructed. Consequently, this work paved the way for designing fine-tuned promoters, in which added UASs could increase the expression level of endogenous promoters that could be selected from developing genomic and transcriptomic data.

4.3.2.2 Targeting signals for secretion and surface display

The *XPR2* prepro region has been the most widely used secretion signal, until it was shown that the *XPR2* pre region alone was sufficient to drive efficient heterologous secretion and occasionally ensured a more complete maturation (Swennen et al. 2002). As a consequence, this smaller signal is now preferably used for directing secretion. Alternative secretion signals have also been used, such as *LIP2* prepro region and a *XPR2/LIP2* prepro hybrid (Nicaud et al. 2002). More recently, it was also shown that the *LIP2* pre region with its dipeptide stretch performed better than the larger *LIP2* prepro (Gasmi et al. 2011). In addition, native secretion signals, especially those originating from fungi or plants, were often shown to be efficient in *Y. lipolytica* (reviewed in the works of Madzak et al. 2013).

Surface display systems have been developed rather recently in *Y. lipolytica.* They are based on transcriptional fusion of the heterologous gene to a sequence encoding the C-terminal part of a cell wall protein, a region acting as a signal for the addition of a glycosylphosphatidylinositol (GPI) anchor. A GPI anchor is a posttranslational modification that fastens the modified secreted protein into the cell wall of eukaryotic cells. The first report of surface display in *Y. lipolytica* was by Yue and coworkers (2008), using the GPI anchor domain of YlCWP1p, a cell wall protein previously described in an article by Jaafar et al. (2004). Alternative GPI anchor domains have also been used: a heterologous one, from *S. cerevisiae* FLO1p (Yang et al. 2009b), and those from newly described *Y. lipolytica* cell wall proteins, YlCWP3p and YlCWP6p (Yuzbasheva et al. 2011).

4.3.2.3 Targeting to microbodies

The possibility of targeting heterologous proteins to cellular microbodies, such as peroxisomes or lysosomes, is of particular interest in metabolic engineering. Targeting to *Y. lipolytica* peroxisomes of a polyhydroxyalkanoate synthase from *Pseudomonas aeruginosa* has been obtained by fusion to the modified C-terminal region (34 amino acids) of glyoxysomal isocitrate lyase from *Brassica napus* (Haddouche et al. 2010). The C-terminal SKM tripeptide from this latter sequence, previously shown to direct efficient peroxisomal targeting in *S. cerevisiae,* was modified into AKI to fit the supposed *Y. lipolytica* tripeptide consensus, deduced from that found in *Candida tropicalis,* a more closely related yeast (Haddouche et al. 2010). More recently, a C-terminal SKL tripeptide was shown to be efficient for targeting green fluorescent protein (GFP) to *Y. lipolytica* peroxisomes (Xue et al. 2013). Targeting to *Y. lipolytica* lysosomes of recombinant human lysosomal enzymes, for ERTs of lysosomal storage diseases, has been obtained by Oxyrane (Belgium; see Table 4.1). They glycoengineered *Y. lipolytica* (through heterologous expression of a glycosidase from *Cellulosimicrobium cellulans*) for the production of lysosomal enzymes bound to high levels of mannose-6-phosphate, a specific sugar structure enabling efficient targeting of these therapeutic enzymes to the lysosomes through interaction with cation-independent mannose-6 phosphate receptors (Tiels et al. 2012). Recently, several heterologous proteins have been targeted to and anchored on the surface of *Y. lipolytica* oleosomes through their fusion to the C-terminal domain of a sesame oleosin (Han et al. 2013). Namely, plant oleosins are naturally embedded into the phospholipid monolayer membrane of plant oleosomes, and their heterologous expression as fusion proteins in *Y. lipolytica* allows the design of functionalized oleosomes that can be easily purified using floating centrifugation. These nanoparticles (mean diameter, 200–300 nm) were found to be highly stable, which allows their application as scaffolds for protein immobilization and display. The proof of concept of codisplaying multiple proteins on *Y. lipolytica* oleosomes was demonstrated by using the high-affinity interaction between cohesin and dockerin domains from the cellulosome scaffolding system of *Clostridium cellulolyticum* (Fierobe et al. 1999). Oleosomes displaying a cohesin domain can be incubated with different dockerin-fusion proteins to form a self-assembled multifunctional nanoparticle able to target a specific cell type and to bring along new catalytic functions to its surface. For proof of concept, cohesin-displaying oleosomes interacted with (i) a GFP-dockerin fusion protein, for cell-targeting by binding to a GFP-antibody surface-displayed on *Y. lipolytica* cells; and (ii) a dockerin-bearing CelA endoglucanase from *C. cellulolyticum*, whose activity was followed by the detection of a red fluorescent compound. These engineered oleosomes were shown to be functional for both cell-targeting and reporting activities (Han et al. 2013). Such tunable multifunctional oleosomes could be useful for various applications, for example, targeted drug delivery, pathogen detection, or self-assembly of nanofactories (functionalized supramolecular nanostructures).

4.3.2.4 Vectors, related methods, and amplification strategies

Similar to other yeasts, the expression/secretion vectors used for transforming *Y. lipolytica* are shuttle vectors containing a bacterial moiety (plasmidic backbone) and a yeast moiety composed of a selection marker, an expression/secretion cassette, and an element for transformation and maintenance into yeast cells. The latter can be either an autonomously replicating sequence (ARS) in replicative vectors, or a sequence that will target or promote integration into the genome in integrative vectors. In *Y. lipolytica*, replicative vectors are not very attractive for heterologous protein production due to their low copy number and high loss frequency, both linked to the colocalization of centromeric and replicative functions in ARSs (Fournier et al. 1993). However, they are useful when transient expression is required, for example, when using marker rescue for performing multiple gene deletions. Namely, an efficient gene knockout method based on *Cre-lox* recombination system consists of integrating a promoter–terminator gene disruption cassette, carrying a selection marker flanked by *loxR–loxP* sequences, which are able to recombine during transient expression of Cre recombinase, allowing efficient marker rescue (Fickers et al. 2003). This method constitutes a powerful tool for genetic engineering of metabolic pathways in *Y. lipolytica*. Integrative vectors are widely used for heterologous production in *Y. lipolytica*. Their integration is generally targeted by linearizing the vector within a region homologous to the genome and occurs by single crossover. Their transformation efficiency can be very high (in the range of 10^5 transformants per microgram of DNA) when using chemical (lithium acetate) methods (Chen et al. 1997; Madzak et al. 2005; Xuan et al. 1988). INRA has developed autocloning vectors from which the bacterial moiety can be removed prior to transformation, enabling the resulting recombinant strain to retain its GRAS status (Nicaud et al. 2000). These autocloning vectors use zeta sequences, which are the long terminal repeats (LTRs) from Ylt1 retrotransposon for targeting or promoting integration into the *Y. lipolytica* genome. In Ylt1-carrying strains, zeta sequences provide at least 100 targeting sites per genome (Schmid-Berger et al. 1994), constituting a powerful tool for multiple homologous integration when combined with a defective selection marker. Surprisingly, zeta sequences also exhibit the unexpected property of enhancing nonhomologous integration when introduced into *Y. lipolytica* strains devoid of Ylt1 retrotransposon, thus providing a method for random genomic integration (patented by Nicaud et al. 2000). Some examples of ready-to-use vectors for intracellular expression, secretion, or surface display of heterologous proteins are shown in Table 4.3.

Some expression/secretion vectors and optimized strains constructed at INRA have been commercially available since 2006, such as the YLEX Expression Kit (see Table 4.1), which can be purchased from Yeastern Biotech Co. (Taiwan) or from their retailer (for Europe, the United States, and Canada), Gentaur (Belgium). The preferred strategy for increasing the copy number of a heterologous gene in *Y. lipolytica* is to perform multiple integrations of a vector (or expression cassette) into the genome, which can only be obtained by using a defective selection marker, such as the *ura3d4* allele (Le Dall et al. 1994). It is noteworthy that the use of *ura3d4* for selection dramatically decreases the transformation efficiency to approximately 10^2 transformants per microgram (Juretzek et al. 2001). High copy number integrants can be obtained not only when using repeated sequences (rDNA or zeta) as targets but also with a single integration site (Juretzek et al. 2001), because multiple homologous integrations almost always occur in tandemly repeated copies, at only one or two sites, irrespective of the number of potential targets (Le Dall et al. 1994). This phenomenon could unfortunately lead to genetic instability problems in the producing strains. In this context, the development of a nonhomologous integration system, by using autocloning vectors carrying zeta sequences into *Y. lipolytica* strains devoid of Ylt1 retrotransposon, has offered new possibilities: randomly integrated copies are more dispersed in the genome, leading to a better stability of high copy number integrants (Nicaud et al. 2000). This method, however, has its own drawbacks: transformation efficiency is reduced further, to approximately 10 transformants per microgram (Juretzek et al. 2001), when random integration effects require the testing of numerous transformants for selecting a good producer strain. Despite such difficulties, this method has been successfully used to produce numerous heterologous proteins, as reviewed in the works of Madzak et al. (2013). However, some producing strains developed at INRA, using random multiple integrations, have recently failed to obtain marketing authorizations due to an insufficient stability level. Even if these strains were able to roughly maintain their copy number, some changes in integration sites were observed over time (Nicaud et al. 2009); the regulations regarding GMO from GMP guidelines require that genomic modifications of the producing strains should be described precisely, and be stable over hundreds of generations. To alleviate these problems, a new strategy for obtaining stable multicopy transformants was recently patented (Nicaud et al. 2009). This new approach provides a genetically engineered strain, deleted for major homologous secreted proteins (to alleviate metabolic burden), in which several loci have been selected for targeting integrative vectors. This strain could be sequentially transformed by a set of vectors carrying different auxotrophic or dominant markers, with the possible use of marker rescue. Successive integrations could be checked by the inactivation of different reporter genes, which generates new phenotypes detectable in plate assays. This method allows us to obtain three to 10 stable copies of a gene of interest, or to coexpress the same number of different new functions for the engineering of *Y. lipolytica* metabolic pathways (Nicaud et al. 2009). Finally, another strategy for increasing heterologous production or for coexpressing several heterologous genes is to insert multiple expression cassettes into a single vector. The limitations of this method are the vector size (probably limited to between 12 and 14 kb) and possible recombinations between repeated sequences if the same promoter/targeting/terminator or even gene sequences are

Table 4.3 A Selection of Ready-to-Use Integrative Vectors for Expression/Secretion/Surface Display of Heterologous Proteins in *Y. lipolytica*

Plasmid type: characteristics/drawbacks					
Function	Name	Expression cassette	Selection marker	Targeting sequence	Reference
Monocopy pBR-based vectors for precise targeting at pBR docking platform: very high transformation efficiency, growth-phase related expression (hp4d promoter)/presence of bacterial sequences					
Intracellular expression	pINA1269	hp4d/MCS/ *XPR2*t	LEU2	pBR322 backbone	(Madzak et al. 2000)
Secretion	pINA1296	hp4d/*XPR2* pre/ MCS/*XPR2*t	LEU2	pBR322 backbone	(Madzak et al. 2000)
Mono/low copy auto-cloning vectors for targeting at zeta sequences or random integration into Ylt1-devoid strains: high transformation efficiency, devoid of bacterial DNA, constitutive (pTEF1) or growth-phase related expression (hp4d) or control of induction (pPOX2)					
Intracellular expression	pINA1312	hp4d/MCS/ *XPR2*t	ura3d1	zeta sequence	(Nicaud et al. 2002)
Intracellular expression	JMP62	p*POX2*/MCS/ *LIP2*t	ura3d1	zeta sequence	(Nicaud et al. 2002)
Intracellular expression	pKOV96	p*TEF1*/MCS/ *LIP2*t	ura3d1	zeta sequence	(Labuschagne et al. 2007)
Secretion	pINA1317	hp4d/*XPR2* pre/ MCS/*XPR2*t	ura3d1	zeta sequence	(Nicaud et al. 2002)
Secretion	JMP61	p*POX2*/*LIP2* prepro/MCS/ *LIP2*t	ura3d1	zeta sequence	(Nicaud et al. 2002)
Surface display	pINA1317-YlCWP110	hp4d/*XPR2* pre/ MCS/*YlCWP1*GPI anchor/*XPR2*t	ura3d1	zeta sequence	(Yue et al. 2008)
Multicopy auto-cloning vectors for targeting at zeta sequences or random integration into Ylt1-devoid strains: devoid of bacterial DNA, growth-related expression (hp4d) or control of induction (pPOX2), amplification of expression/very low transformation efficiency					
Intracellular expression	pINA1292	hp4d/MCS/ *XPR2*t	ura3d4	zeta sequence	(Nicaud et al. 2002)
Intracellular expression	JMP64	p*POX2*/MCS/ *LIP2*t	ura3d4	zeta sequence	(Nicaud et al. 2002)
Secretion	pINA1297	hp4d/*XPR2* pre/ MCS/*XPR2*t	ura3d4	zeta sequence	(Nicaud et al. 2002)
Secretion	JMP63	p*POX2*/*LIP2* prepro/MCS/ *LIP2*t	ura3d4	zeta sequence	(Nicaud et al. 2002)
Surface display	pINA1297-YlCWP110	hp4d/*XPR2* pre/ MCS/*YlCWP1*GPI anchor/*XPR2*t	ura3d4	zeta sequence	(Madzak et al. submitted)

Note: MCS, multiple cloning sites; pINA1269 and pINA1296 are, respectively, pYLEX1 and pYLSC1 in the YLEX kit (Yeastern Biotech Co., Taiwan).

used for different cassettes. Vectors with two expression cassettes are currently used for marker rescue (Fickers et al. 2003), although for transient expression only. More recently, *Y. lipolytica* fatty acid pathways were engineered by the coexpression of two heterologous desaturases, using a tandem double cassette vector with the same hp4d promoter (Chuang et al. 2010; see Section 4.3.3.3). Finally, triple and even quadruple expression cassettes in a single vector were described in several patents (using different promoters and terminator sequences) and, very recently, in an article by Celińska et al. (2013). These authors chose to coexpress three heterologous genes involved in glycerol metabolism using the same regulatory elements: a glycerol-induced *G3P* promoter and an *XPR2*-like terminator. The expression cassettes were shown to be stably maintained in the genome throughout more than 40 passages in selective medium, lyophilization followed by reviving process, and bioreactor cultivation without selective pressure (Celińska et al. 2013).

4.3.2.5 Strains

The E150 strain (*MatB, his1, leu2-270, ura3-302,* and *xpr2-322*) was chosen for the *Y. lipolytica* genome-sequencing project (Dujon et al. 2004). *Y. lipolytica* strains used for heterologous production were selected empirically at first for their growth and secretion capacities, and genetically modified to facilitate their use or to improve yields. A comprehensive list of such *Y. lipolytica* strains was published by Madzak et al. (2004). The most frequently used for heterologous expression are E129 (*MatA, lys11-23, leu2-270, ura3-302,* and *xpr2-322*) and essentially Po1d (*MatA, leu2-270, ura3-302,* and *xpr2-322*) and its derivatives. Po1d presents many interesting features: a high level of secretion, the deletion of AEP, which was a potent threat for secreted heterologous proteins, two auxotrophies due to nonrevertant mutations and the production of recombinant invertase (*ura3-302* allele corresponds to *ura3::pXPR2:SUC2*; namely, disruption of *URA3* by *SUC2* from *S. cerevisiae* under the control of p*XPR2*), which allows it to metabolize sucrose (Nicaud et al. 1989). These characteristics have been retained in a series of derivatives that were further adapted for heterologous protein production (Madzak et al. 2000, 2004): Po1f, Po1g, and Po1h were deleted for acid extracellular protease (ΔAXP: *axp1-2* allele), the other source of secreted proteasic activity. Po1g received an integrated pBR322 docking platform to ease the integration of pBR-based vectors. Finally, Po1g and Po1h retain only one auxotrophy, respectively, Leu$^-$ and Ura$^-$. Some *Y. lipolytica* strains, like E129 and E150, carry Ylt1 retrotransposons in their genome, when others, like the Po1 series of strains, are devoid of this transposable element. All these strains are available from INRA's CIRM (International Center for Microbial Resources) Yeasts Library (http://www7.inra.fr/cirmlevures/page.php?page=home&lang=en). Other derivatives of Po1d have been optimized for applications linked to lipid metabolism, such as the MTLY60 strain, deleted for extracellular lipases (*MatA, ura3-302, leu2-270, xpr2-322,* Δ*lip2* Δ*lip7,* and Δ*lip8*), which has been further fitted with a zeta docking platform to produce the JMY1212 strain, used in high-throughput expression platforms (Bordes et al. 2007; see Section 4.3.3.1). More recently, "obese" strains were derived from Po1d by deleting *YlGUT2*, encoding glycerol-3-phosphate dehydrogenase and, optionally, the acyl-coA oxidase genes *YlPOX1* to *POX6* (implicated in the β-oxidation pathway), have been patented at INRA (Nicaud et al. 2000). These mutant strains exhibit a threefold and fourfold increase, respectively, in lipid accumulation, which makes them interesting tools for the production of valuable lipid products (Beopoulos et al. 2008; see Section 4.3.3.4). Finally, some wild-type strains have been selected for their interesting properties (high lipid or protein content) from a library of 78 marine *Y. lipolytica* strains (Ocean University of China, Qingdao) and have been genetically engineered for use as SCO (Zhao et al. 2010), SCP (Cui et al. 2011), or citric acid producers (Liu et al. 2010), illustrating the potentialities offered by exploiting natural *Y. lipolytica* biodiversity.

4.3.3 Examples of processes and products of recombinant Y. lipolytica *strains*

As already exemplified in the commercial and industrial applications presented in Table 4.1, genetic engineering of *Y. lipolytica* for the production of valuable compounds has developed in two directions: on the one hand, using it as a host for the production of heterologous proteins of interest and, on the other hand, modifying its metabolism for use as a whole cell biocatalyst (microbial factory). Moreover, the relatively recent development of surface display and lysosome targeting methods opened new possibilities of using arming *Y. lipolytica* or derived nanoparticles for molecular recognition applications. For example, a *Y. lipolytica* strain harboring hemolysin from *Vibrio harveyi* could be used as a live vaccine for the fish-farming industry (Yue et al. 2008). However, it remains critical to know if, despite its GRAS status, the use of GM *Y. lipolytica* for whole cell applications will be socially acceptable, especially in the environmental domain and in the food industry.

4.3.3.1 Y. lipolytica *as a high-throughput expression platform for protein engineering*

Industrial processes generally require optimizing enzymatic properties using protein engineering and directed evolution methods. For this purpose, a high-throughput expression platform was developed for *Y. lipolytica* using the JMY1212 strain (INRA/INSA Toulouse/Toulouse University, France collaboration; Bordes et al. 2007). It was applied to the engineering of *C. antarctica* lipase B (CalB), one of the enzymes most widely used in industrial biocatalysis: large libraries of CalB mutants were created and screened, identifying new variants with higher catalytic efficiencies (Emond et al. 2010).

4.3.3.2 Glycoengineering of Y. lipolytica *for the production of humanized pharmaceutical proteins*

Yeasts modify glycoproteins with high mannose-type *N*-glycans, which can reduce the protein half-life *in vivo* and could be immunogenic. Since the genetics of *N*-glycosylation of proteins in *Y. lipolytica* were developed (Barnay-Verdier et al. 2004, 2008), some research groups from South Korea and Belgium focused on engineering the *Y. lipolytica* metabolism to provide "humanized" strains that are able to produce glycoproteins with human-compatible *N*-linked oligosaccharides for therapeutic applications. Song and coworkers (2007) have shown that a strain deleted for α-1,6-mannosyltransferase *YlOCH1* synthesized only the core oligosaccharide Man8GlcNAc2, whereas wild-type strains synthesized oligosaccharides with heterogeneous sizes (up to Man12GlcNAc2). More recently, the same South Korean research groups showed that *YlMPO1* was necessary for mannosylphosphorylation of *N*-linked oligosaccharides in secreted proteins (Park et al. 2011). They constructed a Δ*Yloch1* Δ*Ylmpo1* double mutant as a host for producing glycoproteins lacking yeast-specific hypermannosylation and mannosylphosphorylation. Another strategy was developed by the Belgian research groups, which engineered *N*-glycan biosynthesis in *Y. lipolytica* so that it produces *N*-glycoproteins homogenously carrying a Man3GlcNAc2 core. After patenting their work, the steps involved were recently fully described (De Pourcq et al. 2012a): (i) disruption of *YlALG3* α-1,3-mannosyltransferase, (ii) overexpression of *YlALG6* glucosyltransferase, (iii) overexpression of the heterodimeric *Apergillus niger* glucosidase II, and (iv) overexpression of a *Y. lipolytica*-optimized ER-targeted *T. reesei* α-1,2-mannosidase. Interestingly, the resulting core Man3GlcNAc2 is common to all mammalian *N*-glycan structures, and it can be further modified *in vitro* into any complex type of *N*-glycan, using the appropriate glycosyltranferases and sugar-nucleotide donors (De Pourcq et al. 2012a). The same research group also engineered *Y. lipolytica* to produce human-type *N*-glycoproteins homogenously carrying either Man8GlcNAc2 or Man5GlcNAc2. For this purpose, they inactivated both yeast-specific Golgi α-1,6-mannosyltransferases YlOch1p and YlMnn9p, yielding a strain producing homogeneous Man8GlcNAc2 *N*-glycans. Further introduction of a *Y. lipolytica*-optimized ER-targeted *T. reesei* α-1,2-mannosidase yielded a strain producing proteins homogeneously glycosylated with Man5GlcNAc2 (De Pourcq et al. 2012b). These new expression platforms will allow the development of *Y. lipolytica* as a competitive heterologous host for pharmaceutical applications.

4.3.3.3 Y. lipolytica *as whole-cell biocatalyst for SCO production*

Genetic engineering of *Y. lipolytica* for use as a whole-cell biocatalyst has essentially focused on its lipid metabolism pathways. This oleaginous yeast possesses a very effective lipid uptake system, enabling the efficient transport of hydrophobic substrates, from protrusions on the cell surface to catalytic sites within the cells through specific channels (Fickers et al. 2005). *Y. lipolytica* can accumulate lipids to more than 50% of its cell dry weight and, among oleaginous yeasts, it has the highest proportion of linoleic acid (more than 50% of fatty acids; Beopoulos et al. 2009). There is a strong economic interest in the production of SCO enriched in essential fatty acids, namely, unsaturated fatty acids that are not synthesized by mammals but are essential for health and must be supplied by the diet. Engineering *Y. lipolytica* ω-3/ω-6 biosynthetic pathways target the commercial niche of SCO enriched with ω-3 fatty acids, such as EPA, docosahexaenoic acid (DHA), and α-linolenic acid (ALA), or ω-6 fatty acids such as arachidonic acid (AA) and γ-linolenic acid (GLA), for use as dietary supplements in feed or food. Damude and coworkers (2006) have obtained a GM *Y. lipolytica* strain accumulating ALA up to 28% of its cell dry weight, by expressing a bifunctional Δ12/ω3 desaturase from *Fusarium moniliformis*. More recently, efficient GLA production, accounting for 20% of the total lipids and 44% of triacylglycerol fraction, was obtained by coexpression of Δ6 and Δ12 desaturases from *Mortierella alpina* under the control of hp4d promoter (Chuang et al. 2010). After patenting their process for the production of EPA-rich SCO (see Table 4.1), DuPont has recently described the numerous steps involved in the metabolic engineering of their production strain, which accumulated EPA to 15% of its dry cell weight and 57% of its lipids (Xue et al. 2013). Briefly, to produce EPA from endogenous LA whereas avoiding GLA build-up, the Δ-9 pathway was selected by overexpressing several heterologous enzymes (Δ-9 elongases, Δ-8 desaturases, Δ-5 desaturases, Δ-17 desaturases, Δ12 desaturases, and C16/18 elongases) and further improvements were brought by knocking out several genes (*YlPEX10*, involved in the import of peroxisomal matrix proteins; *YlLIP1*, encoding lipase 1; and *YlSCP2*, encoding a sterol carrier protein).

4.3.3.4 Y. lipolytica *as whole-cell biocatalyst for carotenoid production*

Microbia (see Table 4.1) and DuPont are independently developing *Y. lipolytica* as a production host for carotenoids, organic pigments used as natural coloring, and antioxidant agents for food and feed. Ye and coworkers (2012) from DuPont disclosed some information on *Y. lipolytica* engineering for the production of lycopene (an important intermediate in the biosynthesis of carotenoids), involving the expression of bacterial *crtE*, *crtB*, and *crtI* genes. Very recently, the production of unprecedented lycopene levels for a eukaryotic host was reported by a group from Dresden University. These authors used an "obese" strain (Δ*YlGUT2*, Δ*YlPOX1* to *POX6*) developed at INRA, in which they expressed *crtB* and *crtI* genes from *Pantoea ananatis* and overexpressed the rate-limiting genes for isoprenoid biosynthesis in *Y. lipolytica*, *YlGGS1*, and *YlHMG1* (Matthäus et al. 2014).

4.3.3.5 *Future prospects*

The production of biofuel, namely, of biosynthetic hydrocarbons from engineered microorganisms, in replacement of petroleum-derived fuels, constitutes a developing research area. In that context, Blazeck and coworkers (2013) recently engineered *Y. lipolytica* to produce pentane from linoleic acid by strongly overexpressing a soybean lipoxygenase (Blazeck et al. 2013). This work constitutes the first report of pentane microbial production and demonstrates that *Y. lipolytica* is a promising host for short-chain *n*-alkane synthesis. This yeast is also expected to become a source for many other valuable lipid-derived compounds such as wax esters (industrial lubricants), polyhydroxyalkanoates (PHAs), and free hydroxylated fatty acids (HFAs, used as chiral building blocks in fine chemicals synthesis). PHAs are linear polyesters produced by bacterial fermentation of sugar or lipids, which are biodegradable and can be used for producing bioplastics. Recombinant *Y. lipolytica* strains expressing the PHA synthase gene from *P. aeruginosa* in their peroxisomes have been shown to produce PHAs in various yields, depending on their *POX* (acyl-CoA oxidases) genotype (Haddouche et al. 2010). It was further shown that redirecting the fatty acid flux toward β-oxidation, by deleting neutral lipid synthesis pathways (*YlLRO1*, *DGA1*, *DGA2*, and *ARE1* acyltransferase genes), increased PHA levels (up to more than 7% of cell dry weight; Haddouche et al. 2011). Developing *Y. lipolytica* into a versatile and high-throughput microbial factory for the production of industrially valuable lipid-derived compounds (wax esters, isoprenoid-derived compounds, PHAs, and HFAs) constitutes the purpose of the European Union-sponsored LipoYeasts project (Sabirova et al. 2011). This consortium used specific enzymatic pathways from hydrocarbonoclastic bacteria, and from other hydrocarbon and lipid-degrading bacteria from various lipid-contaminated environments, to redirect *Y. lipolytica* versatile lipid metabolisms toward high-value lipid products. The main metabolic routes to be optimized will be β- and ω-oxidations, together with lipid accumulation. The results of this work can be found on the consortium web page (http://www.lipoyeasts.ugent.be/).

4.4 Hansenula polymorpha

H. polymorpha is a thermotolerant species capable of growing at record high temperatures of up to 50°C (Cabeç-Silva et al. 1984; Ishchuk et al. 2009; Reinders et al. 1999). The maximal growth temperature of this species is the absolute recorded among all known yeast species. It is only 10°C below the maximal growth temperature of all known eukaryotic organisms (Middelhoven 2002; Reinders et al. 1999). Cells of *H. polymorpha*, grown under increased temperatures, accumulate trehalose as a thermoprotector (Reinders et al. 1999). Thermotolerance could be further elevated by the disruption of the *ATH1* gene, coding for acid trehalose, or by the overexpression of *HSP16* and *HSP104* genes coding for two heat-shock proteins (Ishchuk et al. 2009).

H. polymorpha is widely used by scientists and in industrial biotechnology, reaching in this respect *P. pastoris* among the methylotrophic yeasts. Along with this latter, *H. polymorpha* is the favorite organism in yeast cell biology, especially for studying peroxisome biogenesis and degradation. *H. polymorpha* is the only species among the methylotrophic yeasts used to produce bioethanol from lignocellulosics and glycerol (Dmytruk et al. 2013; Hong et al. 2010; Ishchuk et al. 2009; Ryabova et al. 2003). Additionally, *H. polymorpha* is known to be the most efficient yeast producer of glutathione (Ubiyvovk et al. 2011a).

4.4.1 *Available strains*

The most popular strains of *H. polymorpha* are DL-1, CBS4732, and NCYC495. There have been lengthy discussions on the reclassification of *H. polymorpha.* Alternative names have been proposed, such as *Pichia angusta* (Barnett et al. 2000; Kurtzman et al. 1998), *Torulopsis methanothermo, Hansenula angusta,* or *Ogataea polymorpha* (Middelhoven 2002; Sudbery 2003; Suh et al. 2010; Teunison et al. 1960; Yamada et al. 1994). It has to be pointed out that strains CBS4732 and NCYC495 are closely related and intercrossed whereas strain DL-1 is unable to cross with either of the abovementioned strains. According to the latest classification, strains CBS4732 and NCYC495 belong to the species *O. polymorpha,* whereas strain DL-1 belongs to the separate species *Ogataea parapolymorpha.* Additional species are also known (*Ogataea angusta;* Suh et al. 2010; Kurtzman 2011; Naumova et al. 2013). In this review, the commonly accepted name *H. polymorpha* is used.

4.4.2 *Genetic methods and genome research*

Using an inbreeding program, a genetic cell line of *H. polymorpha* derived from strain CBS4732 was isolated. This presents an efficient hybridization and massive sporulation of four-spored asci, thus, allowing, tetrad analysis (Lahtchev et al. 2002). Auxotrophic and other types of mutants have been isolated from this genetic cell line. Most of the markers analyzed showed Mendelian segregation in tetrad analysis. Strain DL-1, although unable to hybridize, seems to be suitable for molecular genetics experiments. It is characterized by high growth rate and exhibits a high frequency of homologous recombination (Kang et al. 2001). Transformation of *H. polymorpha* was originally developed for *ura3* mutants with *URA3* gene as selectable marker (Merckelbach et al. 1993). Besides this, *leu2* mutant and heterologous *LEU2* gene from *S. cerevisiae* have been proposed (Tikhomirova et al. 1986, 1988). Several dominant markers, such as geneticin (G418) and zeocin resistance for transformant selection have been adapted for *H. polymorpha* (Dmytruk et al. 2008a,b). Transformation is carried out using the LiCl-based method, protoplasts in the presence of polyethylene glycol (Tikhomirova et al. 1988) or electroporation (Faber et al. 1994). Methods for gene disruption and knock out in *H. polymorpha,* including those with the use of modified Cre-loxP system, have been described (Kang et al. 2002; Qian et al. 2009). Methods of multicopy gene integrations in the genome of *H. polymorpha* have also been developed (Agaphonov et al. 1999; Krasovska et al. 2013; Sohn et al. 1996, 1999).

Pulse field electrophoresis of *H. polymorpha* displays six chromosomes in both strains (CDS4732 and DL-1), with sizes ranging from 0.9 to 1.9 Mb, although the electrophoretic patterns of chromosomes in both strains is quite different (Ramezani-Rad et al. 2003; Kunze et al. 2009). Pulse electrophoresis of chromosomes from nine different strains of *H. polymorpha* showed a strong difference in chromosome numbers, which ranged from two to six depending on the strains (Marri et al. 1993).

Until now, the genomes of the three abovementioned strains have been sequenced. However, data from strains CDS4732 and DL-1 are not public and belong to a German private company, Rhein Biotech GmbH (Ramezani-Rad et al. 2003) and the Korean Institute KRIBB (Kang et al. 2005, personal communication). Data on the genome sequence of strain NCYC495 is publicly available (http://genome.jgi-psf.org/Hanpol/Hanpol.home.html). The genome of *H. polymorpha* NCYC 495 is 8.97 Mb in size and contains 5162 open reading frames.

In strain CBS4732, 8.73 Mb of DNA is sequenced, which contains 5848 open reading frames coding for proteins larger than 80 amino acids. In addition, 389 open reading frames were identified as coding for proteins containing less than 100 amino acids and 4771 open reading frames (81.6%) showed homology to known proteins. Average gene density is one gene per 1.5 Kb, whereas the protein contains 440 amino acids on average. In total, 91 introns were identified in the *H. polymorpha* genome. Functional annotation showed the following gene distribution depending on putative functions: 4% encode genes involved in energy metabolism, 3% in signaling, 6% in protein synthesis, 4% in stress response, 9% in transport, 9% in cell cycle and DNA metabolism, 17% in protein processing, 13% in transcription, and 19% in metabolism.

4.4.3 *Genetic control of methanol induction and glucose catabolite repression in* H. polymorpha

In *H. polymorpha,* and similarly to other methylotrophic yeasts, methanol strongly induces synthesis of the enzymes involved in primary methanol metabolism localized both in cytosol (formaldehyde and formate dehydrogenases, dihydroxyacetone kinase, and fructose-1,6-bisphosphatase) and in peroxisomes (alcohol oxidase, catalase, and dihydroxyacetone synthase). Methanol also induces peroxisome growth and proliferation (Eggeling et al. 1980; van der Klei et al. 2006; Veenhuis et al. 1983). Shifting methanol-grown cells to glucose or ethanol-based media represses the synthesis of these enzymes and additionally causes autophagic peroxisome degradation or pexophagy (Dunn et al. 2005; van der Klei et al. 2006). Transcriptome analysis of *H. polymorpha* showed that 2 h after shifting glucose-grown cells to methanol, 1184 genes out of approximately 6000 showed a twofold increased expression level, whereas 1246 other genes presented a twofold reduction of their expression level (van Zutphen et al. 2010). Maximal induction was observed for the genes of the central regulator of methanol metabolism *MPP1* (394-fold) and *FMD*, coding for formate dehydrogenase (347-fold). A strong induction was also observed for genes coding for enzymes of fatty acid β-oxidation, glyoxylic shunt and carnitine mitochondrial transporter (111-fold increase). The *MPP1* gene identified belongs to the family of $Zn(II)_2Cys_6$ zinc cluster transcriptional activators (Leão-Helder et al. 2003). The mutant Δ*mpp1* failed to grow on methanol as the sole carbon and energy source, and was characterized by a strong decrease in the level of peroxins and enzymes of methanol catabolism. However, the mechanism of transcription activation by Mpp1p remains unknown. *H. polymorpha MPP1* shows a close similarity with transcription factors from other species of methylotrophic yeasts, namely, *P. pastoris MXR1* (Lin-Cereghino et al. 2006) and *Candida boidinii TRM1* (Sasano et al. 2008), which are also specifically required for the expression of genes involved in methanol metabolism and peroxisome biogenesis.

Carbon catabolite repression is the process of differential inhibition of enzymes of less favored carbon substrates (which support lower growth rate) in the medium containing more favored (supporting higher growth rate) substrates (Schuller 2003; Stasyk et al. 2003). In *S. cerevisiae*, catabolite repression is controlled by transcription repressors *MIG1* and *MIG2*, as well as transcription activators and different protein kinases (Gancedo 2008; Turcotte et al. 2010). An important role in catabolite repression in *S. cerevisiae* belongs to the *HXK2* gene encoding hexokinase II (Kraakman et al. 1999). In *H. polymorpha*, catabolite repression is induced by glucose and ethanol. The latter compound is involved only in the repression of enzymes of methanol metabolism whereas glucose additionally represses enzymes involved in the catabolism of alternative substrates, such as maltose (Alamäe et al. 2003; Liiv et al. 2001). There are several publications on *H. polymorpha* mutants that are capable of synthesizing enzymes of methanol metabolism and peroxins in medium containing glucose as the sole carbon source (Alamäe et al. 1998; Hodgkins et al. 1993; Parpinello et al. 1998; Roggenkamp 1988; Stasyk et al. 1997a). Only some of the corresponding genes and mutations were identified. In one such mutant, the corresponding mutation was recessive and monogenic, and was designated as *glr1* (Parpinello et al. 1998). These mutations led to the synthesis of enzymes of methanol metabolism, corresponding mRNAs and peroxisome proliferation in medium containing glucose but not methanol. Repression by ethanol and pexophagy were not affected. A double mutant of *H. polymorpha,* which was defective in hexokinase and glucokinase, was unable to catabolize glucose and showed defects in catabolite repression of enzymes from methanol metabolism and maltase. Thus, sugar phosphorylation is necessary for repression (Karp et al. 2003; Kramarenko et al. 2000; Laht et al. 2002).

Mutants defective in glucose catabolite repression were also isolated among strains resistant to 2-deoxyglucose in a methanol medium (Stasyk et al. 1997b). Genetic analysis led to the identification of the monogenic recessive mutation *gcr1.* Subsequently, the wild-type *GCR1* allele and mutant *gcr1-2* allele were cloned and sequenced. Additionally, knock-out mutant Δ*gcr1* was isolated (Stasyk et al. 2004). These mutants were characterized by pleiotropic defects of metabolism, namely, by constitutive synthesis of peroxisomal enzymes of methanol metabolism (alcohol oxidase, catalase) and constitutive presence of peroxisome in glucose medium (but not in ethanol medium), a decrease in the intracellular level of glycolysis intermediates, and defects in catabolite repression of cytosolic enzymes formaldehyde, formate dehydrogenases, and α-glucosidase. *GCR1* gene displayed strong homology to yeast and fungal genes coding for hexose transporters and sensors. Gcr1 protein is localized on the cytoplasmic membrane. It was suggested that Gcr1 protein fulfills transport and sensing functions in glucose catabolite repression but not pexophagy (Stasyk et al. 2004). Subsequently, two new genes (*HXS1* and *HXT1*), coding for glucose sensors and glucose transporters, respectively, have been identified in *H. polymorpha* (Stasyk et al. 2008). Thus, *H. polymorpha* possesses several hexose sensors and transporters involved in catabolite repression but not in pexophagy. To identify the other components involved in catabolite repression in *H. polymorpha*, orthologues of transcriptional repressor *S. cerevisiae* genes *MIG1, MIG2,* and *TUP1,* as well as their knockout mutants, were isolated. Surprisingly, it was found that in single Δ*mig1,* double Δ*mig1* Δ*mig2* as well as in Δ*tup1* mutants, glucose catabolite repression was only slightly affected (Stasyk et al. 2007). At the same time, the abovementioned mutations strongly impaired pexophagy induced by both glucose and ethanol. These data suggest substantial differences in mechanisms of catabolite repression between *S. cerevisiae* and *H. polymorpha.* It was also suggested that in *H. polymorpha* catabolite repression is mediated by orthologues of *S. cerevisiae* transcriptional regulators *MIG1, MIG2,* and *TUP1.*

4.4.4 H. polymorpha *as a system for the expression of own and heterologous proteins*

As mentioned previously, *H. polymorpha* is, along with *P. pastoris,* a favorite system for the expression of heterologous proteins. For expression, mitotically stable integrative vectors are used (Gellissen et al. 1997). Some strains allow multicopy integration with maximally up to 100 copies of integrated plasmids per genome represented as tandem repeats (Gatzke et al. 1995; Gellissen 2000; Kang et al. 2002). In other strains, multicopy integration does not occur spontaneously. Two methods of multicopy integrant selection in such strains have been developed based on the use of heterologous yeast auxotrophic genes, namely, the

S. cerevisiae URA3 gene and the *P. pastoris ADE1* gene with a shortened native promoter. Sequential use of both selection markers produced stable transformants containing up to 30 integration cassettes with HBsAg gene encoding hepatitis B surface antigen (Krasovska et al. 2013). In this case, vectors contained *FMD, MOX, TPS1* promoters of formate dehydrogenase, alcohol oxidase, and trehalose-6-phosphate synthetase genes. Also, recombination in the corresponding chromosomal genes occurred (Kunze et al. 2009). It is interesting to note that the production of proteins, especially of secreted ones, often shows no correlation with the copy number of the integrated gene (Gellissen et al. 1992). For instance, maximal expression of the urinary plasminogen activator u-PA and of human serum albumin HSA in DL-1 strain was achieved by integration of one to two copies of expression vector (Kang et al. 2001), whereas for oversynthesis of other proteins integration of high copy number of the expression cassette is necessary (Krasovska et al. 2013). Directed integration in *H. polymorpha* needs much longer homologous sequences as compared with *S. cerevisiae* (González et al. 1999; van Dijk et al. 2001). Vectors containing a set of several subtelomer ARS sequences seemed to be capable of homologous genome integration resulting in the integration of single or multiple tandem repeats into corresponding telomere genome regions (Sohn et al. 1999; Kim et al. 2003). A set of vectors for the integration of heterologous sequences into the ribosomal DNA locus of *H. polymorpha* was constructed (Steinborn et al. 2005, 2006; Kunze et al. 2009).

Several strong constitutive and regulatory promoters have been proposed for heterologous expression (Gellissen et al. 1997; Kang et al. 2002; Song et al. 2003). *MOX* and *FMD* promoters are strong, inducible by methanol, and repressible by glucose. *TPS1* promoter is constitutive, however, its expression in regulated by temperature (Amuel et al. 2000). In combination with the high copy number of integrated cassettes, the abovementioned promoters could drive the production of large amounts of heterologous proteins. For example, *FMD* promoter-regulated expression of secreted phytase provided a yield of 13.5 g/L (Mayer et al. 1999). Rarely, *MAL1* promoter induced by maltose (Liiv et al. 2001; Alamäe et al. 2003); nitrate-inducible promoters *YNT1, YNI1,* and *YNR1* (Avila et al. 1998); *PHO1* promoter from acid phosphatase gene repressed by phosphate (Phongdara et al. 1998); or *FLD* promoter induced by methanol (Baerends et al. 2002) are used. The list of constitutive promoters used includes *ACT* (Kang et al. 2001), *GAP* (Heo et al. 2003; Voronovsky et al. 2005; Dmytruk et al. 2008a,b), *PMA1* (Cox et al. 2000), and heterologous strong constitutive promoter *TEF1* from *A. adeninivorans* (Steinborn et al. 2006).

For the secretion of heterologous protein and its delivery to certain cell compartments, different signal sequences were successfully exploited. To provide efficient secretion, several leader sequences have been used, such as that of *PHO1* gene (Phongdara et al. 1998), *Schwanniomyces occidentalis GAM1* (Brake et al. 1984; van Dijk et al. 2000). However, the most popular one is the MFα1 pre-pro-leader sequence of *S. cerevisiae* (Brake et al. 1984; Gellissen 2000; Krasovska et al. 2007). To deliver proteins to peroxisomes, *PTS1* and *PTS2* targeting signals were used (van Dijk et al. 2000). Glycosyl phosphatidyl inositol motifs were used for the delivery of heterologous proteins, such as glucose oxidase, on the cell surface (Kim et al. 2002). Expression vectors of *H. polymorpha* contain prokaryotic and eukaryotic DNA fragments (Gellissen et al. 1997). Native circular or linearized vectors are used for transformation. The following loci were successfully used for homologous integration: *MOX/TRP1* (Agaphonov et al. 1995), *URA3* (González et al. 1999), *LEU2* (Agaphonov et al. 1999), *GAP* promoter region (Heo et al. 2003), ARS sequence (Sohn et al. 1996; Agaphonov et al. 1999), and rDNA cluster (Klabunde et al. 2002, 2003). The known vector system CoMed, with broad host spectrum, contains individual modules consisting of expression cassettes with efficient promoters, selective markers, and rDNA sequences for integration (Steinborn et al. 2006).

4.4.5 *Examples of processes and products based on the recombinant strains of* H. polymorpha

4.4.5.1 *Production of heterologous protein*

H. polymorpha is used for the industrial production of hepatitis B surface antigen (HBsAg), human interferon IFNa-2a, insulin, hexose oxidase, and phytase (Böer et al. 2007). Design and optimization of the fermentation procedure depend significantly on the characteristics of the host cell, targeted localization of the aimed product and, most importantly, on the promoter used. Most often, medium with simple synthetic components including trace elements and suitable nitrogen source, which provide efficient expression of the gene of interest and high biomass yield, are used. Total fermentation time varies from 60 to 150 h. Based on the characteristics of the most often used promoters (*FMD* and *MOX*), glycerol, methanol, glucose, and their combinations are normally used as carbon sources. For the expression of secretory heterologous proteins in *H. polymorpha*, such as hirudin, glycerol was used as a carbon source (Avgerinos et al. 2001; Weydemann et al. 1995). The recombinant strain contained 40 copies of an expression cassette containing *MFα1* pre-pro secretory sequence and hirudin structural gene expressed under control of *MOX* promoter. Hirudin production was activated by a decrease in the initial glycerol concentration and by maintaining it at an optimal level by pO_2-controlled glycerol feeding. Fermentation started by the addition of 3% glycerol and after its consumption, pO_2-controlled glycerol feeding started to maintain its concentration in the range of 0.05% to 0.3%, thus leading to derepression of the *MOX* promoter.

For the production of HBsAg, a nutrition regime based on the changes between two carbon sources was used (Brocke et al. 2005). The producing strain possessed a high copy number of expression cassettes containing the coding sequence for small surface antigen (S-antigen) under control of methanol-regulating promoters. Fermentation was conducted in 50 L fermenters. Cells with high antigen concentrations were obtained using a two-fermenter cascade, with a first seeding fermenter of 5 L and a production fermenter of 50 L. The first stage was similar to that used for hirudin production. Cultivation started in glycerol medium using a fed-batch regime. Then, methanol was added leading to an HBsAg accumulation to up to several grams per liter. It was suggested that methanol is necessary during the production phase, in particular, for lipoprotein synthesis and membrane proliferation. To isolate HBSAg, cells were disrupted and inclusion bodies were purified using a multistep procedure involving adsorption, ion-exchange chromatography, ultrafiltration, and ultracentrifugation (Brocke et al. 2005; Schaefer et al. 2007). The assumption that methanol is specifically necessary for HBsAg synthesis (due to the activation of lipoprotein synthesis and membrane proliferation) contradicts recent observations that *AOX* promoter-controlled HBsAg synthesis in *gcr1* mutant is activated by glucose in the absence of methanol (Krasovska et al. 2013).

H. polymorpha seems to be an especially productive organism for phytase synthesis (Mayer et al. 1999; Papendieck et al. 2005). A production yield of 13.5 g/L of phytase was obtained with a strain expressing the phytase gene under control of the *FMD* promoter cultivated in a medium with glucose starvation. The secreted product was purified by centrifugation, filtration, and ultrafiltration, which provided high yields (up to 92%; Mayer et al. 1999).

Known systems of heterologous expression in *H. polymorpha* based on methanol-inducible promoters are not free of serious drawbacks. Thus, the use of toxic and highly inflammable methanol for promoter induction is a real challenge. Methanol concentrations have to be maintained at a strictly defined level because high concentrations of this alcohol are toxic for the cell (Gellissen et al. 2005). In addition, methanol utilization requires intense aeration rates due to the high oxygen requirement of alcohol oxidase (Jenzelewski 2002). It is known that the *MOX* promoter of *H. polymorpha* is not strictly dependent on methanol because of its sufficient derepression when glycerol or glucose are limited in the medium (Hellwig et al. 2005). However, for maximal promoter activity, methanol is necessary. The mutants *H. polymorpha gcr1* and their derivatives EAO, producing alcohol oxidase constitutively in glucose medium without methanol, were isolated. The alcohol oxidase activity in EAO mutants was much higher than that in methanol medium (Gonchar et al. 1998; Moroz et al. 2000; Stasyk et al. 1997a, 2004). The nature of mutations leading to the EAO phenotype remains unidentified. Expression of heterologous genes of *A. niger* glucose oxidase *GOD*, S-antigen HBsAg and secreted mini-proinsulin under control of the *MOX* promoter in the EAO strain of *H. polymorpha*, resulted in transformants producing recombinant proteins in the medium without methanol. Moreover, the highest promoter induction was achieved in the medium containing xylose (Krasovska et al. 2007). Expression of HBsAg in a strain with multicopy-integrating cassettes with this gene further increased the production level. Deletion of the *PEX3* gene coding for peroxine involved in the early step of peroxisome formation substantially increased the production of HBsAg in glucose medium as compared with the parental EAO strain. Maximal production of HBsAg in the Δ*pex3* strain was nearly in the range of 8% to 9% of total cell protein (Krasovska et al. 2013). Optimization and scaling-up for heterologous protein synthesis in the mentioned system has not yet been carried out.

In the case of the production of toxic proteins or small peptides, which undergo rapid cytosolic degradation, the products could be targeted to peroxisomes using peroxisome-targeting signals PTS1 or PTS2. In such a way, transformants of *H. polymorpha* accumulating in peroxisomes' small functional peptides as insulin-like human growth factor II (IGF-II) and *Xenopus laevis* magainin II have been obtained (Faber et al. 1996).

Some efforts have been made to improve heterologous protein secretion. It was shown that an increase of *HpCNE1* gene expression, coding for membrane chaperone protein synthesis, led to an improvement in the secretion of several tested glycoproteins and of one nonglycosylated protein by twofold to fourfold (Klabunde et al. 2007; Qian et al. 2009). Improvements in the secretion of the urokinase type of plasminogen activator were achieved with mutations in *PMT1* and *PMR1* genes coding for protein *O*-mannosyl transferase and Golgi ion pump, respectively (Agaphonov et al. 2005, 2007). Secretion of this protein was also activated by mutations in the *RET1* gene coding for α-COP, the subunit COPI of the protein complex involved in protein translocation in the Golgi apparatus (Chechenova et al. 2004), as well as genetic defects in the vacuole protein sorting (vps) pathway due to mutations in the genes *PEP3, VPS8, VPS10, VPS17,* and *VPS35* (Agaphonov et al. 2005).

Sometimes, genetic manipulations led to the secretion of normally intracellular proteins such as alcohol oxidase, which is normally a peroxisomal enzyme. However, expression of the alcohol oxidase gene *AOX* under control of its own promoter and fused with *MFα1* pre-pro-secretory resulted in the accumulation of monomeric protein in the endoplasmic reticulum. However, substitution of the *MFα1* signaling sequence for *S. cerevisiae ISS* invertase signaling sequence provided synthesis of the intact enzymatically active protein secreted into the cultivation medium (van Der Heide et al. 2007). High levels of protein secretion were obtained by increasing the copy number of expression cassettes. Thus, the heterologous *ISS* signal appeared to be stronger than the native PTS1 signal for peroxisome targeting.

Normally, in *H. polymorpha* and similarly to *S. cerevisiae* and other yeast species, hyperglycosylation of glycoproteins occurs as compared with the glycosylation level in animal and human cells. *H. polymorpha* strains lacking hyperglycosylation similar to those for *P. pastoris* (Hamilton et al. 2007; Li et al. 2006) have been isolated. For this, the *HpALG3* gene, coding for dolychol phosphomannose-dependent α-1,3-mannosyl transferase was deleted in the background of the *Hpoch1* mutant with a defect in yeast-specific synthesis of the outer mannose chain, which additionally expresses the heterologous gene of α-1,2-mannosidase from *Aspergillus saitoi* (Oh et al. 2008). Such recombinant strains produced glycoproteins containing *N*-glycan of trimannosyl type ($Man_3GlcNAc_2$), which is the typical core for different human glycans.

4.4.5.2 *Production of low-molecular weight biologically active compounds in* H. polymorpha

In addition to heterologous proteins, *H. polymorpha* seemed to be a promising producer for some low-molecular weight compounds, both native and heterologous. These compounds include antibiotic penicillin, tripeptide glutathione, and ethanol. The corresponding producers were constructed by a combination of methods of classic selection and metabolic engineering.

4.4.5.2.1 Penicillin. The market for β-lactam antibiotics, penicillins, and cephalosporin consists of approximately 40% of the total antibiotic market, with an overall market of nearly 55 billion US dollars (Demain 2007; Kresse et al. 2007). Several years ago, the market for penicillin was at 8.2 billion US dollars (Demain 2007). Industrial production of penicillin is based on specially constructed strains of mycelial fungus *Penicillium chrysogenum* (Kresse et al. 2007). Use of a multicellular mycelial organism generates some difficulties for large-scale cultivation; therefore, introduction and efficient expression of penicillin biosynthesis genes into cells of an evolutionarily related organism, the unicellular yeast *H. polymorpha,* has undoubted scientific and applied interest. Penicillin biosynthesis involves four enzymatic steps, the first two being located in the cytosol whereas the last two occur in peroxisomes (Evers et al. 2004). First, the *P. chrysogenum* enzyme successfully expressed in *H. polymorpha* was the peroxisomal isopenicillin *N*-acyl transferase catalyzing the penultimate reaction of penicillin biosynthesis (Lutz et al. 2005). The enzyme was localized in yeast peroxisomes. Subsequently, the last enzyme of the pathway, phenylacetyl-CoA ligase (which is also located in yeast peroxisomes) was expressed (Gidijala et al. 2007). Finally, strains of *H. polymorpha* producing heterologous first and second enzymes of penicillin biosynthesis were constructed (Gidijala et al. 2008, 2009). A strain of *H. polymorpha* that expressed all four enzymes of penicillin biosynthesis was constructed. The amount of penicillin secreted by this strain was similar to that of the wild-type *P. chrysogenum* NRRL1951 strain (nearly 1 mg/L; Gidijala et al. 2009). This work opens up opportunities to increase penicillin production levels in *H. polymorpha* up to industrial levels as well as to construct on the base of *H. polymorpha* from the producers of other β-lactam antibiotics and pharmacologically important peptides. Thus, one may assume that, in the future, *H. polymorpha* will become an industrial producer of antibiotics and immunosuppressive and cytostatic peptides.

4.4.5.2.2 Glutathione. Glutathione (γ-glutamylo-L-cysteinyloglycine) is a natural tripeptide and simultaneously is the main nonprotein thiol in all eukaryotic and many prokaryotic organisms (Penninckx 2000). Glutathione is the main redox buffer because the ease of interconversion between oxidized and reduced forms of glutathione maintains cell redox status (Ostergaard et al. 2004). In addition, glutathione plays a major role in cell response to stress induced by substrate starvation, heavy metals, xenobiotics, and free radicals as well as participating in sulfur storage, regulation of gene expression, and cell signaling (Bachhawat et al. 2009; Pócsi et al. 2004; Sies 1999). More than 90% of glutathione is located in cells in reduced form (Pócsi et al. 2004). Both forms of glutathione, reduced thiol and oxidized disulfide, are widely used in medicine, the cosmetic industry, and as a food additive (Li et al. 2004). Being an active ingredient of food, medicines, and cosmetics, glutathione relieves the effects of harmful oxidizers, removes toxic products, and enhances the action of cosmetics regenerating skin cells. The oxidized form of glutathione is used as a cryoprotector and immunomodulator (Chatterjee et al. 2001; Kozhemyakin et al. 2008). It was found that an increase in glutathione cell content biosynthesis in *H. polymorpha* due to overexpression of *GSH2* and *MET4* genes coding for the first enzyme of glutathione biosynthesis and the central regulator of sulfur metabolism, respectively, strongly activates glucose (but not xylose) alcoholic fermentation (Grabek-Lejko et al. 2011).

Methylotrophic yeasts are considered to be a rich source of glutathione because the metabolism of the products of alcohol oxidase reaction, formaldehyde, and hydrogen peroxide depends on glutathione. Thus, oxidation of glutathione in cytosol is catalyzed by NAD-dependent and glutathione-dependent formaldehyde dehydrogenase (Sahm 1977), whereas detoxification of hydrogen peroxide depends on the activity of Pmp20p, glutathione peroxidase, localized in the vacuolar membrane (Horiguchi et al. 2001; Yano et al. 2009). It is known that methanol induces an increase in intracellular glutathione concentration (Ubiivovk et al. 1986), which is especially noticeable in *H. polymorpha* mutants defective in formaldehyde dehydrogenase (Sibirny et al. 1990). However, the regulatory mechanisms involved in glutathione synthesis and degradation in methylotrophic yeasts remained unknown for a long time. Glutathione auxotrophic mutants of *H. polymorpha*, divided into two complementation groups designated as *gsh1* and *gsh2*, have been isolated (Ubiivovk et al. 1999). Using a *gsh2-1* mutant, the existence of glutathione-dependent and independent systems of xenobiotic detoxification were demonstrated (Ubiyvovk et al. 2003). The gene *GSH2* was cloned by functional complementation of *gsh2-1* mutant. It seems that this latter encodes γ-glutamylo-L-cysteine synthetase, in other words, this is a homologue of the *GSH1* gene of *S. cerevisiae* (Ubiyvovk et al. 2002). The *H. polymorpha GSH1* gene was also cloned by functional complementation and was found to be the homologue of the gene *S. cerevisiae MET1*, coding for *S*-adenosyl-L-methionine uroporphyrinogen III methyltransferase, which is involved in siroheme biosynthesis (Ubiyvovk et al. 2011b). The gene *H. polymorpha GGT1*, which seemed to be the homologue of the *S. cerevisiae CIS2* gene, coding for γ-glutamyl transpeptidase was cloned and a Δ*ggt1* knockout strain defective in glutathione degradation was isolated (Ubiyvovk et al. 2006). It was found that Δ*ggt1* and Δ*met1* (Δ*gsh1*) mutants are unable, in contrast with the *H. polymorpha* wild-type strain, to accumulate Cd^{2+} ions inside cells, whereas the cells of the Δ*gsh2* mutant accumulated increased amounts of this ion (Blazhenko et al. 2006). It was found that *H. polymorpha* synthesizes only one type of cadmium-binding chelator, namely, glutathione, and does not synthesize compounds of the phytochelatine type. It is suggested that the gene *GSH1/MET1* participates in maturation, and gene *GSH2* is involved in metabolism of the intracellular complex of Cd^{2+} ions with glutathione.

To construct *H. polymorpha* mutants overproducing glutathione, the strains overexpressing either *GSH2*, coding for the first enzyme of glutathione synthesis, or *MET4*, which is the central positive regulator of glutathione biosynthesis, were constructed (Kang et al.

2002; Thomas et al. 1992; Ubiyvovk et al. 2011a). The production of intracellular and extracellular glutathione was studied in the constructed strains depending on cultivation parameters (dissolved oxygen pressure, pH, and mixing rate) and on the type of carbon source (methanol or glucose). Under optimal conditions during cultivation in fed-batch mode with glucose medium, recombinant strains accumulated 900 to 2300 mg of total intracellular glutathione per liter of culture broth. Strain cultivation in the same fed-batch mode in methanol medium resulted in the accumulation of 250 mg of extracellular glutathione per liter of culture broth (Ubiyvovk et al. 2011a). The obtained titers exceed maximal data for glutathione synthesis in *S. cerevisiae*, even after the addition of exogenous amino acids to the medium (Wang et al. 2007). Thus, *H. polymorpha* seems to be a promising organism for the industrial production of this important tripeptide.

4.4.5.2.3 H. polymorpha *as a promising producer of fuel ethanol from lignocellulose.* *H. polymorpha* is a promising organism for high-temperature alcoholic fermentation. All eight strains of *H. polymorpha* that were tested were capable of fermenting glucose, xylose, mannose, maltose, and cellobiose, whereas galactose and L-arabinose practically did not support the growth of this yeast (Dmytruk et al. 2013; Ryabova et al. 2003). The optimal temperature for fermentation was 37°C to 40°C, although fermentation was quite intensive even at 45°C to 48°C, which is the absolute temperature recorded for eukaryotic alcoholic fermentation. At 50°C, fermentation is strongly suppressed. However, an increase of intracellular trehalose content due to knockout of the acid trehalase gene *ATH1* or overexpression of the heat shock proteins Hsp16 and Hsp104 allowed normal xylose fermentation at 50°C (Ishchuk et al. 2009). Fermentation was most active under limited aeration and starvation for flavins, which are necessary for cell respiration. *H. polymorpha* was more tolerant of ethanol compared with the most effective natural xylose-fermenting yeast, *Pichia stipitis*; however, it was more susceptible relative to *S. cerevisiae* (Ryabova et al. 2003). Ethanol tolerance could be further increased due to overexpression of the heterologous gene *MPR1* (Ishchuk et al. 2010) or *ETT1*'s own gene (Ishchuk et al. in preparation; Abbas et al. 2013). Using an auxonographic selection method, *H. polymorpha* mutants accumulating increased amounts of ethanol in xylose medium were isolated (Grabek-Lejko et al. 2006). Recombinant strains of this organism, which express amylolytic and xylanolytic enzymes and directly ferment starch and xylan to ethanol, have been isolated (Voronovsky et al. 2009).

H. polymorpha could be a promising organism for simultaneous saccharification and fermentation (SSF) processes. In the SSF process, pretreated lignocellulose is hydrolyzed by cellulases and hemicellulases and, to avoid product inhibition of hydrolysis by the liberated monosaccharides, the vessels also contain microorganisms that efficiently ferment produced sugars to ethanol (Olofsson et al. 2008). Because cellulases and hemicellulases express maximal activity under temperatures in the range of 50°C to 60°C, the microorganisms used in the SSF process, in addition to effective alcoholic fermentation of major sugars of lignocellulosic hydrolysates, have to survive and actively ferment at high temperatures (Merino et al. 2007). *H. polymorpha* is perfectly suited to this requirement as it belongs to the very few thermotolerant yeast species capable of xylose fermentation and could be the organism of choice. However, it is not free from several drawbacks. Most importantly, ethanol yield and productivity from xylose in wild-type strains of *H. polymorpha* are very low. However, these features could be improved through classic selection and metabolic engineering. Three approaches have been used previously for the construction of *H. polymorpha* strains with improved characteristics of ethanol production from xylose. In one line of investigation, *XYL1*'s own gene coding for xylose reductase (XR), and two paralogs of xylitol dehydrogenase (XDH), *XYL2A* and *XYL2B*, were deleted in the collection strain CBS4732, and bacterial *xylA* gene from *E. coli* or *Streptomyces coelicolor* were expressed (Voronovsky et al. 2005). The corresponding transformants possessed the activity of xylose isomerase and grew on xylose, but the amount of accumulated ethanol was very low (both transformants and wild-type cells accumulated a maximum of 0.15 g/L of ethanol). Overexpression of *E. coli xylA* together with own *XYL3* coding for xylulokinase (XK) increased ethanol production. However, maximal ethanol accumulation did not exceed 0.6 g/L at 48°C (Dmytruk et al. 2008a,b). In another work, XR was engineered by site-specific mutagenesis to reduce affinity toward NADPH, according to the approach developed for *Candida tenuis* (Petschacher et al. 2005). Consequently, genes coding for modified XR (*XYL1m*), native XDH (*XYL2*), and XK (*XYL3*) were overexpressed in strain CBS4732 and resulted in twofold higher ethanol accumulation in corresponding transformants reaching 1.3 g/L of ethanol (Dmytruk et al. 2008a,b). In another line of investigations, the wild-type strain NCYC495 (the currently sequenced strain) was used initially because we showed that it seemed to be a more efficient xylose fermenter relative to strain CBS4732. The mutant 2EthOH$^-$, unable to use ethanol as a sole carbon source, was isolated from strain NCYC495 and characterized by threefold elevated ethanol accumulation. Subsequently, *PDC1* gene coding for pyruvate decarboxylase (PDC) was cloned and overexpressed in strain 2EthOH$^-$. The best-selected transformants accumulated 2.5 g/L of ethanol at 48°C (Ishchuk et al. 2008). However, the ethanol production achieved is still very low and has to be substantially increased before the strain meets the requirements for an SSF process.

Recently, more efficient *H. polymorpha* high-temperature ethanol producers from xylose have been isolated (Kurylenko et al. submitted for publication). For this, a combination of the methods of metabolic engineering and classic selection were applied. The mutant 2EtOH$^-$ (Ishchuk et al. 2008) was used for overexpression of the genes *XYL1m, XYL2, XYL3,* and *PDC1.* The best-selected transformant was used for isolation of the mutants that were resistant to the anticancer drug 3-bromopyruvate, which is known to inhibit glycolysis (Cardaci et al. 2012; Ganapathy-Kanniappan et al. 2010; Shoshan 2012). The best obtained strain showed a 15-fold enhancement in ethanol synthesis from xylose as compared with the wild-type strain, accumulating up to 10.0 g/L of ethanol at 45°C, which, after correction for ethanol evaporation, corresponds to the calculated concentration of 20 g/L.

Thus, accumulation of ethanol from xylose increased by more than 15 times as compared with the *H. polymorpha* wild-type strain. The maximal observed level of ethanol produced from xylose by the best-isolated strains (near 10 g/L at 45°C, which approximately corresponds to the 20 g/L value calculated after correction for evaporation, resulting in ethanol yields of 0.18 and 0.35 g/g xylose, respectively) makes *H. polymorpha* quite close to known promising organisms for use in the SSF process. Still, these results are lower than ethanol production in mesophilic xylose-fermenting organisms such as *P. stipitis* (0.35–0.44 g/g xylose) and *Spathaspora passalidarum* (0.42 g/g xylose; Jeffries et al. 2007; Long et al. 2012) but higher than those in the best engineered strain of thermotolerant yeast *Kluyveromyces marxianus* (0.31 g/g xylose under anaerobic conditions at 45°C; Wang et al. 2013). To be industrially feasible, ethanol yield in *H. polymorpha* has to be further increased to be close to the theoretical maximum. It is desirable to manipulate with new targets for metabolic engineering (e.g., xylose transport, pentose phosphate pathway), which could be successful for the further increase of ethanol yield from xylose during high-temperature alcoholic fermentation of this promising organism.

4.5 Conclusion

In this chapter, we provide information and data that clearly highlights that the nonconventional yeasts *P. pastoris*, *Y. lipolytica*, and *H. polymorpha* present many advantages over other model organisms such as *E. coli* or even mammalian cells for the production of heterologous proteins or other valuable compounds. Indeed, these yeasts are able to grow at high cell densities in low-cost culture media, to express heterologous genes at high levels, to synthesize and secrete the corresponding proteins at high yield, and to perform human-like posttranslational modifications.

Acknowledgment

This work was supported in part by a Wallonie-Bruxelles-Pologone bilateral project (WBI grant SOR-2013-115126) grant to Andriy Sibirny and Patrick Fickers.

References

Abbas, C. A., Sibirny, A. A., Voronovsky, A. Y., and Ishchuk, O. P. "Alcoholic xylose fermentation at high temperatures by the thermotolerant yeast *Hansenula polymorpha*." US Patent 20,130,084,616, 2013.

Agaphonov, M. O., Beburov, M. Y., Ter-Avanesyan, M. D., and Smirnov, V. N. "A disruption-replacement approach for the targeted integration of foreign genes in *Hansenula polymorpha*." *Yeast* 11 (1995): 1241–1247.

Agaphonov, M. O., Plotnikova, T. A., Fokina, A. V., Romanova, N. V., Packeiser, A. N., Kang, H. A., and Ter-Avanesyan, M. D. "Inactivation of the *Hansenula polymorpha* PMR1 gene affects cell viability and functioning of the secretory pathway." *FEMS Yeast Research* 7 (2007): 1145–1152.

Agaphonov, M. O., Sokolov, S. S., Romanova, N. V., Sohn, J. H., Kim, S. Y., Kalebina, T. S. et al. "Mutation of the protein-O-mannosyltransferase enhances secretion of the human urokinase-type plasminogen activator in *Hansenula polymorpha*." *Yeast* 22 (2005): 1037–1047.

Agaphonov, M., Trushkina, P., Sohn, J. H., Choi, E. S., Rhee, S. K., and Ter-Avanesyan, M. "Vectors for rapid selection of integrants with different plasmid copy numbers in the yeast *Hansenula polymorpha* DL1." *Yeast* 15 (1999): 541–551.

Alamäe, T., and Liiv, L. "Glucose repression of maltase and methanol-oxidizing enzymes in the methylotrophic yeast *Hansenula polymorpha*: Isolation and study of regulatory mutants." *Folia Microbiologica* 43 (1998): 443–452.

Alamäe, T., Pärn, P., Viigand, K., and Karp, H. "Regulation of the *Hansenula polymorpha* maltase gene promoter in *H. polymorpha* and *Saccharomyces cerevisiae*." *FEMS Yeast Research* 4 (2003): 165–173.

Amuel, C., Gellissen, G., Hollenberg, C., and Suckow, M. "Analysis of heat shock promoters in *Hansenula polymorpha*: The TPS1 promoter, a novel element for heterologous gene expression." *Biotechnology and Bioprocess Engineering* 5 (2000): 247–252.

Avgerinos, G., Turner, B., Gorelick, K., Papendieck, A., Weydemann, U., and Gellissen, G. "Production and clinical development of a *Hansenula polymorpha*-derived pegylated hirudin." *Seminar in Thrombosis and Hemostasis* 27 (2001): 357–372.

Avila, J., Gonzalez, C., Brito, N., and Siverio, J. M. "Clustering of the YNA1 gene encoding a Zn(II)2Cys6 transcriptional factor in the yeast *Hansenula polymorpha* with the nitrate assimilation genes *YNT1, YNI1* and *YNR1*, and its involvement in their transcriptional activation." *Biochemical Journal* 335 (1998): 647–652.

Bachhawat, A., Ganguli, D., Kaur, J., Kasturia, N., Thakur, A., Kaur, H. et al. "Glutathione production in yeast." In T. Satyanarayana and G. Kunze (Eds.), *Yeast Biotechnology: Diversity and Applications* (Vol. 259, pp. 259–280). Netherlands: Springer Science, Business Media B.V., 2009.

Baerends, R., Sulter, G., Jeffries, T., Cregg, J., and Veenhuis, M. "Molecular characterization of the *Hansenula polymorpha* FLD1 gene encoding formaldehyde dehydrogenase." *Yeast* 19 (2002): 37–42.

Bankar, A., Kumar, A., and Zinjarde, S. "Environmental and industrial applications of *Yarrowia lipolytica*." *Applied Microbiology and Biotechnology* 84 (2009): 847–865.

Barnard, G., Kull, A., Sharkey, N., Shaikh, S., Rittenhour, A., Burnina, I. et al. "High-throughput screening and selection of yeast cell lines expressing monoclonal antibodies." *Journal of Industrial Microbiology and Biotechnology* 37 (2010): 961–971.

Barnay-Verdier, S., Beckerich, J. M., and Boisramé, A. "New components of *Yarrowia lipolytica* Golgi multi-protein complexes containing the alpha-1,6-mannosyltransferases YlMnn9p and YlAnl1p." *Current Genetics* 54 (2008): 313–323.

Barnay-Verdier, S., Boisramé, A., and Beckerich, J. M. "Identification and characterization of two alpha-1,6-mannosyltransferases, Anl1p and Och1p, in the yeast *Yarrowia lipolytica*." *Microbiology* 150 (2004): 2185–2195.

Barnett, R. J., Payne, R. W., and Yarrow, D. *Yeast: Characteristics and Identification* (3rd Edition). Cambridge University Press, 2000.

Barth, G. (Ed.). *Yarrowia Lipolytica—Genetics, Genomics, and Physiology*. Heidelberg, Germany: Springer, 2013a.

Barth, G. (Ed.). *Yarrowia Lipolytica—Biotechnological Applications*. Heidelberg, Germany: Springer, 2013b.

Barth, G., and Gaillardin, C. "Yarrowia lipolytica." In K. Wolf (Ed.), *Nonconventional Yeasts in Biotechnology: A Handbook* (pp. 313–388). Heidelberg, Germany: Springer-Verlag, 1996.

Battista, M. C., Bergamini, G., Campanini, F., Landini, M. P., and Ripalti, A. "Intracellular production of a major cytomegalovirus antigenic protein in the methylotrophic yeast *Pichia pastoris*." *Gene* 176 (1996): 197–201.

Beck, A., Cochet, O., and Wurch, T. "GlycoFi's technology to control the glycosylation of recombinant therapeutic proteins." *Expert Opinion on Drug Discovery* 5 (2010): 95–111.

Beopoulos, A., Cescut, J., Haddouche, R., Uribelarrea, J. L., Molina-Jouve, C., and Nicaud, J. M. "*Yarrowia lipolytica* as a model for bio-oil production." *Progress in Lipid Research* 48 (2009): 375–387.

Beopoulos, A., Mrozova, Z., Thevenieau, F., Le Dall, M.-T., Hapala, I., Papanikolaou, S. et al. "Control of lipid accumulation in the yeast *Yarrowia lipolytica*." *Applied and Environmental Microbiology* 74 (2008): 7779–7789.

Berrin, J. G., Williamson, G., Puigserver, A., Chaix, J. C., McLauchlan, W., and Juge, N. "High-level production of recombinant fungal Endo-β-1,4-xylanase in the methylotrophic yeast *Pichia pastoris*." *Protein Expression and Purification* 19 (2000): 179–187.

Blanchin-Roland, S., Cordero Otero, R., and Gaillardin, C. "Two upstream activation sequences control the expression of the *XPR2* gene in the yeast *Yarrowia lipolytica*." *Molecular and Cellular Biology* 14 (1994): 327–338.

Blazeck, J., Liu, L., Knight, R., and Alper, H. "Heterologous production of pentane in the oleaginous yeast *Yarrowia lipolytica*." *Journal of Biotechnology* 165 (2013): 184–194.

Blazeck, J., Liu, L., Redden, H., and Alper, H. "Tuning gene expression in *Yarrowia lipolytica* by a hybrid promoter approach." *Applied and Environmental Microbiology* 77 (2011): 7905–7914.

Blazhenko, O., Zimmermann, M., Kang, H., Bartosz, G., Penninckx, M., Ubiyvovk, V., and Sibirny, A. "Accumulation of cadmium ions in the methylotrophic yeast *Hansenula polymorpha*." *Biometals* 19 (2006): 593–599.

Böer, E., Steinborn, G., Matros, A., Mock, H. P., Gellissen, G., and Kunze, G. "Production of interleukin-6 in *Arxula adeninivorans*, *Hansenula polymorpha* and *Saccharomyces cerevisiae* by applying the wide-range yeast vector (CoMed™) system to simultaneous comparative assessment." *FEMS Yeast Research* 7 (2007): 1181–1187.

Boisramé, A., Kabani, M., Beckerich, J. M., Hartmann, E., and Gaillardin, C. "Interaction of Kar2p and Sls1p is required for efficient co-translational translocation of secreted proteins in the yeast *Yarrowia lipolytica*." *The Journal of Biological Chemistry* 273 (1998): 30903–30908.

Bordes, F., Fudalej, F., Dossat, V., Nicaud, J. M., and Marty, A. "A new recombinant protein expression system for high-throughput screening in the yeast *Yarrowia lipolytica*." *Journal of Microbiological Methods* 70 (2007): 493–502.

Brake, A. J., Merryweather, J. P., Coit, D. G., Heberlein, U. A., Masiarz, F. R., Mullenbach, G. T. et al. "Alpha-factor-directed synthesis and secretion of mature foreign proteins in *Saccharomyces cerevisiae*." *Proceedings of the National Academy of Sciences of the United States of America* 81 (1984): 4642–4646.

Brierley, R. A., Bussineau, C., Kosson, R., Melton, A., and Siegel, R. S. "Fermentation development of recombinant *Pichia pastoris* expressing the heterologous gene: Bovine lysozyme." *Annals of the New York Academy of Sciences* 589 (1990): 350–362.

Brocke, P., Schaefer, S., Melber, K., Jenzelewski, V., Müller, F., Dahlems, U. et al. "Recombinant hepatitis B vaccines: Disease characterization and vaccine production." In G. Gellisen (Ed.), *Production of Recombinant Proteins* (pp. 319–359). Weinhem: Wiley-VCH Verlag GmbH & Co. KGaA, 2005.

Buckholz, R., and Gleeson, M. "Yeast systems for the commercial production of heterologous proteins." *Nature Biotechology* 9 (1991): 1067–1072.

Cabeç-Silva, C., and Madeira-Lopes, A. "Temperature relations of yield, growth and thermal death in the yeast *Hansenula polymorpha*." *Zeitschrift für allgemeine Mikrobiologie* 24 (1984): 129–132.

Çalık, P., İnankur, B., Soyaslan, E., Şahin, M., Taşpınar, H., Açık, E., and Bayraktar, E. "Fermentation and oxygen transfer characteristics in recombinant human growth hormone production by *Pichia pastoris* in sorbitol batch and methanol fed-batch operation." *Journal of Chemical Technology and Biotechnology* 85 (2010): 226–233.

Cardaci, S., Desideri, E., and Ciriolo, M. R. "Targeting aerobic glycolysis: 3-bromopyruvate as a promising anticancer drug." *Journal of Bioenergetics and Biomembranes* 44 (2012): 17–29.

Çelik, E., and Çalık, P. "Production of recombinant proteins by yeast cells." *Biotechnology Advances* 30 (2012): 1108–1118.

Çelik, E., Çalık, P., and Oliver, S. "Fed-batch methanol feeding strategy for recombinant protein production by *Pichia pastoris* in the presence of co-substrate sorbitol." *Yeast* 26 (2009): 473–484.

Çelik, E., Çalık, P., and Oliver, S. "Metabolic flux analysis for recombinant protein production by *Pichia pastoris* using dual carbon sources: Effects of methanol feeding rate." *Biotechnology and Bioengineering* 105 (2010): 317–329.

Celińska, E., and Grajek, W. "A novel multigene expression construct for modification of glycerol metabolism in *Yarrowia lipolytica*." *Microbial Cell Factories* 12 (2013): 102.

Cereghino, G., Cereghino, J., Ilgen, C., and Cregg, J. "Production of recombinant proteins in fermenter cultures of the yeast *Pichia pastoris*." *Current Opinion in Biotechnology* 13 (2002): 329–332.

Cereghino, G., Cereghino, J., Sunga, A., Johnson, M., Lim, M., Gleeson, M., and Cregg, J. "New selectable marker/auxotrophic host strain combinations for molecular genetic manipulation of *Pichia pastoris*." *Gene* 263 (2001): 159–169.

Cereghino, J., and Cregg, J. "Heterologous protein expression in the methylotrophic yeast *Pichia pastoris*." *FEMS Microbiology Reviews* 24 (2000): 45–66.

Chatterjee, S., de Lamirande, E., and Gagnon, C. "Cryopreservation alters membrane sulfhydryl status of bull spermatozoa: Protection by oxidized glutathione." *Molecular Reproduction and Development* 60 (2001): 498–506.

Chechenova, M., Romanova, N., Deev, A., Packeiser, A., Smirnov, V., Agaphonov, M., and Ter-Avanesyan, M. "C-terminal truncation of α-COP affects functioning of secretory organelles and calcium homeostasis in *Hansenula polymorpha*." *Eukaryotic Cell* 3 (2004): 52–60.

Chen, D., Beckerich, J. M., and Gaillardin, C. "One-step transformation of the dimorphic yeast *Yarrowia lipolytica*." *Applied Microbiology and Biotechnology* 48 (1997): 232–235.

Chuang, L. T., Chen, D. C., Nicaud, J. M., Madzak, C., Chen, Y. H., and Huang, Y. S. "Co-expression of heterologous desaturase genes in *Yarrowia lipolytica*." *New Biotechnology* 27 (2010): 277–282.

Clare, J. J., Rayment, F. B., Ballantine, S. P., Sreekrishna, K., and Romanos, M. A. "High-level expression of tetanus toxin fragment C in *Pichia pastoris* strains containing multiple tandem integrations of the gene." *Nature Biotechnology* 9 (1991): 455–460.

Cos, O., Ramón, R., Montesinos, J., and Valero, F. "Operational strategies, monitoring and control of heterologous protein production in the methylotrophic yeast *Pichia pastoris* under different promoters: A review." *Microbial Cell Factories* 5 (2006): 17.

Cox, H., Mead, D., Sudbery, P., Eland, R., Mannazzu, I., and Evans, L. "Constitutive expression of recombinant proteins in the methylotrophic yeast *Hansenula polymorpha* using the PMA1 promoter." *Yeast* 16 (2000): 1191–1203.

Crawford, K., Zaror, I., Bishop, R. J., and Innis, M. A. "*Pichia* secretory leader for protein expression." US Patent 6,410,264, 2002.

Cregg, J., Barringer, K., Hessler, A., and Madden, K. "*Pichia pastoris* as a host system for transformations." *Molecular and Cellular Biology* 5 (1985): 3376–3385.

Cregg, J., Cereghino, J., Shi, J., and Higgins, D. "Recombinant protein expression in *Pichia pastoris*." *Molecular Biotechnology* 16 (2000): 23–52.

Cregg, J., and Russell, K. "Transformation." In D. Higgins and J. Cregg (Eds.), *Pichia Protocols* (pp. 27–39). Totowa: NJ, Humana Press, 1998.

Cregg, J., Vedvick, T., and Raschke, W. "Recent advances in the expression of foreign genes in *Pichia pastoris*." *Nature Biotechnology* 11 (1993): 905–910.

Cui, W., Wang, Q., Zhang, F., Zhang, S. C., Chi, Z. M., and Madzak, C. "Direct conversion of inulin into single cell protein by the engineered *Yarrowia lipolytica* carrying inulinase gene." *Process Biochemistry* 46 (2011): 1442–1448.

d'Anjou, M., and Daugulis, A. "Mixed-feed exponential feeding for fed-batch culture of recombinant methylotrophic yeast." *Biotechnology Letters* 22 (2000): 341–346.

Daly, R., and Hearn, M. "Expression of heterologous proteins in *Pichia pastoris*: A useful experimental tool in protein engineering and production." *Journal of Molecular Recognition* 18 (2005): 119–138.

Damasceno, L., Huang, C., and Batt, C. "Protein secretion in *Pichia pastoris* and advances in protein production." *Applied Microbiology and Biotechnology* 93 (2012): 31–39.

Damaso, M., Almeida, M., Kurtenbach, E., Martins, O., Pereira, N., Andrade, C., and Albano, R. "Optimized expression of a thermostable xylanase from *Thermomyces lanuginosus* in *Pichia pastoris*." *Applied and Environmental Microbiology* 69 (2003): 6064–6072.

Damude, H., Zhang, H., Farrall, L., Ripp, K., Tomb, J. M., Hollerbach, D., and Yadav, N. "Identification of bifunctional delta12/omega3 fatty acid desaturases for improving the ratio of omega3 to omega6 fatty acids in microbes and plants." *Proceedings of the National Academy of Sciences of the United States of America* 103 (2006): 9446–9451.

Davidow, L. S., and Dezeeuw, J. R. "Process for transformation of *Yarrowia lipolytica*." EP0138508, 1985.

De Pourcq, K., Tiels, P., Van Hecke, A., Geysens, S., Vervecken, W., and Callewaert, N. "Engineering *Yarrowia lipolytica* to produce glycoproteins homogeneously modified with the universal Man3GlcNAc2 N-glycan core." *PloS One* 7 (2012a): e39976.

De Pourcq, K., Vervecken, W., Dewerte, I., Valevska, A., Van Hecke, A., and Callewaert, N. "Engineering the yeast *Yarrowia lipolytica* for the production of therapeutic proteins homogeneously glycosylated with $Man_8GlcNAc_2$ and $Man_5GlcNAc_2$." *Microbial Cell Factories* 11 (2012b): 53.

De Schutter, K., Lin, Y. C., Tiels, P., Van Hecke, A., Glinka, S., Weber-Lehmann, J. et al. "Genome sequence of the recombinant protein production host *Pichia pastoris*." *Nature Biotechnology* 27 (2009): 561–566.

Demain, A. "The business of biotechnology." *Industrial Biotechnology* 3 (2007): 269–283.

Ding, S. J., Ge, W., and Buswell, J. "Secretion, purification and characterisation of a recombinant Volvariella volvacea endoglucanase expressed in the yeast *Pichia pastoris*." *Enzyme and Microbial Technology* 31 (2002): 621–626.

Dmytruk, K., and Sibirny, A. "Metabolic engineering of the yeast *Hansenula polymorpha* for the construction of efficient ethanol producers." *Cytology and Genetics* 47 (2013): 329–342.

Dmytruk, O., Dmytruk, K., Abbas, C., Voronovsky, A., and Sibirny, A. "Engineering of xylose reductase and overexpression of xylitol dehydrogenase and xylulokinase improves xylose alcoholic fermentation in the thermotolerant yeast *Hansenula polymorpha*." *Microbial Cell Factories* 7 (2008a): 21.

Dmytruk, O., Voronovsky, A., Abbas, C., Dmytruk, K., Ishchuk, O., and Sibirny, A. "Overexpression of bacterial xylose isomerase and yeast host xylulokinase improves xylose alcoholic fermentation in the thermotolerant yeast *Hansenula polymorpha*." *FEMS Yeast Research* 8 (2008b): 165–173.

Dragosits, M., Stadlmann, J., Albiol, J., Baumann, K., Maurer, M., Gasser, B. et al. "The effect of temperature on the proteome of recombinant *Pichia pastoris*." *Journal of Proteome Research* 8 (2009): 1380–1392.

Drocourt, D., Calmels, T., Reynes, J. P., Baron, M., and Tiraby, G. "Cassettes of the Streptoalloteichus hindustanus ble gene for transformation of lower and higher eukaryotes to phleomycin resistance." *Nucleic Acids Research* 18 (1990): 4009.

Dujon, B., Sherman, D., Fischer, G., Durrens, P., Casaregola, S., Lafontaine, I. et al. "Genome evolution in yeasts." *Nature* 430 (2004): 35–44.

Dunn, W., Cregg, J., Kiel, J., Klei, I., Oku, M., Sakai, Y. et al. "Pexophagy: The Selective autophagy of peroxisomes." *Autophagy* 1 (2005): 75–83.

Eckart, M., and Bussineau, C. "Quality and authenticity of heterologous proteins synthesized in yeast." *Current Opinion in Biotechnology* 7 (1996): 525–530.
Eggeling, L., and Sahm, H. "Regulation of alcohol oxidase synthesis in *Hansenula polymorpha*: Oversynthesis during growth on mixed substrates and induction by methanol." *Archives of Microbiology* 127 (1980): 119–124.
Eiden-Plach, A., Zagorc, T., Heintel, T., Carius, Y., Breinig, F., and Schmitt, M. "Viral preprotoxin signal sequence allows efficient secretion of green fluorescent protein by *Candida glabrata, Pichia pastoris, Saccharomyces cerevisiae,* and *Schizosaccharomyces pombe.*" *Applied and Environmental Microbiology* 70 (2004): 961–966.
Emond, S., Akoh, C., Nicaud, J. M., Marty, A., Monsan, P., André, I., and Remaud-Siméon, M. "New efficient recombinant expression system to engineer *Candida antarctica* lipase B." *Applied and Environmental Microbiology* 76 (2010): 2684–2687.
Evers, M. E., Trip, H., van den Berg, M. A., Bovenberg, R. A. L., and Driessen, A. J. M. "Compartmentalization and transport in β-lactam antibiotics biosynthesis." In A. A. Brakhage (Ed.), *Molecular Biotechnolgy of Fungal beta-Lactam Antibiotics and Related Peptide Synthetases* (Vol. 88, pp. 111–135). Berlin, Heidelberg: Springer, 2004.
Faber, K., Haima, P., Harder, W., Veenhuis, M., and Ab, G. "Highly-efficient electrotransformation of the yeast *Hansenula polymorpha.*" *Current Genetics* 25 (1994): 305–310.
Faber, K. N., Westra, S., Waterham, H. R., Keizer-Gunnink, I., Harder, W., Ab, G., and Veenhuis, M. "Foreign gene expression in *Hansenula polymorpha.* A system for the synthesis of small functional peptides." *Applied Microbiology and Biotechnology* 45 (1996): 72–79.
Fickers, P., Benetti, P. H., Waché, Y., Marty, A., Mauersberger, S., Smit, M. S., and Nicaud, J. M. "Hydrophobic substrate utilisation by the yeast *Yarrowia lipolytica,* and its potential applications." *FEMS Yeast Research* 5 (2005): 527–543.
Fickers, P., Le Dall, M. T., Gaillardin, C., Thonart, P., and Nicaud, J. M. "New disruption cassettes for rapid gene disruption and marker rescue in the yeast *Yarrowia lipolytica.*" *Journal of Microbiological Methods* 55 (2003): 727–737.
Fierobe, H., Pagès, S., Bélaïch, A., Champ, S., Lexa, D., and Bélaïch, J. P. "Cellulosome from *Clostridium cellulolyticum*: Molecular study of the dockerin/cohesin interaction." *Biochemistry* 38 (1999): 12822–12832.
Fournier, P., Abbas, A., Chasles, M., Kudla, B., Ogrydziak, D. M., Yaver, D. et al. "Colocalization of centromeric and replicative functions on autonomously replicating sequences isolated from the yeast *Yarrowia lipolytica.*" *Proceedings of the National Academy of Sciences of the United States of America* 90 (1993): 4912–4916.
Ganapathy-Kanniappan, S., Vali, M., Kunjithapatham, R., Buijs, M., Syed, L. H., Rao, P. P. et al. "3-bromopyruvate: A new targeted antiglycolytic agent and a promise for cancer therapy." *Current Pharmaceutical Biotechnology* 11 (2010): 510–517.
Gancedo, J. "The early steps of glucose signalling in yeast." *FEMS Microbiology Reviews* 32 (2008): 673–704.
Gao, M. J., Li, Z., Yu, R. S., Wu, J. R., Zheng, Z. Y., Shi, Z. P. et al. "Methanol/sorbitol co-feeding induction enhanced porcine interferon-α production by *P. pastoris* associated with energy metabolism shift." *Bioprocess and Biosystems Engineering* 35 (2012): 1125–1136.
Gasmi, N., Fudalej, F., Kallel, H., and Nicaud, J. M. "A molecular approach to optimize hIFN α2b expression and secretion in *Yarrowia lipolytica.*" *Applied Microbiology and Biotechnology* 89 (2011): 109–119.
Gatzke, R., Weydemann, U., Janowicz, Z. A., and Hollenberg, C. P. "Stable multicopy integration of vector sequences in *Hansenula polymorpha.*" *Applied Microbiology and Biotechnology* 43 (1995): 844–849.
Gellissen, G. "Heterologous protein production in methylotrophic yeasts." *Applied Microbiology and Biotechnology* 54 (2000): 741–750.
Gellissen, G., and Hollenberg, C. "Application of yeasts in gene expression studies: A comparison of *Saccharomyces cerevisiae, Hansenula polymorpha* and *Kluyveromyces lactis*—A review." *Gene* 190 (1997): 87–97.
Gellissen, G., Janowicz, Z., Weydemann, U., Melber, K., Strasser, A., and Hollenberg, C. "High-level expression of foreign genes in *Hansenula polymorpha.*" *Biotechnology Advances* 10 (1992): 179–189.

Gellissen, G., Kunze, G., Gaillardin, C., Cregg, J., Berardi, E., Veenhuis, M., and Van Der Klei, I. "Nex yeast expression platforms base on methylotrophic *Hansenula polymorpha* and *Pichia pastoris* and on *Arxula adeninivorans* and *Yarrowia lipolytica*—A comparison." *FEMS Yeast Research* 5 (2005): 1079–1096.
Gidijala, L., Bovenberg, R., Klaassen, P., van der Klei, I., Veenhuis, M., and Kiel, J. "Production of functionally active *Penicillium chrysogenum* isopenicillin N synthase in the yeast *Hansenula polymorpha*." *BMC Biotechnology* 8 (2008): 1–8.
Gidijala, L., Kiel, J., Douma, R., Seifar, R., van Gulik, W., Bovenberg, R. et al. "An engineered yeast efficiently secreting penicillin." *PLoS One* 4 (2009): e8317.
Gidijala, L., Van Der Klei, I., Veenhuis, M., and Kiel, J. "Reprogramming *Hansenula polymorpha* for penicillin production: Expression of the *Penicillium chrysogenum* pcl gene." *FEMS Yeast Research* 7 (2007): 1160–1167.
Goncalves, A. M., Pedro, A., Maia, C., Sousa, F., Queiroz, J., and Passarinha, L. "*Pichia pastoris*: A recombinant microfactory for antibodies and human membrane proteins." *Journal of Microbiology and Biotechnology* 23 (2013): 587–601.
Gonchar, M., Maidan, M., Moroz, O., Woodward, J., and Sibirny, A. "Microbial O2- and H2O2-electrode sensors for alcohol assays based on the use of permeabilized mutant yeast cells as the sensitive bioelements." *Biosensors and Bioelectronics* 13 (1998): 945–952.
González, C., Perdomo, G., Tejera, P., Brito, N., and Siverio, J. "One-step, PCR-mediated, gene disruption in the yeast *Hansenula polymorpha*." *Yeast* 15 (1999): 1323–1329.
Grabek-Lejko, D., Kurylenko, O., Sibirny, V., Ubiyvovk, V., Penninckx, M., and Sibirny, A. "Alcoholic fermentation by wild-type *Hansenula polymorpha* and *Saccharomyces cerevisiae* versus recombinant strains with an elevated level of intracellular glutathione." *Journal of Industrial Microbiology and Biotechnology* 38 (2011): 1853–1859.
Grabek-Lejko, D., Ryabova, O., Oklejewicz, B., Voronovsky, A., and Sibirny, A. "Plate ethanol-screening assay for selection of the *Pichia stipitis* and *Hansenula polymorpha* yeast mutants with altered capability for xylose alcoholic fermentation." *Journal of Industrial Microbiology and Biotechnology* 33 (2006): 934–940.
Groenewald, M., Boekhout, T., Neuvéglise, C., Gaillardin, C., van Dijck, P., and Wyss, M. "*Yarrowia lipolytica*: Safety assessment of an oleaginous yeast with a great industrial potential." *Critical Reviews in Microbiology* 40 (2014): 187–206.
Guo, J. P., and Ma, Y. "High-level expression, purification and characterization of recombinant *Aspergillus oryzae* alkaline protease in *Pichia pastoris*." *Protein Expression and Purification* 58 (2008): 301–308.
Guo, W., González-Candelas, L., and Kolattukudy, P. E. "Cloning of a novel constitutively expressed pectate lyase gene pelB from *Fusarium solani* f. sp. pisi (*Nectria haematococca*, mating type VI) and characterization of the gene product expressed in *Pichia pastoris*." *Journal of Bacteriology* 177 (1995): 7070–7077.
Haddouche, R., Delessert, S., Sabirova, J., Neuvéglise, C., Poirier, Y., and Nicaud, J. M. "Roles of multiple acyl-CoA oxidases in the routing of carbon flow towards β-oxidation and polyhydroxyalkanoate biosynthesis in *Yarrowia lipolytica*." *FEMS Yeast Research* 10 (2010): 917–927.
Haddouche, R., Poirier, Y., Delessert, S., Sabirova, J., Pagot, Y., Neuvéglise, C., and Nicaud, J. M. "Engineering polyhydroxyalkanoate content and monomer composition in the oleaginous yeast *Yarrowia lipolytica* by modifying the β-oxidation multifunctional protein." *Applied Microbiology and Biotechnology* 91 (2011): 1327–1340.
Hamilton, S., Davidson, R., Sethuraman, N., Nett, J., Jiang, Y., Rios, S. et al. "Humanization of yeast to produce complex terminally sialylated glycoproteins." *Science* 313 (2006): 1441–1443.
Hamilton, S., and Gerngross, T. "Glycosylation engineering in yeast: The advent of fully humanized yeast." *Current Opinion in Biotechnology* 18 (2007): 387–392.
Han, Y., and Lei, X. G. "Role of glycosylation in the functional expression of an *Aspergillus niger* phytase (phyA) in *Pichia pastoris*." *Archives of Biochemistry and Biophysics* 364 (1999): 83–90.
Han, Z., Madzak, C., and Su, W. W. "Tunable nano-oleosomes derived from engineered *Yarrowia lipolytica*." *Biotechnology and Bioengineering* 110 (2013): 702–710.
Hazeu, W., Bruyn, J., and Bos, P. "Methanol assimilation by yeasts." *Archives für Mikrobiologie* 87 (1972): 185–188.

Heimo, H., Palmu, K., and Suominen, I. "Human placental alkaline phosphatase: Expression in *Pichia pastoris*, purification and characterization of the enzyme." *Protein Expression and Purification* 12 (1998): 85–92.

Hellwig, S., Stöckmann, C., Gellissen, G., and Büchs, J. "Comparative fermentation." In G. Gellissen (Ed.), *Production of Recombinant Proteins* (pp. 287–317). Weinheim, Germany: Wiley-VCH Verlag GmbH & Co. KGaA, 2005.

Heo, J. H., Hong, W. K., Cho, E. Y., Kim, M. W., Kim, M., Kim, C. et al. "Properties of the Hansenula polymorpha-derived constitutive *GAP* promoter, assessed using an *HSA* reporter gene." *FEMS Yeast Research* 4 (2003): 175–184.

Heyland, J., Fu, J., Blank, L., and Schmid, A. "Quantitative physiology of Pichia pastoris during glucose-limited high-cell density fed-batch cultivation for recombinant protein production." *Biotechnology and Bioengineering* 107 (2010): 357–368.

Higgins, D., Busser, K., Comiskey, J., Whittier, P. S., Purcell, T. J., and Hoeffler, J. P. "Small vector for expression based on dominant drug resistance with direct multicopy selection." In D. R. Higgins and J. Cregg (Eds.), *Methods in Molecular Biology: Pichia Pastoris* (pp. 41–53). Totowa, NJ: Humana Press, 1998.

Hodgkins, M., Sudbery, P., Mead, D., Ballance, D., and Goodey, A. "Expression of the glucose oxidase gene from *Aspergillus niger* in *Hansenula polymorpha* and its use as a reporter gene to isolate regulatory mutations." *Yeast* 9 (1993): 625–635.

Hong, S. P., Seip, J., Walters-Pollak, D., Rupert, R., Jackson, R., Xue, Z., and Zhu, Q. "Engineering *Yarrowia lipolytica* to express secretory invertase with strong FBA1(IN) promoter." *Yeast* 29 (2012): 59–72.

Hong, W., Kim, C., Heo, S. Y., Luo, L., Oh, B. R., and Seo, J. W. "Enhanced production of ethanol from glycerol by engineered *Hansenula polymorpha* expressing pyruvate decarboxylase and aldehyde dehydrogenase genes from *Zymomonas mobilis*." *Biotechnology Letters* 32 (2010): 1077–1082.

Horiguchi, H., Yurimoto, H., Kato, N., and Sakai, Y. "Antioxidant system within yeast peroxisome: Biochemical and physiological characterisation of CbPmp20 in the methylottrophic yeast *Candida boidini*." *Journal of Biological Chemistry* 276 (2001): 14279–14288.

Inan, M. "*Pichia pastoris* PIR1 secretion signal peptide for recombinant protein expression and *Pichia pastoris PIR1* and *PIR2* anchor domain peptides for recombinant surface display." US Patent 7,741,075, 2010.

Inan, M., Chiruvolu, V., Eskridge, K., Vlasuk, G., Dickerson, K., Brown, S., and Meagher, M. "Optimization of temperature–glycerol–pH conditions for a fed-batch fermentation process for recombinant hookworm (*Ancylostoma caninum*) anticoagulant peptide (AcAP-5) production by *Pichia pastoris*." *Enzyme and Microbial Technology* 24 (1999): 438–445.

Inan, M., and Meagher, M. "Non-repressing carbon sources for alcohol oxidase (AOX1) promoter of *Pichia pastoris*." *Journal of Bioscience and Bioengineering* 92 (2001): 585–589.

Invitrogen. "Pichia fermentation process guidelines." 2000. Available at http://tools.lifetechnologies.com/content/sfs/manuals/pichiaferm_prot.pdf.

Ishchuk, O., Abbas, C., and Sibirny, A. "Heterologous expression of *Saccharomyces cerevisiae MPR1* gene confers tolerance to ethanol and l-azetidine-2-carboxylic acid in *Hansenula polymorpha*." *Journal of Industrial Microbiology and Biotechnology* 37 (2010): 213–218.

Ishchuk, O., Voronovsky, A., Abbas, C., and Sibirny, A. "Construction of *Hansenula polymorpha* strains with improved thermotolerance." *Biotechnology and Bioengineering* 104 (2009): 911–919.

Ishchuk, O., Voronovsky, A., Stasyk, O., Gayda, G., Gonchar, M., Abbas, C., and Sibirny, A. "Overexpression of pyruvate decarboxylase in the yeast *Hansenula polymorpha* results in increased ethanol yield in high-temperature fermentation of xylose." *FEMS Yeast Research* 8 (2008): 1164–1174.

Jaafar, L., and Zueco, J. "Characterization of a glycosylphosphatidylinositol-bound cell-wall protein (GPI-CWP) in *Yarrowia lipolytica*." *Microbiology* 150 (2004): 53–60.

Jacobs, P., Geysens, S., Vervecken, W., Contreras, R., and Callewaert, N. "Engineering complex-type N-glycosylation in *Pichia pastoris* using GlycoSwitch technology." *Nature Protocols* 4 (2008): 58–70.

Jeffries, T., Grigoriev, I. V., Grimwood, J., Laplaza, J. M., Aerts, A., Salamov, A. et al. "Genome sequence of the lignocellulose-bioconverting and xylose-fermenting yeast *Pichia stipitis*." *Nature Biotechnology* (2007): 319–326.

Jenzelewski, V. "Fermentation and primary product recovery." In G. Gellisen (Ed.), *Hansenula Polymorpha* (pp. 156–174). Weinheim: Wiley-VCH Verlag GmbH & Co. KGaA, 2002.

Jeong, K. J., Jang, S. H., and Velmurugan, N. "Recombinant antibodies: Engineering and production in yeast and bacterial hosts." *Biotechnology Journal* 6 (2011): 16–27.

Jia, B., Liu, W., Yang, J., Ye, C., Xu, L., and Yan, Y. "*Burkholderia cepacia* lipase gene modification and its constitutive and inducible expression in *Pichia pastoris*." *Acta Microbiologica Sinica* 50 (2010): 1194–1201.

Jorda, J., Jouhten, P., Camara, H., Maaheimo, H., Albiol, J., and Ferrer, P. "Metabolic flux profiling of recombinant protein secreting *Pichia pastoris* growing on glucose: Methanol mixtures." *Microbial Cell Factories* 11 (2012): 57.

Jungo, C., Marison, I., and von Stockar, U. "Mixed feeds of glycerol and methanol can improve the performance of *Pichia pastoris* cultures: A quantitative study based on concentration gradients in transient continuous cultures." *Journal of Biotechnology* 128 (2007a): 824–837.

Jungo, C., Schenk, J., Pasquier, M., Marison, I. W., and von Stockar, U. "A quantitative analysis of the benefits of mixed feeds of sorbitol and methanol for the production of recombinant avidin with *Pichia pastoris*." *Journal of Biotechnology* 131 (2007b): 57–66.

Juretzek, T., Le Dall, M., Mauersberger, S., Gaillardin, C., Barth, G., and Nicaud, J. M. "Vectors for gene expression and amplification in the yeast *Yarrowia lipolytica*." *Yeast* 18 (2001): 97–113.

Juretzek, T., Wang, H. J., Nicaud, J. M., Mauersberger, S., and Barth, G. "Comparison of promoters suitable for regulated overexpression of beta-galactosidase in the alkane-utilizing yeast *Yarrowia lipolytica*." *Biotechnology and Bioprocess Engineering* 5 (2000): 320–326.

Kang, H. A., and Gellissen, G. "*Hansenula polymorpha*." In G. Gellissen (Ed.), *Production of Recombinant Proteins* (pp. 111–142). Weinheim: Wiley-VCH Verlag GmbH & Co. KGaA, 2005.

Kang, H. A., Rhee, S. K., Sohn, M. J., Sibirny, A. A., and Ubiyvovk, V. M. "Gamma-glutamylcysteine synthetase and gene coding for the same." Korea Patent 10,2002,0073219, 2002.

Kang, H., Kang, W., Hong, W. K., Kim, M. W., Kim, J. Y., Sohn, J. H. et al. "Development of expression systems for the production of recombinant human serum albumin using the MOX promoter in *Hansenula polymorpha* DL-1." *Biotechnology and Bioengineering* 76 (2001): 175–185.

Karp, H., Järviste, A., Kriegel, T., and Alamäe, T. "Cloning and biochemical characterization of hexokinase from the methylotrophic yeast *Hansenula polymorpha*." *Current Genetics* 44 (2003): 268–276.

Kato, S., Ishibashi, M., Tatsuda, D., Tokunaga, H., and Tokunaga, M. "Efficient expression, purification and characterization of mouse salivary α-amylase secreted from methylotrophic yeast, *Pichia pastoris*." *Yeast* 18 (2001): 643–655.

Khandekar, S., Silverman, C., Wells-Marani, J., Bacon, A., Birrell, H., Brigham-Burke, M. et al. "Determination of carbohydrate structures N-Linked to soluble CD154 and characterization of the interactions of CD40 with CD154 expressed in *Pichia pastoris* and Chinese hamster ovary cells." *Protein Expression and Purification* 23 (2001): 301–310.

Kim, S. Y., Sohn, J. H., Bae, J. H., Pyun, Y. H., Agaphonov, M., Adamczak, M., and Choi, E. S. "Efficient library construction by in vivo recombination with a telomere-originated autonomously replicating sequence of *Hansenula polymorpha*." *Applied and Environmental Microbiology* 69 (2003): 4448–4454.

Kim, S. Y., Sohn, J. H., Pyun, Y. R., and Choi, E. S. "A cell surface display system using novel GPI-anchored proteins in *Hansenula polymorpha*." *Yeast* 19 (2002): 1153–1163.

Kim, T., and Lei, X. G. "Expression and characterization of a thermostable serine protease (TfpA) from *Thermomonospora fusca* YX in *Pichia pastoris*." *Applied Microbiology and Biotechnology* 68 (2005): 355–359.

Kimura, M., Takatsuki, A., and Yamaguchi, I. "Blasticidin S deaminase gene from *Aspergillus terreus* (BSD): A new drug resistance gene for transfection of mammalian cells." *Biochimica et Biophysica Acta (BBA)—Gene Structure and Expression* 1219 (1994): 653–659.

Klabunde, J., Diesel, A., Waschk, D., Gellissen, G., Hollenberg, C., and Suckow, M. "Single-step co-integration of multiple expressible heterologous genes into the ribosomal DNA of the methylotrophic yeast *Hansenula polymorpha*." *Applied Microbiology and Biotechnology* 58 (2002): 797–805.

Klabunde, J., Kleebank, S., Piontek, M., Hollenberg, C., Hellwig, S., and Degelmann, A. "Increase of calnexin gene dosage boosts the secretion of heterologous proteins by *Hansenula polymorpha*." *FEMS Yeast Research* 7 (2007): 1168–1180.

Klabunde, J., Kunze, G., Gellissen, G., and Hollenberg, C. "Integration of heterologous genes in several yeast species using vectors containing a *Hansenula polymorpha*-derived rDNA-targeting element." *FEMS Yeast Research* 4 (2003): 185–193.
Kobayashi, K., Kuwae, S., Ohya, T., Ohda, T., Ohyama, M., Ohi, H. et al. "High-level expression of recombinant human serum albumin from the methylotrophic yeast *Pichia pastoris* with minimal protease production and activation." *Journal of Bioscience and Bioengineering* 89 (2000a): 55–61.
Kobayashi, K., Kuwae, S., Ohya, T., Ohda, T., Ohyama, M., and Tomomitsu, K. "Addition of oleic acid increases expression of recombinant human serum albumin by the *AOX2* promoter in *Pichia pastoris*." *Journal of Bioscience and Bioengineering* 89 (2000b): 479–484.
Kottmeier, K., Ostermann, K., Bley, T., and Rödel, G. "Hydrophobin signal sequence mediates efficient secretion of recombinant proteins in *Pichia pastoris*." *Applied Microbiology and Biotechnology* 91 (2011): 133–141.
Kozhemyakin, L., and Balasovski, M. "Methods for production of the oxidized glutathione composite with CIS-diamminedichloroplatinium and pharmaceutical compositions based thereof regulating metabolism, proliferation, differentiation and apoptotic mechanism for normal and transformed cells." US Patent 7,371,411, 2008.
Kraakman, L. S., Winderickx, J., Thevelein, J. M., and De Winde, J. H. "Structure-function analysis of yeast hexokinase: Structural requirements for triggering cAMP signalling and catabolite repression." *Biochemical Journal* 343 Pt 1 (1999): 159–168.
Kramarenko, T., Karp, H., Järviste, A., and Alamäe, T. "Sugar repression in the methylotrophic yeast *Hansenula polymorpha* studied by using hexokinase-negative, glucokinase-negative and double kinase-negative mutants." *Folia Microbiologica* 45 (2000): 521–529.
Krasovska, O., Stasyk, O., Nahorny, V., Stasyk, O., Granovski, N., Kordium, V. et al. "Glucose-induced production of recombinant proteins in *Hansenula polymorpha* mutants deficient in catabolite repression." *Biotechnology and Bioengineering* 97 (2007): 858–870.
Krasovska, O., Stasyk, O., and Sibirny, A. "Stable overproducer of hepatitis B surface antigen in the methylotrophic yeast *Hansenula polymorpha* due to multiple integration of heterologous auxotrophic selective markers and defect in peroxisome biogenesis." *Applied Microbiology and Biotechnology* 97 (2013): 9969–9979.
Kresse, H., Belsey, M. J., and Rovini, H. "The antibacterial drugs market." *Nature Reviews Drug Discovery* 6 (2007): 19–20.
Kunze, G., Kang, H., and Gellissen, G. "*Hansenula polymorpha* (*Pichia angusta*): Biology and applications." In T. Satyanarayana and G. Kunze (Eds.), *Yeast Biotechnology: Diversity and Applications* (pp. 47–64). Netherlands: Springer, 2009.
Kurtzman, C. "A new methanol assimilating yeast, Ogataea parapolymorpha, the ascosporic state of *Candida parapolymorpha*." *Antonie van Leeuwenhoek* 100 (2011): 455–462.
Kurtzman, C., and Fell, J. W. *The Yeasts, A Taxonomic Study* (4th Edition). Amsterdam: Elsevier Science, 1998.
Labuschagne, M., and Albertyn, J. "Cloning of an epoxide hydrolase-encoding gene from Rhodotorula mucilaginosa and functional expression in *Yarrowia lipolytica*." *Yeast* 24 (2007): 69–78.
Laht, S., Karp, H., Kotka, P., Järviste, A., and Alamäe, T. "Cloning and characterization of glucokinase from a methylotrophic yeast *Hansenula polymorpha*: Different effects on glucose repression in *H. polymorpha* and *Saccharomyces cerevisiae*." *Gene* 296 (2002): 195–203.
Lahtchev, K. L., Semenova, V. D., Tolstorukov, I. I., Klei, I., and Veenhuis, M. "Isolation and properties of genetically defined strains of the methylotrophic yeast *Hansenula polymorpha* CBS4732." *Archives of Microbiology* 177 (2002): 150–158.
Larsen, M., Bornscheuer, U., and Hult, K. "Expression of *Candida antarctica* lipase B in *Pichia pastoris* and various *Escherichia coli* systems." *Protein Expression and Purification* 62 (2008): 90–97.
Le Dall, M. T., Nicaud, J. M., and Gaillardin, C. "Multiple-copy integration in the yeast *Yarrowia lipolytica*." *Current Genetics* 26 (1994): 38–44.
Leão-Helder, A. N., Krikken, A. M., van der Klei, I. J., Kiel, J., and Veenhuis, M. "Transcriptional down-regulation of peroxisome numbers affects selective peroxisome degradation in *Hansenula polymorpha*." *Journal of Biological Chemistry* 278 (2003): 40749–40756.
Li, H., Sethuraman, N., Stadheim, T., Zha, D., Prinz, B., Ballew, N. et al. "Optimization of humanized IgGs in glycoengineered *Pichia pastoris*." *Nature Biotechnology* 24 (2006): 210–215.

Li, J. F., Hong, Y. Z., Xiao, Y. Z., Xu, Y. H., and Fang, W. "High production of laccase B from *Trametes* sp. in *Pichia pastoris*." *World Journal of Microbiology and Biotechnology* 23 (2007): 741–745.

Li, S., Sing, S., and Wang, Z. "Improved expression of *Rhizopus oryzae* α-amylase in the methylotrophic yeast *Pichia pastoris*." *Protein Expression and Purification* 79 (2011): 142–148.

Li, Y., Wei, G., and Chen, J. "Glutathione: A review on biotechnological production." *Applied Microbiology and Biotechnology* 66 (2004): 233–242.

Liang, S., Li, C., Ye, Y., and Lin, Y. "Endogenous signal peptides efficiently mediate the secretion of recombinant proteins in *Pichia pastoris*." *Biotechnology Letter* 35 (2013): 97–105.

Liiv, L., Pärn, P., and Alamäe, T. "Cloning of maltase gene from a methylotrophic yeast, *Hansenula polymorpha*." *Gene* 265 (2001): 77–85.

Lin-Cereghino, G., Godfrey, L., de la Cruz, B., Johnson, S., Khuongsathiene, S., Tolstorukov, I. et al. "Mxr1p, a key regulator of the methanol utilization pathway and peroxisomal genes in *Pichia pastoris*." *Molecular and Cellular Biology* 26 (2006): 883–897.

Linnevers, C., Mcgrath, M., Armstrong, A., Mistry, F., Barnes, M., Klaus, J. et al. "Expression of human cathepsin K in *Pichia pastoris* and preliminary crystallographic studies of an inhibitor complex." *Protein Science* 6 (1997): 919–921.

Liu, H., Tan, X., Russell, K., Veenhuis, M., and Cregg, J. "PER3, a gene required for peroxisome biogenesis in *Pichia pastoris*, encodes a peroxisomal membrane protein involved in protein import." *Journal of Biological Chemistry* 270 (1995): 10940–10951.

Liu, X. Y., Chi, Z., Liu, G. L., Wang, F., Madzak, C., and Chi, Z.-M. "Inulin hydrolysis and citric acid production from inulin using the surface-engineered *Yarrowia lipolytica* displaying inulinase." *Metabolic Engineering* 12 (2010): 469–476.

Loewen, M. C., Liu, X., Davies, P. L., and Daugulis, A. J. "Biosynthetic production of type II fish antifreeze protein: Fermentation by *Pichia pastoris*." *Applied Microbiology and Biotechnology* 48 (1997): 480–486.

Loira, N., Dulermo, T., Nicaud, J. M., and Sherman, D. J. "A genome-scale metabolic model of the lipid-accumulating yeast *Yarrowia lipolytica*." *BMC Systems Biology* 6 (2012): 35.

Long, T. M., Su, Y. K., Headman, J., Higbee, A., Willis, L. B., and Jeffries, T. W. "Cofermentation of glucose, xylose, and cellobiose by the beetle-associated yeast *Spathaspora passalidarum*." *Applied and Environmental Microbiology* 78 (2012): 5492–5500.

Lu, L., Zhao, M., Liang, S.-C., Zhao, L.-Y., Li, D.-B., and Zhang, B.-B. "Production and synthetic dyes decolourization capacity of a recombinant laccase from *Pichia pastoris*." *Journal of Applied Microbiology* 107 (2009): 1149–1156.

Lutz, M. V., Bovenberg, R., van der Klei, I. J., and Veenhuis, M. "Synthesis of Penicillium chrysogenum acetyl-CoA:isopenicillin N acyltransferase in *Hansenula polymorpha*: First step towards the introduction of a new metabolic pathway." *FEMS Yeast Research* 5 (2005): 1063–1067.

Macauley-Patrick, S., Fazenda, M., McNeil, B., and Harvey, L. "Heterologous protein production using the *Pichia pastoris* expression system." *Yeast* 22 (2005): 249–270.

Madzak, C., and Beckerich, J. "Heterologous protein expression and secretion in *Yarrowia lipolytica*." In G. Barth (Ed.), *Yarrowia Lipolytica: Biotechnological Application* (pp. 1–76). Heidelberg, Germany: Springer, 2013.

Madzak, C., Blanchin-Roland, S., Cordero Otero, R., and Gaillardin, C. "Functional analysis of upstream regulating regions from the *Yarrowia lipolytica XPR2* promoter." *Microbiology* 145 (1999): 75–87.

Madzak, C., Blanchin-Roland, S., and Gaillardin, C. "Upstream activating sequences and recombinant promoter sequences functional in Yarrowia and vectors containing them." European Patent EP0747484, 1996.

Madzak, C., Gaillardin, C., and Beckerich, J. M. "Heterologous protein expression and secretion in the non-conventional yeast *Yarrowia lipolytica*: A review." *Journal of Biotechnology* 109 (2004): 63–81.

Madzak, C., Nicaud, J. M., and Gaillardin, C. "*Yarrowia lipolytica*." In G. Gellissen (Ed.), *Production of Recombinant Proteins* (pp. 163–189). Weinhem: Wiley-VCH Verlag GmbH & Co. KGaA, 2005.

Madzak, C., Tréton, B., and Blanchin-Roland, S. "Strong hybrid promoters and integrative expression/secretion vectors for quasi-constitutive expression of heterologous proteins in the yeast *Yarrowia lipolytica*." *Journal of Molecular Microbiology and Biotechnology* 2 (2000): 207–216.

Marri, L., Rossolini, G. M., and Satta, G. "Chromosome polymorphisms among strains of *Hansenula polymorpha* (syn. *Pichia angusta*)." *Applied and Environmental Microbiology* 59 (1993): 939–941.
Matthäus, F., Ketelhot, M., Gatter, M., and Barth, G. "Production of Lycopene in the non-carotenoid producing yeast *Yarrowia lipolytica*." *Applied and Environmental Microbiology* 80 (2014) 1660–1669.
Mayer, A. F., Hellmuth, K., Schlieker, H., Lopez-Ulibarri, R., Oertel, S., Dahlems, U. et al. "An expression system matures: A highly efficient and cost-effective process for phytase production by recombinant strains of *Hansenula polymorpha*." *Biotechnology and Bioengineering* 63 (1999): 373–381.
Menendez, J., Valdes, I., and Cabrera, N. "The ICL1 gene of *Pichia pastoris*, transcriptional regulation and use of its promoter." *Yeast* 20 (2003): 1097–1108.
Merckelbach, A., Gödecke, S., Janowicz, Z., and Hollenberg, C. "Cloning and sequencing of the ura3 locus of the methylotrophic yeast *Hansenula polymorpha* and its use for the generation of a deletion by gene replacement." *Applied Microbiology and Biotechnology* 40 (1993): 361–364.
Merino, S., and Baratti, J. "Progress and challenges in enzyme development for biomass utilization." In L. Olsson (Ed.), *Biofuels* (Vol. 108, pp. 95–120). Berlin, Heidelberg: Springer, 2007.
Middelhoven, W. J. "Identification of yeasts present in sour fermented foods and fodders." *Molecular Biotechnology* 21 (2002): 279–292.
Mochizuki, S., Hamato, N., Hirose, M., Miyano, K., Ohtani, W., Kameyama, S. et al. "Expression and characterization of recombinant human antithrombin III in *Pichia pastoris*." *Protein Expression and Purification* 23 (2001): 55–65.
Montesino, R., Nimtz, M., Quintero, O., García, R., Falcón, V., and Cremata, J. "Characterization of the oligosaccharides assembled on the *Pichia pastoris*—Expressed recombinant aspartic protease." *Glycobiology* 9 (1999): 1037–1043.
Moroz, O. M., Gonchar, M. V., and Sibirny, A. "Efficient bioconversion of ethanol to acetaldehyde using a novel mutant strain of the methylotrophic yeast *Hansenula polymorpha*." *Biotechnology and Bioengineering* 68 (2000): 44–51.
Müller, S., Sandal, T., Kamp-Hansen, P., and Dalbøge, H. "Comparison of expression systems in the yeasts *Saccharomyces cerevisiae*, *Hansenula polymorpha*, *Klyveromyces lactis*, *Schizosaccharomyces pombe* and *Yarrowia lipolytica*. Cloning of two novel promoters from *Yarrowia lipolytica*." *Yeast* 14 (1998): 1267–1283.
Naumova, E. S., Dmitruk, K. V., Kshanovskaya, B. V., Sibirny, A. A., and Naumov, G. I. "Molecular identification of the industrially important strain *Ogataea parapolymorpha*." *Microbiology (Moscow)* 82 (2013): 453–458.
Nett, J., and Gerngross, T. "Cloning and disruption of the PpURA5 gene and construction of a set of integration vectors for the stable genetic modification of *Pichia pastoris*." *Yeast* 20 (2003): 1279–1290.
Nett, J., Hodel, N., Rausch, S., and Wildt, S. "Cloning and disruption of the *Pichia pastoris* ARG1, *ARG2*, *ARG3*, *HIS1*, *HIS2*, *HIS5*, *HIS6* genes and their use as auxotrophic markers." *Yeast* 22 (2005): 295–304.
Nicaud, J. M., Fabre, E., and Gaillardin, C. "Expression of invertase activity in *Yarrowia lipolytica* and its use as a selective marker." *Current Genetics* 16 (1989): 253–260.
Nicaud, J. M., Fudalej, F., Neuvéglise, C., and Beckerich, J. M. "Procédé d'intégration ciblée de multicopies d'un gène d'intérêt dans une souche de *Yarrowia*." French Patent FR2927089, 2009.
Nicaud, J. M., Gaillardin, C., Adamczak, M., and Pignède, G. "Procédé de transformation nonhomologue de *Yarrowia lipolytica*." French Patent FR2782733, 2000.
Nicaud, J. M., Madzak, C., van den Broek, P., Gysler, C., Duboc, P., Niederberger, P., and Gaillardin, C. "Protein expression and secretion in the yeast *Yarrowia lipolytica*." *FEMS Yeast Research* 2 (2002): 371–379.
Niu, H., Jost, L., Sassi, H., Daukandt, M., Rodriguez, C., and Fickers, P. "A quantitative study of methanol/sorbitol co-feeding process of a *Pichia pastoris* Mut+/pAOX1-lacZ strain." *Microbial Cell Factories* 12 (2013): 33.
Ogrydziak, D., Demain, A., and Tannenbaum, S. "Regulation of extracellular protease production in *Candida lipolytica*." *Biochimica and Biophysica Acta* 497 (1977): 525–538.
Ogunjimi, A., Chandler, J., Gooding, C., Recinos, III, A., and Choudary, P. "High-level secretory expression of immunologically active intact antibody from the yeast *Pichia pastoris*." *Biotechnology Letters* 21 (1999): 561–567.

Oh, B. R., Park, J., Kim, M. W., Cheon, S. A., Kim, E. J., Moon, H. Y. et al. "Glycoengineering of the methylotrophic yeast *Hansenula polymorpha* for the production of glycoproteins with trimannosyl core N-glycan by blocking core oligosaccharide assembly." *Biotechnology Journal* 3 (2008): 659–668.

Ohi, H., Okazaki, N., Uno, S., Miura, M., and Hiramatsu, R. "Chromosomal DNA patterns and gene stability of *Pichia pastoris.*" *Yeast* 14 (1998): 895–903.

Olofsson, K., Bertilsson, M., and Lidén, G. "A short review on SSF—An interesting process option for ethanol production from lignocellulosic feedstocks." *Biotechnology for Biofuels* 1 (2008): 1–14.

Ostergaard, H., Tachibana, C., and Winther, J. R. "Monitoring disulfide bond formation in the eukaryotic cytosol." *Journal of Cell Biology* 166 (2004): 337–345.

Pan, R., Zhang, J., Shen, W., Tao, Z., Li, S., Yan, X. "Sequential deletion of Pichia pastoris genes by a self-excisable cassette." *FEMS Yeast Research* 11 (2011): 292–298.

Papendieck, A., Dahlems, U., and Gellissen, G. "Technical enzyme production and whole-cell biocatalysis: Application of *Hansenula polymorpha.*" In G. Gellisen (Ed.), *Hansenula Polymorpha* (pp. 255–271). Weinheim: Wiley-VCH Verlag GmbH & Co. KGaA, 2005.

Park, J. M., Song, Y., Cheon, S. A., Kwon, O., Oh, D. B., Chen, Y. H. et al. "Essential role of YlMPO1, a novel *Yarrowia lipolytica* homologue of *Saccharomyces cerevisiae MNN4*, in mannosylphosphorylation of N- and O-linked glycans." *Applied and Environmental Microbiology* 77 (2011): 1187–1195.

Parpinello, G., Berardi, E., and Strabbioli, R. "A Regulatory mutant of *Hansenula polymorpha* exhibiting methanol utilization metabolism and peroxisome proliferation in glucose." *Journal of Bacteriology* 180 (1998): 2958–2967.

Paulová, L., Hyka, P., Branská, B., Melzoch, K., and Kovar, K. "Use of a mixture of glucose and methanol as substrates for the production of recombinant trypsinogen in continuous cultures with *Pichia pastoris* Mut+." *Journal of Biotechnology* 157 (2012): 180–188.

Penheiter, A., Klucas, R., and Sarath, G. "Purification and characterization of a soybean root nodule phosphatase expressed in *Pichia pastoris.*" *Protein Expression and Purification* 14 (1998): 125–130.

Penninckx, M. "A short review on the role of glutathione in the response of yeasts to nutritional, environmental, and oxidative stresses." *Enzyme and Microbial Technology* 26 (2000): 737–742.

Petschacher, B., Leitgeb, S., Kavanagh, K. L., Wilson, D. K., and Nidetzky, B. "The coenzyme specificity of *Candida tenuis* xylose reductase (AKR2B5) explored by site-directed mutagenesis and X-ray crystallography." *Biochemical Journal* 385 (2005): 75–83.

Phongdara, A., Merckelbach, A., Keup, P., Gellissen, G., and Hollenberg, C. P. "Cloning and characterization of the gene encoding a repressible acid phosphatase (PHO1) from the methylotrophic yeast *Hansenula polymorpha.*" *Applied Microbiology and Biotechnology* 50 (1998): 77–84.

Pignède, G., Wang, H. J., Fudalej, F., Seman, M., Gaillardin, C., and Nicaud, J. M. "Autocloning and amplification of LIP2 in *Yarrowia lipolytica.*" *Applied and Environmental Microbiology* 66 (2000): 3283–3289.

Pócsi, I., Prade, R., and Penninckx, M. "Glutathione, altruistic metabolite in fungi." *Advances in Microbial Physiology* 49 (2004): 1–76.

Potgieter, T., Cukan, M., Drummond, J., Houston-Cummings, N., Jiang, Y., Li, F. et al. "Production of monoclonal antibodies by glycoengineered *Pichia pastoris.*" *Journal of Biotechnology* 139 (2009): 318–325.

Qian, W., Song, H., Liu, Y., Zhang, C., Niu, Z., Wang, H., and Qiu, B. "Improved gene disruption method and Cre-loxP mutant system for multiple gene disruptions in *Hansenula polymorpha.*" *Journal of Microbiological Methods* 79 (2009): 253–259.

Raemaekers, R., de Muro, L, Gatehouse, J., and Fordham-Skelton A. P. "Functional phytohemagglutinin (PHA) and Galanthus nivalis agglutinin (GNA) expressed in Pichia pastoris." *European Journal of Biochemistry* 265 (1999): 394–403.

Ramezani-Rad, M., Hollenberg, C., Lauber, J., Wedler, H., Griess, E., Wagner, C. et al. "The *Hansenula polymorpha* (strain CBS4732) genome sequencing and analysis." *FEMS Yeast Research* 4 (2003): 207–215.

Ramón, R., Ferrer, P., and Valero, F. "Sorbitol co-feeding reduces metabolic burden caused by the overexpression of a Rhizopus oryzae lipase in *Pichia pastoris.*" *Journal of Biotechnology* 130 (2007): 39–46.

Reinders, A., Romano, I., Wiemken, A., and De Virgilio, C. "The thermophilic yeast *Hansenula polymorpha* does not require trehalose synthesis for growth at high temperatures but does for normal acquisition of thermotolerance." *Journal of Bacteriology* 181 (1999): 4665–4668.

Roggenkamp, R. "Constitutive appearance of peroxisomes in a regulatory mutant of the methylotrophic yeast *Hansenula polymorpha.*" *Molecular and General Genetics* 213 (1988): 535–540.

Romanos, M. "Advances in the use of *Pichia pastoris* for high-level gene expression." *Current Opinion in Biotechnology* 6 (1995): 527–533.

Romanos, M., Scorer, C., and Clare, J. "Foreign gene expression in yeast: A review." *Yeast* 8 (1992): 423–488.

Rosenfeld, S. "Use of *Pichia pastoris* for expression of recombinant proteins." In M. C. S. J. C. Glorioso (Ed.), *Methods in Enzymology* (Vol. 306, pp. 154–169). Pasadena, USA: Academic Press, 1999.

Rosenfeld, S., Nadeau, D., Tirado, J., Hollis, G., Knabb, R., and Jia, S. "Production and purification of recombinant hirudin expressed in the methylotrophic yeast *Pichia pastoris.*" *Protein Expression and Purification* 8 (1996): 476–482.

Ryabova, O. B., Chmil, O. M., and Sibirny, A. "Xylose and cellobiose fermentation to ethanol by the thermotolerant methylotrophic yeast *Hansenula polymorpha.*" *FEMS Yeast Research* 4 (2003): 157–164.

Sabirova, J., Haddouche, R., Van Bogaert, I., Mulaa, F., Verstraete, W., Timmis, K. et al. "The "LipoYeasts" project: Using the oleaginous yeast *Yarrowia lipolytica* in combination with specific bacterial genes for the bioconversion of lipids, fats and oils into high-value products." *Microbial Biotechnology* 4 (2011): 47–54.

Sahm, H. "Metabolism of methanol by yeasts." *Advances in Biochemical Engineering* 6 (1977): 77–103.

Sasano, Y., Yurimoto, H., Yanaka, M., and Sakai, Y. "Trm1p, a Zn(II)2Cys6-type transcription factor, Is a master regulator of methanol-specific gene activation in the methylotrophic yeast *Candida boidinii.*" *Eukaryotic Cell* 7 (2008): 527–536.

Schaefer, S., Piontek, M., Ahn, S. J., Papendieck, A., Janowicz, Z. A., and Gellissen, G. "Recombinant hepatitis B vaccines-characterization of the viral disease and vaccine production." In G. Gellisen (Ed.), *Hansenula Polymorpha. Biology and Applications* (pp. 245–274). Weinheim: Wiley-VCH Verlag GmbH, 2007.

Schmid-Berger, N., Schmid, B., and Barth, G. "Ylt1, a highly repetitive retrotransposon in the genome of the dimorphic fungus *Yarrowia lipolytica.*" *Journal of Bacteriology* 176 (1994): 2477–2482.

Schuller, H. J. "Transcriptional control of nonfermentative metabolism in the yeast *Saccharomyces cerevisiae.*" *Current Genetics* 43 (2003): 139–160.

Scorer, C., Clare, J., McCombie, W., Romanos, M., and Sreekrishna, K. "Rapid selection using G418 of high copy number transformants of *Pichia pastoris* for high-level foreign gene expression." *Nature Biotechnology* (1994): 181–184.

Sears, I., O'Connor, J., Rossanese, O., and Glick, B. "A versatile set of vectors for constitutive and regulated gene expression in *Pichia pastoris.*" *Yeast* 14 (1998): 783–790.

Shen, S., Sulter, G., Jeffries, T., and Cregg, J. "A strong nitrogen source-regulated promoter for controlled expression of foreign genes in the yeast *Pichia pastoris.*" *Gene* 216 (1998): 93–102.

Shoshan, M. "3-bromopyruvate: Targets and outcomes." *Journal of Bioenergetics and Biomembranes* 44 (2012): 7–15.

Sibirny, A., Ubiyvovk, V., Gonchar, M., Titorenko, V., Voronovsky, A., Kapultsevich, Y., and Bliznik, K. "Reactions of direct formaldehyde oxidation to CO2 are non-essential for energy supply of yeast methylotrophic growth." *Archives of Microbiology* 154 (1990): 566–575.

Sies, H. "Glutathione and its role in cellular functions." *Free Radical Biology and Medicine* 27 (1999): 916–921.

Soderholm, J., Bevis, B., and Glik, B. "Vector for pop-in/pop-out gene replacement in *Pichia pastoris.*" *BioTechniques* 31 (2001): 306–310, 312.

Sohn, J., Choi, E., Kang, H., Rhee, J., Agaphonov, M., Ter-Avanesyan, M., and Rhee, S. "A dominant selection system designed for copy-number-controlled gene integration in *Hansenula polymorpha* DL-1." *Applied Microbiology and Biotechnology* 51 (1999): 800–807.

Sohn, J. H., Choi, E. S., Kim, C. H., Agaphonov, M. O., Ter-Avanesyan, M. D., Rhee, J. S., and Rhee, S. K. "A novel autonomously replicating sequence (ARS) for multiple integration in the yeast *Hansenula polymorpha* DL-1." *Journal of Bacteriology* 178 (1996): 4420–4428.

Song, H., Li, Y., Fang, W., Geng, Y., Wang, X., Wang, M., and Qiu, B. "Development of a set of expression vectors in *Hansenula polymorpha*." *Biotechnology Letters* 25 (2003): 1999–2006.

Song, Y., Choi, M. H., Park, J., Kim, M. Y., Kim, E. J., Kang, H. A., and Kim, J. Y. "Engineering of the yeast *Yarrowia lipolytica* for the production of glycoproteins lacking the outer-chain mannose residues of N-glycans." *Applied and Environmental Microbiology* 73 (2007): 4446–4454.

Sreekrishna, K., Nelles, L., Potenz, R., Cruze, J., Mazzaferro, P., Fish, W. et al. "High-level expression, purification, and characterization of recombinant human tumor necrosis factor synthesized in the methylotrophic yeast *Pichia pastoris*." *Biochemistry* 28 (1989): 4117–4125.

Stasyk, O., Maidan, M., Stasyk, O., Van Dijck, P., Thevelein, J., and Sibirny, A. "Identification of hexose transporter-like sensor HXS1 and functional hexose transporter *HXT1* in the methylotrophic yeast *Hansenula polymorpha*." *Eukaryotic Cell* 7 (2008): 735–746.

Stasyk, O., Stasyk, O., Komduur, J., Veenhuis, M., Cregg, J., and Sibirny, A. "A Hexose transporter homologue controls glucose repression in the methylotrophic yeast *Hansenula polymorpha*." *Journal of Biological Chemistry* 279 (2004): 8116–8125.

Stasyk, O., Van Zutphen, T., Ah Kang, H., Stasyk, O., Veenhuis, M., and Sibirny, A. "The role of *Hansenula polymorpha MIG1* homologues in catabolite repression and pexophagy." *FEMS Yeast Research* 7 (2007): 1103–1113.

Stasyk, O. V., Ksheminskaya, G. P., Kulachkovskii, A. R., and Sibirnyi, A. A. "Mutants of the methylotrophic yeast *Hansenula polymorpha* with impaired catabolite repression." *Mikrobiologiya* 66 (1997a): 755–760.

Stasyk, O. V., Ksheminskaya, G. P., Kulachkovskii, A. R., and Sibirnyi, A. A. "Mutants of the methylotrophic yeast *Hansenula polymorpha* with impaired catabolite repression." *Microbiology (Moscow)* 66 (1997b): 631–636.

Stasyk, O. V., and Sybirnyi, A. A. "Molecular mechanisms of catabolic repression in yeast." *Mikrobiolohichnyi Zhurnal (Kiev)* 65 (2003): 84–103.

Steinborn, G., Böer, E., Scholz, A., Tag, K., Kunze, G., and Gellissen, G. "Application of a wide-range yeast vector (CoMed™) system to recombinant protein production in dimorphic *Arxula adeninivorans*, methylotrophic *Hansenula polymorpha* and other yeasts." *Microbial Cell Factories* 5 (2006): 1–13.

Steinborn, G., Gellissen, G., and Kunze, G. "Assessment of *Hansenula polymorpha* and *Arxula adeninivorans*-derived rDNA-targeting elements for the design of *Arxula adeninivorans* expression vectors." *FEMS Yeast Research* 5 (2005): 1047–1054.

Stratton, J., Chiruvolu, V., and Meagher, M. "High cell-density fermentation." In D. Higgins and J. Bhagat (Eds.), *Pichia Protocols* (pp. 107–120). Totowa, NJ: Humana Press, 1998.

Sudbery, P. "*Hansenula polymorpha*: Biology and application." *Yeast* 20 (2003): 1307–1308.

Suh, S. O., and Zhou, J. J. "Methylotrophic yeasts near *Ogataea* (*Hansenula*) *polymorpha*: A proposal of *Ogataea angusta* comb. nov. and *Candida parapolymorpha* sp. nov." *FEMS Yeast Research* 10 (2010): 631–638.

Sunga, A., and Cregg, J. "The *Pichia pastoris* formaldehyde dehydrogenase gene (FLD1) as a marker for selection of multicopy expression strains of *P. pastoris*." *Gene* 330 (2004): 39–47.

Swennen, D., Paul, M. F., Vernis, L., Beckerich, J. M., Fournier, A., and Gaillardin, C. "Secretion of active anti-Ras single-chain Fv antibody by the yeasts *Yarrowia lipolytica* and *Kluyveromyces lactis*." *Microbiology* 148 (2002): 41–50.

Szöőr, B., Gross, S., and Alphey, L. "Biochemical characterization of recombinant Drosophila type 1 serine/threonine protein phosphatase (PP1c) roduced in *Pichia pastoris*." *Archives of Biochemistry and Biophysics* 396 (2001): 213–218.

Talmont, F., Sidobre, S., Demange, P., Milon, A., and Emorine, L. "Expression and pharmacological characterization of the human μ-opioid receptor in the methylotrophic yeast *Pichia pastoris*." *FEBS Letters* 394 (1996): 268–272.

Tanaka, H., Okuno, T., Moriyama, S., Muguruma, M., and Ohta, K. "Acidophilic xylanase from Aureobasidium pullulans: Efficient expression and secretion in *Pichia pastoris* and mutational analysis." *Journal of Bioscience and Bioengineering* 98 (2004): 338–343.

Teunison, D. J., Hall, H., and Wickerham, L. "*Hansenula angusta*, an excellent species for demonstration of the coexistence of haploid and diploid cells in homothalic yeast." *Mycologia* 52 (1960): 184–188.

Thomas, D., Jacquemin, I., and Surdin-Kerjan, Y. "*MET4*, a leucine zipper protein, and centromere-binding factor 1 are both required for transcriptional activation of sulfur metabolism in *Saccharomyces cerevisiae*." *Molecular and Cellular Biology* 12 (1992): 1719–1727.
Thongekkaew, J., Ikeda, H., Masaki, K., and Iefuji, H. "An acidic and thermostable carboxymethyl cellulase from the yeast Cryptococcus sp. S-2: Purification, characterization and improvement of its recombinant enzyme production by high cell-density fermentation of *Pichia pastoris*." *Protein Expression and Purification* 60 (2008): 140–146.
Thor, D., Xiong, S., Orazem, C., Kwan, A. C., Cregg, J., Cereghino, J., and Cereghino, G. "Cloning and characterization of the *Pichia pastoris MET2* gene as a selectable marker." *FEMS Yeast Research* 5 (2005): 935–942.
Tiels, P., Baranova, E., Piens, K., De Visscher, C., Pynaert, G., Nerinckx, W. et al. "A bacterial glycosidase enables mannose-6-phosphate modification and improved cellular uptake of yeast-produced recombinant human lysosomal enzymes." *Nature Biotechnology* 30 (2012): 1225–1231.
Tikhomirova, L. P., Ikonomova, R. N., and Kuznetsova, E. N. "Evidence for autonomous replication and stabilization of recombinant plasmids in the transformants of yeast *Hansenula polymorpha*." *Current Genetics* 10 (1986): 741–747.
Tikhomirova, L. P., Ikonomova, R. N., Kuznetsova, E. N., Fodor, I., Bystrykh, L. V., Aminova, L. R., and Trotsenko, Yu. A. "Transformation of methylotrophic yeast *Hansenula polymorpha*: Cloning and expression of genes." *Journal of Basic Microbiology* 28 (1988): 343–351.
Tull, D., Gottschalk, T., Svendsen, I., Kramhøft, B., Phillipson, B., Bisgård-Frantzen, H. et al. "Extensive N-Glycosylation reduces the thermal stability of a recombinant alkalophilic *Bacillus* α-amylase produced in *Pichia pastoris*." *Protein Expression and Purification* 21 (2001): 13–23.
Turcotte, B., Liang, X. B., Robert, F., and Soontorngun, N. "Transcriptional regulation of nonfermentable carbon utilization in budding yeast." *FEMS Yeast Research* 10 (2010): 2–13.
Ubiivovk, R. M., and Trotsenko, Yu. A. "Regulation of formaldehyde level in the methylotrophic yeast *Candida boidinii*." *Mikrobiologiya* 55 (1986): 181–185.
Ubiivovk, V. M., Telegus, Y. V., and Sibirnyi, A. A. "Isolation and characterization of glutathione-deficient mutants of the methylotrophic yeast *Hansenula polymorpha*." *Microbiology (Moscow)* 68 (1999): 26–31.
Ubiyvovk, V., Ananin, V., Malyshev, A., Kang, H., and Sibirny, A. "Optimization of glutathione production in batch and fed-batch cultures by the wild-type and recombinant strains of the methylotrophic yeast *Hansenula polymorpha* DL-1." *BMC Biotechnology* 11 (2011a): 1–12.
Ubiyvovk, V., Blazhenko, O., Gigot, D., Penninckx, M., and Sibirny, A. "Role of gamma-glutamyltranspeptidase in detoxification of xenobiotics in the yeasts *Hansenula polymorpha* and *Saccharomyces cerevisiae*." *Cell Biology International* 30 (2006): 665–671.
Ubiyvovk, V., Maszewski, J., Bartosz, G., and Sibirny, A. "Vacuolar accumulation and extracellular extrusion of electrophilic compounds by wild-type and glutathione-deficient mutants of the methylotrophic yeast *Hansenula polymorpha*." *Cell Biology International* 27 (2003): 785–789.
Ubiyvovk, V., Nazarko, T., Stasyk, O., Sohn, M. J., Kang, H. A., and Sibirny, A. "GSH2, a gene encoding γ-glutamylcysteine synthetase in the methylotrophic yeast *Hansenula polymorpha*." *FEMS Yeast Research* 2 (2002): 327–332.
Ubiyvovk, V. M., Blazhenko, O. V., Zimmermann, M., Sohn, M. J., and Kang, H. A. "Cloning and functional analysis of the *GSH1/MET1* gene complementing cysteine and glutathione auxotrophy of the methylotrophic yeast *Hansenula polymorpha*." *Ukrainskii Biokhimicheskii Zhurnal* 83 (2011b): 67–81.
van Der Heide, M., Leão, A., Van der Klei, I., and Veenhuis, M. "Redirection of peroxisomal alcohol oxidase of *Hansenula polymorpha* to the secretory pathway." *FEMS Yeast Research* 7 (2007): 1093–1102.
van der Klei, I., Yurimoto, H., Sakai, Y., and Veenhuis, M. "The significance of peroxisomes in methanol metabolism in methylotrophic yeast." *Biochimica et Biophysica Acta* 1763 (2006): 1453–1462.
van Dijk, R., Faber, K., Kiel, J., Veenhuis, M., and van der Klei, I. "The methylotrophic yeast *Hansenula polymorpha*: A versatile cell factory." *Enzyme and Microbial Technology* 26 (2000): 793–800.
van Dijk, R., Faber, K. N., Hammond, A. T., Glick, B. S., Veenhuis, M., and Kiel, J. A. "Tagging *Hansenula polymorpha* genes by random integration of linear DNA fragments (RALF)." *Molecular Genetics and Genomics* 266 (2001): 646–656.

Van Heeke, G., Ott, T., Strauss, A., Ammaturo, D., and Bazer, F. "High yield expression and secretion of the ovine pregnancy recognition hormone interferon-τ by *Pichia pastoris*." *Journal of Interferon and Cytokine Research* 16 (1996): 119–126.
van Zutphen, T., Baerends, R., Susanna, K., de Jong, A., Kuipers, O., Veenhuis, M., and Van Der Klei, I. "Adaptation of *Hansenula polymorpha* to methanol: A transcriptome analysis." *BMC Genomics* 11 (2010): 1.
Veenhuis, M., van Dijken, J. P., and Harder, W. "The significance of peroxisomes in the metabolism of one-carbon compounds in yeast." *Advance in Microbial Physiology* 24 (1983): 1–82.
Vervecken, W., Kaigorodov, V., Callewaert, N., Geysens, S., Vusser, K., and Contreras, R. "In vivo Synthesis of mammalian-like, hybrid-type N-Glycans in *Pichia pastoris*." *Applied and Environmental Microbiology* 70 (2004): 2639–2646.
Vogl, T., and Glieder, A. "Regulation of *Pichia pastoris* promoters and its consequences for protein production." *New Biotechnology* 30 (2013): 385–404.
Voronovsky, A., Rohulya, O., Abbas, C., and Sibirny, A. "Development of strains of the thermotolerant yeast *Hansenula polymorpha* capable of alcoholic fermentation of starch and xylan." *Metabolic Engineering* 11 (2009): 234–242.
Voronovsky, A., Ryabova, O., Verba, O., Ishchuk, O., Dmytruk, K., and Sibirny, A. "Expression of xylA genes encoding xylose isomerases from *Escherichia coli* and *Streptomyces coelicolor* in the methylotrophic yeast *Hansenula polymorpha*." *FEMS Yeast Research* 5 (2005): 1055–1062.
Wang, R., Li, L., Zhang, B., Gao, X., Wang, D., and Hong, J. "Improved xylose fermentation of *Kluyveromyces marxianus* at elevated temperature through construction of a xylose isomerase pathway." *Journal of Industrial Microbiology and Biotechnology* 40 (2013): 841–854.
Wang, Y., Liang, Z. H., Zhang, Y. S., Yao, S. Y., Xu, Y., Tang, Y. H. et al. "Human insulin from a precursor overexpressed in the methylotrophic yeast *Pichia pastoris* and a simple procedure for purifying the expression product." *Biotechnology and Bioengineering* 73 (2001): 74–79.
Wang, Z., Tan, T., and Song, J. "Effect of amino acids addition and feedback control strategies on the high-cell-density cultivation of *Saccharomyces cerevisiae* for glutathione production." *Process Biochemistry* 42 (2007): 108–111.
Wang, Z., Wang, Y., Zhang, D., Li, J., Hua, Z., Du, G., and Chen, J. "Enhancement of cell viability and alkaline polygalacturonate lyase production by sorbitol co-feeding with methanol in *Pichia pastoris* fermentation." *Bioresource Technology* 101 (2010): 1318–1323.
Waterham, H., Digan, M. E., Koutz, P., Benjamin, S., and Cregg, J. "Isolation of the *Pichia pastoris* glyceraldehyde-3-phosphate dehydrogenase gene and regulation and use of its promoter." *Gene* 186 (1997): 37–44.
Weiβ, M., Haase, W., Michel, H., and Reiländer, H. "Expression of functional mouse 5-HT5A serotonin receptor in the methylotrophic yeast *Pichia pastoris:* Pharmacological characterization and localization." *FEBS Letters* 377 (1995): 451–456.
Weydemann, U., Keup, P., Piontek, M., Strasser, A. W. M., Schweden, J., Gellissen, G., and Janowicz, Z. A. "High-level secretion of hirudin by *Hansenula polymorpha*—Authentic processing of three different preprohirudins." *Applied Microbiology and Biotechnology* 44 (1995): 377–385.
Williams, K., Jiang, J., Ju, J., and Olsen, D. "Novel strategies for increased copy number and expression of recombinant human gelatin in *Pichia pastoris* with two antibiotic markers." *Enzyme and Microbial Technology* 43 (2008): 31–34.
Xiong, A. S., Yao, Q.-H., Peng, R. H., Zhang, Z., Xu, F., Liu, J. G. et al. "High level expression of a synthetic gene encoding *Peniophora lycii* phytase in methylotrophic yeast *Pichia pastoris*." *Applied Microbiology and Biotechnology* 72 (2006): 1039–1047.
Xiong, A.-S., Yao, Q.-H., Peng, R.-H., Han, P.-L., Cheng, Z.-M., and Li, Y. "High level expression of a recombinant acid phytase gene in *Pichia pastoris*." *Journal of Applied Microbiology* 98 (2005): 418–428.
Xiong, R., Chen, J., and Chen, J. "Secreted expression of human lysozyme in the yeast *Pichia pastoris* under the direction of the signal peptide from human serum albumin." *Biotechnology and Applied Biochemistry* 51 (2008): 129–134.
Xuan, J. W., Fournier, P., and Gaillardin, C. "Cloning of the *LYS5* gene encoding saccharopine dehydrogenase from the yeast *Yarrowia lipolytica* by target integration." *Current Genetics* 14 (1988): 15–21.

Xue, Z., Sharpe, P., Hong, S. P., Yadav, N., Xie, D., Short, D. et al. "Production of omega-3 eicosapentaenoic acid by metabolic engineering of *Yarrowia lipolytica*." *Nature Biotechnology* 31 (2013): 734–740.
Yamada, Y., Maeda, K., and Mikata, K. "The phylogenetic relationships of the hat-shaped ascospore-forming, nitrate-assimilating *Pichia* Species, formerly classified in the genus *Hansenula* SYDOW et SYDOW, based on the partial sequences of 18S and 26S ribosomal RNAs (Saccharomycetaceae): The proposals of three new genera, Ogataea, Kuraishia, and Nakazawaea." *Bioscience, Biotechnology and Biochemistry* 58 (1994): 1245–1257.
Yang, C. H., Huang, Y. C., Chen, C. Y., and Wen, C. Y. "Expression of *Thermobifida fusca* thermostable raw starch digesting alpha-amylase in *Pichia pastoris* and its application in raw sago starch hydrolysis." *Journal of Industrial Microbiology and Biotechnology* 37 (2010): 401–406.
Yang, J., Jiang, W., and Yang, S. "mazF as a counter-selectable marker for unmarked genetic modification of *Pichia pastoris*." *FEMS Yeast Research* 9 (2009a): 600–609.
Yang, X. S., Jiang, Z. B., Song, H. T., Jiang, S. J., Madzak, C., and Ma, L.-X. "Cell-surface display of the active mannanase in *Yarrowia lipolytica* with a novel surface-display system." *Biotechnology and Applied Biochemistry* 54 (2009b): 171–176.
Yano, T., Takigami, E., Yurimoto, H., and Sakai, Y. "Yap1-regulated glutathione redox system curtails accumulation of formaldehyde and reactive oxygen species in methanol metabolism of *Pichia pastoris*." *Eukaryotic Cell* 8 (2009): 540–549.
Ye, J., Ly, J., Watts, K., Hsu, A., Walker, A., McLaughlin, K. et al. "Optimization of a glycoengineered *Pichia pastoris* cultivation process for commercial antibody production." *Biotechnology Progress* 27 (2011): 1744–1750.
Ye, R., Sharpe, P., and Zhu, Q. "Bioengineering of oleaginous yeast *Yarrowia lipolytica* for lycopene production." In J. L. Barredo (Ed.), *Microbial Carotenoids from Fungi* (pp. 153–159). New York, USA: Humana Press, 2012.
Yokoyama, S. "Protein expression systems for structural genomics and proteomics." *Current Opinion in Chemical Biology* 7 (2003): 39–43.
Yu, M., Lange, S., Richter, S., Tan, T., and Schmid, R. "High-level expression of extracellular lipase Lip2 from *Yarrowia lipolytica* in *Pichia pastoris* and its purification and characterization." *Protein Expression and Purification* 53 (2007): 255–263.
Yue, L., Chi, Z., Wang, L., Liu, J., Madzak, C., Li, J., and Wang, X. "Construction of a new plasmid for surface display on cells of *Yarrowia lipolytica*." *Journal of Microbiological Methods* 72 (2008): 116–123.
Yuzbasheva, E., Yuzbashev, T., Laptev, I., Konstantinova, T., and Sineoky, S. "Efficient cell surface display of Lip2 lipase using C-domains of glycosylphosphatidylinositol-anchored cell wall proteins of *Yarrowia lipolytica*." *Applied Microbiology and Biotechnology* 91 (2011): 645–654.
Zalai, D., Dietzsch, C., Herwig, C., and Spadiut, O. "A dynamic fed batch strategy for a *Pichia pastoris* mixed feed system to increase process understanding." *Biotechnology Progress* 28 (2012): 878–886.
Zhang, W., Potter, K., Plantz, B., Schlegel, V., Smith, L., and Meagher, M. "*Pichia pastoris* fermentation with mixed-feeds of glycerol and methanol: Growth kinetics and production improvement." *Journal of Industrial Microbiology and Biotechnology* 30 (2003): 210–215.
Zhang, W., Sinha, J., Smith, L., Inan, M., and Meagher, M. "Maximization of production of secreted recombinant proteins in *Pichia pastoris* fed-batch fermentation." *Biotechnology Progress* 21 (2005): 386–393.
Zhao, C. H., Cui, W., Liu, X. Y., Chi, Z. M., and Madzak, C. "Expression of inulinase gene in the oleaginous yeast *Yarrowia lipolytica* and single cell oil production from inulin-containing materials." *Metabolic Engineering* 12 (2010): 510–517.
Zhu, T., You, L., Gong, F., Xie, M., Xue, Y., Li, Y., and Ma, Y. "Combinatorial strategy of sorbitol feeding and low-temperature induction leads to high-level production of alkaline β-mannanase in *Pichia pastoris*." *Enzyme and Microbial Technology* 49 (2011): 407–412.
Zhuge, B., Du, G. C., Shen, W., Zhuge, J., and Chen, J. "Expression of a *Bacillus subtilis* pectate lyase gene in *Pichia pastoris*." *Biochemical Engineering Journal* 40 (2008): 92–98.
Zwart, K., Veenhuis, M., van Dijken, J. P., and Harder, W. "Development of amine oxidase-containing peroxisomes in yeast during growth on glucose in the presence of methylamine as the sole source of nitrogen." *Archives of Microbiology* 126 (1980): 117–126.

chapter five

Induced anhydrobiosis

Powerful method for preservation of industrial microorganisms

Armando Hernández García

Contents

5.1 *Introduction*

The preservation of microorganisms has been a challenge since the beginning of mankind. Some strains of beneficial microorganisms are involved in the production of dairy, bakery, spirits, alcohol, vaccines, antibiotics, enzymes, silage, vinegar, and others (Uzunova-Doneva and Donev 2005). There are several methods for the preservation of industrial microorganisms, for example, subcultivation, use of mineral oils, water–salt solutions, cryogenic conservation, and drying (Uzunova-Doneva and Donev 2005). Among these methods, the preservation of microorganisms by desiccation has been the preferred method for long-term storage (Morgan et al. 2006).

Nowadays, there are big culture collections that depend on drying methods for preserving their microorganisms. In addition, there are many industrial applications of microorganisms preserved in aliquots. Microbiological quality control is necessary in the pharmaceutical, food, and beverage industries. It is also evident in clinical microbiology, which is performed by using samples of microbial strains acting as positive controls. These strains are preserved through drying, which increases their stability for transportation at room temperature. Moreover, emerging fields such as the probiotic and biocontrol industries require the conservation of certain microorganisms (Garcia 2011; Morgan et al. 2006).

Anhydrobiosis is the state at which an organism stops its vital functions temporarily due to partial or total desiccation. This state is characterized by the extreme reduction of mensurable metabolism (Rebecchi et al. 2007). A quantitative definition of the total desiccation states that the water content in an anhydrobiote is less than 0.1 g of free water per gram of dry cell weight (Alpert 2005).

Concerning the industrial importance of anhydrobiosis, the worldwide market in the stabilization of cells and cell products is approximately $500 billion (Garcia 2011; Potts et al. 2005). Moreover, induced anhydrobiosis is applied in the preservation of reference strains used in the quality control of pharmaceutical and food industries, the conservation of reference strains from culture collections, and the generation of biotechnological products in the emerging fields of probiotics and biocontrol (Morgan et al. 2006).

In the fields of probiotics and biocontrol, an intense amount of work has been performed during the last decade using the technologies of freeze-drying and spray-drying (Ananta et al. 2005; Arunsiri et al. 2003; Carvalho et al. 2004; Corcoran et al. 2006; Fravel 2005; Hernández et al. 2006, 2007; Kuang et al. 2010; Lauten et al. 2010; Nag and Das 2012; Strasser et al. 2009; Ziadi et al. 2005). These technologies show relevant results in the use of bacterial dehydration as a powerful tool for future applications in agriculture, food industry, and medicine.

According to recent reports, to achieve high survival rates, a combination of physiology and technology should be considered (Garcia 2011; Fu and Chen 2011). Thus, sample preparation and optimal drying conditions should be rigourosuly selected for long-term stabilization of desiccated cells (Garcia 2011; Julca et al. 2012).

Considering the technological approach, although freeze-drying is the most widely used technology, it is, however, the most expensive one (Fu and Chen 2011; Hernández et al. 2007). Thus, nowadays, spray-drying and fluidized bed drying are being evaluated by many reseachers worldwide as potential drying techniques to be used as alternatives for freeze-drying in the preservation of several microbial species (Fu and Chen 2011; Morgan et al. 2006).

With respect to the mechanisms of desiccation tolerance, most studies have been performed mainly on prokaryotes (Potts 1994). Although, up to date, little is known about such mechanisms. Research suggests that small changes of energy are involved in the decrease of viability, and that those cells sensitive to desiccation could lose an important fraction of free water (Potts et al. 2005).

Among the methodologies used to improve desiccation tolerance, we found: (1) selection—training through multiple exposure to stress; (2) accumulation of intracellular protective agents—through uptake or through genetic engineering; (3) use of extracellular protective agents—as additives or secretion through genetic engineering; and (4) manipulation of cell metabolism/physiology—before exposure to stress (Potts et al. 2005).

In this chapter, we focus our attention on the induced anhydrobiosis phenomenon, and the practical implications of the drying processes in the preservation of microorganisms.

5.2 *Anhydrobiosis in microorganisms*

Desiccation tolerance is generally defined as the ability of an organism to reach desiccation up to a state of equilibrium with moderately or extremely dry air, and then the recovery of its normal functions after rehydration (Alpert 2005). According to França et al. (2007), the term *desiccation tolerance* is not the same as *drought tolerance*. Drought is related to the moderate dehydration of a microorganism, whereas desiccation is related to the extreme dehydration of cells when the hydration shell of molecules is gradually lost (França et al. 2007).

Most microorganisms with industrial applications are bacteria; as a consequence, the anhydrobiosis phenomenon has been studied mainly in prokaryotes (Garcia 2011; Potts et al. 2005). The minimal water content or hydric potential for survival in these desiccation-tolerant cells is extremely low. Thus far, a decrease of up to 2% (w/w) in the water content of bacterial cells has been reported (Potts 1994).

5.3 Mechanisms of desiccation damage

Microbial cells, under desiccation, are subjected to several stress processes once they reach the anhydrobiosis state. One of the most important processes related to the desiccated state is oxidative stress. Little is known in relation to the damage caused by oxidative stress as a consequence of dehydration. Although oxygen is a vital compound for aerobic organisms, this substance could be harmful in excess (França et al. 2007). Figure 5.1 shows the mechanisms of desiccation damage. The chemical damage caused by free radicals is one of the main reasons for desiccation lesions. The water stress increases reactive oxygen species (ROS) formation, which produces lipid peroxidation, protein denaturation, and damage in nucleic acids with drastic consequences in the general metabolism (Hansen et al. 2006).

Aerobic organisms use oxygen as electron acceptors. Nevertheless, during respiration, this compound could be partially reduced by forming ROS, such as superoxide anions (O_2^-), hydrogen peroxide (H_2O_2), and hydroxyl radicals (OH•). The balance between the ROS production and the cellular defenses determines the degree of oxidative stress (França et al. 2007). According to Pereira et al. (2003), yeast cells increase the oxidation process by more than 10-fold when they are dehydrated.

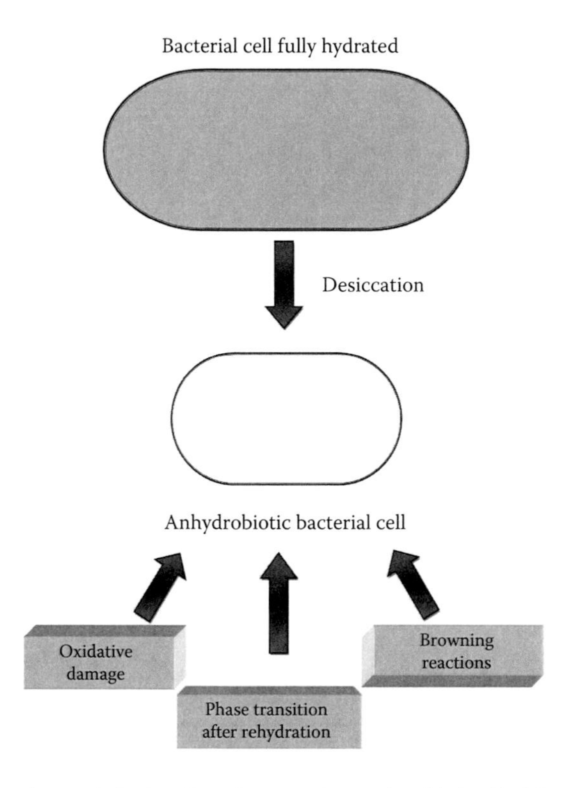

Figure 5.1 Main mechanisms of desiccation damage in a microbial cell. A bacterial cell is shown as the model.

Changes in the water content of the cells could produce disfunctions in specific enzymes, or promote the development of chemical reactions that normally would not occur in a fully hydrated cell. Thus, under normal metabolic conditions, free radicals would be trapped by the antioxidant defense systems. However, under water stress, these mechanisms could be inactive (França et al. 2007).

Once the cell membranes are dehydrated, they are more susceptible to ROS attack (Crowe et al. 1989). These radicals frequently cause an extensive peroxidation and de-esterification of membrane lipids for dehydration levels of up to 50% (Senaratna et al. 1987). Furthermore, an increase in the packing density of the polar heads of the membrane phospholipids takes place, leading to the strengthening of the van der Waals interactions between the carbon chains (Crowe et al. 1992). Consequently, the phase transition temperature (T_m) increases significantly, and lipids are in gel phase at room temperature. Then, when these lipids are rehydrated, they undergo a phase transition producing extensive leakage and cell death (Potts 1994).

Desiccation-tolerant cells have mechanisms for decreasing the T_m value, and thus preventing phase transition. A decrease in the T_m value could be reached by increasing the unsaturation degree of fatty acid membranes (Hoekstra et al. 2001). However, desiccated cells using this mechanism are more susceptible to lipid oxidation (França et al. 2007; Hoekstra et al. 2001). In addition, the polar heads of the phospholipids generate ROS through Fe^{2+} auto-oxidation (Motta and Sechi 1992).

Protein dehydration induces significant conformational changes that were detected by Fourier transform infrared spectroscopy (Prestrelski et al. 1993). The drying, by itself, could produce protein denaturation, which causes a loss of biological activity after rehydration (França et al. 2007). On the other hand, protein oxidation catalyzed by metals is an additional consequence of cellular desiccation. Oxygen reduction directly produces H_2O_2 or superoxide as an intermediary, which reacts with Fe^{2+} to yield Fe^{3+} and oxygen. The ion Fe^{2+} binds a metal-binding site in the protein, and the complex protein-Fe^{2+} reacts with H_2O_2 to generate ferryl ions, leading to reactions with protein side chains (França et al. 2007; Potts 1994).

After oxidative modification, proteins become more sensitive to proteolysis and may be inactivated or decrease their biological activity (Potts 1994). Some amino acid residues are converted to carbonyl derivatives and the carbonylated proteins accumulate in the desiccated cells (Fredrickson et al. 2008; Yin and Chen 2005). In fact, most recent experimental evidences indicate that protein oxidation seems to be mainly responsible for the desiccation damage in bacterial cells (Fredrickson et al. 2008). Protection of proteins against ROS damage is equally important (or more critical) for maintaining ancient bacteria in a viable state than DNA repair.

According to Potts et al. (2005), drastic changes occur in bacterial populations exposed to air desiccation. Among these changes, the most remarkable are related to increases in surface and saline precipitation, as well as changes in shape, color, and texture. Also, important changes occur at the cellular level, affecting the bacterial cell physically and chemically. Consequently, physiological changes will be observed. For instance, the thickness of the cell wall is decreased due to the packing of the peptidoglycan layers after dehydration (Hernández 2009). Simultaneously, the intracellular salt concentration is increased, producing an agglomeration of biopolymers. In addition, intracellular viscosity is increased causing low rates of diffusion, enzymatic activity is reduced, and finally, growth stops.

For the survival of microbial cells, DNA must be fully hydrated, which ensures its chemical stability (Potts 1994; Potts et al. 2005). Thus, changes in the DNA hydration patterns could affect several molecular processes, for example, replication, transcription, and protein synthesis, leading to the disruption of cellular equilibrium. After the desiccation process, many harmful reactions occur; browning reactions could affect the stability of DNA and proteins. In fact, the Maillard reaction produces melanoidins through condensation between carbonyl groups of saccharides and the amino group of proteins and nucleic acids (Potts et al. 2005). This chemical change in DNA could generate abnormalities that inhibit the copying process performed by polymerases. Other undesired reactions such as Haber–Weiss and Fenton reactions produce OH•, this radical is the most reactive species among ROS, causing oxidation of pyrimidines, enhancing mutation rates, and leading to cell death (Kranner and Birtic 2005).

5.4 Mechanisms of desiccation tolerance

Some radiation-tolerant bacterial species, for example, *Deinococcus radiodurans* and cyanobacteria such as *Nostoc commune* and *Chroococcidiopsis* spp. are considered anhydrobiotes (Mattimore and Battista 1996; Potts et al. 2005; Rebecchi et al. 2007). *N. commune* did not show significant damage in DNA molecules after 13 years of desiccation (Shirkey et al. 2003). Also, tolerant cyanobacteria in the desert of Negev can recover up to 50% of its normal activity in photosystem II in only 2 min after rehydration (Harel et al. 2004). Recently, Slade and Radman (2011) have illustrated the remarkable correlation between desiccation tolerance and resistance to ionizing radiation in *D. radiodurans*. Considering the data presented by these authors, the survival curves for both gamma radiation and desiccation show identical shapes for survival rates between 100% and 10%. This evidence suggests the hypothesis that desiccation tolerance and radiation resistance could be molecularly regulated by similar mechanisms.

In general, Gram-positive bacteria are the most desiccation-tolerant bacteria (Potts et al. 2005). Many Gram-positive species accumulate Mn^{2+}, have a high Mn^{2+}/Fe^{2+} ratio, and are resistant to radiation (Daly et al. 2004). On the other hand, cyanobacteria, despite having a similar cell wall to Gram-negative bacteria, show phylogenetic affinity to Gram-positive species, being markedly tolerant of both desiccation and gamma radiation (Billi et al. 2000; Potts 1994). Recently, the intracellular accumulation Mn^{2+} has been proposed as the most probable mechanism involved in the protection of proteins from oxidative damage after desiccation (Fredrickson et al. 2008). These researchers found a strong correlation between the intracellular levels of Mn and the desiccation tolerance in several bacterial genera.

In Gram-negative bacteria such as *Escherichia coli* and *Pseudomonas putida*, the induction of desiccation tolerance was achieved using trehalose and hydroxyectoine (de Castro et al. 2000; Manzanera et al. 2002, 2004; Tunnaclife et al. 2001). In addition, high radioactivity tolerance has lead to the discovery of new bacterial species resistant to desiccation in radioactive work areas (Venkateswaran et al. 2003).

In relation to the mechanisms of desiccation tolerance (Figure 5.2), during the drying process, loss of water is balanced by reversible interactions with other molecules (Rebecchi et al. 2007). Thus, biomolecules and structures are protected, and they retain their native conformation after rehydration (Wolkers et al. 2002).

One of the survival strategies against desiccation is the accumulation of nonreducing disaccharides such as sucrose and trehalose (Clegg 2001; Rebecchi et al. 2007). These sugars have a double function in the desiccation-tolerant organisms. First, the sugar molecules protect the cells and biomolecules by replacing the water normally forming hydrogen bonds. Second, the saccharides are involved in the formation of a vitreous cytoplasmatic matrix (Wolkers et al. 2002).

Although the importance of nonreducing disaccharides has been emphasized for both unicellular and pluricellular organisms, the molecular mechanisms of anhydrobiosis are not clearly understood (Rebecchi et al. 2007). It seems that trehalose does not induce anhydrobyosis

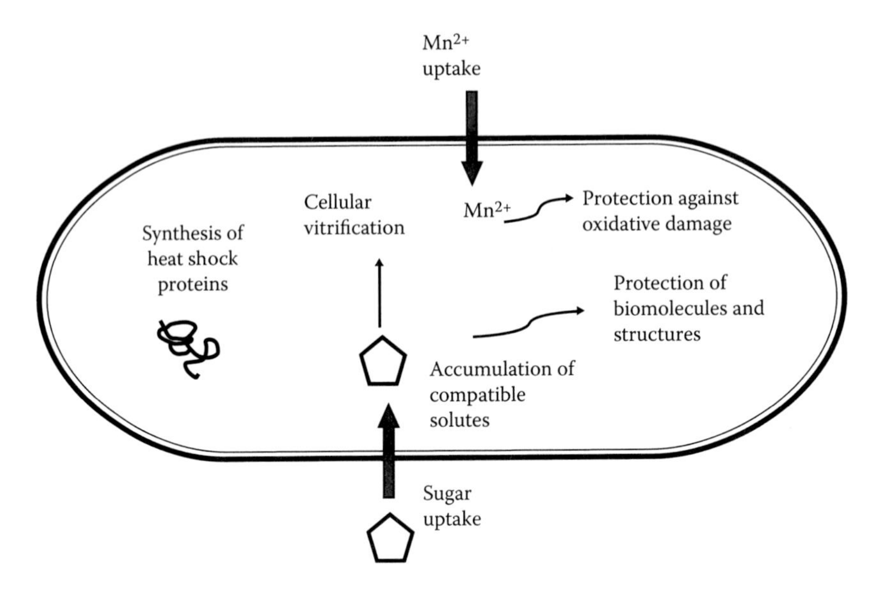

Figure 5.2 Main mechanisms of desiccation tolerance in a microbial cell. A bacterial cell is presented as the model.

by itself. In fact, in some anhydrobiotes, trehalose is not relevant to desiccation and they need other adaptations (Clegg 2001; Kshamata et al. 2003; Tunnacliffe and Lapinski 2003).

In this context, the attention has been focused on the definition of other molecular adaptations required for anhydrobiosis. Several families of stress proteins seem to be the key for understanding the anhydrobiotic mechanisms. During drying, the response to stress is associated with a rapid synthesis of stress proteins (Potts et al. 2005; Rebecchi et al. 2007). Proteins commonly known as "heat shock proteins" (HSP), which are regulated through thermal stress, and seem to have some function in desiccation tolerance (França et al. 2007; Goyal et al. 2005).

The HSP genes are highly conserved in most species in which they have been studied (Feder and Hofmann 1999). As in any biochemical system, these proteins act like molecular chaperones, even in nonstressed cells. They have important functions such as protein biosynthesis, folding, assembly, intracellular localization, secretion, and degradation of other proteins (Feder and Hofmann 1999).

The intracellular proteins are able to compensate for the loss of water through the formation of hydrogen bonds to other molecules. It is very probable that proteins and saccharides interact by hydrogen bonds in the cytoplasm of anhydrobiotes. Thus, stress proteins and sugars could play an important role in the molecular organization of the desiccated cell once they are present in the same cellular compartment (Wolkers et al. 2002).

In bacteria, the regulation of the genes encoding HSP in response to desiccation has been observed, for example, in the cyanobacterium *Anabaena* sp. 7120 (Katoh et al. 2004). In addition, the overexpression of Hsp20 has been detected in *D. radiodurans* (Tanaka et al. 2004).

On the other hand, the so-called late embryogenesis abundant (LEA) proteins play an important role in anhydrobiotes, mainly in plants (Hoekstra et al. 2001). LEA proteins are usually classified into five groups on the basis of amino acid sequence and conserved motif. Recently, LEA proteins have been found in nonplant organisms (Rebecchi et al. 2007). In addition, several genes encoding LEA-like proteins have been identified from the genome analysis of *D. radiodurans*. In fact, the inactivation of the gene encoding for the group-3 LEA reduces the desiccation tolerance of this bacterium (Battista et al. 2001). The group-3 LEA proteins are characterized by 11 amino acid tandem repeats (TAQAAKEKAGE; Ried and Walker-Simmons 1993). More recent studies have shown the high correlation between the expression of proteins with hydrophilic low complexity (LC) regions and desiccation tolerance in several bacterial species, including *D. radiodurans* (Krisco et al. 2010). Presumably, these proteins provide resistance to water stress conditions by avoiding aggregation. However, the exact mechanism remains unclear.

Any damage caused by air-drying is manifested during cell rehydration (Potts 1994). Apparently, there are similarities between the kinds of damage produced during desiccation and during the subsequent rehydration stage. Certainly, during both desiccation and rehydration, a dynamic change in the proteins and nucleic acid contents takes place (Shirkey et al. 2000). Therefore, a question arises in relation to the probable parallelism between the response to some types of stress and the poststress recuperation (Potts et al. 2005).

In this sense, oxidative damage is considered as an important consequence of desiccation stress. For example, Aldsworth (1999) proposed that bacterial cells perceive any stress as an oxidative stress. The decoupling between growth and metabolism generates free radicals, which are lethal to cells (Potts et al. 2005). Under such conditions, the role of superoxide dismutase (SOD) in response to stress is critical, and it could explain why some SODs are excreted by cells (Cannio et al. 2001).

A protein complex Fe-SOD (SodF) is the third more abundant in the desiccated cells of *N. commune*. In this cyanobacterium, SodF could protect cells from the free radicals generated when the extracellular polysaccharide is exposed to UV radiation (Shirkey et al. 2000).

The aconitases are involved in the rapid initial response to oxidative stress in *E. coli* and *Bacillus subtilis* by acting as mRNA binding proteins (Alén and Sonenshein 1999). In this mechanism, the cofactor sensor is the cluster iron-sulfur (redox sensor), which constitutes a positive posttranscriptional self-regulatory switch (Potts et al. 2005).

5.5 *Induced anhydrobiosis as a powerful method for preservation of microorganisms*

In general, the studies related to desiccation tolerance indicate that the stress processes induce several survival strategies offering protection to microorganisms subjected to desiccation. The production of stress proteins during bacterial growth could protect the cells, preparing them from further desiccation processes (Morgan et al. 2006). Thus, the design of experiments to evaluate drying resistance in bacterial strains is important for the industry. Moreover, the genomics of the microorganism subjected to desiccation should be studied to characterize the genes associated with the drying tolerance, which allows the development of strategies for the manipulation of external conditions to maximize the expression of such genes (Potts et al. 2005).

Another approach to induce anhydrobiosis in microorganisms is related to the use of protectants. The protective agents could be added during microorganism growth, or just before the drying process. The type of protectant depends mainly on the microorganism; nevertheless, some protectants seem to be universal (Morgan et al. 2006). These include skimmed milk, serum, trehalose, glycerol, betaine, adonytol, sucrose, glucose, lactose, and polymers such as dextrane and polyethylene glycol (Hubalek 2003).

According to Morgan et al. (2006), the protectant additives could be classified into two groups: (1) those forming amorphous glasses, and (2) the crystallized eutectic salts. The first group includes substances such as carbohydrates, proteins, and polymers. The term "glass" is related to a supersaturated liquid that is thermodynamically unstable and highly viscous (Garcia 2011). In this sense, the glassy additives are the best protectant agents during freeze-drying processes. The glassy state formation induces enough viscosity inside and outside the cells to reduce the molecular mobility to a minimum value (Morgan et al. 2006).

Trehalose and sucrose have been reported as the best membrane stabilizers and proteins in drying several microorganisms through a mechanism of water replacement (Rudolph and Crowe 1985). Both disaccharides preserve the structure and function of the proteins during drying, through stabilization by hydrogen bonds, keeping the tertiary structure in the absence of water (Leslie et al. 1995).

Several studies suggest that trehalose provides higher survival rates than sucrose (Crowe et al. 1998; Gómez Zavaglia et al. 2003; Leslie et al. 1995; Streeter 2003). The scientific explanation for this evidence is focused on the high glass transition temperature (T_g) of the trehalose. The glass transition temperature is the temperature at which glass is transformed to the corresponding liquid phase. Hence, the higher the T_g value, the higher the stability of the dried matrix when the temperature is increased (Morgan et al. 2006). Particularly, anhydrous trehalose has $T_g = 110°C$, keeping a high viscosity at high temperatures. Otherwise, sucrose has $T_g = 65°C$, reducing its potential use with respect to trehalose (Crowe et al. 1998). However, sucrose is cheaper than trehalose.

Previous studies have established the correlation between the high T_g value of trehalose, and the increase in the protection to desiccation by comparison to other protecting carbohydrates with lower T_g values (Sun and Davidson 1998). In addition, trehalose has the ability to form a crystalline dihydrate. Consequently, in the formation of the dehydrated crystal, a small amount of water is absorbed and the matrix stays in a glassy state without decreasing the T_g of the system (Crowe et al. 1998).

Although it is believed that sugars are the major components in glass formation, it has been shown that polypeptides could significantly alter the vitreous properties of sugars. Buitink et al. (2000) found that proteins were more stable above their T_g values than sugars. This fact suggests that proteins contribute higher extension to glass formation in comparison with sugars. Moreover, it could explain why milk serum and skimmed milk are such good protectors against desiccation (Hubalek 2003).

Therefore, it seems that the most efficient protector against desiccation is a mixture of proteins and sugars. Desmond et al. (2002) found that the survival of *Lactobacillus paracasei* significantly increased up to 1000-fold when acacia gum (10% w/w) was added to reconstituted skimmed milk (10% w/v).

On the other hand, rehydration is the most important stage once the microbial cells are subjected to the desiccation process. Rehydration is critical because, according to the rehydration protocol, the survival of the cells will be more or less affected. Hence, the rehydration medium plays a key role. For example, it has been observed that a protective solution used before the drying is a good rehydration solution (Garcia 2011; Julca et al. 2012; Potts et al. 2005). In addition, a complex medium generally has higher osmotic pressure, which diminishes the negative effect of osmotic shock in anhydrobiotic cells (Abe et al. 2009; Ahi et al. 2010; Fu and Chen 2011; Morgan et al. 2006).

The rehydration temperature is another variable to be considered, mainly because rehydration at a temperature below T_m, generally leads to phase transition with the subsequent leakage of intracellular material and lethal damages (Garcia 2011; Julca et al. 2012; Potts et al. 2005). Besides, the rehydration speed should be controlled. Some authors have shown that the survival rate is increased once the cells are slowly rehydrated (Potts 1994; Potts et al. 2005). Nevertheless, if rehydration is extremely slow, the process is not economically feasible (Morgan et al. 2006).

5.6 Technologies used in the drying of microorganisms

There are several methods used in the drying of bacteria, such as freeze-drying, spray-drying, and fluidized bed drying (Morgan et al. 2006). Spray-drying is the predominant process in the dairy industry and could be used for producing large amounts of milk by-products at relatively low cost in comparison to freeze-drying (Lian et al. 2002; Morgan et al. 2006). Using the spray-drying technology, it is possible to obtain granulated dry powders from a cream. This process implies the spraying of the cream into the drying chamber. The air in the chamber is injected at high speed with temperatures up to 200°C (inlet temperature; Fu and Etzel 1995; Silva et al. 2004, 2005; To and Etzel 1997). The little drops formed during the atomization of the cream are dried in a rapid manner through direct contact with the hot air, forming particles or powder granules that reach the final temperature (outlet temperature).

In the last decade, several studies have shown the possibility of using this relatively cheap and highly productive technology in the preservation of probiotic bacteria (Corcoran et al. 2004; Desmond et al. 2002; Gardiner et al. 2000; Silva et al. 2002). Furthermore, some biocontrol bacterial agents have been formulated by spray-drying for their use in agriculture, mainly as wettable powders of small particle size (Arunsiri et al. 2003; Bashan et al. 2002; Hernández et al. 2007).

Table 5.1 shows several examples of the use of spray-dried bacteria in medicine, agriculture and the food industry. For instance, microencapsulation of probiotic bacteria by spray-drying is the most commonly used method for the formulation of such products

Table 5.1 Application of Anhydrobiosis in Bacteria to Several Fields

Drying technology	Use in medicine	Use in agriculture	Use in food industry
Spray-drying	Probiotics to reinforce the body's natural defense mechanisms (Anal and Singh 2007; Corcoran et al. 2006) Therapeutic use of live bacterial cells (pathobiotechnology; Culligan et al. 2009; Prakash and Jones 2005)	Biological control of insects (Arunsiri et al. 2003; Zhou et al. 2004) Biological control of nematodes (Hernández et al. 2007, 2009)	Starters for the production of polysaccharides (Boza et al. 2004) Probiotic food (Ross et al. 2005; Simpson et al. 2005)
Freeze-drying	Treatment of gastrointestinal diseases (e.g., Crohn's disease with human IL-10 expressed in *L. lactis*) (Huyghebaert et al. 2005) Vaccines based on attenuated bacteria (Edwards and Slater 2008; Rexroad et al. 2002)	Biofertilizers for nitrogen fixation (Arraes Pereira et al. 2002; Vriezen et al. 2006) Encapsulation of plant growth-promoting bacteria (Bashan et al. 2002)	Starter cultures for food production (Kiviharju et al. 2005; Ziadi et al. 2005)
Fluidized bed drying	Encapsulation of probiotic bacteria for therapeutic uses (Anal and Singh 2007; Lopez-Rubio et al. 2006)	Biopesticides containing heat-sensitive bacteria (Slininger et al. 2010)	Optimization in the drying of Lactobacilli strains (Mille et al. 2004) Improvement in the preservation of lactic acid bacteria (Strasser et al. 2009)

(Anal and Singh 2007; Kailasapathy 2002). These probiotic bacteria have beneficial effects on patients with gastrointestinal disorders, mainly by improving the immunological system (Anal and Singh 2007; Huyghebaert et al. 2005).

Fluidized bed drying is a method for drying granulated solids. An advantage of this technology is the incorporation of a spray-dryer to the fluidized bed apparatus, which allows the conversion of liquids in powders and the subsequent drying of the granulated powders. This method is recommended for drying heat-sensitive bacterial strains, mainly by microencapsulation (Anal and Singh 2007; Lopez-Rubio et al. 2006). In the drying chamber, temperatures of approximately 40°C are reached, which produces a gentle drying without thermal stress.

Although freeze-drying is the favorite method of drying in the microbiological industry, there are losses in cell viability (Morgan et al. 2006). One of the main causes affecting the loss of viability is the need to freeze the samples to be desiccated. The freezing process and especially the freezing speed could damage the cells during drying (Uzunova-Doneva and Donev 2000). The formation of large ice crystals could lead to damage in the cell membranes, causing mortality in the processed biomass.

The freeze-drying process involves two key stages: primary and secondary drying. Primary drying takes place at low temperatures, when the frozen mixture is sublimed by reduction of the chamber pressure below the ice pressure within the product (Morgan et al. 2006). The secondary stage takes place once all frozen water has been sublimed, and the chamber temperature is slowly increased, producing the sublimation of bound water. The end point of the secondary drying is usually determined by analysis of the residual water content in the final product.

Always the freeze-drying method is cited considering the long-term storage of cellular suspensions containing more than 10^8 cells/mL (Costa et al. 2000; Miyamoto-Shinohara et al. 2006). Thus, creams usually have cell concentrations above this value to ensure high survival rates after the drying process (Garcia 2011; Morgan et al. 2006). The freeze-drying technology has been used in the preservation of bacterial strains for several applications (Table 5.1), mainly in the production of probiotics for the treatment of gastrointestinal disorders.

5.7 *Stability of desiccated microorganisms*

The shelf-life time of a product based on desiccated cells will depend on the method of packing and storage. In desiccated bacterial cells, the mechanisms inducing cell death by oxidation have been acting since the beginning of the drying process (França et al. 2007; Hernández et al. 2009; Potts 1994).

Concerning the analysis of stability in anhydrobiosis state, there are only isolated reports (Achour et al. 2001; Hernández et al. 2007, 2009; Ziadi et al. 2005; Morgan et al. 2006). Most researchers describe the stability of anhydrobiotic cells by using first-order models of microbial death (Achour et al. 2001; Ananta et al. 2005; Ziadi et al. 2005). This means that, according to classical kinetic mechanisms, bacterial inactivation is considered a first-order reaction. However, each bacterial species has its particularities and for both desiccated cells and liquid-phase cells, the thermal death patterns could be dissimilar.

van Boekel Martinus (2002) reported the use of the Weibull model (Equation 5.1) to describe the inactivation of bacterial strains in liquid phase. In this model, S is the survival rate, and α and β are parameters or constants. The parameter α is called the scale parameter, and β is called the shape parameter. If $\beta < 1$, an upward concavity will be observed.

Otherwise, if $\beta > 1$, a downward concavity will be observed. Then, a straight line will be obtained when $\beta = 1$.

$$\log(S) = -\frac{1}{2.3}\left(\frac{t}{\alpha}\right)^{\beta} \tag{5.1}$$

In general, the formation of tails and arms has been observed in thermal death curves for microorganisms exposed to heat in liquid phase; however, this phenomenon has not been reported for anhydrobiotic cells. In the few reports concerning the stability of desiccated cells, most authors fitted the viability data to linear curves, assuming $\beta = 1$, without considering the concavities appearing in survival curves (Achour et al. 2001; Ananta et al. 2005; Ziadi et al. 2005). This could cause mistakes in the estimation of survival for particular zones in the curve.

Recently, Hernández et al. (2009) have reported that the introduction of the term "anhydrobiosis quotient" (Equation 5.2):

$$\varepsilon = \log\left(\frac{Xv}{Hr}100\right) \tag{5.2}$$

To characterize the stability of anhydrobiotic cells, where Xv is the viability or viable cells (colony-forming units [CFU]/mL); and Hr is the residual moisture (%). This function is advantageous because it allows the estimation of stability by using the nonlinear model (Equation 5.3):

$$\varepsilon = \varepsilon_0 - (\alpha.t)^{\beta} \tag{5.3}$$

Considering the concavities of the curves, moreover, it seems to give a reasonable prediction for the stability of desiccated bacterial cells. Nevertheless, attention should be focused on further mathematical modeling of the desiccated state.

5.8 Case study: Anhydrobiotic cells of a biocontrol agent

Considering the effect of biopesticides in agriculture and the environment, here we present an experimental work related to the obtainment of a solid formulation containing the nematocidal bacterium *Tsukamurella paurometabola* C-924. This microorganism was desiccated by spray-drying technology, and the stability of anhydrobiotic cells was evaluated at several temperatures (Hernández et al. 2007, 2009).

Cells of *T. paurometabola* C-924 were cultured in a 300 L fermentor, and then the biomass produced was subjected to spray-drying at several outlet temperatures between 50°C and 62°C to evaluate the effect of outlet temperature on survival rate using sucrose 10% (w/w dry biomass) as a protective agent. All the experiments were done in triplicate; in total, 27 runs were carried out with a Mobile Minor spray dryer. The cream (pH 6) was subdivided in 1-L samples per run. The inlet temperature was 120°C. According to the experimental design, outlet temperature was between 50°C and 60°C.

Survival rate was determined as follows: 0.1 g of powder was suspended by vortexing in 10 mL of NaCl at 9 g/L, and serial dilutions were made to plate. The survival rate was calculated as follows (Equation 5.4):

$$\text{Survival (\%)} = \left(\frac{N}{N_0}\right)100 \tag{5.4}$$

The moisture content of the powders was determined in triplicate by oven-drying at 105°C.

From an analysis of trends in Figure 5.3, it suggests that by increasing sucrose concentration, cells are more protected against thermal stress during dehydration. Data in Figure 5.3 were well-fitted to a reduced quadratic model (Equation 5.5; $r^2 = 0.9992$, $F = 2118.82$, $p < 0.0001$):

$$\text{Survival} = 57.19 - 3.12 \times \text{sucrose}\ (\%) + 1.5 \times \text{sucrose}\ (\%)^2 - 0.74 \times T_{\text{outlet}} \quad (5.5)$$

All the coefficients were significant ($p < 0.05$); mainly, sucrose concentration had a remarkably positive effect on survival. As expected, outlet temperature exerted a negative contribution to survival. High survival rates reflect tolerance to desiccation of *T. paurometabola* C-924 in the presence of sucrose. From evaluation of the model, sucrose exerts a more significant effect on survival than temperature. For example, at 10% (w/w), sucrose survival decreases by only 9% when temperature changes from 50°C to 62°C, meanwhile at 62°C, survival is increased in 73.5% when sucrose concentration changes from 0% to 10% (w/w). In all the experimental points, moisture was less than 10% (w/w), which must ensure high stability of formulation according to Garcia (2011).

From this result, the selected experimental conditions were T_{inlet} 120°C, T_{output} 62°C, and sucrose concentration 10% (w/w). Using these conditions, survival rates higher than 80% were obtained. Cell membranes remained undamaged after desiccation, as evidenced by transmission electron microscopy (Figure 5.4), contributing to cell survival after spray-drying.

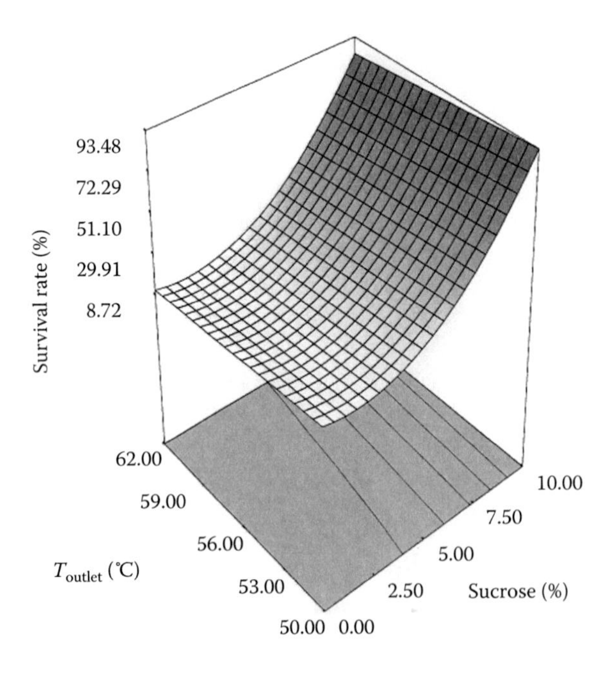

Figure 5.3 Response surface for the survival rate during the spray-drying of *T. paurometabola* C-924. A three-level factorial design was performed using sucrose as an additive. Sucrose concentration was set between 0% and 10% (w/w). Outlet temperature was set between 50°C and 62°C. Data were analyzed using Design Expert 6.0.1.

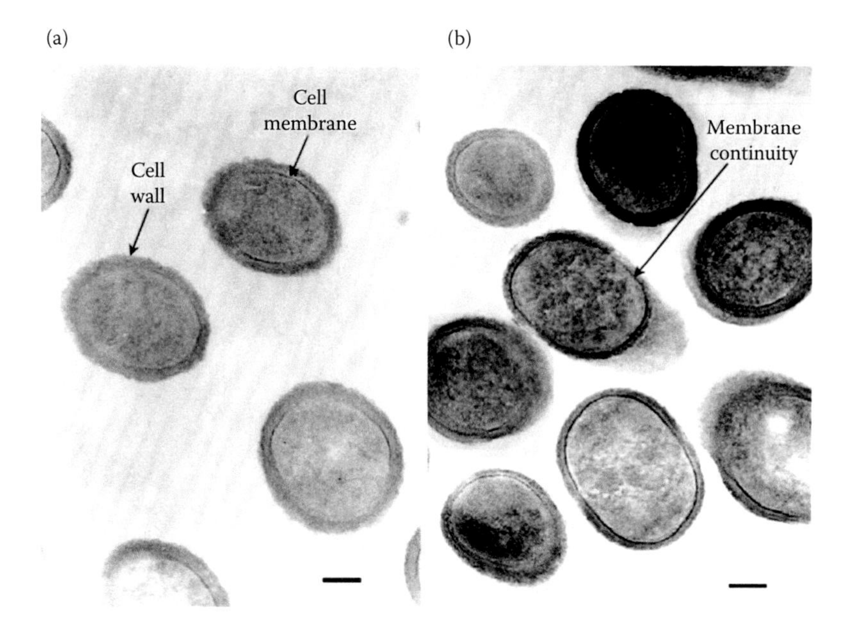

Figure 5.4 Electron micrographs of *T. paurometabola* C-924 cells. (a) Cells before the spray-drying process; (b) cells after desiccation by spray-drying. Bar = 100 nm.

A stability assay of anhydrobiotic cells was performed. Samples were incubated at 4°C, 25°C, and 32°C (under vacuum and without vacuum), respectively, and the inactivation curves were obtained by plotting survival rate versus time.

Shelf-life time (SLT) and half-life time (HLT) are the most useful parameters to describe the stability of a formulation. SLT has been defined as the period in which a biological product can be stored before its biological activity decays (Hernández et al. 2007).

On the other hand, HLT of a bioproduct is the time at which the biological activity of the formulation decreases by 50%. In practice, the most used parameter is SLT. In our case, SLT was defined as the time value at which the cell concentration per gram of the powder (CFU/g), decreases to 1×10^{12} CFU/g. The starting point of viability was 4×10^{12} CFU/g; therefore, in terms of survival, SLT is the time at which survival decreases by 75%.

It was expected that the highest stability was at 4°C for both treatments, and the SLT value was ninefold higher for the powders stored at vacuum. With respect to the powders stored at room temperature (25°C) in both conditions (under vacuum and without vacuum), SLT was twofold higher for powders under vacuum; whereas at 32°C SLT was threefold higher for the powders stored under vacuum. These findings are very similar to those previously obtained by authors who have been working with other bacteria (Achour et al. 2001).

Considering the values of the cell death constant, according to classical kinetics, the Arrhenius plot was performed (Achour et al. 2001). Plotting $\ln(k)$ versus $(1/T)$ was well-fitted to a straight line for the powders stored at vacuum in the interval 4°C to 32°C, where k is the cell death constant and T is the absolute temperature.

$$\ln(k) = -4423\left(\frac{1}{T}\right) + 12.45 \text{ with } r^2 = 0.9435 \tag{5.6}$$

The activation energy calculated from the slope was $\Delta E^{\#} = 8.75$ kcal/mol, which could be considered as an energy related to diffusion controls, in other words, the energy related to those reactions that are not diffusion-limited reactions in a glassy cytoplasm (Potts 1994).

Non-Arrhenius behavior was obtained for the powders exposed to air (data not shown); this phenomenon could be explained on the basis of the glass transition. Vitrification is an extensively studied process (Crowe et al. 1998), and dried bacterial cells in the presence of sugars could form bacterial glasses (Potts 1994). In the glassy state, rates are low, but Arrhenius kinetics nevertheless still apply. When the system is over the glass transition temperature (T_g), non-Arrhenius kinetics are observed, and the dependence of viscoelastic properties on temperature is described by the Williams–Landel–Ferry theory (Potts 1994), which relates the k value to T according to Equation 5.7; where k_g is the death constant at T_g:

$$\log\left(\frac{k_g}{k}\right) = \frac{-C_1(T - T_g)}{C_2 + (T - T_g)} \tag{5.7}$$

Consequently, deviations from Arrhenius law will be found experimentally when the interval of temperatures includes T_g. In addition, T_g decreases when moisture is increased. In these experiments, the bags containing the powders were sealed in the presence of air with high moisture content (>80%); as a consequence, residual moisture of powders was increased, and likely T_g decreased. Based on this analysis, we could hypothesize that T_g value is in the experimental interval of temperatures, explaining the non-Arrhenius behavior obtained.

5.9 Conclusion

Induced anhydrobiosis is one of the most powerful methods used in the preservation of microorganisms. It has been widely applied in the preservation of microbial strains with industrial, agricultural, and biomedical applications. Among the most frequently applied technologies for the stabilization of microbial cells by drying, freeze-drying is the most popular and the most expensive. Although a common methodology for evaluating the stability of anhydrobiotic microbial cells has been developed, further research on mathematical modeling of the desiccated state in microorganisms should be performed to predict the stability of such biological systems.

References

Abe, F., Miyauchi, H., Uchijima, A., Yaeshima, T. and Iwatsuki, K. 2009. Effects of storage temperature and water activity on the survival of bifidobacteria in powder form. *Int. J. Dairy Technol.* 62: 234–239.

Achour, M., Mtimet, N., Cornelius, C. et al. 2001. Application of the accelerated shelf life testing method (ASLT) to study the survival rates of freeze-dried *Lactococcus* starter cultures. *J. Chem. Technol. Biotechnol.* 76: 624–628.

Ahi, M., Hatamipour, M. and Goodarzi, A. 2010. Optimization of leavening activity of baker's yeast during the spray-drying process. *Drying Technol.* 28: 490–494.

Aldsworth, T. G. 1999. Bacterial suicide through stress. *Cell. Mol. Life Sci.* 56: 378–383.

Alén, C. and Sonenshein, A. L. 1999. *Bacillus subtilis* aconitase is an RNA-binding protein. *Proc. Natl. Acad. Sci. U.S.A.* 96: 10412–10417.

Alpert, P. 2005. The limits on frontiers of desiccation-tolerant life. *Integr. Comp. Biol.* 45: 685–695.

Anal, A. K. and Singh, H. 2007. Recent advances in microencapsulation of probiotics for industrial applications and targeted delivery. *Trends Food Sci. Tech.* 18: 240–251.

Ananta, E., Volkert, M. and Knorr, D. 2005. Cellular injuries and storage stability of spray-dried *Lactobacillus rhamnosus* GG. *Int. Dairy J.* 15: 399–409.

Arraes Pereira, P. A., Oliver, A., Bliss, F. A., Crowe, L. and Crowe, J. 2002. Preservation of rhizobia by lyophlization with trehalose. *Pesq. Agropec. Bras.* 37 (6): 831–839.

Arunsiri, A., Suphantarika, M. and Ketunuti, U. 2003. Preparation of spray-dried wettable powder formulations of *Bacillus thuringiensis*-based biopesticides. *J. Econ. Entomol.* 96 (2): 292–299.

Bashan, Y., Hernández, J. P., Leyva, L. A. and Bacilio, M. 2002. Alginate microbeads as inoculant carriers for plant growth-promoting bacteria. *Biol. Fertil. Soils* 35: 359–368.

Battista, J. R., Park, M. J. and McLemore, A. E. 2001. Inactivation of two homologues of proteins presumed to be involved in the desiccation tolerance of plants sensitizes *Deinococcus radiodurans* R1 to desiccation. *Cryobiology* 43: 133–139.

Billi, D., Wright, D. J., Helm, R. F., Prickett, T., Potts, M. and Crowe, J. H. 2000. Engineering desiccation tolerance in *Escherichia coli*. *Appl. Environ. Microbiol.* 66: 1680–1684.

Boza, Y., Barbin, D. and Scamparini, A. P. 2004. Effect of spray-drying on the quality of encapsulated cells of *Beijerinckia* sp. *Process Biochem.* 39: 1275–1284.

Buitink, J., van den Dries, I. J., Hoekstra, F. A., Alberda, M. and Hemminga, M. A. 2000. High critical temperature above Tg may contribute to the stablity of biological systems. *Biophys. J.* 79 (2): 1119–1128.

Cannio, R., Fiorentino, G., Morana, A., Rossi, M. and Bartolucci, S. 2001. Oxygen: Friend or foe? Archaeal superoxide dismutases in the protection of intra- and extracellular oxidative stress. *Front. Biosci.* 5: 768–779.

Carvalho, A. S., Silva, J., Ho, P., Teixeira, P., Malcata, F. X. and Gibbs, P. 2004. Effects of various sugars added to growth and drying media upon thermotolerance and survival throughout storage of freeze-dried *Lactobacillus delbrueckii* ssp. *Bulgaricus*. *Biotechnol. Progr.* 20: 248–254.

Clegg, J. S. 2001. Cryptobiosis—A peculiar state of biological organization. *Comp. Biochem. Physiol.* 128 (6): 613–624.

Corcoran, B. M., Ross, R. P., Fitzgerald, G. F., Dockery, P. and Stanton, C. 2006. Enhanced survival of GroESL-overproducing *Lactobacillus paracasei* NFBC 338 under stressful conditions induced by drying. *Appl. Environ. Microbiol.* 72 (7): 5104–5107.

Corcoran, B. M., Ross, R. P., Fitzgerald, G. and Stanton, C. 2004. Comparative survival of probiotic lactobacilli spray dried in the presence of prebiotic substances. *J. Appl. Microbiol.* 96: 1024–1039.

Costa, E., Usall, J., Teixeidó, N., García, N. and Viñas, I. 2000. Effect of protective agents, rehydration media and initial cell concentration on viability of *Pantoea aglomerans* strain CPA-2 subjected to freeze-drying. *J. Appl. Microbiol.* 89: 793–800.

Crowe, J. H., Hoekstra, F. A. and Crowe, L. M. 1992. Anhydrobiosis. *Annu. Rev. Physiol.* 54: 579–599.

Crowe, J. H., McKersie, B. D. and Crowe, L. M. 1989. Effects of free fatty acids and transition temperature on the stability of dry liposomes. *Biochem. Biophys. Acta* 97 (9): 7–10.

Crowe, J., Carpenter, J. and Crowe, L. 1998. The role of vitrification on anhydrobiosis. *Annu. Rev. Physiol.* 60: 73–103.

Culligan, E. P., Hill, C. and Sleator, R. D. 2009. Probiotics and gastrointestinal disease: Successes, problems and future prospects. *Gut Pathol.* 1: 19–26.

Daly, M. J., Gaidamakova, E. K., Matrosova, V. Y. et al. 2004. Accumulation of Mn(II) in *Deinococcus radiodurans* facilitates gamma-radiation resistance. *Science* 306: 1025–1028.

de Castro, A. G., Lapinski, J. and Tunnacliffe, A. 2000. Anhydrobiotic engineering. *Nat. Biotechnol.* 18: 473.

Desmond, C., Ross, R. P., O'Callaghan, E., Fitzgerald, G. and Stanton, C. 2002. Improved survival of *Lactobacillus paracasei* NFBC in spray-dried powders containing gum acacia. *J. Appl. Microbiol.* 93: 1003–1011.

Edwards, A. D. and Slater, N. K. 2008. Formulation of a live bacterial vaccine for stable room temperature storage results in loss of acid, bile and bile salt resistance. *Vaccine* 26: 5675–5678.

Feder, M. E. and Hofmann, G. E. 1999. Heat-shock proteins, molecular chaperones, and the stress response: Evolutionary and ecological physiology. *Ann. Rev. Physiol.* 61: 243–282.

França, M. B., Panek, A. D. and Eleutherio, E. C. 2007. Oxidative stress and its effects during dehydration. *Comp. Biochem. Phys. Part A* 146: 621–631.

Fravel, D. R. 2005. Commercialization and implementation of biocontol. *Annu. Rev. Phytopathol.* 43: 337–359.

Fredrickson, J. K., Li, S. W., Gaidamakova, E. K. et al. 2008. Protein oxidation: Key to bacterial desiccation resistance? *ISME J.* 2: 393–403.

Fu, N. and Chen, X. D. 2011. Towards a maximal cell survival in convective thermal drying processes. *Food Res. Int.* 44: 1127–1149.

Fu, W. Y. and Etzel, M. R. 1995. Spray drying of *Lactococcus lactis* ssp. *lactis* C2 and cellular injury. *J. Food Sci.* 60: 195–200.

Garcia, A. H. 2011. Anhydrobiosis in bacteria: From physiology to applications. *J Biosci.* 36: 939–950.

Gardiner, G. E., O'Sullivan, E., Kelly, J. et al. 2000. Comparative survival rates of human-derived probiotic *Lactobacillus paracasei* and *L. salivarius* strains during heat treatment and spray drying. *Appl. Environ. Microbiol.* 6 (6): 2605–2612.

Gómez Zavaglia, A., Tymczyszyn, E., De Antoni, G. and Anibal Disalvo, E. 2003. Action of trehalose on the preservation of *Lactobacillus delbrueckii* ssp. *bulgaricus* by heat and osmotic dehydration. *J. Appl. Microbiol.* 95: 1315–1320.

Goyal, K., Walton, L. J., Browne, J. A., Burnell, A. M. and Tunnaclife, A. 2005. Molecular anhydrobiosis: Identifying molecules implicated in vertebrate anhydrobiosis. *Integr. Comp. Biol.* 45: 702–709.

Hansen, J. M., Go, Y. M. and Jones, D. P. 2006. Nuclear and mitochondrial compartmentation of oxidative stress and redox signaling. *Annu. Rev. Pharmacol. Toxicol.* 46: 215–234.

Harel, Y., Ohad, I. and Kaplan, A. 2004. Activation of photosynthesis and resistance to photoinibition in cyanobacteria within biological desert crust. *Plant. Physiol.* 136: 3070–3079.

Hernández, A., Weekers, F., Mena, J., Borroto, C. and Thonart, P. 2006. Freeze-drying of the biocontol agent *Tsukamurella paurometabola* C-924: Predicted stability of formulated powders. *Ind. Biotechnol.* 2 (3): 209–212.

Hernández, A., Weekers, F., Mena, J., Zamora, J., Borroto, C. and Thonart, P. 2007. Culture and spray-drying of *Tsukamurella paurometabola* C-924: Stability of formulated powders. *Biotechnol. Lett.* 29 (11): 1723–1728.

Hernández, A., Zamora, J., González, N., Salazar, E. and Sánchez, M. 2009. Anhydrobosis quotient: A novel approach to evaluate stability in desiccated bacterial cells. *J. Appl. Microbiol.* 107: 436–442.

Hoekstra, F. A., Golovina, E. A. and Buitink, J. 2001. Mechanisms of plant desiccation tolerance. *Trends Plant. Sci.* 6: 431–438.

Hubalek, Z. 2003. Protectants used in the cryopreservation of microorganisms. *Cryobiology* 46: 205–229.

Huyghebaert, N., Vermeire, A., Neirynck, S., Steidler, L., Remaut, E. and Remon, J. P. 2005. Development of an enteric-coated formulation containing freeze-dried, viable recombinant *Lactococcus lactis* for the ileal mucosal delivery of human interleukin-10. *Eur. J. Pharm. Biopharm.* 60: 349–359.

Julca, I., Alaminos, M., González-López, J. and Manzanera, M. 2012. Xeroprotectants for the stabilization of biomaterials. *Biotechnol. Adv.* 30: 1641–1654.

Kailasapathy, K. 2002. Microencapsulation of probiotic bacteria: Technology and potential applications. *Curr. Issues Intest. Microbiol.* 3: 39–48.

Katoh, H., Asthana, R. K. and Ohmori, M. 2004. Gene expression in the cyanobacterium *Anabaena* sp. PCC7120 under desiccation. *Microbiol. Ecol.* 47: 164–174.

Kiviharju, K., Leisola, M. and Eerik, T. 2005. Optimization of a *Bifidobacterium longum* production process. *J. Biotechnol.* 117: 299–308.

Kranner, I. and Birtic, S. 2005. A modulating role for antioxidants in desiccation tolerance. *Integr. Comp. Biol.* 45: 734–740.

Krisko, A., Smole, Z., Debret, G., Nikolic, N. and Radman, M. 2010. Unstructured hydrophilic sequences in prokaryotic proteomes correlate with dehydration tolerance and host association. *J. Mol. Biol.* 402: 775–782.

Kshamata, G., Tisi, L., Basran, A., Browne, J., Burnell, A. and Zurdo, J. 2003. Transition from natively unfolded to folded state induced by desiccation in an anhydrobiotic nematode protein. *J. Biol. Chem.* 278: 12977–12984.

Kuang, S. S., Oliveira, J. C. and Crean, A. M. 2010. Microencapsulation as a tool for incorporating bioactive ingredients into food. *Crit. Rev. Food Sci. Nutr.* 50: 951–968.

Lauten, E. H., Pulliman, B. L., Derousse, J., Bhatta, D. and Edwards, D. A. 2010. Gene expression, bacteria viability and survivability following spray drying of mycobacterium smegmatis. *Materials* 3: 2684–2724.

Leslie, S. B., Israeli, E., Lighthart, B., Crowe, J. H. and Crowe, L. M. 1995. Trehalose and sucrose protect both membranes and proteins in intact bacteria during drying. *Appl. Environ. Microbiol.* 61 (10): 3592–3597.

Lian, W. C., Hsiao, H. C. and Chou, C. C. 2002. Survival of bifidobacteria after spray-drying. *Int. J. Food Microbiol.* 74: 79–86.

Lopez-Rubio, A., Gavara, R. and Lagaron, J. M. 2006. Bioactive packaging: Turning foods into healthier foods through biomaterials. *Trends Food Sci. Tech.* 17: 567–575.

Manzanera, M., de Castro, A. G., Tondervik, A., Rayner-Brandes, M., Strom, A. R. and Tunnaclife, A. 2002. Hydroxyectoine is superior to trehalose for anhydrobiotic engineering of *Pseudomonas putida* KT2440. *Appl. Environ. Microbiol.* 68: 4328–4333.

Manzanera, M., Vilchez, S. and Tunnaclife, A. 2004. High survival and stability rates of *Escherichia coli* dried in hydroxyectoine. *FEMS Microbiol. Lett.* 233: 347–352.

Mattimore, V. and Battista, J. R. 1996. Radioresistance of *Deinococcus radiodurans*: Functions necessary to survive ionizing radiation are also necessary to survive prolonged desiccation. *J. Bacteriol.* 178: 633–637.

Mille, Y., Obert, J. P., Beney, L. and Gervais, P. 2004. New drying process for lactic bacteria based on their dehydration behavior in liquid medium. *Biotechnol. Bioeng.* 88 (1): 71–76.

Miyamoto-Shinohara, Y., Sukenobe, J., Imaizumi, T. and Nakahara, T. 2006. Survival curves for microbial species stored by freeze-drying. *Cryobiology* 52: 27–32.

Morgan, C. A., Herman, N., White, P. A. and Vesey, G. 2006. Preservation of micro-organisms by drying: A review. *J. Microbiol. Methods* 66: 183–193.

Motta, T. B. and Sechi, A. M. 1992. Phospholipid polar heads affect the generation of oxygen active species by Fe^{2+} autooxidation. *Biochem. Int.* 26: 987–994.

Nag, A. and Das, S. 2012. Improving ambient temperature stability of probiotics with stress adaptation and fluidized bed drying. *J. Funct. Foods* 5 (1): 170–177.

Pereira, E. J., Panek, A. D. and Eleutherio, E. C. 2003. Protection against oxidation during dehydration of yeast. *Cell Stress Chaperon* 8: 120–124.

Potts, M. 1994. Desiccation tolerance of prokaryotes. *Microbiol. Rev.* 58 (4): 755–805.

Potts, M., Slaughter, S. M., Hunneke, F., Garst, J. F. and Helm, R. F. 2005. Desiccation tolerance of prokaryotes: Application of principles to human cells. *Integr. Comp. Biol.* 45: 800–809.

Prakash, S. and Jones, M. L. 2005. Artificial cell therapy: New strategies for the therapeutic delivery of live bacteria. *J. Biomed. Biotechnol.* 1: 44–56.

Prestrelski, S. J., Tedeschi, N., Arakawa, T. and Carpenter, J. F. 1993. Dehydration induced conformational transitions in proteins and their inhibition by stabilizers. *Biophys. J.* 65: 661–671.

Rebecchi, L., Altiero, T. and Guidetti, R. 2007. Anhydrobiosis: The extreme limit of desiccation tolerance. *Invertebr. Surv. J.* 4: 65–81.

Rexroad, J., Wiethoff, C. M., Jones, L. S. and Middaugh, C. R. 2002. Lyophilization and the thermostability of vaccines. *Cell. Preserv. Technol.* 1 (2): 91–104.

Ried, J. L. and Walker-Simmons, M. K. 1993. Group 3 late embryogenesis abundant proteins in desiccation-tolerant seedlings of wheat (*Triticum aestivum*). *Plant. Physiol.* 102: 125–131.

Ross, R. P., Desmond, C., Fitzgerald, G. F. and Stanton, C. 2005. Overcoming the technological hurdles in the development of probiotic foods. *J. Appl. Microbiol.* 98: 1410–1417.

Rudolph, A. S. and Crowe, J. H. 1985. Membrane stabilization during freezing: The role of two natural cryoprotectants, trehalose and proline. *Cryobiology* 22: 367–377.

Senaratna, T., McKersie, B. D. and Borochov, A. 1987. Desiccation and free radical mediated changes in plant membranes. *J. Exp. Bot.* 38: 2005–2014.

Shirkey, B., Kovarcik, D. P., Wright, D. J. et al. 2000. Active Fe-SOD and abundant sodFmRNA in Nostoc commune (Cyanobacteria) after years of desiccation. *J. Bacteriol.* 182: 189–197.

Shirkey, B., McMaster, N. J., Smith, S. C., Wright, D. J., Rodriguez, H. and Jaruga, P. 2003. Genomic DNA of *Nostoc commune* (Cyanobacteria) becomes covalently modified during long-term (decades) desiccation but is protected from oxidative damage and degradation. *Nucleic Acids Res.* 31: 2995–3005.

Silva, J., Carvalho, A. S., Ferreira, R. et al. 2005. Effect of the pH of growth on the survival of *Lactobacillus delbrueckii* subsp. *bulgaricus* to stress conditions during spray-drying. *J. Appl. Microbiol.* 98 (3): 775–782.

Silva, J., Carvalho, A. S., Pereira, H., Teixeira, P. and Gibbs, P. A. 2004. Induction of stress tolerance in *Lactobacillus delbrueckii* ssp. *bulgaricus* by the addition of sucrose to the growth medium. *J. Dairy Res.* 71: 121–125.

Silva, J., Carvalho, A. S., Teixeira, P. and Gibbs, P. A. 2002. Bacteriocin production by spray-dried lactic acid bacteria. *Lett. Appl. Microbiol.* 34: 77–81.

Simpson, P. J., Stanton, C., Fitzgerald, G. F. and Ross, R. P. 2005. Intrinsic tolerance of *Bifidobacterium* species to heat and oxygen and survival following spray drying and storage. *J. Appl. Microbiol.* 99: 493–501.

Slade, D. and Radman, M. 2011. Oxidative stress resistance in *Deinococcus radiodurans*. *Microbiol. Mol. Biol. R.* 75: 133–191.

Slininger, P. J., Dunlap, C. A. and Schisler, D. A. 2010. Polysaccharide production benefits dry storage survival of the biocontrol agent *Pseudomonas fluorescens* S11:P:12 effective against several maladies of stored potatoes. *Biocontrol Sci. Techn.* 20 (3): 227–244.

Strasser, S., Neureiter, M., Geppl, M., Braun, R. and Danner, H. 2009. Influence of lyophilization, fluidized bed drying, addition of protectants, and storage on the viability of lactic acid bacteria. *J. Appl. Microbiol.* 107: 167–177.

Streeter, J. G. 2003. Effect of trehalose on survival of *Bradyrhizobium japonicum* during desiccation. *J. Appl. Microbiol.* 95: 484–491.

Sun, W. Q. and Davidson, P. 1998. Protein inactivation in amorphous sucrose and trehalose matrices: Effects of phase separation and crystallization. *Biochem. Biophys. Acta* 1425: 235–244.

Tanaka, M. M., Earl, A. M., Howell, H. A., Park, M. J., Eisen, J. A. and Peterson, S. N. 2004. Analysis of *Deinococus radiodurans* transcriptional response to ionizing radiation reveals novel protein that contribute to extreme radioresistance. *Genetics* 168: 21–33.

To, B. C. S. and Etzel, M. R. 1997. Survival of *Brevibacterium linens* (ATCC 9174) after spray drying, freeze drying or freezing. *J. Food Sci.* 62: 167–169.

Tunnaclife, A., de Castro, A. G. and Manzanera, M. 2001. Anhydrobiotic engineering of bacterial and mammalian cells: Is intracellular trehalose sufficient? *Cryobiology* 43: 124–132.

Tunnacliffe, A. and Lapinski, J. 2003. Resurrecting Van Leeuwenhoek's rotifers: A reappraisal of the role of disaccharides in anhydrobiosis. *Philosophical transactions of the Royal Society of London Series B, Biological Sciences* 358: 1755–1771.

Uzunova-Doneva, T. and Donev, T. 2000. Influence of the freezing rate on the survival of strains of *Saccharomyces cerevisiae* after cryogenic preservation. *J. Cult. Collect.* 3: 78–83.

Uzunova-Doneva, T. and Donev, T. 2005. Anabiosis and conservation of microorganisms. *J. Cult. Collect.* 4 (1): 17–28.

van Boekel Martinus, A. J. 2002. On the use of the Weibull model to describe thermal inactivation of microbial vegetative cells. *J. Food Microbiol.* 74: 139–159.

Venkateswaran, K., Kempf, M., Chen, F., Satomi, M., Nicholson, W. and Kern, R. 2003. *Bacillus nealsonii* sp. *nov.*, isolated from a spacecraft-assembly facility, whose spores are gamma-radiation resistant. *Int. J. Syst. Evol. Micr.* 53: 165–172.

Vriezen, J. A., de Bruijn, F. J. and Nüsslein, K. 2006. Desiccation responses and survival of *Sinorhizobium meliloti* USDA 1021 in relation to growth phase, temperature, chloride and sulfate availability. *Lett. Appl. Microbiol.* 42: 172–178.

Wolkers, W. F., Tablin, F. and Crowe, J. H. 2002. From anhydrobiosis to freeze-drying of eukaryotic cells. *Comp. Biochem. Phys.* 131 A: 535–543.

Yin, D. and Chen, K. 2005. The essential mechanisms of ageing: Irreparable damage accumulation of biochemical side-reactions. *Expl. Gerontol.* 40: 455–465.

Zhou, X., Chen, S. and Yu, Z. 2004. Effects of spray drying parameters on the processing of a fermentation liquor. *Biosyst. Eng.* 88 (2): 193–199.

Ziadi, M., Touhami, Y., Achour, M., Thonart, P. and Hamdi, M. 2005. The effect of heat stress on freeze-drying and conservation of *Lactococcus*. *Biochem. Eng. J.* 24: 141–145.

chapter six

Recent developments in solid-state fermentation

Chinese herbs as substrate

Hongzhang Chen and Guanhua Li

Contents

6.1 Introduction

Chinese herb solid-state fermentation (SSF) processing has a long history; numerous Chinese medicines such as Shenqu and Hongqu are all produced using SSF. In this chapter, the essence, history, and advantages of SSF are introduced, and the application of SSF in the concoction and extraction of herbs is also stated. The key technology used in Chinese herb SSF processes such as pretreatment, microorganisms, and SSF technology are systematically summarized. Finally, the idea of process engineering for Chinese herbs SSF and its ecoindustrial integration was proposed and explained.

6.2 Solid-state fermentation versus submerged fermentation

Before discussing Chinese herb SSF processing, detailed comparisons between SSF and submerged fermentation (SMF) are provided in Table 6.1. From the table, the many advantages of SSF over SMF can be revealed:

1. The culture conditions in SSF are more similar to their natural habitat, which leads to higher product yields (Castilho et al. 2000).
2. The fermentation media per mass of substrate in SSF is smaller than that in SMF, which results in smaller reactors and lower capital costs yet higher product yields.

Table 6.1 The Detailed Comparisons between SSF with SMF

		SSF	SMF
Components	Solid phase	More than 70%	Less than 5%
	Liquid phase	Important part	Main body
	Gas phase	Free oxygen, playing an important role	Dissolved oxygen, limit factor
		Gas phase	Liquid phase
Substrate	Medium	Culture media are simple and low-cost	High quality requirements and high cost
	Nutrients	Nutrients are dissolved in water that is absorbed on substrate, and the distribution is uneven	Nutrients are dissolved in water, and the distribution is uniform
Microorganisms	Growth state	Microorganisms adsorb on or penetrate into the solid substrate	Microorganisms suspend in the culture system
	Strains	Natural enrichment or artificial breeding strains	Pure strains
	Inoculation size	More than 15%	Less than 10%
	Water activity	Low	High
	Environment	Similar to their natural habitat	Not conducive to microbial growth
Fermentation process	Determination and control	Hard	Easy
	Uniformity	Heterogeneity	Homogeneity
	Heat removal	Hard	Easy
Bioreactor	Characteristics	Simple, open	Complex, sealed
	Cost	Operation cost is high	Investment cost is high
	Gas circulation	Oxygen supply and heat removal	Oxygen supply
	Scale-up	Hard	Easy
Product	Concentration	High	Low
	Yield	High	Low
	Wastewater	Seldom	Large
	Extraction	Complicated	Easy

Source: With kind permission from Springer Science+Business Media: Modern Solid State Fermentation—Theory and Practice, 2013, Chen, H.Z.

3. Aeration is much easier and the aeration efficiency is higher in SSF, which is beneficial for those fermentation processes demanding high levels of oxygen (Nagao et al. 2003).
4. Diversity in products, especially for anaerobic mixed SSF (Sindhu et al. 2006).
5. Lower demand on sterility due to the low water activity used in SSF (Hölker et al. 2004).
6. Less water is needed in SSF.

The major challenges in SSF that need to be overcome can be summarized as follows:

1. Scale-up in SSF has long since been a limiting factor because of its inefficient heat and mass transfer.
2. Determination and separation of biomass are also challenges in SSF.

Although there are some challenges such as high cost of product recovery and purification in SSF, an economic evaluation of the overall process should be done to determine its feasibility for a specific purpose (Rodrıguez-Couto and Sanroman 2005; Singhania et al. 2009). This system is suitable for the production of high-value products such as medicines and enzymes, especially those applications in which concentrated products with high titers are required rather than a high degree of purity.

6.3 *History of Chinese herb SSF processing*

In a contemporary Chinese dictionary, traditional Chinese medicine (TCM) is defined as the sum of medicines that are used by herbalist doctors, which includes plants, animals, and a few minerals. For this definition, folk medicine, ethnic medicine, modern Chinese herbs, and even ordinary western medicine were all included. Lei (1998) defined TCM as the generic term for all medicines that have been used by Chinese people in the long history of that nation, which excludes modern Chinese herbs and western herbs. Gao (2007) pointed out that TCM represents not only the medicines but also the traditional syndrome differentiation theory. In conclusion, we thought that TCM were the medicines that were used for the prevention, diagnosis and treatment, and rehabilitation of health, under the guidance of syndrome differentiation theory. Because of the kernel of syndrome differentiation theory of Chinese medicine, TCM cannot be deemed as an empirical health care system. The various aspects of TCM includes Chinese materia medica, acupuncture, and moxibustion, among others. Here, Chinese herbs will be introduced in detail.

Solid-state fermentation originated in ancient China and developed into a vibrant industry in traditional Chinese liquor, rice wine, soy sauce, and vinegar. As early as 4500 years ago, Chinese herbs were processed by using solid fermentation technology to change the original quantity, improve the medicinal efficacy, and expand medication performance. For example, medicated leaven, fermented soybean, and red fermented rice are all fermented products that have been and are still used up to now (Table 6.2). The SSF of Chinese herbs is the cultivation process of beneficial microbes on pretreated herbs; the optimum temperature is approximately 30°C to 37°C, and relative humidity is 70% to 80%.

Solid-state fermentation of Chinese herbs can be divided into two categories:

1. Herbs mixed with other raw materials are used as culture medium, which could provide adequate carbon and nitrogen sources for the microorganisms. At the same time, the microorganisms produce metabolites or transform the herbs' ingredients into beneficial active substances.
2. Medicinal fungi are cultivated on agricultural materials that provide either nutrients or inducers to obtain active metabolites.

Table 6.2 Chronology of Important Events Related to TCM

Period	Content
Ancient times	Herb was washed and cut, and also roasted when fire was discovered, to reduce toxicity and enhance the efficacy
Xia and Zhou	It was the beginning of fermentation of Chinese herb, for example, rice wine, soy sauce, and vinegar were produced
Qin and Han	Several medicinal fungi such as *Ganoderma lucidum*, *Poria cocos*, and *Phellinus baumii* were discovered and their medicinal effects were recorded in "Shen Nong Materia Medica." The theory of TCM came into being
Han and Jin	Many processing technologies and medicinal functions of TCM such as medicated leaven and fermented soybean were recorded in "Febrile Disease"
The North Wei	The SSF technology was developed further
Tang	Pretreatment methods of Chinese herbs were summarized; the theory of TCM was concluded and spread to East Asian countries
Song	The effects of auxiliary materials on Chinese herbs during the SSF were studied and utilized
Jing	Mongolian medicine and Tibetan medicine were absorbed into TCM
Ming	The Ming era was the golden era of TCM research, and the theory of fermentation and concoction had been established completely. The TCM and traditional syndrome differentiation theory were combined more closely
Qing	The theory of fermentation and concoction was enriched and developed

Source: Chen, H.Z. Medicinal Plants, Process Engineering and its Ecological Industry Integration. Science Press, Beijing, China, 2010; Gong, Q.F. The Concoction of Chinese Traditional Medicine. Chinese Medicine Press, Beijing, China, 2007.

Here, we introduce two fermented Chinese herbs:

(a) *Shenqu*—also known as medicated leaven or fry bran koji, was first recorded in the Han Dynasty, and is a traditional Asian fermented medicine. Medicated leaven is a type of SSF product made from flour and food fungus. In China, the documentation recommends the use of medicated leaven to help digestion, eliminate phlegm, and so on.

The traditional SSF methods were summarized as follows: (1) *Prunusmandshurica Koehne* and *Phaseolus calcaratus Roxburg* were ground to powders and mixed with flour; (2) *Artemisia annua* juice, *Xanthium sibiricum* juice, and *Polygonum flaccidum* juice were added; (3) after that, the powders were mixed and kneaded, and formed into a paste; (4) the fermentation proceeded at 30°C to 37°C for 7 to 8 days; and finally, (5) the Shenqu should be kept in a cool and dry place (Nan and Zhu 1992).

(b) *Hongqu*—also known as red fermented rice, red yeast rice, or anka, was first recorded in the Zhou Dynasty, and is a traditional Asian fermented food and folk medicine. Hongqu is a type of SSF product made from steamed rice and food fungus from the *Monascus* genus. Hongqu is rice-shaped, purple or brown-red, crisp, friable, and slightly tartish or tasteless. In China, the documentation recommends the use of Hongqu as a colorant for cooking, Chinese medicine, and fermentation starters to brew red wine.

The traditional SSF methods were summarized as follows: (1) A 1-m-deep pit is made in red soil, and a mat is put around this pit; (2) rice (preimmersed for 2 h in water at 30°C) and *Monascus* are placed into the pit; (3) a big stone is placed over the fermentation medium; (4) fermentation proceeds at approximately 28°C for 3 to

4 days; (5) the shell of the rice turns red, and the core of the rice turns purple-red; (6) the red fermented rice is put into a pot and roasted until the shell becomes black and the core becomes yellow; and finally, (7) Hongqu should be kept in a cool and dry place (Chen and Hu 2005).

6.4 Principle of Chinese herb SSF processing

6.4.1 Natural SSF of herbs

In ancient times, people occasionally vomited or even lost consciousness after eating some plants and animals. However, these symptoms might be alleviated or relieved by eating some other foods. People probably gained the first cogitation of medicine from these daily practices. In other circumstances, people consciously preserved some excess fruits during harvest time, which would begin to ferment if the environment was suitable. The sugars in the fruits were first fermented into ethanol by microbial metabolism, and subsequently, the ethanol was partially transformed into acetic acid. Then, the reaction of ethanol with acetic acid produced some special flavor substances such as ethyl acetate, which would bring in a magical taste. In fact, this is similar to SSF, which occurred during the storage process of Chinese herbs. The herbs would be invaded by some microbes, especially molds such as *Penicillium* and *Aspergillus*, which would produce enzymes and secondary metabolites using it as a carbon and nitrogen source. Some active ingredients in the herbs were modified or transformed by enzyme or microbes during the process. Therefore, the fermented herbs could be used as medicine, and the herbs resulting from different microbes would produce different medical efficacies.

6.4.2 SSF applied to a concoction of Chinese herbs

Herbs should be concocted before being used as medicine, which is characteristic of traditional Chinese herbs, and is also a major feature of the theory behind Chinese medicine. The effects of concoction can be summarized by two aspects: (1) destruction or alteration of the chemical structure of the noxious ingredients in the herbs to release toxicity; and (2) inactivation of the coexisting enzymes in the herbs to ensure the clinical efficacy of the medicine.

The effects of concoction caused by microbes on Chinese herbs can be mainly stated as follows:

1. *Biotransformation of herbs*. The ingredients in herbs can be modified or even transformed into new compounds by various enzymes such as cellulase, laccase, lignin peroxidase, and lipase, which is produced from microbes at room temperature. For example, Pu-erh tea is a well-known traditionally fermented product that is derived from the dry leaves of *Camellia sinensis* var. *assamica* in Yunnan, China. Several investigations have recently confirmed the beneficial effects of Pu-erh tea on health, such as its antioxidant, antimicrobial, and anticancer effects, as well as lowering cholesterol, blood pressure, and blood sugar levels (Hou et al. 2009). Wang et al. (2011) studied the SSF of Pu-erh tea using *Foxia* as a microbe. The SSF proceeded for 20 days at 45°C and 70% relatively humidity. The results showed that the theabrownin, total carbohydrate, polysaccharide, amino acid, and protein contents all increased, whereas the tea polyphenol content decreased sharply. Kluge and coworkers (2012) reported that stereoselective benzylic hydroxylation and C1–C2 epoxidation of

alkylbenzenes were catalyzed by heme-thiolate peroxygenase produced from *Agrocybe aegerita*. Benzylic hydroxylation led exclusively to the (*R*)-1-phenylalkanols. Epoxidation of straight-chain and cyclic styrene derivatives gave a heterogeneous picture. Cranberry pomace contains large amounts of phenolic glycosides, which are important sources of free phenolics that are used as antioxidants, flavorings, and nutraceuticals. Zheng and Shett (2000) used β-glucosidase produced by *Lentinus edodes* to hydrolyze glycosides. The results showed that after 50 days of cultivation, the yield of total free phenolics could reach 0.5 mg/g. The major free phenolics produced from cranberry pomace were gallic acid, chlorogenic acid, *p*-hydroxybenzoic acid, and *p*-coumaric acid. It can be concluded that the enzymes produced by microorganisms were closely related to the compositional changes occurring as well as the quality of the herbs used.

2. *Biological detoxification of herbs.* Some toxic substances in the herbs can be decomposed by the microbes during the SSF. Brand et al. (2000) studied the detoxification of coffee husk in SSF using *Rhizopus*, *Phanerochaete*, and *Aspergillus*. The results showed that *Rhizopus arrizus* can degrade 87% of caffeine and 65% of tannins when fermentation proceeded for 6 days at pH 6.0 and at 60% moisture. When *Phanerochaete chrysosporium* was used, the maximum degradation of caffeine and tannins were 70.8% and 45%, respectively. When fermentation proceeded for 14 days at pH 5.5 and at 65% moisture, *Aspergillus* was used; the best detoxification rates could achieve 92% for caffeine and 65% for tannins (Brand et al. 2000). The results demonstrated that microbes had the ability to detoxify herbs.
3. *The formation of new active compounds.*
   a. Several secondary metabolites could be produced by medicinal fungus
   b. The secondary metabolites formed may interact with the active ingredients of herbs to generate new compounds
   c. Compared with a conventional culture medium (straw or starch), herbs are full of active ingredients, such as polyphenol, which may promote the growth of microbes

Treviño-Cueto and coworkers (2007) used *Larrea tridentata* powder as the sole carbon source and inducer to cultivate *Aspergillus niger*. The results indicated that tannase production reached values of 1040 units/L, and a high content of tannins and gallic acid were obtained during fermentation for 43 h. Lin and Chiang (2008) used *Radix astragali* as the medium to culture *Cordyceps militaris*, which would produce a substance with antitumor activity. These results showed that the antitumor activity of the crude fermentation broth can be enhanced by microorganisms. Besides cordycepin, several other unknown compounds were also responsible for the antitumor activity, which indicated that the novel antitumor activity mechanisms were caused by the addition of *R. astragali*.

6.4.3 SSF applied to the extraction of Chinese herbs

The distributions of active ingredients in Chinese herbs are very complex. Active ingredients in different parts of Chinese herbs may have different pharmacological effects. Therefore, the extraction of active ingredients is an important step. A basic requirement in the extraction process is the maximization of the yield of active ingredients from the herbs while keeping the yield of harmful and ineffective substances at a minimum. Traditional

Table 6.3 Modern Technology Applied to the Extraction of Herbs

Types	Mechanism/Characteristic
Ultrasound-assisted extraction (UAE)	The mechanism is generally attributed to mechanical, cavitation, and thermal effects. These effects can result in the disruption of cell walls, reduction in particle size, and enhancement of mass transfer. The implosion of cavitation bubbles accelerates the eddy diffusion and internal diffusion (Cabaleiro et al. 2013)
Supercritical fluid extraction (SFE)	SFE is based on the use of an extractant in its supercritical state; gas can penetrate into porous solid materials more effectively than liquid. Carbon dioxide is commonly used for this purpose, because it easily reaches supercritical conditions, and moreover, it displays low toxicity (Lang and Wai 2001) and easy separation
Microwave-assisted extraction (MAE)	MAE uses microwave energy to enhance the extraction efficiency. The electric field causes heating via two simultaneous mechanisms, dipolar rotation and ionic conduction. This oscillation produces collisions with surrounding molecules and thus the liberation of thermal energy into the medium (Kaufmann and Christen 2002)

Source: Chen, H.Z. Process Engineering in Plant-Based Products. Nova Science Publishers, Inc., New York, 2009.

extraction methods for herbs include boiling, soaking, percolation, refluxing, sublimation, and others. Although these classic methods have their own advantages, some issues still need to be addressed, such as being time-consuming, having low efficiency, using large amounts of solvent, and having a complicated extraction procedure. Consequently, researchers are still exploring more effective extraction methods. With the development of science and technology, more and more new technologies are being applied to the extraction process of herbs (Table 6.3; Chen 2009).

Active ingredients in herbs wrapped by cellulose, hemicellulose, lignin, and others could cause difficulties in extraction. The dense structure of herbs can be destroyed by the microbe's invasion or effectively degraded by enzymes, which are beneficial for further extraction processes. The SSF of Pu-erh tea was done by Wang and colleagues (2011). The results demonstrated that the surfaces of untreated tea appeared to be smooth and intact, and the structures of the cell, cellulose, and lignin were complete. Yet, after several days of fermentation, their surfaces were covered by microorganisms and the structures of the cells were largely disrupted. The scanning electron micrograph (SEM) morphology also suggests that the structural changes in the tea leaves were related to the growth of microorganisms (Figure 6.1).

6.4.4 *Advantages of Chinese herb SSF processing*

Compared with other methods, the advantages of SSF on Chinese herb processing can be stated as follows:

1. Mild reaction condition. SSF is often carried out under relatively mild conditions such as ambient temperature and pressure; as a result, the active ingredients in the herbs can be kept from destruction as effectively as possible, especially for heat-sensitive aromatic volatile substances, vitamins, and other active ingredients.

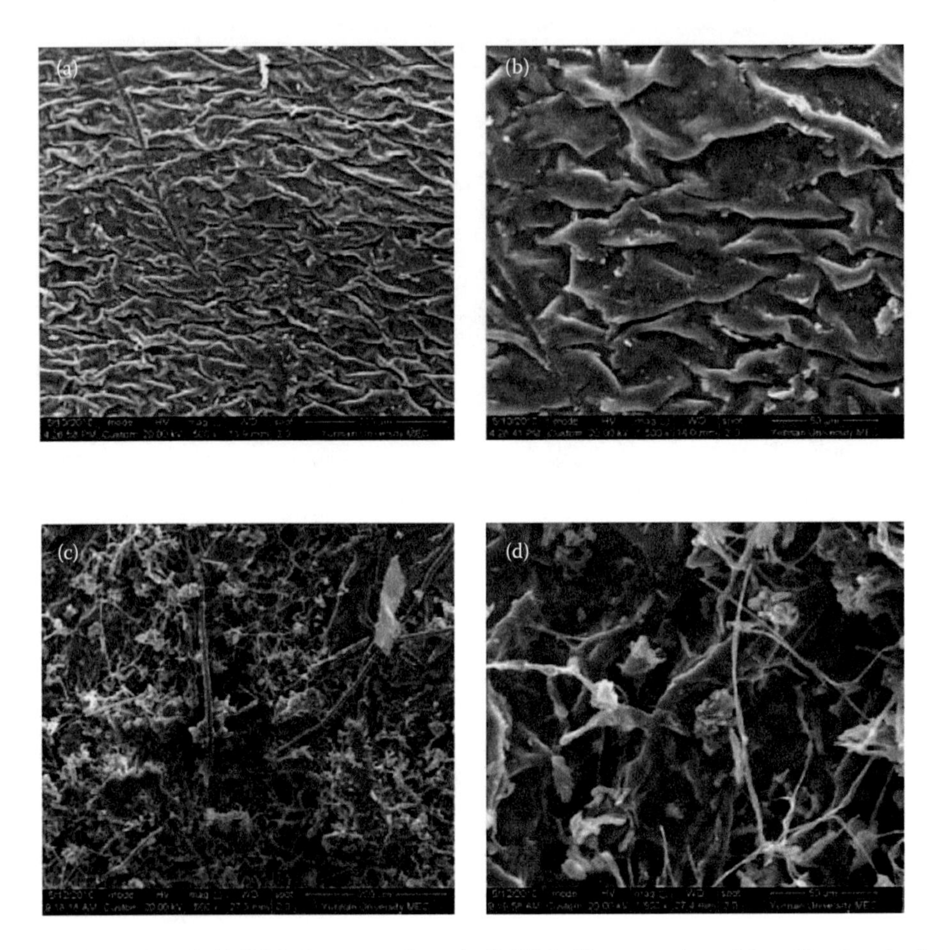

Figure 6.1 SEM of tea. (a, b) SEM of untreated tea. (c, d) SEM of fermented tea. Magnification: (a, c) ×500; (b, d) ×1500. (From Wang, Q.P. et al., *Journal of the Science of Food and Agriculture*, 91: 2412–2418, 2011.)

2. Formation of new Chinese medicine compounds. Some microbes have been identified as traditional folk medicines for their special active metabolites. Microbes could interact with herbs during the SSF. The microbial growth and metabolism could be promoted, and the ingredients in herbs could be modified. As a result, the medicinal effects of Chinese medicine compounds were improved.
3. Being convenient for assimilation and transformation of active ingredients. The macromolecule substances can be degraded into micromolecule substances by microbial metabolism. Compared with macromolecule substances, small active molecule substances easily pass through the blood–brain barrier and combine with the protein of human cells. Thus, they can be faster and more completely assimilated and used by humans.
4. Multilevel utilization of Chinese herbs. The ingredients of herbs are complex; in the traditional process, only several special components have been used, yet many other parts were discarded, which resulted in a waste of resources. During the fermentation process, herbs were repeatedly used and most of the nutrients can be converted. Therefore, SSF is beneficial for comprehensive utilization of herb resources.
5. Last but not least, the fermentation products can be used without any postprocessing.

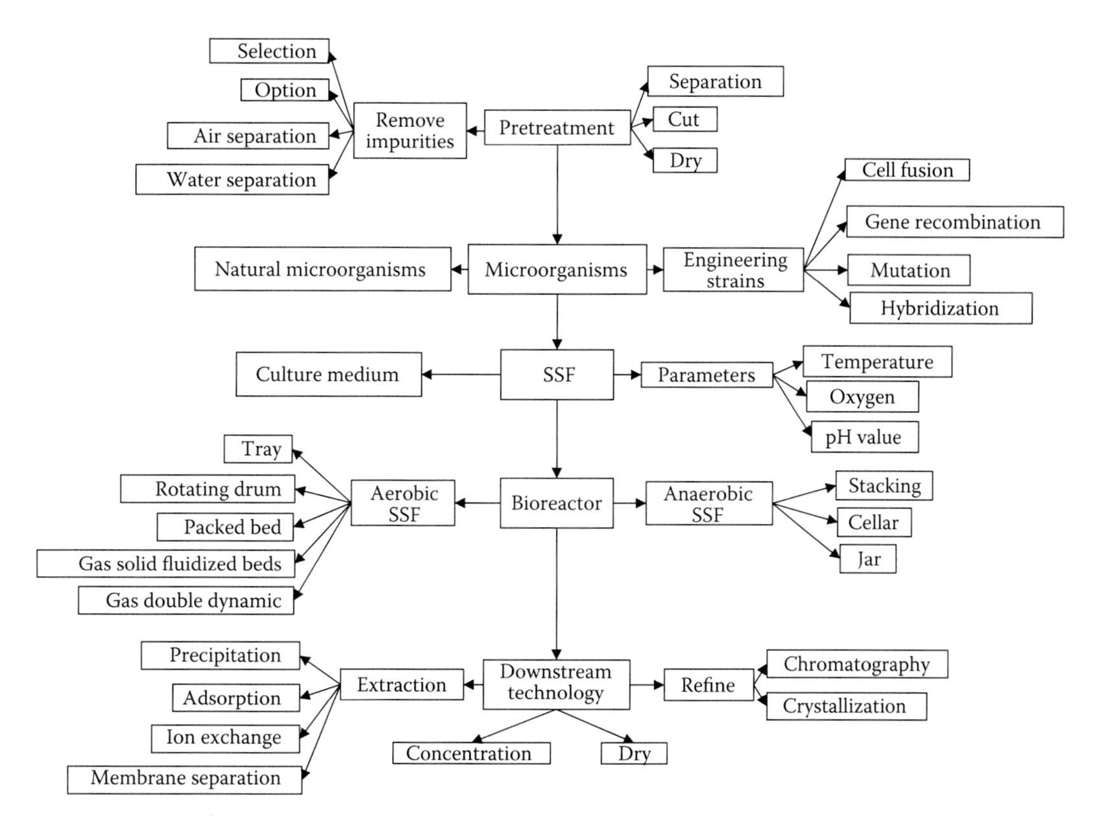

Figure 6.2 SSF on Chinese herb processing. (From Chen, H.Z., *Medicinal Plants, Process Engineering and Its Ecological Industry Integration*. Science Press, Beijing, China, 2010; Gong, Q.F., *The Concoction of Chinese Traditional Medicine*. Chinese Medicine Press, Beijing, China, 2007.)

6.5 Process and key technologies in Chinese herb SSF process

There are five sections involved in SSF of herbs, which are summarized as follows (Figure 6.2).

6.5.1 Pretreatment process of Chinese herbs

6.5.1.1 Selecting and washing

Herbs are often mixed with sediment impurities, and invaded by insects and fungus during the collection, pretreatment, storage, and transportation processes. Therefore, herbs need to be pretreated to reach the standard of medicinal purity and to facilitate the subsequent processes. Selection is the first step of herb processing. A complete herb usually consists of roots, stems, leaves, flowers, and fruit. According to the different ingredients contained in different parts of one herb, the herb should be selectively fractionated to reach uniformity, such as the removal of roots, residual stems, cores, and hull.

The sand and impurities can be separated from herbs, after being passed through different specification sieves. On the other hand, the different parts of herbs can be fractionated based on their size, which is suitable for subsequent fermentations and concoctions (Figure 6.3; Gong 2007).

Air separation is one of the most common methods used in herb processing. Herbs and impurities can be blown off, shaken, and separated by wind according to their different

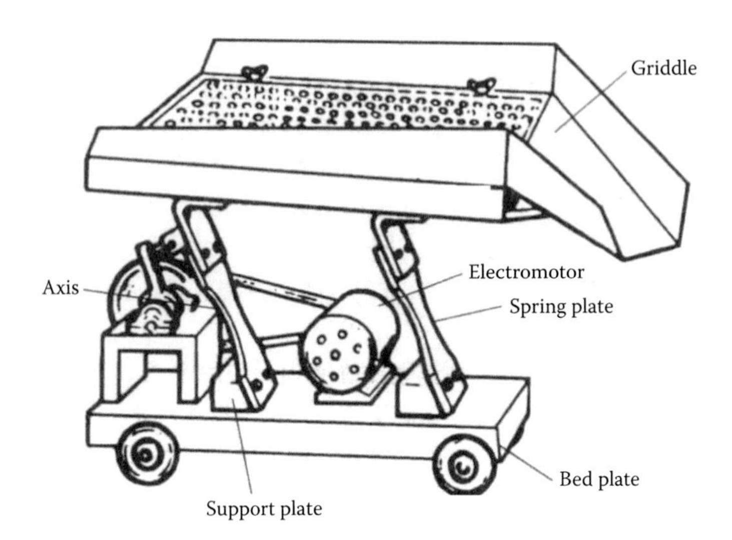

Figure 6.3 Shock screening machine. (From Gong, Q.F., *The Concoction of Chinese Traditional Medicine*. Chinese Medicine Press, Beijing, China, 2007.)

specific gravities. If the sediments, salt, or other impurities cannot be removed by air separation, the herbs should be washed in water.

6.5.1.2 *Cutting process technology*

The particle size of the herbs is also a major factor affecting the performance of SSF. Large particle size can hamper the penetration of fungi into cellulosic biomass and also prevent the diffusion of air, water, and metabolite intermediates into the particles. However, the reduced particle size with a decreased size of interparticle channel may adversely affect interparticle gas circulation (Zadrazil and Puniya 1995). Cutting is the process in which herbs are cut into special specifications to facilitate identification, fermentation, concoction, and extraction. Dried herbs are relatively hard, so soaking pretreatment is the first step that needs to be done before cutting. During the soaking process, herbs are soaked to soften, as a result, herbs are easy to cut and the toxic effects of the herbs can be alleviated. The cutting process can be divided into hand-cutting and machine-cutting. Hand-cutting is flexible, but less efficient, thus it is suitable for fine herb processing. Machine-cutting is fast, highly efficient, and less labor-intensive, thus it is suitable for bulk herb processing. Commonly used machines include chopping knife, cutting machines, and rotary cutting machines (Gong 2007; Figure 6.4).

6.5.1.3 *Drying process technology*

After the cutting process, Chinese herbs should be dried for long-term preservation. The basic requirements being that the herbs should retain their shape, color, smell, and taste. The drying process is usually divided into natural drying and machine drying. For natural drying, herbs are placed in the sun or in a cool but ventilated place. Traditional natural drying technology does not need special equipment, but the drying efficiency is often affected by the weather. For machine drying, herbs are dried by equipment such as a direct hot air dryer, electric dryer, or far infrared dryer, which is clean, weatherproof, and highly efficient.

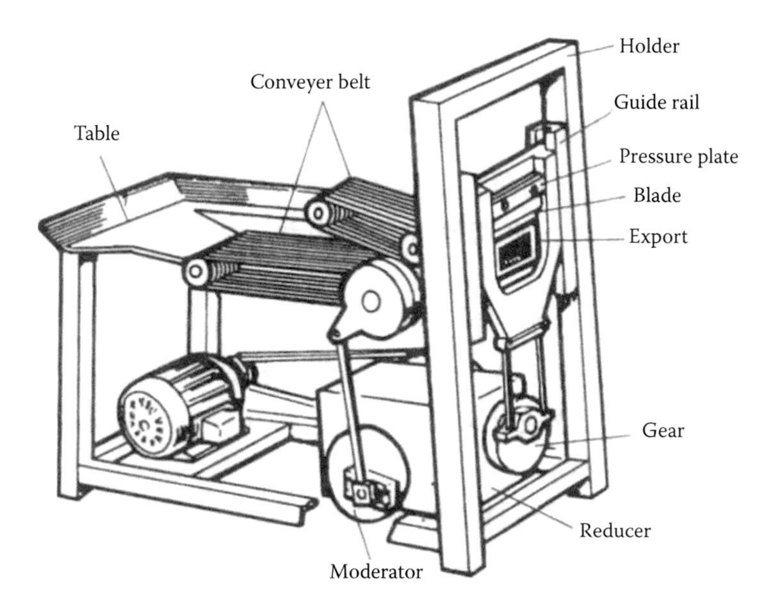

Figure 6.4 Rotary cutting machine. (From Gong, Q.F., *The Concoction of Chinese Traditional Medicine.* Chinese Medicine Press, Beijing, China, 2007.)

6.5.2 SSF microorganisms

Microorganisms constitute the most abundant and diverse set of organisms on earth, and are widely distributed in the soil, water, and air. Microorganisms play a dominant role in performing key biochemical reactions essential to sustaining the biosphere. Generally, it is accepted that more than 80% of the total microorganisms are unknown; however, reactions mediated by both the known and the unknown microorganisms are already used in SSF (Baxter and Cummings 2006). Regardless of the exact nature of the fermentation and concoction process, all fermentation techniques depend on the right microorganisms in the right place with the right environmental conditions for degradation or biotransformation to occur (Iranzo et al. 2001). The right microorganisms are those microbes that have the physiological and metabolic capabilities to utilize and degrade the herbs. Compared with other environments, the conditions in soil are more suitable for the growth of microorganisms, thus the most industrial strains are isolated from the soil. The vast majority of wild-type microorganisms cannot produce commercially acceptable yields of the desired product, so the isolated strains are often modified.

Selection of suitable microorganisms can be deemed as one of the most important criteria in SSF. The requirements of SSF for strains can be stated as follows:

1. A unique characteristic of SSF is having no free water in the medium, so the microbes can grow in low concentrations of moisture
2. The microorganisms are capable of producing a range of extracellular enzymes required for the hydrolysis of complex, polymeric raw solid substrates
3. Growth cycle is as short as possible
4. The culture condition such as temperature, pH, and oxygen can be easily met

5. The microorganisms should grow fast and can inhibit other microbes' reproduction
6. They cannot be pathogenic microorganisms or gene-engineering strains
7. For the microorganisms used in SSF of herbs, there are several additional requirements that need to be stated
   a. The growth of microorganisms cannot be inhibited by the active ingredients in the herbs, and the active ingredients in the herbs cannot be destroyed by microbes either
   b. The microorganisms do not produce harmful substances to humans, and can easily be inactivated

6.5.2.1 *Medicinal fungi*

In "Shen Nong Materia Medica," plenty of fungi were recorded as medicine, which had been used for thousands of years in East Asia, especially in China. Medicinal fungus belonging to higher Basidiomycetes and Ascomycetesares are an immensely rich yet largely untapped resource of useful, accessible, and natural compounds with various biological activities. The medicinal ingredients produced by fungus have been isolated and identified from the fruit bodies, culture mycelium, and culture broth. These various cellular components and secondary metabolites (polysaccharides, proteins and their complexes, phenolic compounds, polyketides, triterpenoids, steroids, alkaloids, nucleotides, etc.) have provided significant opportunities for the treatment of patients, for example, using drugs with antitumor, antioxidant, immunomodulatory, antibacterial, antiparasitic, radical scavenging, antihypercholesterolemia, cardiovascular, antiviral, detoxification, and hepatoprotective effects, among others (Elisashvili 2012). For example, the best known antibiotic, penicillin, discovered from a species of *Penicillium* in the early 1940s, has thus far played an important role in human life.

The active ingredients in medical fungus can be divided into five types: polysaccharides, alkaloids, sterols, terpenes, and polypeptides (Table 6.4).

Wasser (2010) summarized that approximately 700 species of higher Heterobasidiomycetes and Homobasidiomycetes could produce fungal polysaccharides. Approximately 126 medicinal compounds produced by fungus have been reported, and several of these compounds have been proceeded through Phase I, II, and III clinical trials and used extensively and successfully in Asia.

Phellinus Quél. is an important genus in Basidiomycota; approximately 220 species belong to this genus (Larsen and Cobb-Poulle 1990). Dai and coworkers summarized that approximately 70 species of *Phellinus* have been reported in China, and 26 species are recognized as medicinal fungi. Polysaccharides and polyphenols of the *Phellinus* genus are two of the main kinds of medicinal metabolites, and their medicinal functions can be summarized as antitumor, improving immunity, and antioxidant (Dai and Yang 2008; Dai et al. 2010).

6.5.2.2 *Photosynthetic bacteria*

Photosynthetic bacteria (PSB), the common microorganisms in the natural environment, have been applied in the fields of environmental protection and energy production, as well as in the medical, fertilizer, and food industries. Photosynthetic bacteria can produce relatively large amounts of physiological active substances such as vitamin B_{12}, ubiquinone, 5-aminolevulinic acid, porphyrins, and RNA, some of which have been prepared and commercialized (Sasaki et al. 2005; Table 6.5).

Table 6.4 The Active Ingredients Produced from Medical Fungi

Active ingredients	Medicinal metabolites	Functions	Species
Polysaccharide	Hispidin,1,1-distyrylpyrylethan 3,14′-bihispidinyl, hypholomine	b, c	*Inonotus xeranticus, Phellinus linteus* (Jung et al. 2008)
		i, l, n, o	*P. baumii* (Hwang et al. 2007)
	Intracellular polysaccharide, triterpenoid	e, h, k	(Shih et al. 2006)
	Intracellular polysaccharides, Exopolysaccharides	c	*Laetiporus sulphureus* Lung and Huang (2012)
	Polysaccharide fractions	a, g	*G. lucidum* (Zhao et al. 2010)
	Heteropolysaccharide maitake Z-fraction	g	*Grifola frondosa* (Masuda et al. 2010)
Alkaloid	Isoindolinone	g	*Hericium erinaceum* (Noh et al. 2012)
	Sesquiterpenes	g	*Flammulina velutipes* (Xu et al. 2013)
Sterol	Ascorbic acid, flavonoids	c, e	*Phellinus igniarius* (Lung et al. 2010)
	Geranylated aromatic compound	c	*Hericium erinaceum* (Yaoita et al. 2012b)
	Vitamin D_2	g, h, j	*Agaricus bisporus* (Koyyalamudi et al. 2011)
	Vitamin D_4	j	(Phillips et al. 2012)
	Cerevisterol	g	(Froufe et al. 2012)
Terpene	Lanostane triterpenoids	d	*Ganoderma orbiforme* (Isaka et al. 2013)
	Sesquiterpenoids	d	*Russula sanguinea* (Yaoita et al. 2012a)
	(-)-(2S,8R)-8,12-dihydroxy-isolongifolanol	d	*Aspergillus niger* (Sakata et al. 2011)
Polypeptide	Low-molecular weight protein fraction	g	*Grifola frondosa* (Kodama et al. 2010)
		g	(Liang et al. 2011)
		h	*G. lucidum* (Yeh et al. 2010)

Notes: Medicinal functions of fungus: (a) activation of macrophage cell; (b) alleviating septic shock; (c) anti-oxidation; (d) antibacterial; (e) anti-inflammatory; (f) antiplasmodial activity; (g) antitumor activities; (h) improving immunity; (i) lowering serum lipids; (j) preventing and treating cardiovascular diseases; (k) promoting blood circulation; (l) restoring liver; (n) treating anemia; and (o) treating human diabetes.

Zhu et al. (2010) studied active extracts from Liuweidihuang that was fermented by PSB. The results showed that PSB could lead to an increase in endogenous antioxidant enzyme activity and a reduction in the risks of lipid peroxidation accelerated by age-induced free radicals. Li added certain amounts of extract of Chinese herbs such as liquorice, baikal skullcap, and trichosanthis into the medium to cultivate PSB. The results showed that the growth of PSB can be greatly prompted (Sun et al. 2007).

Table 6.5 Active Substances Produced from Photosynthetic Bacteria

Medicinal metabolites	Medicinal functions	Species
Vitamin B_{12}	Treating anemia and neuritis, or used as an eye lotion	*R. spheroids* (Cauthen et al. 1967) *R. spheroids, R. rubrum* (Ohmori et al. 1971) *Propionibacterium* (Sasaki et al. 1998)
Coenzyme Q10	Antioxidative supplement, enhancing immune system, easing hypertension	*R. rubrum* (Tian et al. 2010) *R. palustris* (Fang et al. 2005) *Rhodobacteria* sp. (Lai et al. 2006)
5-Aminolevulinic acid	Treatment of skin cancer	*R. sphaeroides* (Tanaka et al. 1994)
Porphyrin	Treatment of skin cancer	Photosynthetic bacteria (Mongin et al. 1999)

6.5.3 SSF technologies

6.5.3.1 Preparation culture medium with Chinese herbs

Nutrition is the process in which microorganisms get energy and nutrients from the external environment in their lifetime, which also plays basic physiological functions that provide structure substances, energy metabolism regulation substance and the necessary physiological environment to metabolisms (Zhou 2002). The addition of nutrients to the herb medium can generally increase the formation of the fungal biomass and also facilitated fungal colonization in the deeper areas. Microbial basic nutritional elements can be divided into six categories: carbon sources, nitrogen sources, energy, mineral, water, and growth factors (Chen 2013). Carbon sources are major nutrients for microorganisms, and includes organic carbon and inorganic carbon sources. In SSF processes, the carbon source substances can often be used both as nutrients and as inert carrier materials that maintain the microorganisms' growth. Consequently, it is essential to go into the characteristics of the solid substrate during the SSF process, especially for the amplification process. Chen (2013) has paid much attention to the pretreatment of the herbs, which will be discussed in later sections. Nitrogen sources mainly provide nitrogen elements for microbial growth, which are used to synthesize important live protein materials and nucleic acid. Common raw protein materials mainly include bean substances, such as soybean peas, soybean meal, bran, urea, peptone, cicada chrysalis powder, and others. Energy mainly provides initial energy sources for nutrition or for microbial organisms. Inorganic salts and growth factors are two other substances needed for microbial growth. One common characteristic is their lower demand; they are essential to the growth of the microorganisms yet they cannot be synthesized by the microbial organisms themselves. These mainly include a vitamin base, amine, and small-molecule fatty acids. Inorganic salts refer to K, P, S, Ca, Mg, and others. During the fermentation process, the additional carbon and nitrogen sources are often mixtures that include combinations of ingredients. For example, straw is rich in K, P, S, Ca, Cu, Mg, and others, which can provide enough inorganic salt for the growth of microbes (Yu and Chen 2010). Water is a necessary nutritional element for microbial growth and is an important part of the organism itself. The initial water content of the substrate is important for fungal establishment and growth, which also affects the secondary metabolism of the fungus. Previous studies suggested that an initial water content ranging from 70% to 80% was the optimal level for most white rot fungi (Asgher et al. 2006; Wan and Li 2010) Water can assist microbes in transferring metabolic nutrients from the outside into the cell. On the other hand, water molecules also play a role in the maintenance of macromolecular stability, and provide a relatively stable microenvironment. In aerobic SSF, water can be divided into bound water and free water, and water content is an important factor that influences heat and mass transfer. Bound water exists as a thin water film layer, and plays a role in the absorption of nutrients and the desorption of metabolic substances. Free water, commonly expressed by water activity, can be defined as the ratio of solvent fugacity and pure solvent fugacity (Chen 2013; van den Doel et al. 2009).

Compared with ordinary solid substrates, the ingredients in herbs are more complex, and the interactions between herbs are more diverse. Therefore, some of the additional requirements of a culture medium for microbial growth should be stated: (1) to meet the basic growth requirements of the microorganisms, (2) the changes of active ingredients in herbs caused by the microorganisms during fermentation, (3) the effects of the active ingredients on the growth of microbes, (4) the conditions of pretreatment of herbs, and (5) the drying process of the products.

6.5.3.2 Bioreactors of SSF

Various bioreactor types have been used in SSF processes, based on mixing and aeration. They can be divided into four general categories (Figure 6.5; Mitchell et al. 2000).

1. *Unmixed beds without forced aeration*—the bed is static and air is circulated around the bed, but not blown forcefully through it (i.e., tray bioreactors). Each tray contains a thin layer of substrate (~5 to 15 cm deep).
2. *Unmixed beds without forced aeration*—the bed is continuously mixed or mixed intermittently with a frequency of minutes to hours, and air is circulated around the bed, but not blown forcefully through it. The substrate bed is held within a horizontal or near-horizontal drum, which may or may not have baffles yet continuously rotates, such as a rotating drum.
3. *Unmixed beds with forced aeration*—the bed is static and air is blown forcefully though the bed (i.e., packed-bed bioreactors). There is a static bed on top of a perforated plate through which conditioned air is blown, although one interesting variation is the

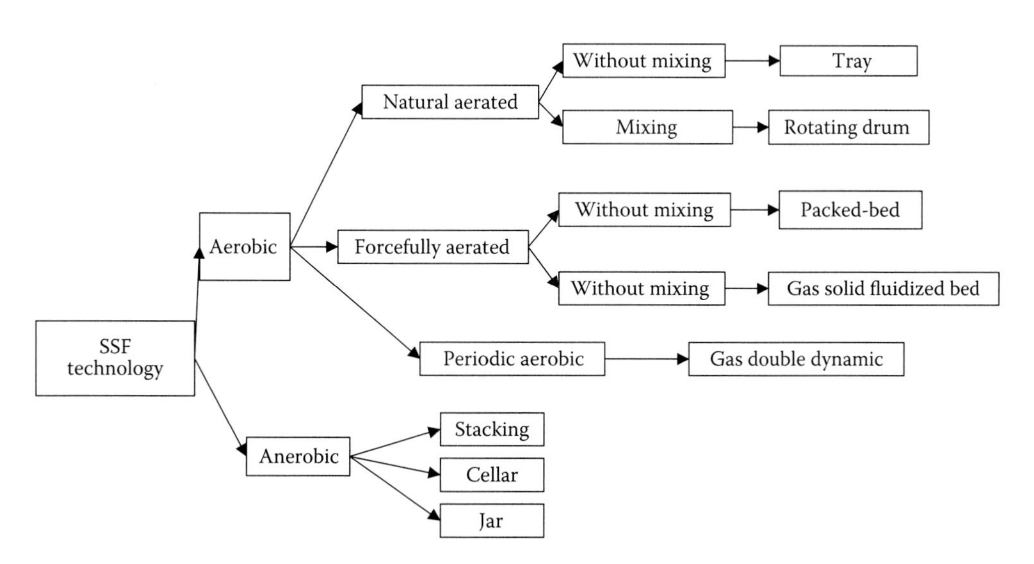

Figure 6.5 The classification of SSF technology used in Chinese herb processing. (From Mitchell, D.A. et al., *Process Biochemistry, 35 (10)*: 1211–1225, 2000.)

introduction of air through a perforated rod inserted into the center of the bed, such as a packed bed.
4. *Mixed beds with forced aeration*—the bed is agitated and air is blown forcefully through the bed (Chen 2013).

Chen and coworkers designed gas double-dynamic SSF technology. In 1998, the large-scale solid-state pure culture fermentation demonstration plant was built. The 110 m^3 bioreactors were used to produce biopesticides. The results showed that economic indicators of this technology were better than traditional SMF. On this basis, the gas double-dynamic SSF technology gradually developed as one of the modern SSF technologies (Chen and Li 2007; Chen 2013; Chen and He 2013).

6.6 *Prospects of Chinese herb SSF processing*

6.6.1 *Overall multistage conversion and directional fractionation of Chinese herbs*

Chinese herbs, as treasures of the Chinese civilization, play important roles in human life, yet most of them are rare species in nature and should be preserved. Due to overexploitation and utilization, the wild resources of Chinese herbs have suffered serious damage. Therefore, the utilization and protection of Chinese herbs should be coupled together; ecological, social, and economic benefits should aim for harmony and unity, which are the core requirements for the modernization of the TCM industry.

With respect to the low extraction efficiency and high separation cost in the traditional Chinese herb processing, the overall multistage conversion and directional fractionation of Chinese herb processing was proposed by Chen (2010). This idea contains two aspects. First, for various kinds of herbs or exploitation targets, different transformation processes should be established. Second, all different active ingredients in herbs should be separated and used based on their characteristics.

6.6.2 *Novel Chinese herb SSF processing based on steam explosion technology*

The herbs are enclosed by a thick and rigid phellem layer, and most active ingredients in herbs are combined with the lignin in the cell wall that is wrapped by the cellulose and hemicellulose, which causes the difficulty in the extraction of the active ingredients. Based on the essential characters of the herbs, Chen and others developed the unpolluted low-pressure steam explosion technology, which has been successfully applied to Chinese herbs processing, clean pulping, and xylan production. For this technology, the pressures are lower than 1.5 MPa, and no chemicals are added (Chen and Liu 2007a; Zhang et al. 2012). Raw herbs are cooked by vapor under certain pressure and temperature conditions, the cell wall is broken, starch and hemicellulose are hydrolyzed into soluble sugars, and the lignin is softened and partially degraded. On the other hand, vapor can penetrate into the pores of materials, and then the pressure is discharged immediately, resulting in a mechanical crushing effect. That is to say, the steam explosion technology is coupled with chemical hydrolysis and physical crushing (Chen and Liu 2007b).

The effects of steam explosion technology on the extraction of ephedrine from the ephedra herb were studied. The optimum steam explosion condition was 35% moisture at 1.5 MPa for 3 min, and then the pressure was discharged immediately. Under these conditions, the extraction yield of ephedrine was 0.345%, which was 243% higher than the control. The pretreated ephedra herbs were observed by electron microscopy; the results showed that the holes of the untreated ephedra surfaces were scarce and small, yet the holes of the pretreated ephedra surfaces were large and dense, and cellulose fibers were exposed and became sparse (Figure 6.6; Chen 2006; Chen and Peng 2012).

Resveratrol (3,5,4′-trihydroxystilbene) is a naturally occurring phytoalexin produced by some spermatophytes, such as *Polygonum cuspidatum*. As a phenolic compound, resveratrol contributes to the antioxidant potential and may play a role in the prevention of human cardiovascular diseases. Resveratrol has also been shown to modulate the metabolism of lipids, and to inhibit the oxidation of low-density lipoproteins and the aggregation of platelets. Moreover, as a phytoestrogen, resveratrol may provide cardiovascular protection (Adrian and Jeandet 2012). *P. cuspidatum* is a perennial herb, its roots are full of resveratrol and is commonly known in TCM. The combination pretreatment of steam explosion with biological fermentation was established to transform polydatin into resveratrol. The steam explosion condition was 0.8 to 1.6 MPa for 1 to 6 min, the pretreated sample underwent enzymatic hydrolysis for 36 h, and the fermentation condition was 28°C for 5 days (Chen and Peng 2010).

Steam explosion is conducive to the extraction of the active ingredient from herbs and the growth of microorganisms in SSF. Therefore, the unpolluted low-pressure steam explosion technology has broad application prospects in the extraction, concoction, and detoxification of herbs.

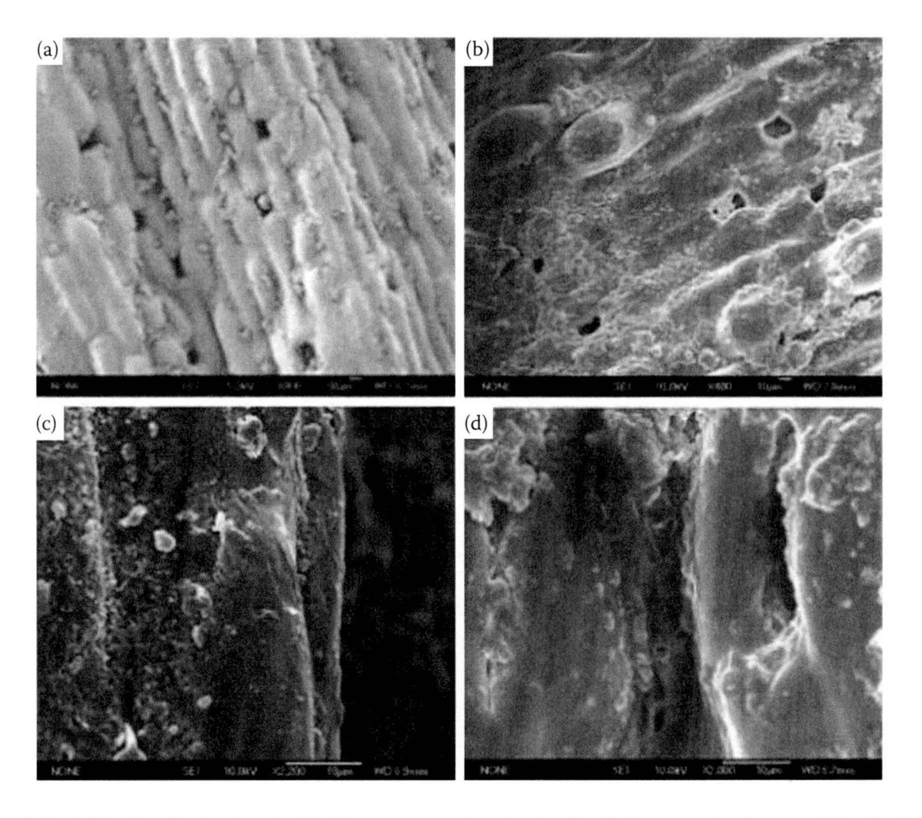

Figure 6.6 SEM of ephedra herb (a) The untreated ephedra herb (magnification, ×400); (b) the pretreated ephedra herb (magnification, ×400); (c) the untreated ephedra herb (magnification, ×2200); (d) the pretreated ephedra herb (magnification, ×2000). (From Chen, H.Z. and Peng, X.W., *Progress in Chemistry*, *24 (9)*: 1857–1864, 2012.)

6.6.3 Process engineering of Chinese herb SSF and its ecoindustrial integration

In traditional Chinese herb processing, the characteristics and interactions between operating units were often neglected. Each operating unit was relatively independent, lacking coordination and complementarity, which lead to low efficiency and high production costs. For example, in the basic process of traditional concoction, only one active ingredient was utilized and the other parts of the herb were often discarded as waste, or used for mushroom cultivation or feed.

From the perspective of engineering and processing, Chen proposed an ecological industry chain thinking that was based on steam explosion technology and SSF technology. The key points of ecological industry chain thinking were selective fractionation and economic functional utilization. Namely, based on the characteristics of the herb, its components were fractionated and used. Various technologies were introduced, each operating unit was analyzed deeply, and multiple processes were coupled (Chen 2009, 2010). In this process, resource utilization, energy conversion, and environmental protection were organically combined; the ecological industry chain, including clean production, technology integration, and product diversification, was established ultimately.

Pueraria (*Radix puerariae*) is a perennial leguminous plant, which was first recorded in Shennong's Chinese Materia Medica. It is rich in starch and isoflavones. *R. puerariae* is listed as both food and medicine by the Ministry of Public Health of China. More than 30 types of Pueraria are found around the world, and most of them are distributed in China, Japan, and Southeast Asia.

R. puerariae is composed of starch, cellulose, protein, isoflavones, and a small amount of fat, pectin, tannins, and alkaloid. *R. puerariae* can be used for the extraction of flavones, or as a starch-type energy source plant to produce ethanol. Consequently, *R. puerariae* has become one of the focuses of economic crop development projects in China, and its planting area is huge (Figure 6.7).

Chen invented a new process for coproducing ethanol and *R. puerariae* flavones as well as fibers from *R. puerariae*. Finally, a classified transformation and a comprehensive and clean utilization of *R. puerariae* components technology was established. This technology is composed of the following steps: (1) *R. puerariae* was pretreated by steam explosion; (2) ethanol was produced through simultaneous saccharified SSF; (3) *R. puerariae* flavones were extracted from the fermentation residue; and (4) fibers were separated from residues (Chen et al. 2007; Chen and Qiu 2010; Chen and Li 2013).

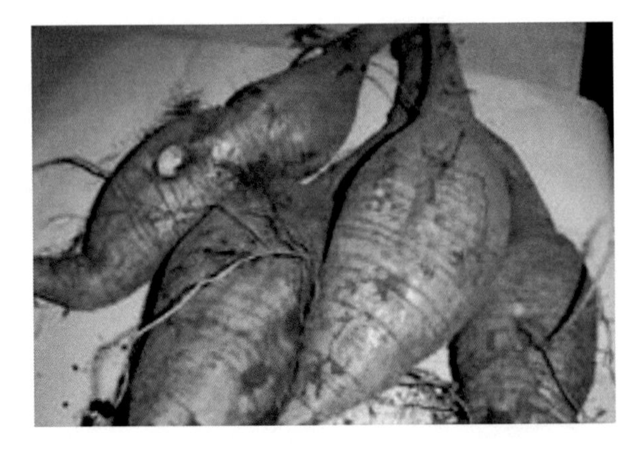

Figure 6.7 Pueraria (*R. puerariae*).

The concrete steps can be stated as follows (Figure 6.8):

1. *R. puerariae* was steam explosion pretreated under a steam pressure of 0.5 to 1.0 MPa for 2 to 4 min
2. SSF for ethanol production
   a. The steam-exploded *R. puerariae* was sterilized in the SSF reactor. Then the glucoamylase, $(NH_4)_2SO_4$, KH_2PO_4, and the activated yeast were added to the sterilized materials.
   b. The motor of the airlock was turned on; the sterilely inoculated *R. puerariae* substrate was slowly delivered into the fermentation pot. When the materials were fed for fermentation, CO_2 entered through the gas inlet and replaced the air inside the pot. The temperature was maintained at 35 ± 1°C in the fermentation pot.
   c. The gas was circulated by the CO_2 entering through the gas inlet at the bottom of the gas extraction pot, and ethanol was carried away from the substrate.
   d. The mixed gas of CO_2 and ethanol passed into active carbon absorbing columns.

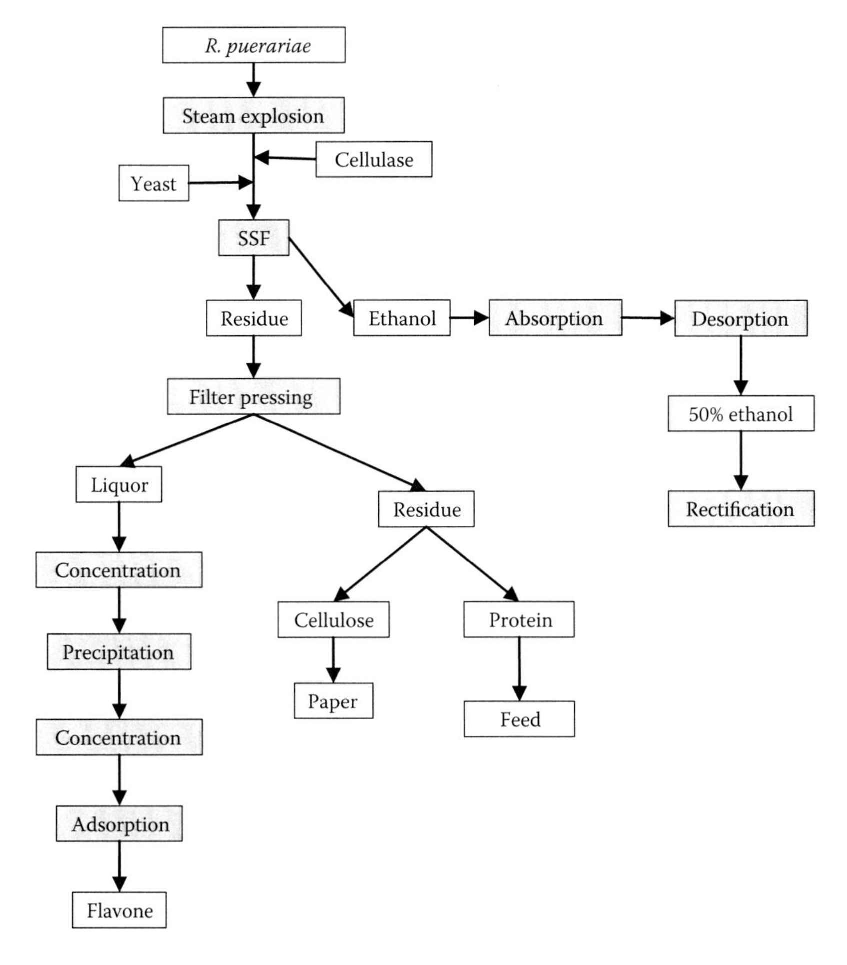

Figure 6.8 The process of a technical system for cleanly coproducing fuel ethanol, flavones, and fibers by steam explosion and SSF of *R. puerariae*. (From Chen, H.Z. et al., Process for comprehensively utilizing steam exploded *radix puerariae* and device therefor. US Patents: US8333999 B2, 2007.)

   e. The adsorbed columns were heated to 90°C for recovery of ethanol.
   f. The ethanol was recovered and obtained after condensation.
   g. The *R. puerariae* residue was filter-pressed.
3. 95% ethanol was added to the residue for the extraction of isoflavones.
4. The extraction liquid was filtered, and the concentrated liquid was passed onto macroporous adsorbing resins for separation of flavonoids.

Fibrous residues were dried, followed by mechanical classification into a fiber portion or a protein portion with a ratchet.

6.7 Conclusions

Solid-state fermentation has been widely applied to the concoction and extraction of Chinese herbs due to the advantages of mild reaction condition, formation of new compounds, and multilevel utilization. Microorganisms, especially medicinal fungus, and modern SSF technologies should be introduced into the processing of Chinese herbs. Process engineering of Chinese herb SSF and its eco-industrial integration based on SSF and steam explosion were proposed, and will become one of the main directions of traditional Chinese herb development in the future.

Acknowledgments

This research received financial support from the National Basic Research Program of China (973 Project, no. 2011CB707401), the National High Technology Research and Development Program of China (863 Program, SS2012AA022502), and the National Key Project of Scientific and Technical Supporting Program of China (no. 2011BAD22B02).

References

Adrian, M. and Jeandet, P. 2012. Effects of resveratrol on the ultrastructure of *Botrytis cinerea conidia* and biological significance in plant/pathogen interactions. *Fitoterapia, 83 (8)*: 1345–1350.

Asgher, M., Asad, M.J. and Legge, R.L. 2006. Enhanced lignin peroxidase synthesis by *Phanerochaete Chrysosporium* in solid state bioprocessing of a lignocellulosic substrate. *World Journal of Microbiology and Biotechnology, 22 (5)*: 449–453.

Baxter, J. and Cummings, S.P. 2006. The current and future applications of microorganism in the bioremediation of cyanide contamination. *Antonie Van Leeuwenhoek, 90 (1)*: 1–17.

Brand, D., Pandey, A., Roussosand, S. and Soccol, C.R. 2000. Biological detoxification of coffee husk by filamentous fungi using a solid state fermentation system. *Enzyme and Microbial Technology, 27 (1–2)*: 127–133.

Cabaleiro, N., Calle, I., Bendichoand, C. and Lavilla, I. 2013. Current trends in liquid-liquid and solid-liquid extraction for cosmetic analysis: A review. *Analytical Methods, 5 (2)*: 323–340.

Castilho, L.R., Medronho, R.A. and Alves, T.L.M. 2000. Production and extraction of pectinases obtained by solid state fermentation of agroindustrial residues with *Aspergillus niger*. *Bioresource Technology, 71 (1)*: 45–50.

Cauthen, S.E., Pattisonand, J. and Lascelles, J. 1967. Vitamin B_{12} in photosynthetic bacteria and methionine synthesis by *Rhodopseudomonas spheroides*. *Biochemical Journal, 102 (3)*: 774.

Chen, F.S. and Hu, X.Q. 2005. Study on red fermented rice with high concentration of monacolin K and low concentration of citrinin. *International Journal of Food Microbiology, 103 (3)*: 331–337.

Chen, H.Z. and Yuan, Y.T. 2006. A method on the extraction of ephedrine by air steam explosion technology. Chinese Patent: CN200410090699.1.

Chen, H.Z., Fu, X.G. and Wang, W.D. 2007. Process for comprehensively utilizing steam exploded *radix puerariae* and device therefore. US Patents: US8333999 B2.
Chen, H.Z. and Li, Z.H. 2007. Gas dual-dynamic solid state fermentation technique and apparatus. US Patents: 7183074B2.
Chen, H.Z. and Liu, L.Y. 2007a. *Principle and Application of Steam Explosion Technology*. Chemical Industry Press, Beijing, China, pp. 8–14.
Chen, H.Z. and Liu, L.Y. 2007b. Unpolluted fractionation of wheat straw by steam explosion and ethanol extraction. *Bioresource Technology, 98 (3)*: 666–676.
Chen, H.Z. 2009. *Process Engineering in Plant-Based Products*. Nova Science Publishers, Inc., New York.
Chen, H.Z. 2010. *Medicinal Plants, Process Engineering and Its Ecological Industry Integration*. Science Press, Beijing, China.
Chen, H.Z. and Peng, X.W. 2010. A method on transformation of polydatin into resveratrol by combination of Steam explosion and biology. Chinese Patent: CN201010128286.3.
Chen, H.Z. and Qiu, W.H. 2010. Key technologies for bioethanol production from lignocellulose. *Biotechnology Advance, 28 (5)*: 556–562.
Chen, H.Z. and Peng, X.W. 2012. Steam explosion technology applied to high-value utilization of herb medicine resources. *Progress in Chemistry, 24 (9)*: 1857–1864.
Chen, H.Z. 2013. *Modern Solid State Fermentation-Theory and Practice*. Springer, Germany.
Chen, H.Z. and He, Q. 2013. A novel structured bioreactor for solid-state fermentation. *Bioprocess and Biosystems Engineering, 36 (2)*: 223–230.
Chen, H.Z. and Li, G.H. 2013. An industrial level system with nonisothermal simultaneous solid state saccharification, fermentation and separation for ethanol production. *Biochemical Engineering Journal, 74*: 121–126.
Dai, Y.C. and Yang, Z.L. 2008. A revised checklist of medicinal fungi in China. *Mycosystema, 27 (6)*: 801–824.
Dai, Y.C., Zhou, L.W., Cui, B.K., Chen, Y.Q. and Decock, C. 2010. Current advances in Phellinus sensu lato: Medicinal species, functions, metabolites and mechanisms. *Applied Microbiology and Biotechnology, 87 (5)*: 1587–1593.
Elisashvili, V. 2012. Submerged cultivation of medicinal mushrooms: Bioprocesses and products (review). *International Journal of Medicinal Mushrooms, 14 (3)*: 25–59.
Fang, L.C., Huang, X.F., Du, Z.H., Yuan, J., Wei, H., Cheng, H. and Liu, Y. 2005. Isolation and identification of a photosynthetic bacteria producing coenzyme Q10. *Acta Microbiologica Sinica, 45 (5)*: 772–775.
Froufe, H., Abreu, R., Barros, L. and Ferreira, I. 2012. Docking studies to evaluate mushrooms low molecular weight compounds as inhibitors of the anti-apoptotic protein BCL-2. *Planta Medica, 78 (11)*: 1463.
Gao, X.M. 2007. *Chinese Pharmacy*. People's Medical Publishing House, Beijing, China.
Gong, Q.F. 2007. *The Concoction of Chinese Traditional Medicine*. Chinese Medicine Press, Beijing, China.
Hölker, U., Höfer, M. and Lenz, J. 2004. Biotechnological advantages of laboratory-scale solid-state fermentation with fungi. *Applied Microbiology and Biotechnology, 64 (2)*: 175–186.
Hou, Y., Shao, W.F., Xiao, R., Xu, K.L., Ma, Z.Z., Johnstone, B.H. and Du, Y.S. 2009. Pu-erh tea aqueous extracts lower atherosclerotic risk factors in a rat hyperlipidemia model. *Experimental Gerontology, 44 (6)*: 434–439.
Hwang, H.J., Kim, S.W. and Yun, J.W. 2007. Modern biotechnology of *Phellinus baumii*—from fermentation to proteomics. *Food Technology and Biotechnology, 45 (3)*: 306–318.
Iranzo, M., Sainz-Pardo, I., Boluda, R., Sanchez, J. and Mormeneo, S. 2001. The use of microorganisms in environmental remediation. *Annals of Microbiology, 51 (2)*: 135–144.
Isaka, P., Chinthanom, M., Kongthong, S., Srichomthong, K. and Choeyklin, R. 2013. Lanostane triterpenes from cultures of the Basidiomycete *Ganoderma orbiforme* BCC 22324. *Phytochemistry, 87*: 133–139.
Jung, J.Y., Lee, I.K., Seok, S.J., Lee, H.J., Kim, Y.H. and Yun, B.S. 2008. Antioxidant polyphenols from the mycelial culture of the medicinal fungi *Inonotus xeranticus* and *Phellinus linteus*. *Journal of Applied Microbiology, 104 (6)*: 1824–1832.
Kaufmann, B. and Christen, P. 2002. Recent extraction techniques for natural products: Microwave-assisted extraction and pressurised solvent extraction. *Phytochemical Analysis, 13 (2)*: 105–113.

Kluge, M., Ullrich, R., Scheibner, K. and Hofrichter, M. 2012. Stereoselective benzylic hydroxylation of alkylbenzenes and epoxidation of styrene derivatives catalyzed by the peroxygenase of *Agrocybe aegerita*. *Green Chemistry, 14 (2)*: 440–446.

Kodama, N., Mizuno, S., Nanba, H. and Saito, N. 2010. Potential antitumor activity of a low-molecular-weight protein fraction from *Grifola frondosa* through enhancement of cytokine production. *Journal of Medicinal Food, 13 (1)*: 20–30.

Koyyalamudi, S.R., Jeong, S.C., Pang, G., Teal, A. and Biggs, T. 2011. Concentration of vitamin D in white button mushrooms (*Agaricus bisporus*) exposed to pulsed UV light. *Journal of Food Composition and Analysis, 24 (7)*: 976–979.

Lai, J.T., Kuo, C.M. and Liao, B.C.C. 2006. Investigation of response surface methodology on the coenzyme Q10 production by using photosynthetic bacteria. The 2006 Annual Meeting.

Lang, Q.Y. and Wai, C.M. 2001. Supercritical fluid extraction in herbal and natural product studies—A practical review. *Talanta, 53 (4)*: 771–782.

Larsen, M.J. and Cobb-Poulle, L.A. 1990. *Phellinus (Hymenochaetaceae): A Survey of the World Taxa.* Fungiflora: Olso, Norway *206* 170–178.

Lei, Z.Q. 1998. *Chinese Pharmacy*. Shanghai Science and Technology Press, Shanghai, China.

Liang, Y., Chen, Y., Liu, H., Luan, R., Che, T., Jiang, S., Xie, D. and Sun, H. 2011. The tumor rejection effect of protein components from medicinal fungus. *Biomedicine and Preventive Nutrition, 1 (4)*: 245–254.

Lin, Y.W. and Chiang, B.H. 2008. Anti-tumor activity of the fermentation broth of *Cordyceps militaris* cultured in the medium of *Radix astragali*. *Process Biochemistry, 43 (3)*: 244–250.

Lung, M.Y., Tsai, J.C. and Huang, P.C. 2010. Antioxidant properties of edible basidiomycete *Phellinus igniarius* in submerged cultures. *Journal of Food Science, 75 (1)*: 18–24.

Lung, M.Y. and Huang, W.Z. 2012. Antioxidant properties of polysaccharides from *Laetiporus sulphureus* in submerged cultures. *African Journal of Biotechnology, 11 (23)*: 6350–6358.

Masuda, Y., Ito, K., Konishi, M. and Nanba, H. 2010. A polysaccharide extracted from *Grifola frondosa* enhances the anti-tumor activity of bone marrow-derived dendritic cell-based immunotherapy against murine colon cancer. *Cancer Immunology, Immunotherapy, 59 (10)*: 1531–1541.

Mitchell, D.A., Krieger, N.D., Stuart, M. and Pandey, A. 2000. New developments in solid-state fermentation II. Rational approaches to the design, operation and scale-up of bioreactors. *Process Biochemistry, 35 (10)*: 1211–1225.

Mongin, O., Schuwey, A., Vallot, M.A. and Gossauer, A. 1999. Synthesis of a macrocyclic porphyrin hexamer with a nanometer-sized cavity as a model for the light-harvesting arrays of purple bacteria. *Tetrahedron Letters, 40 (48)*: 8347–8350.

Nagao, N., Matsuyama, T., Yamamoto, H. and Toda, T. 2003. A novel hybrid system of solid state and submerged fermentation with recycle for organic solid waste treatment. *Process Biochemistry, 39 (1)*: 37–43.

Nan, Y.S. and Zhu, C.H. 1992. Studies on the technological process for preparing medicated Leaven. *China Journal of Chinese Materia Medica, 17 (8)*: 471–474.

Noh, H., Kim, K., Park, H., Kim, G., Lee, S., Kim, S., Choi, S. and Lee, K. 2012. Bioactive isoindolinone alkaloid from *Hericium erinaceum*. *Planta Medica, 78 (11)*: PJ128.

Ohmori, H., Ishitani, H., Sato, K., Shimizu, S. and Fukui, S. 1971. Vitamin B dependent glutamate mutase activity in photosynthetic bacteria. *Biochemical and Biophysical Research Communications, 43 (1)*: 156–162.

Phillips, K.M., Horst, R.L., Koszewski, N.J. and Simon, R.R. 2012. Vitamin D4 in mushrooms. *PloS One, 7 (8)*: e40702.

Rodrıguez-Couto, S. and Sanroman, M.A. 2005. Application of solid-state fermentation to ligninolytic enzyme production. *Biochemcal Enginering Journal, 22 (3)*: 211–219.

Sakata, K., Utsunomiya, H., Tokuda, A., Ichinose, M. and Miyazawa, M. 2011. Production of a new terpenoid from biotransformation of (-)-isolongifolanol by *Aspergillus niger* and suppression of SOS-inducing activity. *Biocatalysis and Biotransformation, 29 (5)*: 212–216.

Sasaki, K., Tanaka, T. and Nagai, S. 1998. Use of photosynthetic bacteria for the production of SCP and chemicals from organic wastes. In: *Bioconversion of Waste Materials to Industrial Products*, Martin, A.M. (ed.). Springer: New York.

Sasaki, K., Watanabe, M., Suda, Y., Ishizuka, A. and Noparatnaraporn, N. 2005. Applications of photosynthetic bacteria for medical fields. *Journal of Bioscience and Bioengineering, 100 (5)*: 481–488.
Shih, I.L., Pan, K. and Hsieh, C.Y. 2006. Influence of nutritional components and oxygen supply on the mycelial growth and bioactive metabolites production in submerged culture of *Antrodia cinnamomea*. *Process Biochemistry, 41 (5)*: 1129–1135.
Sindhu, I., Chhibber, S., Capalash, N. and Sharma, P. 2006. Production of cellulase-free xylanase from *Bacillus megaterium* by solid state fermentation for biobleaching of pulp. *Current Microbiology, 53 (2)*: 167–172.
Singhania, R.R., Patel, A.K., Soccol, C.R. and Pandey, A. 2009. Recent advances in solid-state fermentation. *Biochemical Engineering Journal, 44 (1)*: 13–18.
Sun, T., Li, L.S. and Hu, X.L. 2007. Influence of Chinese herbal medicines on the growth of photosynthetic bacteria. *Transactions of Oceanology and Limnology, S0*: 91–96.
Tanaka, T., Sasaki, K., Noparatnaraporn, N. and Nishio, N. 1994. Utilization of volatile fatty acids from the anaerobic digestion liquor of sewage sludge for 5-aminolevulinic acid production by photosynthetic bacteria. *World Journal of Microbiology and Biotechnology, 10 (6)*: 677–680.
Tian, Y., Yue, T., Yuan, Y., Soma, P.K., Williams, P.D., Machado, P.A., Fu, H., Kratochvil, R., Wei, J.C. and Lo, Y.M. 2010. Tobacco biomass hydrolysate enhances coenzyme Q10 production using photosynthetic *Rhodospirillum rubrum*. *Bioresource Technology, 101 (20)*: 7877–7881.
Treviño-Cueto, B., Luis, M., Contreras-Esquivel, J., Rodríguez, R., Aguilera, A. and Aguilar, C. 2007. Gallic acid and tannase accumulation during fungal solid state culture of a tannin-rich desert plant (*Larrea tridentata*). *Bioresource Technology, 98 (3)*: 721–724.
van den Doel, L.R., Mohoric, A., Vergeldt, F.J., van Duynhoven, J., Blonk, H., van Dalen, G., van As, H. and van Vliet, L.J. 2009. Mathematical modeling of water uptake through diffusion in 3D inhomogeneous swelling substrates. *AIChE Journal, 55 (7)*: 1834–1848.
Wan, C. and Li, Y. 2010. Microbial pretreatment of corn stover with *Ceriporiopsis subvermispora* for enzymatic hydrolysis and ethanol production. *Bioresource Technology, 101 (16)*: 6398–6403.
Wang, Q.P., Peng, C.X. and Gong, J.S. 2011. Effects of enzymatic action on the formation of theabrownin during solid state fermentation of Pu-erh tea. *Journal of the Science of Food and Agriculture, 91*: 2412–2418.
Wasser, S.P. 2010. Medicinal mushroom science: History, current status, future trends, and unsolved problems. *International Journal of Medicinal Mushrooms, 12 (1)*: 1–16.
Xu, Z.Y., Wu, Z.A. and Bi, K.S. 2013. A novel norsesquiterpene alkaloid from the mushroom-forming fungus *Flammulina velutipes*. *Chinese Chemical Letters, 24 (1)*: 57–58.
Yaoita, Y., Hiraob, M., Kikuchi, M. and Machida, K. 2012a. Three new lactarane sesquiterpenoids from the mushroom *Russula sanguinea*. *Natural Product Communications, 7 (9)*: 1133–1135.
Yaoita, Y., Yonezawa, S., Kikuchi, M. and Machida, K. 2012b. A new geranylated aromatic compound from the mushroom *Hericium erinaceum*. *Natural Product Communications, 7 (4)*: 527.
Yeh, C.H., Chen, H.C., Yang, J.J., Chuang, W.I. and Sheu, F. 2010. Polysaccharides PS-G and protein LZ-8 from Reishi (*Ganoderma lucidum*) exhibit diverse functions in regulating murine macrophages and T lymphocytes. *Journal of Agricultural and Food Chemistry, 58 (15)*: 8535–8544.
Yu, B. and Chen, H.Z. 2010. Effect of the ash on enzymatic hydrolysis of steam-exploded rice straw. *Bioresource Technology, 101 (23)*: 9114–9119.
Zadrazil, F. and Puniya, A.K. 1995. Studies on the effect of particle size on solid-state fermentation of sugarcane bagasse into animal feed using white-rot fungi. *Bioresource Technology, 54 (1)*: 85–87.
Zhang, Y.Z., Fu, X.G. and Chen, H.Z. 2012. Pretreatment based on two-step steam explosion combined with an intermediate separation of fiber cells—Optimization of fermentation of corn straw hydrolysates. *Bioresource Technology, 121*: 100–104.
Zhao, L., Dong, Y., Chen, G. and Hu, Q. 2010. Extraction, purification, characterization and antitumor activity of polysaccharides from *Ganoderma lucidum*. *Carbohydrate Polymers, 80 (3)*: 783–789.
Zheng, Z. and Shett, K. 2000. Solid-state bioconversion of phenolics from *cranberry pomace* and role of *Lentinus edodes* β-glucosidase. *Journal of Agricultural and Food Chemistry, 48 (3)*: 895–900.
Zhou, D.Q. 2002. *Text Book of Microbiology*. Higher Education Press, Beijing, China, pp. 82–100.
Zhu, Y., Xia, S.Q., Wu, Y.J. and Zhao, Y. 2010. Anti-aging activities of extracts from Liuweidihuang decoction metabolized by photosynthetic bacteria. *Liaoning Journal of Traditional Chinese Medicine, 10*: 1997–1999.

chapter seven

Microbial biofuels production

Anoop Singh and Poonam Singh Nigam

Contents

7.1 Introduction

Increasing energy demand due to the steep rise in industrialization and motorization leads to an increase in crude oil prices, which is directly affected by global economic activity (He et al. 2010). Today, fossil fuels take up 80% of the primary energy consumed in the world, of which 58% is consumed by the transport sector alone (Escobar et al. 2009). The sources for these fossil fuels are becoming exhausted; in addition, the consumption of fossil fuels to fulfill this energy demand has also been found to be a major contributor to greenhouse gas (GHG) emissions (Prasad et al. 2007a; Singh et al. 2010a). It has many negative effects including climate change, receding of glaciers, rise in sea levels, loss of biodiversity, and others, which led to moves toward alternative, renewable, sustainable, efficient, and cost-effective energy sources with fewer emissions (Gullison et al. 2007; Prasad et al. 2010b; Singh et al. 2010b). Among many energy alternatives, biofuels are the most environment-friendly energy source (Nigam and Singh 2011). A worrying statistic is that, the global oil and gas production is approaching its maximum production level and the world is now finding one new barrel of oil for every four it consumes (Aleklett and Campbell 2003). Hence, as an alternative to fossil fuels, biofuels have been portrayed as a future leading supplier of energy sources that have the ability to increase the security of supply, reduce the amount of vehicle emissions, and provide a steady income for farmers.

Biofuels are referred to gas, liquid, and solid fuels predominantly produced from biomass. A variety of fuels can be produced from biomass such as ethanol, methanol, biodiesel, Fischer–Tropsch diesel, hydrogen, and methane (Demirbas 2008). Renewable and carbon-neutral biofuels are necessary for environmental and economic sustainability (Nigam and Singh 2011). Biofuels are important because they replace petroleum fuels. Between 1980 and 2005, worldwide production of biofuels increased by an order of magnitude from 4.4 to 50.1 billion liters with further dramatic increases in the future (Armbruster and Coyle 2006; Licht 2008; Murray 2005). Different countries have adopted different measures to introduce biofuels. The economics of each fuel vary with location, feedstock, and several other factors. Political agendas and environmental concerns also play a crucial role in the production and utilization of biofuels. There are a number of technologies that exist and several that are under development for the production of biofuels such as fermentation of sugar substrates, catalytic technology to convert ethanol to mixed hydrocarbon, hydrolysis of cellulose, biobutanol by fermentation, transesterification of natural oils and fats to biodiesel, hydrocracking of natural oils and fats, pyrolysis and gasification of various biological materials, and others. This chapter is an effort to summarize the microbial biofuel production system.

7.2 Microbial biofuels

Advancements in current research have shown that some microbial species such as yeasts, fungi, and microalgae can be used as potential sources of biomass for the production of biodiesel, which can be biosynthesized to store large amounts of fatty acids (Xiong et al. 2008). The microbial lipids, similar to vegetable oils, mainly contain palmitic, stearic, oleic, and linoleic acid with unsaturated fatty acids amounting to approximately 64% of the total fatty acid content (Nigam and Singh 2011; Singh et al. 2011a,b). The accumulation of lipid within microbial cells could be efficiently enhanced with the addition of various sugars to the pretreated molasses. The lipid content increased to more than 50% of the cell mass (Chen et al. 1992; Fakas et al. 2007).

Recent interest in the use of oleaginous microalgae as a biodiesel feedstock has grown considerably. This is mainly due to high oil yields from algal biomass (5000–100,000 L/ha per year), the ability to capture CO_2, and the capability to grow algae on abandoned or unproductive land using brackish, salt, or wastewaters instead of freshwater (Levine et al. 2010). Some microalgae respond to certain chemical and physical stimuli through the accumulation of intracellular triglycerides (Hu et al. 2008). A biodiesel production process that obviates biomass drying and organic solvent use for oil extraction could lead to significant energy and cost savings (Singh and Olsen 2011). Attempts to combine extraction with acid-catalyzed transesterification in one step have been successful with dry algal biomass, but the reaction is severely inhibited by water (Ehimen et al. 2010; Johnson and Wen 2009).

Microbial oil can be produced from sulfuric acid-treated rice straw hydrolysate (SARSH) by cultivation of the microorganism *Trichosporon fermentans*. Fermentation of SARSH without detoxification gave a poor lipid yield of 0.17% w/v (1.7 g/L; Huang et al. 2009). They also worked on the improvement process to increase the yield and found that the detoxification pretreatment, including overliming, concentration, and adsorption by Amberlite XAD-4 improved the fermentability of SARSH significantly. The pretreatment process helped in increasing the lipid yield by removing the inhibitors in SARSH. Moreover, besides SARSH, *T. fermentans* was also found to be capable of metabolizing other sugars such as mannose, galactose, or cellobiose available in the hydrolysates of other natural lignocellulosic materials used as the single carbon source. Huang and coworkers concluded that the organisms studied could be used as a promising strain for microbial oil production. Levine et al. (2010) have developed a two-step, catalyst-free biodiesel production process involving intracellular lipid hydrolysis coupled with supercritical *in situ* transesterification.

Zhu et al. (2008) have conducted a study on the production of microbial biofuel from waste molasses and concluded that lipids produced in microbial biomass can be used for biodiesel production. In this study, they have optimized the growth medium components for culture cultivation and studied the effects of culture conditions on microbial biomass and lipid production by a microbial strain of *T. fermentans*. The optimal nitrogen source, carbon source, and C/N molar ratio for the best lipid yields were found to be peptone, glucose, and 163, respectively. The most favorable initial pH of the cultivation medium and temperature were 6.5 and 25°C, respectively. *T. fermentans* could be cultivated in a medium consisting of waste molasses from the sugar industry. Zhu and coworkers recorded 12.8 g/L lipid yield from the bioconversion of waste molasses consisting of 15% total sugar concentration (w/v) at pH 6.0 by *T. fermentans*. The microbial oil with an acid value of 5.6 mg KOH/g could be effectively transesterified to produce biodiesel following a process of base catalysis after the removal of free fatty acids, and subsequently, a high methyl ester yield of 92% could be obtained (Zhu et al. 2008). The ability of yeast to grow well on pretreated lignocellulosic biomass could efficiently enhance the lipid accumulation, and provides a promising option for the production of economically and environmentally sound microbial oil from agricultural residues (Prasad et al. 2007a,b; Nigam and Singh 2011).

Multiple approaches are currently being researched for the use of microorganisms in the production of various biofuel, for example, bioalcohols, biohydrogen, biodiesel, and biomethane from multiple feedstocks (Elshahed 2010).

7.2.1 Bioalcohols

Bioalcohols are produced mainly using two approaches, that is, direct fermentation and indirect fermentation. Direct fermentation involves the identification of starting plant material, isolation and development of bacterial and fungal strains, and design of appropriate protocols for efficient conversion of plant material to sugar monomers, and then sugars are converted to ethanol by yeasts or genetically engineered bacterial strains. Indirect fermentation involves pyrolysis of the feedstock, followed by the conversion of the produced gas to alcohol using acetogenic bacteria (Elshahed 2010; Klasson et al. 1992).

Different feedstocks are used for the production of bioalcohols (mainly ethanol). The composition of plant materials varies among plant species. In general, the major polymers within lignocellulosic biomass is cellulose (35%–50%), followed by hemicellulose (20%–35%), and lignin (10%–25%; Liu et al. 2008). Therefore, different kinds of microorganisms, enzymes, incubation conditions, and engineering schemes are required for efficient depolymerization (Elshahed 2010). Celluloses and hemicelluloses can be degraded anaerobically, but lignin requires pretreatment for better degradation. Pretreatment is also required to increase the surface area of cellulose and hemicellulose exposed for microbial or enzymatic degradation. Various pretreatment approaches are employed, such as the use of alkaline peroxidases, concentrated acids, dilute acids, alkali, alkali peroxidases, wet oxidation, steam explosion, ammonia fiber explosion, liquid hot water, or organic solvent treatments (Prasad et al. 2007a; Wyman 1994).

The general steps for producing ethanol include pretreatment of substrates, saccharification process to release the fermentable sugars from polysaccharides, fermentation of released sugars, and finally a distillation step to separate ethanol (Figure 7.1). Pretreatment is designed to facilitate the separation of cellulose, hemicellulose, and lignin, so that complex carbohydrate molecules constituting the cellulose and hemicellulose can be broken down by enzyme-catalyzed hydrolysis into their constituent simple sugars (Nigam and Singh 2011). The degradation of starch, cellulose, or hemicellulose of any plant material yields hexoses and pentoses that need to be fermented to ethanol. Several fermentation schemes, for example, mixed acid fermentation by enteric bacteria and hetereolactic acid fermentation by some lactic acid bacteria are known to produce ethanol. Two groups of microorganisms, yeast *Saccharomyces cerevisiae* and bacterium *Zymomonas mobilis*, naturally produce 2 mol of ethanol per mole of hexose during fermentation (Elshahed 2010).

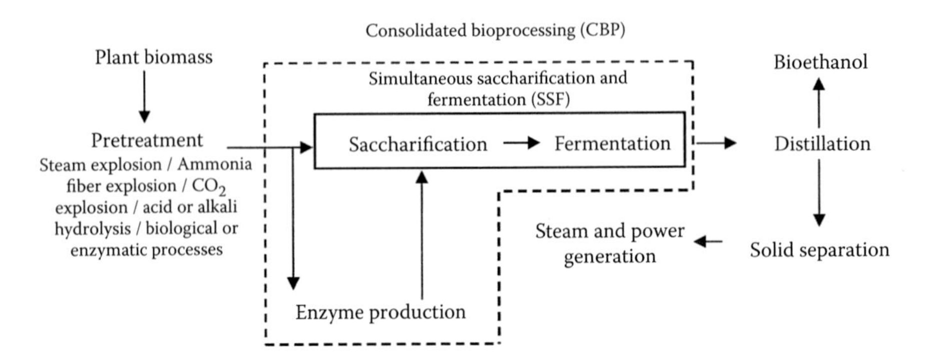

Figure 7.1 Various processes for the production of bioethanol from biomass. (From Nigam, P.S. and Singh, A., *Progress in Energy and Combustion Science* 37, 52–68, 2011.)

The use of *S. cerevisiae* for ethanol conversion from hexoses is one of the best perfected industrial processes. The capability of strains for efficient simultaneous uptake of multiple sugars can be improved through genetic manipulation of sugar transporters. This could direct laboratory evolution results in near-stoichiometric production of ethanol from glucose (Liu et al. 2008). The availability of the genome sequences of *S. cerevisiae* and the advancement of genetic systems allows for continuous genetic manipulations and strain improvements (Goffeau et al. 1996). Because hemicellulose is always present in plant material with cellulose, a microorganism capable of efficiently and simultaneously metabolizing both sugars is the best option to degrade the plant material.

The genetically modified strains efficiently degrade xylose and are shown to work well with pure substrates as well as sugars released by enzymatic treatment of plant materials (Ho et al. 1998; Kuyper et al. 2005; Prasad et al. 2007a; Sedlak and Ho 2004; Wisselink et al. 2009).

The indirect fermentation approach pyrolyzed (burned) the plant material to produce Syngas, which consists primarily of CO, CO_2, and hydrogen, and converted this to ethanol by acetogenic bacteria (Elshahed 2010). Acetogens convert C1 compounds into C2 products, mainly acetate (Müller 2003). Acetogens are usually Gram-positive bacteria belonging to the class Clostridia within the phylum *Firmicutes* (Leadbetter et al. 1999). Several acetogenic *Clostridia* are very efficient for anaerobic fermentation when grown on hexose sugars and produce various products, for example, acetate, butyrate, and ethanol. All plant material or nonplant wastes that could be pyrolized could theoretically be used in such an approach (Gulati et al. 1996). The approach makes use of all plant components, including lignin, which is generally not utilized in direct fermentation approaches, and can use mixed plant flora within a batch (Tilman et al. 2006). The key technical difficulties include relatively low growth rates and low product concentration in the aqueous phase when compared with yeast fermentation (Tanner 2008).

7.2.2 Biodiesel

Biodiesel could be produced from plant lipids, animal fats, and used cooking oils by esterification of triglycerides with methanol (Fukuda et al. 2001). Microbial species such as yeasts, fungi, and algae can be used as potential sources for the production of biofuels. They are rich in oil content and provide multiple advantages over the plant lipids because these microorganisms can grow extremely rapidly and oil content in microalgae can exceed 80%

by weight of dry biomass (Table 7.1; Metting 1996; Singh et al. 2011c; Spolaore et al. 2006). Microalgae commonly double their biomass within 24 h, and biomass doubling times during exponential growth are commonly as short as 3.5 h (Elshahed 2010).

Microalgae can produce lipids, proteins, and carbohydrates in large amounts over short periods of time and these products can be used for the production of biofuels. Microalgae have the ability to take up CO_2 from the atmosphere as well as discharge gases and soluble carbonates. Microalgae can tolerate and use substantially higher levels of CO_2 (up to 150,000 parts per million by volume) compared with higher plants (Brown 1996; Wang et al. 2008). Algal cells are veritable miniature biochemical factories, and appear more photosynthetically efficient than terrestrial plants due to their very efficient CO_2 fixation. The ability of algae to fix CO_2 has been proposed as a method of removing CO_2 from flue gases from power plants, and thus can be used to reduce the emission of greenhouse gases simultaneous with biodiesel production (Singh et al. 2011c, 2012; Singh and Olsen 2011).

Microalgae that are able to survive heterotrophic and exogenous carbon sources offer prefabricated chemical energy, which the cells often store as lipid droplets (Ratledge 2004). Naturally, lipid accumulation increases under certain conditions. Thus, it is important to keep in mind those factors that lead to the accumulation of lipids in algae, such as nutrients, sunlight, and others (Schenk et al. 2008).

Table 7.1 Oil Content of Microorganisms

Microorganism	Oil content (% dry weight)
Microalgae	
Botryococcus braunii	25–75
Cylindrotheca sp.	16–37
Scenedesmus dimorphus	16–40
Chlorella emersonii	28–32
Schizochytrium sp.	50–77
Nitzschia sp.	45–47
Nannochloropsis sp.	31–68
Neochloris oleoabundans	35–54
Yeasts	
Candida curvata	58
Cryptococcus albidus	65
Lipomyces starkeyi	64
Rhodotorula glutinis	72
Bacteria	
Arthrobacter sp.	>40
Bacillus alcalophilus	18–24
Rhodococcus opacus	24–25
Acinetobacter calcoaceticus	27–38
Fungi	
Aspergillus oryzae	57
Mortierella isabellina	86
Humicola lanuginosa	75
Mortierella vinacea	66

Source: Meng, J. et al., *Renewable Energy* 34: 1–5, 2009; Singh, A. et al., *Bioresource Technology* 102: 10–16, 2011; Singh, A. et al., *Bioresource Technology* 102: 26–34, 2011.

The best performing microalgal strains can be obtained by screening a wide range of naturally available isolates, and the efficiency of those can be improved by selection, adaptation, and genetic engineering (Singh et al. 2011a). Liu and Saha (2008) showed that high iron concentrations could induce considerable lipid accumulation in the marine strain *Chlorella vulgaris*, and also suggested that some metabolic pathways related to the lipid accumulation in *C. vulgaris* are probably modified by high levels of iron concentration in the initial medium. Genetic engineering of key enzymes in specific fatty acid production pathways within lipid biosynthesis is a promising target for the improvement of both quantity and quality of lipids. The lipid quality is very important for biodiesel production, as the alkyl ester content dictates the stability and performance of the fuel, and this is also an important factor in meeting international fuel standards (Singh et al. 2011a).

Three distinct algae production mechanisms, photoautotrophic, heterotrophic, and mixotrophic can be used, all of which follow natural growth processes (Singh et al. 2011c). Nitrogen starvation is the most effective method for improving microalgae lipid accumulation. Also, it gradually changes the lipid composition from free fatty acids to triacylglycerol, which a more useful product for biodiesel production. The biodiesel production from algal biomass involves chemical reactions such as transesterification of the extracted oil. Manipulation of metabolic pathways can redirect cellular function toward the synthesis of specific products and even expand the processing capabilities of microalgae (Meng et al. 2009; Singh et al. 2011c).

Singh et al. (2011b) concluded, in a review on mechanism and challenges in commercialization of algal biofuels, that the integration of microalgae cultivation with fish farms, food processing facilities, wastewater treatment plants, and others, will offer the possibility for waste remediation through recycling of organic matter and at the same time provide the low-cost nutrient supply required for algal biomass cultivation.

7.2.3 *Biohydrogen*

Hydrogen gas is seen as a strong candidate for future energy carrier by virtue of the fact that it is renewable, does not evolve the greenhouse gas CO_2 in combustion, it liberates large amounts of energy per unit of weight in combustion, and it is easily converted to electricity by fuel cell (www.oilgae.com/algae/pro/hyd/hyd.html, 2012). Thus, biohydrogen can be considered as the cleanest biofuel because it is oxidized to water with no gas emission in the production and consumption process.

Progress in the late 1990s contributed to a breakthrough in terms of sustainable hydrogen production. However, the problem is with the way it was produced (Rathore and Singh 2013). Although hydrogen biogas can be efficiently produced at the laboratory level, there is no known commercially operating hydrogen from biomass production facility in the world today (Zhu and Beland 2006). Hydrogen has long been known to be produced as a final end product of fermentation, or as a side product in photosynthesis in multiple groups of microorganisms (Elshahed 2010).

Nature has created biological reactions for hydrogen production. Sunlight is used for the oxidation of water by oxygenic photosynthesis reactions, and electrons are used for the generation of H_2 by hydrogenase enzymes (Rathore and Singh 2013). Oxygen adversely affects hydrogenase function and acts as a suppressor of hydrogenase gene expression (Florin et al. 2001; Ghirardi et al. 2000; Happe and Kaminski 2002). Hans Gaffron discovered hydrogen metabolism in unicellular green algae (Gaffron 1940). The first scientific investigation of hydrogen evolution by microalgae was reported by Gaffron and Rubin (1942). They demonstrated that after a period of dark anaerobic "adaptation," the green alga

Scenedesmus obliquus produces hydrogen in the dark at low rates, and hydrogen production is greatly stimulated in the light, although only for relatively brief periods. Other noteworthy observations were that uncouplers and low CO_2 concentrations stimulated light-driven hydrogen production in green microalgae. Gaffron observed that under anaerobic conditions, the green alga *S. obliquus* can use H_2 as an electron donor in the CO_2 fixation process in the dark, and evolve H_2 in the light (Gaffron 1944; Gaffron and Rubin 1942).

Algae structures are primarily for energy conversion without any development beyond cells. Their simple structure developments allow them to adapt to prevailing environmental conditions and prosper in the long term (Rathore and Singh 2013). Interest in green algae emanates from the fact that, in principle, they can use the highly efficient process of photosynthesis to produce hydrogen from sunlight and water as the most abundant of natural resources (Melis and Happe 2004).

Hydrogen has the largest energy content per weight of any known fuel, and can be produced by various means (Hallenbeck and Benemann 2002; Levin et al. 2004). Three main processes are commonly used for the production of biohydrogen. The most direct approach involves using photosynthetic microorganisms, for example, *Cyanobacteria* and green algae for biohydrogen production (Elshahed 2010). These microorganisms have the ability to split water, that is, produce electrons and oxygen from one molecule of water using sunlight as an energy source. The electrons are used for energy production through the electron transport chain as well as biomass production and sugar production using anabolic reactions, and they could also be converted to hydrogen by the action of hydrogenase enzymes (Elshahed 2010; Prince and Kheshgi 2005). A major problem with this approach is the extreme oxygen sensitivity of hydrogenases involved in hydrogen production. Therefore, the two processes, photolysis and hydrogen production, need to be temporarily uncoupled (Elshahed 2010).

The second approach uses nitrogenase enzymes in anoxygenic photoheterotrophic microorganisms for hydrogen production. The function of nitrogenase is to fix atmospheric N_2 gas to ammonia to be incorporated into the cell's biomass. Thus, enabling nitrogen-fixing microorganisms to grow in the absence of organic or inorganic nitrogen sources in growth media could lead to hydrogen production. However, nitrogenase enzymes are also capable of producing hydrogen from electrons and protons in the absence of oxygen and in the presence of light (Elshahed 2010).

The third approach is the production of hydrogen using fermentative bacteria from organic substrates, for example, sugar and lingocellulosic biomass, as well as industrial, residential, and farming wastes for anaerobic fermentation. Several groups of microorganisms are known to produce hydrogen as an end product of fermentation, for example, *Escherichia coli*, Enterobacteraerogenes, and *Clostridium butyricum* (Elshahed 2010).

The main hydrogen production pathways include electrolysis of water and thermocatalytic reformation of hydrogen-rich compounds, and usually require high-energy inputs obtained from nonrenewable resources (Levin et al. 2004). Biological production of hydrogen solves this problem by using microorganisms to convert biomass or solar energy into hydrogen gas (Das and Veziroglu 2001; Hallenbeck and Benemann 2002). Although Gaffron and Rubin (1942) considered both water and carbohydrates as the source of electrons for H_2 evolution, they favored the latter as the main source. Spruit (1958) found evidence supporting water as the main electron donor in such a reaction. A schematic representation of the photosynthetic reactions in the chloroplast thylakoid membrane of green algae is demonstrated in Figure 7.2. In photosynthesis, excitation energy from the sun is transferred from antenna pigments of light-harvesting complexes of photosystem II to photoactive chlorophyll molecule (P680) of the reaction center. As a result, strongly oxidizing cation radical

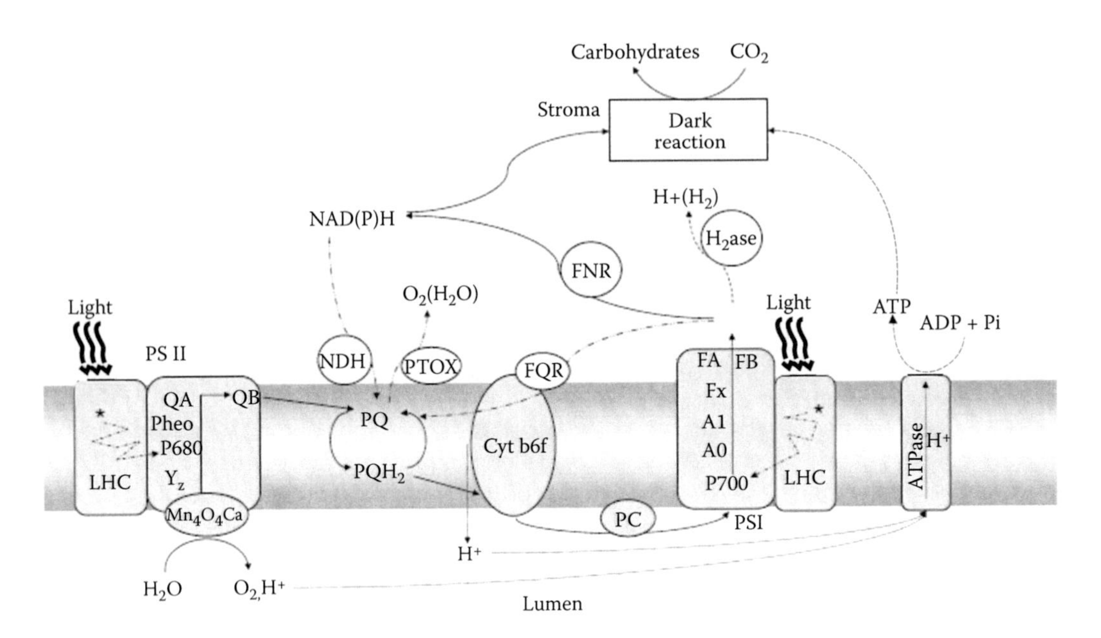

Figure 7.2 A scheme of the photosynthetic reactions in the chloroplast thylakoid membrane of green algae. Dash–dot line shows excitation migration from the antenna to the reaction center. The major electron transfer pathways are indicated by solid lines. The pathways involved in chlororespiration, cyclic electron flow around PS I, and the Mehler reaction are shown by dash–dot–dot lines. The dashed line designates the oxygen-sensitive electron transport route induced under anaerobic conditions. Reactions coupled to the generation of proton gradient and to the ATP synthesis are depicted by dotted lines. (From Antal, T.K. et al., *Applied Microbiology and Biotechnology* 89:3–15, 2011.)

$P680^+$ is formed, which catalyzes the oxidation of water in the oxygen-evolving complex of PS II through a series of redox-active components including the tyrozine Z residue (YZ), and the Mn_4O_4Ca cluster located at the luminal side of the thylakoid membrane (Barber 2008; Goussias et al. 2002; Kern and Renger 2007). Electrons released from $P680^+$ were accepted by a negatively charged radical of the Mg-free chlorophyll pigment pheophytin (Pheo). The reduced Pheo quickly passes an extra electron to the PS II primary quinone molecule (QA). The QA is tightly bound to PS II and acts as a single-electron carrier. From QA, an electron is transferred to the secondary quinone molecule (QB), a two-electron and two-proton acceptor. In double reduced and protonated form, QBH2 is loosely bound to PS II and thus it can be exchanged with the oxidized QB from the plastoquinone (PQ) pool. After that, an electron is transported further to PS I through the PQ pool, cytochrome b6f complex (cyt b6f), and plastocyanin (PC). Subsequent to light excitation, PS I passes an electron to the soluble protein ferredoxin (Fd) through a series of electron carriers including chlorophylls P700 and A0, quinone A1, and iron–sulfur clusters FX, FA, and FB. In the stroma, an electron is transferred from the reduced Fd to the $NADP^+$ with the formation of NADPH by a process catalyzed by the ferredoxin-NADP-reductase (FNR). In chloroplast, NADPH is used as a reducing power for fixation of carbon dioxide in the dark reaction (Bassham et al. 1950). This electron pathway corresponds to the linear electron transport (Antal et al. 2011). However, electrons can be alternatively redirected from the reduced Fd backward to the PQ pool through the putative ferredoxin-quinone-reductase (FQR) in the so-called cyclic electron transport (CET) around PS I (Moss and Bendall 1984). Such CET can be induced when FNR becomes inactive or when carbon fixation proceeds at a slow rate, which takes place in dark-adapted plants or under different environmental stresses (Breyton et al. 2006; Golding and Johnson 2003).

Production of H_2 from algae is mainly based on the photolysis of water molecules in the process of light reaction during photosynthesis. Hydrogen production is a property of many phototrophic organisms (Appel and Schulz 1998; Asada and Miyake 1999; Boichenko et al. 2001; Weaver et al. 1980), and the list of H_2 producers includes several hundred species from different genera of both prokaryotes and eukaryotes (Boichenko and Hoffmann 1994). Oxygenic photosynthetic organisms, such as plants, algae (green, red, brown, and yellow), and cyanobacteria use water as a source of electrons and protons in photosynthesis. Among these organisms, only green microalgae and cyanobacteria have been shown to sustain hydrogen production (Hall et al. 1995; Melis et al. 2000; Sakurai et al. 2004). The enzyme mediating H_2 production in green algae is the reversible (or bidirectional) hydrogenase that catalyzes the following ferredoxin (Fd)-linked reaction in the absence of ATP input (Boichenko and Hoffman 1994):

$$2H^+ + 2e^- \xleftrightarrow{\text{Hydrogenase}} H_2$$

The available H_2 energy production processes from algal biomass can be divided into two general categories: thermochemical and biological processes. Combustion, pyrolysis, liquefaction, and gasification are the four thermochemical processes. Direct biophotolysis, indirect biophotolysis, biological water–gas shift reaction, photofermentation, and dark fermentation are the five biological processes (Ni et al. 2006).

Hydrogen was a vital energy source for organisms during the early stages of our planet, but under reducing atmospheric conditions, that is, in the absence of a substantial amount of oxygen (Rathore and Singh 2013). This process gradually lost its importance with the development of the photosynthetic machinery that was able to exploit light energy more efficiently. Hydrogen-dependent processes therefore lost their central role as a necessity for the survival of most cells. In consequence, there was no strong evolutionary pressure for the design of oxygen-resistant hydrogenases and the increasing oxygen content of the atmosphere produced by the water-splitting process led to these enzymes being switched off (Horner et al. 2002).

Nowadays, hydrogenases still exist in bacteria and microalgae. But their genes are mostly activated under anaerobic conditions and their main function is twofold, either to provide an alternative electron source to aid survival under suboptimal conditions or to capture electrons as a kind of security valve to prevent dangerous overreduction of the electron transport chain (Appel and Schulz 1998). Among the three principally different types of hydrogenases in nature—[Fe only]-type, [NiFe]-type, and [Fe-S cluster-free]-type hydrogenase—the [Fe only]-type are usually the simplest hydrogenases (only one subunit in the case of *Chlamydomonas*) and the most active hydrogenases known (up to 2,000 H_2/s) present in green algae (Girbal et al. 2005; Happe et al. 2002; Lyon et al. 2004).

Recent advances with respect to the identification of genes involved in maturation and regulation of [FeFe] H_2ase (Girbal et al. 2005; McGlynn et al. 2008; Posewitz et al. 2004) and [NiFe] H_2ase (Ludwig et al. 2009; Schubert et al. 2007). The development of heterologous expression systems for both classes of enzyme (Girbal et al. 2005; King et al. 2006; Lenz et al. 2005; Posewitz et al. 2004; Sybirna et al. 2008) make it feasible to biochemically characterize H_2 ases by heterologous expression in organisms that do not possess endogenous H_2 ase machinery. Recently, there has been considerable progress in identifying relevant bioenergy genes and pathways in microalgae, and powerful genetic techniques have been developed to engineer some strains through the targeted disruption of endogenous genes or transgene expression (Rathore and Singh 2013). Some studies indicated that mutagenesis can be used to decrease the O_2 sensitivity of the hydrogenase and thus eventually lead to a system that produces H_2 under aerobic conditions (Flynn et al. 1999, 2002; Ghirardi et al. 1997; McTavish et al. 1995; Seibert et al. 2001). One of the most significant advances in algal genetics is the development of improved gene silencing strategies in *Chlamydomonas reinhardtii*. High-throughput artificial miRNA (amiRNA) techniques for gene knockdown, which are highly specific and stable, were recently reported (Molnar et al. 2009; Zhao et al. 2009).

7.3 Conclusion

The primary focus for the production of microbial biofuel depends on the feasibility of the microbial strains. Various microbial strains were found to be very rich in lipid content. The microorganisms can be used for the production of different biofuels such as bioethanol, biodiesel, and biohydrogen. The whole biomass of microalgae can be converted to biofuels with the biorefinery concept. Metabolic and genetic engineering can play a major role to make these production technologies more economical and sustainable.

References

Aleklett K, Campbell CJ (2003) The peak and decline of world oil and gas production. *Miner Energy* 18:35–42.

Antal TK, Krendelev TE, Rubin AB (2011) Acclimation of green algae to sulfur deficiency: Underlying mechanisms and application for hydrogen production. *Appl Microbiol Biotechnol* 89:3–15.

Appel J, Schulz R (1998) Hydrogen metabolism in organisms with oxygenic photosynthesis: Hydrogenases as important regulatory devices for a proper redox poising? *J Photochem Photobiol* 47:1–11.

Armbruster WJ, Coyle WT (2006) *Pacific Food System Outlook 2006–2007: The Future Role of Biofuels*. Singapore: Pacific Economic Cooperation Council. Available at http://www.pecc.org/food/pfso-singapore2006/PECC_Annual_06_07.pdf.

Asada Y, Miyake J (1999) Photobiological hydrogen production. *J Biosci Bioeng* 88:1–6.

Barber J (2008) Crystal structure of the oxygen-evolving complex of photosystem II. *Inorg Chem* 47:1700–1710.

Bassham J, Benson A, Calvin M (1950) The path of carbon in photosynthesis. *J Biol Chem* 185:781–787.

Boichenko VA, Greenbaum E, Seibert M (2001) Hydrogen production by photosynthetic microorganisms. In: Archer MD, Barber J, editors. *Photoconversion of Solar Energy: Molecular to Global Photosynthesis*, vol. 2. London: Imperial College Press.

Boichenko VA, Hoffmann P (1994) Photosynthetic hydrogen-production in prokaryotes and eukaryotes: Occurrence, mechanism, and functions. *Photosynthetica* 30:527–552.

Breyton C, Nandha B, Johnson GN, Joliot P, Finazzi G (2006) Redox modulation of cyclic electron flow around photosystem I in C3 plants. *Biochemist* 45:13465–13475.

Brown LM (1996) Uptake of carbon dioxide from flue gas by microalgae. *Energy Convers Manage* 37:1363–1367.

Chen J, Ishiii T, Shimura S, Kirimura K, Usami S (1992) Lipase production by Trichosporon fermentans WU-C12, a newly isolated yeast. *J Ferm Bioeng* 5:412e4.

Das D, Veziroglu TN (2001) Hydrogen production by biological processes: A survey of literature. *Int J Hydrogen Energy* 26:13–28.

Demirbas A (2008) Comparison of transesterification methods for production of biodiesel from vegetable oils and fats. *Energy Convers Manage* 49:125–130.

Ehimen EA, Sun ZF, Carrington CG (2010) Variables affecting the in situ transesterification of microalgae lipids. *Fuel* 89:677–684.

Elshahed MS (2010) Microbiological aspects of biofuel production: Current status and future directions. *J Adv Res* 1:103–111.

Escobar JC, Lora ES, Venturini OJ, Yanez EE, Castillo EF, Almazan O (2009) Biofuels: Environment, technology and food security. *Renew Sustain Energy Rev* 13:1275–1287.

Fakas S, Galiotou-Panayotou M, Papanikolaou S, Komaitis M, Aggelis G (2007) Compositional shifts in lipid fractions during lipid turnover in Cunninghamella echinulata. *Enzyme Microbiol Technol* 40:1321e7.

Florin L, Tsokoglou A, Happe T (2001) A novel type of iron hydrogenase in the green alga Scenedesmus obliquus is linked to the photosynthetic electron transport chain. *J Biol Chem* 276:6125–6132.

Flynn T, Ghirardi ML, Seibert M (1999) Isolation of *Chlamydomonas* mutants with improved oxygen-tolerance. In: Division of Fuel Chemistry Preprints of Symposia, 218th ACS National Meeting, vol. 44: 846–850.

Flynn T, Ghirardi ML, Seibert M (2002) Accumulation of O_2-tolerant phenotypes in H_2-producing strains of *Chlamydomonas reinhardtii* by sequential applications of chemical mutagenesis and selection. *Int J Hydrogen Res* 27:1421–1430.

Fukuda H, Kondo A, Noda H (2001) Biodiesel fuel production by transesterification of oils. *J Biosci Bioeng* 92(5):405–416.

Gaffron H (1940) Carbon dioxide reduction with molecular hydrogen in green algae. *Am J Bot* 27:273–283.

Gaffron H (1944) Photosynthesis, photoreduction and dark reduction of carbon dioxide in certain algae. *Biol Rev Cambridge Philos Soc* 19:1–20.

Gaffron H, Rubin J (1942) Fermentative and photochemical production of hydrogen in algae. *J Gen Physiol* 26:219–240.

Ghirardi ML, Togasaki RK, Seibert M (1997) Oxygen sensitivity of algal H_2-production. *Appl Biochem and Biotechnol* 63–65:141–151.

Ghirardi ML, Kosourov SN, Tsygankov AA, Seibert M (2000) Two phase photobiological algal H2-production system. Proceedings of the 2000 DOE Hydrogen Program Review. NREL/CP- 70-28890.

Girbal L, von Abendroth G, Winkler M, Benton PMC, Meynial-Salles I, Croux C, Peters JW, Happe T, Soucaille P (2005) Homologous and heterologous overexpression in Clostridium acetobutylicum and characterization of purified clostridial and algal Fe-only hydrogenases with high specific activities. *Appl Environ Microbiol* 71:2777–2781.

Goffeau A, Barrell G, Bussey H, Davis RW, Dujon B, Feldmann H et al. (1996) Life with 6000 genes. *Science* 274(5287):546–567.

Golding AJ, Johnson GN (2003) Down regulation of linear and activation of cycling electron flow during drought. *Planta* 218:107–114.

Goussias C, Boussac A, Rutherford AW (2002) Photosystem II and photosynthetic oxidation of water: An overview. *Phil Trans R Soc Lond* B357:1369–1381.

Gulati M, Kohlmann K, Ladisch MR, Hespell R, Bothast RJ (1996) Assessment of ethanol production options for corn products. *Bioresour Technol* 58(3):253–264.

Gullison RE, Frumhoff PC, Canadell JG, Field CB, Nepstad DC, Hayhoe K et al. (2007) Tropical forests and climate policy. *Science* 316:985–986.

Hall DO, Markov SA, Watanabe Y, Rao K (1995) The potential applications of cyanobacterial photosynthesis for clean technologies. *Photosynth Res* 46:159–167.

Hallenbeck PC, Benemann JR (2002) Biological hydrogen production; fundamentals and limiting processes. *Int J Hydrogen Energy* 27:1185–1193.

Happe T, Kaminski A (2002) Differential regulation of the Fehydrogenase during anaerobic adaptation in the green alga *Chlamydomonas reinhardtii*. *Eur J Biochem* 269:1022–1032.

Happe T, Winkler M, Hemschemeier A, Kaminski A (2002) Hydrogenases in green algae: Do they save the algae's life and solve our energy problems? *Trends Plant Sci* 7:246–250.

He Y, Wang S, Lai KK (2010) Global economic activity and crude oil prices: A cointegration analysis. *Energy Econ*. 32: 868–876.

Ho NWY, Chen Z, Brainard AP (1998) Genetically engineered Saccharomyces yeast capable of effective cofermentation of glucose and xylose. *Appl Environ Microbiol* 64(5):1852–1859.

Horner D, Heil B, Happe T, Embley M (2002) Iron hydrogenases, ancient enzymes in modern eukaryotes. *Trends Biochem Sci* 27:148–153.

Hu Q, Sommerfeld M, Jarvis E, Ghirardi M, Posewitz M, Seibert M et al. (2008) Microalgal triacylglycerols as feedstocks for biofuel production: Perspectives and advances. *Plant J* 54:621–639.
Huang C, Zong MH, Hong W, Liu QP (2009) Microbial oil production from rice straw hydrolysate by Trichosporon fermentans. *Bioresour Technol* 100:4535–4538.
Johnson MB, Wen Z (2009) Production of biodiesel fuel from the Microalga *Schizochytrium limacinum* by direct transesterification of algal biomass. *Energy Fuels* 23:5179–5183.
Kern J, Renger G (2007) Photosystem II: Structure and mechanism of the water: Plastoquinone oxidoreductase. *Photosynth Res* 94:183–202.
King PW, Posewitz MC, Ghirardi ML, Seibert M (2006) Functional studies of [FeFe] hydrogenase maturation in an *Escherichia coli* biosynthetic system. *J Bacteriol* 188:2163–2172.
Klasson KT, Ackerson MD, Clausen EC, Gaddy JL (1992) Bioconversion of synthesis gas into liquid or gaseous fuels. *Enzyme Microb Technol* 14(8):602–608.
Kuyper M, Toirkens MJ, Diderich JA, Winkler AA, VanDijken JP, Pronk JT (2005) Evolutionary engineering of mixed-sugar utilization by a xylose-fermenting Saccharomyces cerevisiae strain. *FEMS Yeast Res* 5(10):925–934.
Leadbetter JR, Schmidt TM, Graber JR, Breznak JA (1999) Acetogenesis from H_2 plus CO_2 by spirochetes from termite guts. *Science* 283(5402):686–689.
Lenz O, Gleiche A, Strack A, Friedrich B (2005) Requirements for heterologous production of a complex metalloenzyme: The membrane-bound [NiFe] hydrogenase. *J Bacteriol* 187:6590–6595.
Levin DB, Pitt L, Love M (2004) Biohydrogen production: Prospects and limitations to practical application. *Int J Hydrogen Energy* 29:173–185.
Levine RB, Pinnarat T, Savage PE (2010) Biodiesel production from wet algal biomass through in situ lipid hydrolysis and supercritical transesterification. *Energy Fuels* 24:5235–5243.
Licht FO (2008) *World Ethanol and Biofuels Report*. Kent, UK: Agra Informa Ltd. Available at http://www.agra-net.com/portal/puboptions.jsp?Option¼menu&pubId¼ag072.
Liu ZL, Saha BC, Slininger PJ (2008) Lignocellulosic biomass conversion to ethanol by Saccharomyces. In: Harwood CS, Demain AL, Wall JD, editors. *Bioenergy*. Washington, DC: ASM Press.
Liu ZL, Saha BC (2008) Effect of iron on growth and lipid accumulation in Chlorella vulgaris. *Bioresour Technol* 99:4717–4722.
Ludwig M, Schubert T, Zebger I, Wisitruangsakul N, Saggu M, Strack A, Lenz O, Hildebrandt P, Friedrich B (2009) Concerted action of two novel auxiliary proteins in assembly of the active site in a membrane-bound [NiFe] hydrogenase. *J Biol Chem* 284:2159–2168.
Lyon EJ, Shima S, Buurmn G, Chowdhuri S, Btschauer A, Steinbach K, Thauer RK (2004) UV-A/blue-light inactivation of the "metal-free" hydrogenase (Hmd) from methanogenic archaea. *Eur J Biochem* 271:195–204.
McGlynn SE, Shepard EM, Winslow MA, Naumov AV, Duschene KS, Posewitz MC, Broderick WE, Broderick JB, Peters JW (2008) HydF as a scaffold protein in [FeFe] hydrogenase H-cluster biosynthesis. *FEBS Lett* 582:2183–2187.
McTavish H, Sayavedra-Soto LA, Arp DJ (1995) Substitutions of *Azotobacter vinelandii* hydrogenase small-subunit cysteines by serines can create insensitivity to inhibition by O_2 and preferentially damages H_2 oxidation over H_2 evolution. *J Bacteriol* 177:3960–3964.
Melis A, Happe T (2004) Trails of green alga hydrogen research—from Hans Gaffron to new frontiers. *Photosynth Res* 80:401–409.
Melis A, Zhang L, Foestier M, Ghirardi ML, Seibert M (2000) Sustained photobiological hydrogen gas production upon reversible inactivation of oxygen evolution in the green alga *Chlamydomonas reinhardtii*. *Plant Physiol* 122:127–133.
Meng J, Yang X, Xu L, Zhang Q, Nie XM (2009) Biodiesel production from oleaginous microorganisms. *Renew Energy* 34:1–5.
Metting Jr FB (1996) Biodiversity and application of microalgae. *J Ind Microbiol Biotechnol* 17(5–6):477–489.
Molnar A, Bassett A, Thuenemann E, Schwach F, Karkare S, Ossowski S, Weigel D, Baulcombe D (2009) Highly specific gene silencing by artificial microRNAs in the unicellular alga *Chlamydomonas reinhardtii*. *Plant J* 58:165–174.
Moss DA, Bendall DS (1984) Cyclic electron transport in chloroplasts. The Q-cycle and the site of action of antimycin. *Biochim Biophys Acta* 767:389–395.

Müller V (2003) Energy conservation in acetogenic bacteria. *Appl Environ Microbiol* 69(11):6345–6353.
Murray D (2005) *Ethanol's Potential: Looking Beyond Corn*. Washington DC: Earth Policy Institute. Available at http://www.earthpolicy.org/Updates/2005/Update49.htm.
Ni M, Leung DYC, Leung MKH, Sumathy K (2006) An overview of hydrogen production from biomass. *Fuel Process Technol* 87:461–472.
Nigam PS, Singh A (2011) Production of liquid biofuels from renewable resources. *Prog Energy Combust Sci* 37:52–68.
Posewitz MC, King PW, Smolinski SL, Zhang L, Seibert M, Ghirardi ML (2004) Discovery of two novel radical Sadenosylmethionine proteins required for the assembly of an active [Fe] hydrogenase. *J Biol Chem* 279:25711–25720.
Prasad S, Singh A, Jain N, Joshi HC (2007a) Ethanol production from sweet sorghum syrup for utilization as automotive fuel in India. *Energy Fuel* 21(4):2415–2420.
Prasad S, Singh A, Joshi HC (2007b) Ethanol as an alternative fuel from agricultural, industrial and urban residues. *Resour Conserv Recycl* 50:1–39.
Prince RC, Kheshgi HS (2005) The photobiological production of hydrogen: Potential efficiency and effectiveness as a renewable fuel. *Crit Rev Microbiol* 31(1):19–31.
Rathore D, Singh A (2013) Biohydrogen production from micro algae. In: Gupta VK, Tuohy MG, editors. *Biofuel Technologies Recent Developments*. Berlin: Springer-Verlag, pp. 317–333.
Ratledge C (2004) Fatty acid biosynthesis in microorganisms being used for single cell oil production. *Biochimie* 86:807–815.
Sakurai H, Masukawa H, Dawar S, Yoshino F (2004) Photobiological hydrogen production by cyanobacteria utilizing nitrogenase systems-present status and future development. In: Miyake J, Igarashi Y, Rögner M, editors. *Biohydrogen III. Renewable Energy System by Biological Energy Conversion*. Amsterdam: Elsevier, pp. 84–93.
Schenk PM, Thomas-Hall SR, Stephens E, Marx UC, Mussgnug JH, Posten C, Kruse O, Hankamer B (2008) Second generation biofuels: High-efficiency microalgae for biodiesel production. *Bioenerg Res* 1:20–43.
Schubert T, Lenz O, Krause E, Volkmer R, Friedrich B (2007) Chaperones specific for the membrane-bound [NiFe]-hydrogenase interact with the Tat signal peptide of the small subunit precursor in Ralstonia eutropha H16. *Mol Microbiol* 66:453–467.
Sedlak M, Ho NWY (2004) Production of ethanol from cellulosic biomass hydrolysates using genetically engineered Saccharomyces yeast capable of cofermenting glucose and xylose. *Appl Biochem Biotechnol* 114(1–3):403–416.
Seibert M, Flynn T, Ghirardi ML (2001) Strategies for improving oxygen tolerance of algal hydrogen production. In: Miyake J, Matsunaga T, San Pietro A, editors. *Biohydrogen II*. Oxford, UK: Pergamon Press, pp. 67–77.
Singh A, Pant D, Korres NE, Nizami AS, Prasad S, Murphy JD (2010a) Key issues in life cycle assessment of ethanol production from lignocellulosic biomass: Challenges and perspectives. *Bioresour Technol* 101(13):5003–5012.
Singh A, Smyth BM, Murphy JD (2010b) A biofuel strategy for Ireland with an emphasis on production of biomethane and minimization of land-take. *Renew Sustain Energy Rev* 14(1):277–288.
Singh A, Olsen SI (2011) Critical analysis of biochemical conversion, sustainability and life cycle assessment of algal biofuels. *Appl Energy* 88:3548–3555.
Singh A, Nigam PS, Murphy JD (2011a) Renewable fuels from algae: An answer to debatable land based fuels. *Bioresour Technol* 102:10–16.
Singh A, Nigam PS, Murphy JD (2011b) Mechanism and challenges in commercialisation of algal biofuels. *Bioresour Technol* 102:26–34.
Singh A, Olsen SI, Nigam PS (2011c) A viable technology to generate third generation biofuel. *J Chem Technol Biotechnol* 86(11):1349–1353.
Singh A, Pant D, Olsen SI, Nigam PS (2012) Key issues to consider in microalgae based biodiesel. *Energy Educ Sci Technol Part A: Energy Sci Res* 29(1):687–700.
Spolaore P, Joannis Cassan C, Duran E, Isambert A (2006) Commercial applications of microalgae. *J Biosci Bioeng* 101(2):87–96.
Spruit CJP (1958) Simultaneous photoproduction of hydrogen and oxygen by *Chlorella*. *Mededel Landbouwhogeschool Wageningen* 58:1–17.

Sybirna K, Antoine T, Lindberg P, Fourmond V, Rousset M, Mejean V, Bottin H (2008) *Shewanella oneidensis*: A new and efficient system for expression and maturation of heterologous [Fe–Fe] hydrogenase from *Chlamydomonas reinhardtii*. *BMC Biotechnol* 8:73.

Tanner RS (2008) Production of ethanol from synthesis gas. In: Wall JD, Harwood CS, Demain AL, editors. *Bioenergy*. Washington, DC: ASM Press.

Tilman D, Hill J, Lehman C (2006) Carbon-negative biofuels from low-input high-diversity grassland biomass. *Science* 314(5805):1598–1600.

Wang Y, Wu H, Zong MH (2008) Improvement of biodiesel production by lipozyme TL IM-catalyzed methanolysis using response surface methodology and acyl migration enhancer. *Bioresour Technol* 99:7232–7237.

Weaver PF, Lien S, Seibert M (1980) Photobiological production of hydrogen. *Solar Energy* 24:3–45.

Wisselink HW, Toirkens MJ, Wu Q, Pronk JT, VanMaris AJA (2009) Novel evolutionary engineering approach for accelerated utilization of glucose, xylose and arabinose mixtures by engineered Saccharomyces cerevisiae strains. *Appl Environ Microbiol* 75 (4):907–914.

Wyman CE (1994) Ethanol from lignocellulosic biomass: Technology, economics and opportunities. *Bioresour Technol* 50(1):3–15.

Xiong W, Li X, Xiang J, Wu O (2008) High-density fermentation of microalga Chlorella protothecoides in bioreactor for microbiodiesel production. *Appl Microb Biotechnol* 78:29–36.

Zhao T, Wang W, Bai X, Qi Y (2009) Gene silencing by artificial microRNAs in *Chlamydomonas*. *Plant J* 58:157–164.

Zhu H, Beland M (2006) Evaluation of alternative methods of preparing hydrogen producing seeds from digested wastewater sludge. *Int J Hydrogen Energy* 31:1980–1988.

Zhu LY, Zong MH, Wu H (2008) Efficient lipid production with *T. fermentas* and its use for biodiesel preparation. *Bioresour Technol* 99:7881–7885.

chapter eight

Microbial production of organic acids

Sarafadeen Olateju Kareem and Temitope Banjo

Contents

8.1 Introduction

Organic acids are commercially valuable products obtained through microbial fermentation. Generally, microbial fermentation is an aspect of microbial biotechnology that involves the use of microorganisms to metabolize various substrates to produce a particular metabolite. Microbial fermentation was dated to 6000 BC when the art was being used to produce food and beverages. In 1850, Louis Pasteur provided evidence of microbial involvement in beer fermentation when he discovered that yeasts were fermenting sugar into ethanol. He later turned his attention toward vinegar production and other acids of microbial origin.

Many organic acids possess a long chain of carbons attached to a carboxyl group and can be produced through microbial bioprocesses with the exception of a few that are produced commercially. Fermentation processes play a major role in the production of most organic acids such as citric, gluconic, itaconic, lactic, fumaric, and malic acids. Most organic acids are produced as intermediate metabolites of the tricarboxylic acid (TCA) cycle whereas other acids can be derived indirectly from the Krebs cycle such as itaconic acid, or directly from glucose. Acetic and lactic acids are formed as end products from pyruvate or ethanol. Many bacteria and fungi have the potential to produce a variety of organic acids with high yields. However, fungi have an intrinsic ability to accumulate many organic acids because of their ability to thrive at acidic pH from 2 to 5 (Goldberg et al. 2006).

Bioconversion of carbon substrate especially glucose is exploited in large-scale commercial production of a number of organic acids such as citric, gluconic, and itaconic acids. Other acids produced on a smaller scale are lactic acid, malic acid, and kojic acids. This chapter contains a review of the microbial production of organic acids, their biochemistry, upstream processes, fermentation processes, and downstream processes.

8.2 Organic acids: An overview

8.2.1 Acetic acid

Acetic acid is also known as ethanoic acid. It is a colorless organic compound that, in its undiluted form, is also called glacial acetic acid. Acetic acid is the main component of vinegar (4%–8% acetic acid). Vinegar is s a weak acid with a distinctive sour taste and pungent smell. It is often produced by fermentation and subsequent oxidation of ethanol. It is an important chemical reagent and industrial chemical, mainly used in the production of cellulose acetate for photographic film, and polyvinyl acetate for wood glue as well as synthetic fibers and fabrics. In the food industry, acetic acid is used as a condiment.

Acetic acid is used in the production of vinyl acetate polymers, which are components in paints and adhesives. Vinegar is typically 4% to 18% acetic acid by mass. Vinegar is used directly as a condiment, and in the pickling of vegetables and other foods (Bala 2003).

8.2.2 Ascorbic acid

Ascorbic acid or vitamin C is a weak sugar acid, white, soluble, and structurally related to glucose. In biological systems, it can be found only at low pH (acidic), but in neutral solutions above pH 5, it is predominantly found in ionized form (ascorbate). Ascorbate is easily and reversibly oxidized to dehydro-L-ascorbic acid, forming the ascorbyl radical as an intermediate (Schmidt and Moser 1985).

Ascorbic acid enhances iron absorption by forming soluble ferrous–ascorbate complexes that are readily absorbed in the cell. Ascorbic acid is involved in the development of connective tissue cells. It modulates neurotransmitter systems in the brain and stimulates

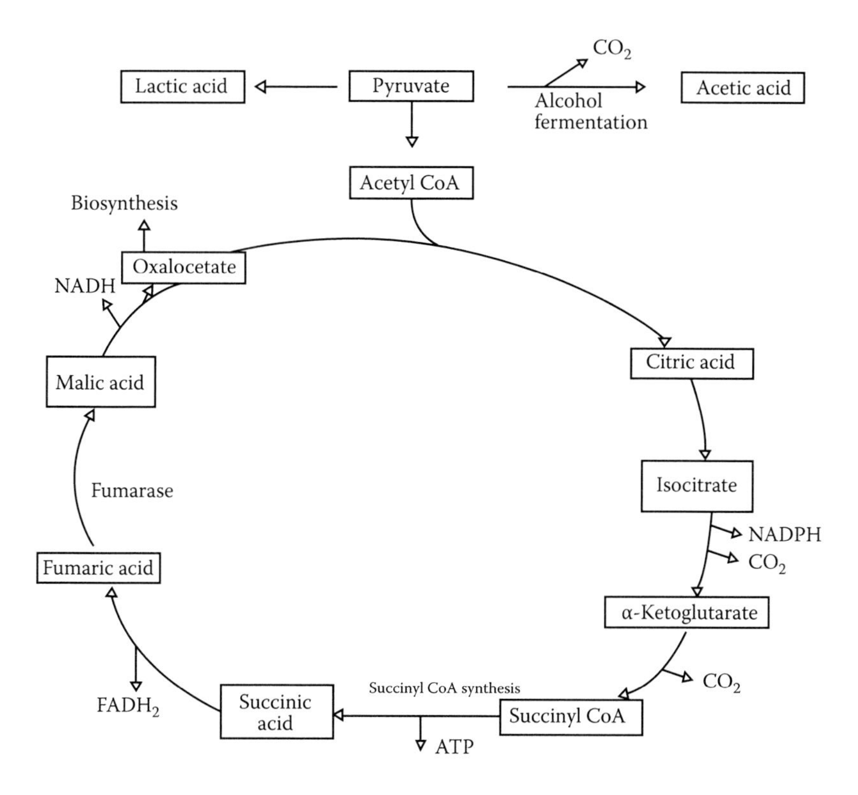

Figure 8.1 Biosynthesis of organic acids through the pyruvate and TCA cycle.

collagen synthesis, which is a requirement for muscle and bone formation (Ball 2004; Hallberg et al. 1996).

8.2.3 Citric acid

Citric acid is a TCA (2-hydroxy-1,2,3-propane tricarboxylic acid), is soluble in water, and has a pleasant taste. It is ubiquitous in nature and exists as an intermediate in the TCA cycle. It is naturally found in fruits such as lemons, oranges, pineapples, and pears as well as in the seeds of different vegetables, and in animal bone, muscle, and blood. The conversion of glucose or sucrose to citric acid by *Aspergillus niger* involves both the glycolytic pathway and the TCA cycle (Figure 8.1). Citric acid synthesis by *A. niger* from sucrose or sugar substrate has been attributed to the presence of the following enzymes; phosphofructokinase, pyruvate kinase, pyruvate decarboxylase, citrate synthase, aconitase and isocitrate dehydrogenase (Kapoor et al. 1983). Citric acid is the most important organic acid used in food industries as food preservatives and anticoagulant. The free acid is employed in pharmaceuticals as an acidulant and to enhance the flavor of syrups, solutions, and elixirs (Ali 2004; Sauer et al. 2007).

8.2.4 Fumaric acid

Fumaric acid is a naturally occurring organic acid that was first isolated from the plant *Fumaria officinalis*. Fumaric acid is also known as (*E*)-2-butenedioic acid or *trans*-1,2-ethylenedicarboxylic acid. Many microorganisms produce fumaric acid in small amounts because it is a key intermediate in the citrate cycle. Fumaric acid is primarily an intermediate of the TCA cycle (Figure 8.1), but it is also involved in other metabolic pathways. Fumaric acid is used as food and in feed additives. It is also used in the production of polyester resins (Kautola and Linko 1989; Kenealy et al. 1986; Sauer et al. 2007).

8.2.5 Gluconic acid

Gluconic acid or pentahydroxycarboxylic acid is a noncorrosive, nonvolatile, nontoxic, and mild organic compound. The annual worldwide production capacity for gluconic acid and its derivatives is estimated to be 60,000 tonnes. The bulk of production (85%) is in the form of sodium gluconate and other alkali gluconate salts. In the oxidation of glucose to gluconic acid, beta-D glucose is oxidized to *d*-D-gluconolactone in the presence of oxygen. The gluconolactone is subsequently hydrolyzed nonenzymatically to D-gluconic acid. Gluconic acid and its salts are prepared exclusively by the oxidation of glucose or glucose-containing raw material through a simple dehydrogenation reaction catalyzed by glucose oxidase. Glucose is generally used as a carbon source for the microbial production of gluconic acid (Sankpal and Kulkarni 2002).

Gluconic acid is used in the chemical, pharmaceutical, food, beverage, textile, and other industries. In the food industry, it is used as a food additive and acidity regulator. It also serves as a cement additive in the construction industry because it enhances cement's resistance and stability under extreme climatic conditions (Ramachandran et al. 2006).

8.2.6 Lactic acid

Lactic acid is a carboxylic acid with an adjacent hydroxyl group. It was first isolated in 1780 by the Swedish chemist Carl Scheele (VickRoy 2000). Lactic acid is miscible with water or ethanol, and is hygroscopic. Lactic acid is chiral and has two optical isomers. One is known as L-(+)-lactic acid or (*S*)-lactic acid and the other, its mirror image, is D-(−)-lactic acid or (*R*)-lactic acid (Tay and Yang 2002). Bacterial species of *Lactobacillus* are major producers of lactic acid and are able to hydrolyze cheap raw materials such as starchy and cellulosic materials into fermentable sugars before lactic acid production. This hydrolysis can be carried out simultaneously with fermentation by amylase-producing *Lactobacillus amylophilus* and *Lactobacillus amylovorus* (Hofvendahl and Hahn-Hägerdal 2000).

Lactic acid has many potential applications in the food, cosmetic, pharmaceutical, and chemical industries. It is used in food to enhance flavor, improve microbial quality, and regulate pH. Moreover, lactic acid is used commercially in processed meats as a preservative and as an acidulant in salads, dressings, and beverages. Other uses of lactic acid include cosmetic applications as moisturizers and pH regulators (Hofvendahl and Hahn-Hägerdal 2000).

8.2.7 Malic acid

Malic acid is a dicarboxylic acid. It was first isolated from apple juice by Carl Wilhelm Scheele in 1785. The main pathway for malic acid synthesis is the condensation of oxaloacetate and acetyl-coenzyme A (acetyl-CoA) to citric acid, followed by its oxidation to malate through the TCA cycle (Figure 8.1). If acetyl-CoA is generated by pyruvate dehydrogenase and oxaloacetate by pyruvate carboxylase, the conversion of glucose to malate through this oxidative pathway results in the release of two molecules of CO_2. Malic acid is used in the treatment of the symptoms of fibromyalgia (Sauer et al. 2007).

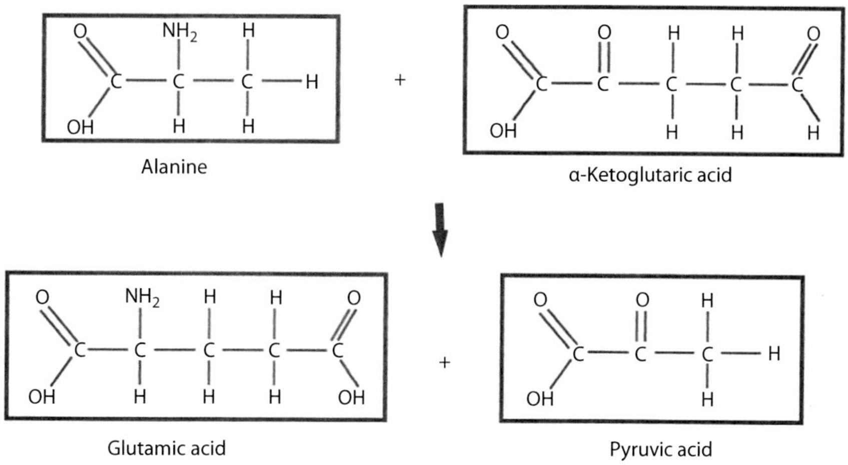

Alanine + α-Ketoglutaric acid

Glutamic acid + Pyruvic acid

Figure 8.2 Transamination reaction for the production of pyruvic acid.

8.2.8 Pyruvic acid

Pyruvic acid is an organic acid, a ketone as well as the simplest of the α-keto acids. It is the end product of the anaerobic portion of glycolysis and is a central metabolite to several metabolic pathways. In aerobic respiration, pyruvate is converted to acetyl-CoA and carbon dioxide by the enzyme pyruvate carboxylase. Under limited amounts of oxygen, pyruvate is converted to lactic acid. It can further be converted to carbohydrates (gluconeogenesis), fatty acids, alanine, and ethanol.

Because glycolysis is the primary metabolic pathway used by microorganisms to generate energy, glucose is a widely used substrate for pyruvic acid synthesis. Prominent pathways for the production of pyruvic acid include glycolysis and transamination processes (Figure 8.2; Bao et al. 2011).

Glycolysis is the anaerobic catabolism of glucose to pyruvate. In this process, glucose passes through a series of metabolic reactions to generate two molecules of pyruvic acid. Transamination involves the transfer of an amine group from one molecule to another catalyzed by transaminases. Amino acids, which can be converted after several steps through transamination into pyruvic acid, include alanine, serine, cysteine, and glycine. In addition, pyruvic acid is also a substrate for the enzymatic production of amino acids such as L-tryptophan, L-tyrosine, D-/L-alanine, and L-dihydroxyphenylalanine through the transamination process.

Pyruvate has also been used in the production of crop protection agents, polymers, cosmetics, and food additives. Calcium pyruvate is useful in the food industry as a fat burner because it accelerates the metabolism of fatty acids in the human body. Clinical studies have demonstrated that pyruvate promotes fat and weight loss significantly, improves exercise endurance capacity, and reduces cholesterol levels effectively (Li et al. 2001; Stanko et al. 1994).

8.2.9 Succinic acid

Succinic acid is another name for amber acid, which has been used in Europe as a natural antibiotic and general curative for centuries. Succinic acid is also a natural constituent of plant and animal tissues. Succinic acid is produced as an intermediate metabolite in the TCA cycle (Figure 8.1). Five key enzymes have been reported for succinic acid production.

These include phosphoenolpyruvate carboxykinase, malate dehydrogenase, malic enzyme, fumarase, and fumarate reductase. The higher CO_2 level resulted in an increased succinic acid production at the expense of ethanol and formic acid.

Succinic acid is known as a powerful antioxidant and has also been shown to stimulate neural system recovery and boost the immune system. In food and beverages, it is used as a flavoring agent and preservative. In pharmaceuticals, it is used in the preparation of active calcium succinate, which acts as an anticarcinogenic agent and as a deodorant for the removal of fishy odors (Sauer et al. 2007).

8.3 Microorganisms involved in organic acid production

Commercial production of many organic acids is usually achieved through fermentation processes. Citric, gluconic, itaconic, and lactic acids are manufactured through large-scale bioprocesses, whereas oxalic, fumaric, and malic acids have been produced in relatively small-scale bioprocesses (Table 8.1). A few other organic acids have been explored for the development of novel processes. The largest commercial quantities of fungal organic acids are those of citric acid and gluconic acid, both of which are prepared through the fermentation of glucose or sucrose by *A. niger*.

Many species of fungi have been evaluated for the production of organic acid (Kareem et al. 2010; Magnuson and Lasure 2004). The selection of a suitable fungus strain is critical

Table 8.1 Organic Acids at a Glance

Organic acid	Chemical formula	Substrate	Microorganisms	References
Acetic	$C_2H_4O_2$	Fermented grains, malt, rice, potato mash, sorghum, apple cider	*Acetobacter* sp., *G. oxydans*, *C. lentocellum*	Sauer et al. 2007
Ascorbic	$C_6H_8O_6$	Glucose	*Aspergillus flavus*, *A. tarmani*, *Acetobacter* sp.	Ball 2004; Sauer et al. 2007
Citric	$C_6H_8O_7$	Glucose, sucrose, cassava whey, dairy whey, pineapple waste	*A. niger*, *A. flavus*, *Candida* spp., *Saccharomyces* spp.	Kareem et al. 2010
Fumaric	$C_4H_4O_4$	Glucose, sucrose, xylose, potato starch	*Rhizopus* sp., *Mucor* sp.	Sauer et al. 2007
Gluconic	$C_6H_{12}O_7$	Glucose, corn starch, cane molasses	*A. niger*	Roukas 2005
Lactic	$C_3H_6O_3$	Starch, glucose	*Lactobacillus* sp., *L. lactis*, *Enterobacter faecalis*, *Rhizopus arrhizus*	Wee et al. 2006
Malic	$C_4H_6O_5$	Glucose	*A. flavus*, *Saccharomyces cerevisae*, *Zygosaccharomyces rouxii*	Sauer et al. 2007
Pyruvic	$C_3H_4O_3$	Glucose	*Proteus*, *Salmonella*, *Escherichia coli*, *Erwinia*, *S. cerevisae*	Bao et al. 2011; Prescott et al. 2005
Succinic	$C_4H_6O_4$	Dairy whey, wood hydrolysate, glucose, lactose, arabinose, maltose, sorbitol	*A. niger*, *A. fumigatus*, *Corynebacterium glutamicum*, *E. faecalis*, *A. succinogenes*	Sauer et al. 2007

to the production of citric acid. Strains with high sporulation, good growth in the substrates, short fermentation times, and high organic acid yields are used in industrial scale operations. Such strains are also expected to have long stability, and resistance to other microorganisms.

A number of microorganisms are involved in the production of acetic acid, which have different pathways or syntheses of production. These microorganisms include *Acetobacter* sp., *Gluconobacter oxydans*, and *Clostridium lentocellum*. Microorganisms that can produce lactic acid can be divided into two groups, bacteria and fungi. Prominent among bacterial species are *Lactobacillus delbrueckii* and *Enterococcus faecalis* (Sauer et al. 2007).

8.4 Upstream processes

8.4.1 Selection of microbial strains

Optimum production of organic acids in a fermentation medium is strongly dependent on the type, nature, and strain of microorganisms. Strain selection is a critical and primary stage of industrial microbiology. The search for a microbial strain with high commercial yields can be obtained from the natural environment or through strain improvement techniques. Strains can be selected after subjecting the genetic material to physical or chemical mutagenic agents or any other genetic manipulation. Mutant strains are able to utilize cheap substrates with high organic acid yield under reduced fermentation periods. They are also stable and resistant to unfavorable conditions. The size of the inocula is also a critical factor in organic acid yield because excess inoculum concentration leads to high biomass concentration and lower organic acid yield, whereas low amounts of inoculum lead to longer fermentation times (Yigitolu 1992).

8.4.2 Immobilization

Immobilization is a technique in which the biomass is immobilized onto the support. It enables repetitive use of the high biomass to carry out biochemical reactions rapidly leading to process economy and stability. Immobilization seems to be an attractive method for accomplishing high cell densities to achieve rapid carbohydrate conversion to organic acids. Immobilization can be carried out in the following ways:

i. *Adsorption*: this process involves the binding of microbial cells directly to water-insoluble carriers such as polysaccharide derivatives, synthetic polymers, and glass. It involves electrostatic interactions between the charged support and the charged cell. The major drawback of this method is cell leakage from the matrix.
ii. *Entrapment*: this involves direct entrapment of microbial cells into polymer matrices such as agar, gelatin, calcium alginate, and others. This process is associated with a diffusion limitation in the transfer of either substrate or product through the matrix.
iii. *Covalent linkage*: this technique involves the linkage of microbial cells or biocatalysts to the carrier through reactive groups on the matrix. The coupling reagents such as glutaraldehyde, diisocyanate, and others, are used to introduce a space group on the carrier surface which subsequently interacts with reactive groups on the cell surface. The use of this technique is limited by partial loss in activity.
iv. *Microencapsulation*: it involves the immobilization of cells or enzymes within semi-permeable microcapsules. These capsules are prepared from organic polymers such as cellulose, polystyrene, polyamine glycol, and others. This is an easy and economic technique; however, cell leakages also occur through microcapsules.

Generally the success of an immobilization process is determined by the type of support, cell retention, stabilization of cells or mycelia, and the amount of biomass. Immobilized cells are preferably used in the bioprocessing of organic acids because of cell stability, constant decrease in medium viscosity, enhanced nutrients and oxygen transfer (Anwar et al. 2009; Kareem et al. 2013; Najafpour 2007).

8.4.3 *Production medium and physicochemical conditions*

A suitable production medium for organic acid production must provide adequate carbon and nitrogen sources. Generally, microbial strains grow well in medium containing carbohydrates ranging from simple sugars to complex sugars. Simple sugars such as glucose, fructose, and maltose are efficiently utilized because organic acids are synthesized through the glycolytic pathway and the TCA cycle. However, complex carbohydrate substrates from agrowastes have also been exploited for organic acid biosynthesis.

Nitrogen is also another factor in the biosynthesis of organic acids because of its role in cell metabolism. Commonly used nitrogen sources are ammonium sulfate or ammonium nitrate (0.25%–0.5%). Adequate balance in C/N ratio is very critical to optimum biosynthesis of organic acids by microorganisms. Many divalent metals such as zinc, manganese, iron, copper, and magnesium at low concentrations (0.02%–0.025%) are known to play significant roles in the biosynthesis of many organic acids (Kareem et al. 2010).

8.4.3.1 *Moisture content*

Another major factor of great influence on the production of organic acid is moisture. Adequate moisture is required for the production of biomass and desired metabolites because it provides nutrients in soluble form for microorganisms. An insufficient moisture level gives a lower degree of swelling and higher water tension, and then reduces the solubility of the nutrients. Whereas excessive moisture levels decrease porosity, changes particle structure, promotes the development of stickiness, reduces gas volume and exchange, and decreases diffusion, which results in lower oxygen transfer.

8.4.3.2 *Effect of temperature*

Temperature is an important factor in the fermentation process (because microbial growth is confined to a narrow range of optimal temperatures). High organic acid yields are obtainable at an optimum growth temperature. Lower or higher temperature values to the optimum always exhibit signs of adverse growth and metabolic production. For instance, several researchers have reported that temperatures between 25°C and 30°C favored adequate sporulation and citric acid production by *A. niger.* Low temperatures lead to slow germination of the fungi and slow metabolic activity, but high temperatures lead to enzyme denaturation and reduced cell viability (Yigitolu 1992).

8.4.3.3 *Control of pH*

The success of microbial production of organic acids is critically dependent on the pH of the fermentation medium. The pH is known to influence important microbial growth and the synthesis of their metabolites. It affects the breakdown of substrates and the permeability of cell membranes, vis-a-vis substrate intake and product release. Optimum yields of many organic acids are obtainable at the optimum pH for microbial growth. It is important to maintain a constant pH during fermentation to ensure optimum yield and reduce the risk of contamination (Kareem et al. 2010). It is also important to note that organic acid production is accompanied by a progressive decrease in the pH of the fermentation broth during the fermentation period.

8.4.3.4 Aeration and agitation

Aeration is a critical factor in the growth and performance of microbial cells by improving mass transfer characteristics with respect to substrate use and product formation. It fulfills four main functions in the fermentation processes: (i) maintain aerobic conditions, (ii) remove carbon dioxide, (iii) regulate the substrate temperature, and (iv) regulate the moisture level.

The amount of oxygen required in a fermentation setup depends on the type and nature of microorganisms (Kamzolova et al. 2003). In organic acid production, an adequate aeration rate is known to enhance the yield and reduce fermentation time. The construction of appropriate aeration devices in fermenters will fulfill the high oxygen demand, which depends on the viscosity of the fermentation broth. Sterile air serves as a source of oxygen intake in fermentation processes. The oxygen is normally dissolved in the liquid phase, which is measured by the amount of dissolved oxygen. In an aerobic fermentation, the bioreactor is usually equipped with a dissolved oxygen meter (Raimbault 1997). In addition to uniform distribution of oxygen, agitation is required to improve rheology, mass, and heat transfer of the fermentation medium.

8.5 Fermentation processes

8.5.1 Surface fermentation

The first fermentation system employed for the production of organic acid on an industrial scale utilized a surface fermentation process. *A. niger* forms a mycelium layer on the liquid surface of the aluminum or stainless steel trays, which are stacked in fermentation rooms supplied with filtered air that serves both to supply oxygen and to control the temperature of fermentation. These techniques make use of the fact that citric acid production occurs in cells not in the active stage of growth (Yigitolu 1992).

Surface fermentation is a simple technique that is easy to control and implement because it needs no aeration or agitation of the fermentation broth. The separation of citric acid from the mycelium is easy because the microorganism is not dispersed into the medium. Only the temperature and humidity of the fermentation chamber need controlling. It can be carried out easily in small plants and therefore is suitable for developing economies. With surface fermentation, the fermentation broth is concentrated due to the high evaporation rate during fermentation. Thus, expenses and losses during recovery and purification are low. However, fermentation time is long and therefore productivity is low (Kapoor et al. 1983).

8.5.2 Submerged fermentation

In submerged fermentation, the microorganism is grown in the fermentation broth. Submerged fermentation has the following advantages: lower total investment costs, higher yields of citric acid, improved process control, reduced fermentation time, reduced labor costs, simpler operations, and easier maintenance of aseptic conditions on an industrial scale. However, the use of submerged fermentation is limited by its high energy consumption and vulnerability to contamination (Kapoor et al. 1983).

8.5.3 Solid-state fermentation

Solid substrate fermentation involves the use of insoluble material as a substrate, both as a physical support and as a source of nutrients for the growth of microorganisms. The solid substrate material can be biodegradable but must provide suitable support for fungal growth and the exchange of gases. For citric acid production, oxygen acts as a direct regulator. As compared with submerged fermentation, solid substrate fermentation offers many advantages: limited consumption of water, low heat transfer capacity with easy aeration, high surface exchange air/substrate, low energy consumption for the process, lower water usage and wastewater treatment costs, limited contamination risks because of a lower moisture level, and simplified nutrient media composition.

On the other hand, the disadvantages of solid-state fermentation include the following: difficulty in controlling process parameters such as pH, heat, and nutrient conditions; problems with heat build-up; higher product impurity level; increased amount of solid waste to manage (including the fungal biomass); and higher product recovery costs (Raimbault 1997).

8.6 Downstream processes

Fermentation broths are complex mixtures of cells, fungal mycelia and spores, partially converted materials, desired products, and unwanted metabolites. These particles must be separated from the product through a series of unit operations known as downstream processing. The primary aim of downstream processing is to efficiently and safely recover the target product to the required specification while maximizing recovery yield and minimizing the process cost. It often accounts for up to 60% of the total production costs, excluding the cost of the purchased raw materials, and determines the economic feasibility of the process. The target products may be the cells (biomass), components within the fermentation broth (extracellular), or those trapped in cells (intracellular). It is often desirable to choose a suitable strain of microorganism that produces extracellular rather than intracellular products for easy recovery. If the product is intracellular, the cells must be ruptured to release the product after which extraction or purification is performed.

The desired product can be recovered by processing the cell or spent medium to a high degree of purity. The level of purity that must be achieved is usually determined by the prospective use of the products such as for food, health, research, and industrial purposes.

There are some fermentation factors that affect downstream processes such as the following:

1. Type and nature of the microorganisms. Some properties of microorganisms such as morphology, flocculation characteristics, size, cell wall rigidity, mycelia formation, and sporulation can affect homogenization, sedimentation, and filtration processes.
2. Presence of media impurities and unwanted metabolites can also add to downstream processing costs. Other fermentation additives and partially or unconverted materials may also interfere with the purification steps.
3. The fermentation substrate is critical to the downstream process. A cheap substrate with many impurities may increase the downstream process cost whereas a more expensive but purer substrate may eventually cut the overall cost.
4. Process economics. Simple and available process equipment may be more cost-effective than sophisticated and complex techniques that might increase the overall cost. The process is normally kept to a minimum to reduce the cost and prevent the overall yield

loss of multistage purification processes. Therefore, the specific unit step in downstream processes is influenced by the economics of the process, the desired degree of purity, the unit yield attainable at each step, and the safety of the product (Najafpour 2007).

Bioseparation processes make use of many separation techniques commonly utilized in the chemical process industries. The first step in downstream processing is the separation of insoluble particles from the fermentation broth. The selection of a separation technique depends on the characteristics of the solids and the liquid medium, which can be accomplished by filtration or centrifugation. Filtration separates particles by forcing the fluid through a filtering medium on which the solids are deposited. The choice of filtration methods depends on the range of particle sizes removed and the pressure differences.

The fermentation broth has to be processed and pass through several stages for separation and purification. The product requires a sequence of operations for high purification. The usual steps are listed below.

Separation of biomass. This stage involves a solid–liquid separation to harvest cells from the spent medium and each fraction is subjected to further treatment depending on whether the target product is intracellular or extracellular. Several factors such as size or morphology of organisms, specific gravity, viscosity, and rheological property of the spent medium can influence the choice of solid–liquid methods. Separation techniques include filtration, sedimentation, and centrifugation.

Cell disruption. It is an appropriate method for the isolation and purification of intracellular products. Available methods for cell disruption can be classified into two groups: (i) mechanical and (ii) nonmechanical. Cell wall destruction is achieved by mechanical methods using solid-shear (bead mill) and liquid-shear forces (high-pressure homogenizer). Nonmechanical methods such as chemical and enzymatic disruption are mild and specific for cell wall degradation. Generally, increases in the viscosity of the suspension are noted as a challenge in cell disruption due to the liberation of DNA and other cellular inclusions.

Concentration of broth. This stage involves the isolation of microbial products through solvent extraction, absorption, precipitation, and ultrafiltration. During this process, target product concentration increases considerably.

Product-specific purification. To obtain a product of high purity, further product concentration techniques such as fractional precipitation, chromatography, and adsorption are used to remove some tiny impurities and further concentrate the product.

Crystallization and drying. Isolated product will be first crystallized and later dried by drum drying, spray drying, and freeze drying. Processes of centrifugation, freeze drying/lyophilization, or organic solvent removal are commonly used. The last step is aimed at presenting the desired product in a form suitable for final formulation (Najafpour 2007).

8.7 Prospects of microbial fermentation of organic acids

Organic acids are vital groups of chemicals that can be produced using fermentation processes. The natural microbial metabolites are intermediates in major metabolic pathways. Because of their functional groups, organic acids are useful as starting materials for the chemical industry. Some organic acids such as fumaric, malic, and succinic acids have the potential to replace petroleum-based chemicals such as maleic anhydride. The need for a sustainable technological development is shifting attention away from petroleum to the use of renewable resources through fermentation processes.

Fermentation technology has allowed the use of indigenously available and cheap substrates for the production of commercially valuable organic acids for sustainable development. Continuous success in the production of organic acids through advances in fermentation technology are envisaged to provide a sustainable industrial bioprocess. Such developments will lead to the evolution of biorefineries to ensure the production of biofuels as well as the building block chemicals from biomass.

8.8 *Case study: Comparative production of citric acid by immobilized* A. niger

Citric acid is an important metabolite produced by fungal fermentation, and is widely used in the food and pharmaceutical industries (Kareem et al. 2010). A large number of microorganisms including fungi, bacteria, and yeasts have been employed for citric acid production. Among the mentioned strains, the fungus *A. niger* has remained the organism of choice for commercial production because it produces more citric acid per time unit. The main advantages of using *A. niger* are its ease of handling, its ability to ferment a variety of cheap raw materials, and high yields (El-Holi and Al-Delamy 2003).

Several raw materials such as hydrocarbons, starch materials, and molasses have been employed as substrates for commercial submerged citric acid production, although citric acid is mostly produced from starch or sucrose-based media using submerged fermentation. Generally, citric acid is produced by fermentation using inexpensive raw materials such as starch hydrolysate, sugar cane broth, and by-products like sugar cane and beet molasses (Yigitolu 1992).

A variety of agro-industrial residues and by-products have also been investigated with solid-state fermentation techniques for their potential to be used as substrates for citric acid production such as coffee husk, wheat bran, apple pomace, pineapple waste, kiwifruit peel, grape pomace, citrus waste, and others (Haq et al. 2004).

The immobilization of microbial cells in polymeric matrices has been reported to enhance the microbial production of organic acids due to the speed and ease of recovery of the cells and ability to reuse the cells. The technique also makes repeated batch and continuous processes possible by allowing a consistent decrease in medium viscosity, and enhanced nutrient and oxygen transfer (Kareem et al. 2013). Cassava starch is abundantly available in the tropics as a cheap substrate. Therefore, this study compares citric acid production by immobilized *A. niger* cells.

8.8.1 *Materials and methods*

8.8.1.1 *Immobilization of* A. niger

Spores of *A. niger* (4.0×10^3) were mixed with sodium alginate solution (2%) in a 1:1 ratio. The alginate mixture was added dropwise into a calcium chloride solution (0.2 M) with continuous shaking at 4°C. As soon as the drop of enzyme–alginate solution mixed with the $CaCl_2$ solution, Na^+ ions of sodium alginate were replaced by Ca^{2+} ions from the $CaCl_2$ solution, which finally formed Ca-alginate beads. The beads thus formed were washed three to four times with deionized water and finally with 50 m Tris-HCl buffer at pH 7.5. These beads were dried and weighed for further studies.

8.8.1.2 *Citric acid production*

Cassava starch (30 g) was added into 200 mL Erlenmeyer flasks and supplemented with nitrogen sources. Distilled water (200 mL) was added and the initial pH of the medium was adjusted to 6.0. The flask was cotton-plugged and autoclaved at 121°C for 15 min. After cooling to room temperature, each medium was inoculated with immobilized Ca-alginate beads and incubated at 30°C in a rotary shaking incubator for 5 days. After incubation, the medium was filtered, and then the filtrate was used for the estimation of citric acid, residual sugar content, and pH.

8.8.1.3 Citric acid determination

Citric acid was determined titrimetrically (Association of Official Analytical Chemists [AOAC] 1995) by using 0.1 M NaOH and phenolphthalein as an indicator, and calculated as a percentage according to the formula:

$$\text{Citric acid g/L}\frac{\text{Titer} \times 0.0064 \times 100 \times 10}{10\text{ mL}} \text{ (where 0.0064 denotes citric acid factor)}$$

8.8.1.4 Biomass, residual sugars, and pH determination

Biomass, sugar, and pH values were determined according to AOAC (1995). To determine the biomass, the whole fungal culture was filtered with sterile filter paper dried at 105°C to a constant weight. Results were expressed in grams per kilogram. Sugar was determined using a refractometer and pH was measured using a pH meter. Each analysis was conducted in triplicate.

8.8.2 Results and discussion

This study presents citric acid production using immobilized *A. niger* on raw cassava starch. The results showed that 26.84 g/L of citric acid was obtained in a medium containing cassava starch alone whereas the addition of nitrogen supplements enhanced citric acid production with the maximum citric acid yield (42.2 g/L) obtained in a cassava starch medium supplemented with peptone (Table 8.2). The results indicated that *A. niger* first hydrolyzed starch to glucose units to provide utilizable substrate for glycolysis and subsequent generation of citric acid in the TCA cycle. Higher citric acid accumulation by *A. niger* in medium containing high concentrations of glucose has been attributed to high hexokinase activity in the glycolytic pathway (Ali 2004). The results also indicate that nitrogen constituents not only enhance microbial growth but also stimulate organic acid synthesis. Kareem and coworkers (2010) reported that fermentation media for citric acid production should consist of substrates necessary for the growth of microorganisms; primarily carbon, nitrogen, and phosphorus sources.

Table 8.2 Effect of Nitrogen Supplements on Citric Acid Production from Cassava Starch by *A. niger*

Supplements	Citric acid (g/L)
KH_2PO_4	32.81
$(NH_4)_2PO_4$	36.32
Peptone	42.12
Control	26.84

Note: Data are means of triplicates, SD = 0.052.

The effect of alcohol on citric acid production is shown in Table 8.3. The results showed that methanol at 1% concentration gave an inductive effect on citric acid production. This observation concurred with the results of Kareem and coworkers (2010) in which the addition of methanol to pineapple waste medium led to an increase in citric acid production by *A. niger*. The inductive effect of methanol for citric acid production may be due to the reduction of the inhibitory effects of metal ions (Kiel et al. 1981). The addition of alcohols to the fermentation media had been reported to increase citric acid yield by filamentous fungi (El-Holi and Al-Delamy 2003). However, in this study, methanol had a more inductive effect on citric acid production than ethanol. The exact mechanism of the alcohol effect, however, is unexplained, although it has been postulated that the addition of methanol increases the tolerance of fungi to Fe^{2+}, Zn^{2+}, and Mn^{2+}.

The results of the effect of agitation on citric acid production by immobilized *A. niger* showed that a higher citric acid yield of 58.1 g/L was obtained under agitated conditions at the fourth day of fermentation (Table 8.4). Agitation and aeration are important in submerged fermentation, which is extremely important for adequate mixing, mass transfer, and heat transfer. In addition, agitation maintains homogeneous chemical and physical conditions in the culture by continuous mixing (Mantzouridou et al. 2002).

The effects of fermentation period on citric acid production, pH, and sugar consumption are presented in Figure 8.3. A steady increase in citric acid production was observed during the fermentation period, with an optimum yield (62.2 g/kg) after fermentation for 96 h, whereas a decrease in the consumption of sugar was observed. Previous reports have shown that the production of citric acid on pineapple waste approximately paralleled the consumption of sugar (Kareem et al. 2010). Based on the amount of fermentable sugar

Table 8.3 Effect of Alcohol on Citric Acid Production from Cassava Starch by Immobilized *A. niger*

	Citric acid (g/L)	
Time (days)	Methanol	Ethanol
0	42.12	42.12
1	55.30	45.10
2	50.03	52.21
3	59.20	35.11

Table 8.4 Effect of Agitation on Citric Acid Production from Cassava Starch by *A. niger*

	Citric acid (g/L)	
Time (days)	Agitation	Static condition
1	15.1	10.2
2	36.2	28.4
3	42.3	34.3
4	58.1	51.2
5	51.2	46.2

Note: Data are means of triplicate determinations, SD = 0.033.

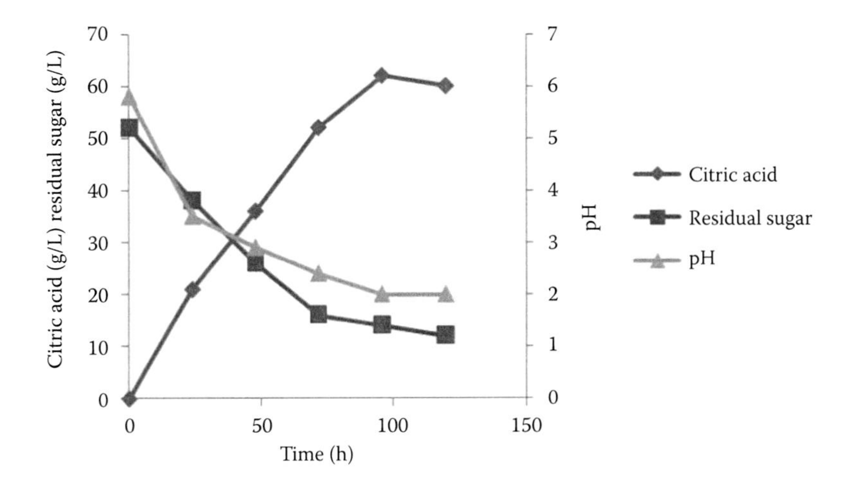

Figure 8.3 Effect of fermentation period on citric acid production, pH and sugar consumption.

consumed, the yield of citric acid was 90.62% under optimum fermentation conditions. A decrease in pH values from 5.0 to 2.2 was noted after 96 h of fermentation. The pH value maintained at the beginning of fermentation was important for a specific biomass formation. The progressive decrease in pH was noted as incubation time increased, which was due to the formation and accumulation of citric acid.

Comparison between fermentation with free cells and immobilized cells of *A. niger* (Figure 8.4) showed that immobilized *A. niger* cells gave an enhanced citric acid production and fermentation efficiency compared with free cells as a result of decreased medium viscosity and enhanced oxygen and nutrient transfer (Kareem et al. 2013). In addition, immobilized cells can be reused up to 20 repeated cycles. This will ensure the optimization of the citric acid bioprocesses and therefore reduce process cost.

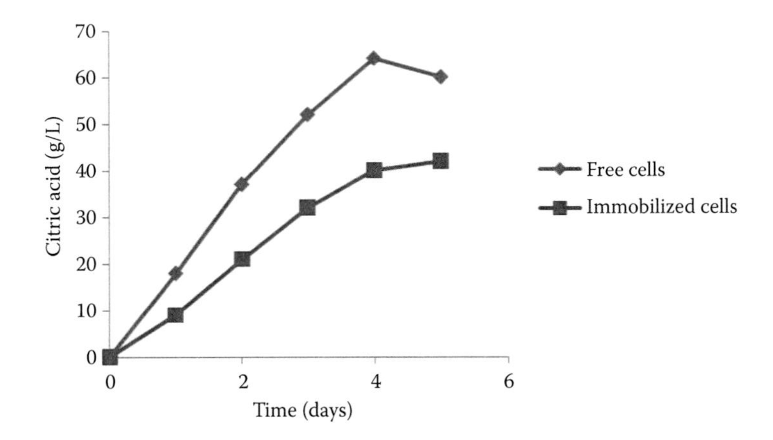

Figure 8.4 Comparative citric acid production by free and immobilized *A. niger* cells.

8.9 Conclusion

Organic acids are important building block chemicals and have a promising global market. Their biosynthesis through fermentation processes can make them available for global use and as well widening their application potential. The production of these commercially valuable organic acids is largely dependent on the choice of a suitable organism capable of utilizing cheap or locally available substrate with a view to reducing the process cost. Downstream processing is also critical to the production of organic acids of high purity. When adequate attention is given to the biosynthesis of these organic metabolites, the use of petrochemicals will be drastically eliminated and greenhouse gas emissions will be reduced.

References

Ali, S. 2004. Studies on the submerged fermentation of citric acid by *Aspergillus niger* in a stirred fermentor, PhD Thesis. University of Punjab, Lahore, Pakistan, pp. 114–115.

Anwar, A., Qader, S.A.U., Raiz, A., Iqbal, S. and Azhar, A. 2009. Calcium alginate: A support material for immobilization of proteases from newly isolated strain of *Bacillus subtilis* KIBGE-HAS. *World Applied Sciences Journal, 7 (10)*: 1281–1286.

AOAC. 1995. *Official Methods of Analysis*, 16th edition. Association of Official Analytical Chemist, Washington, DC.

Bala, S. 2003. Acetic acid. In: *Chemicals Economic Handbook*. SRI International, Menlo Park, CA, pp. 602–5000.

Ball, G.F.M. 2004. *Vitamin C; Their Role in the Human Body*. Blackwell Publishing, Oxford, UK, pp. 394–432.

Bao, X., Yue, L. and Gao, X. 2011. Characterization of *Streptococcus oligofermentus* sucrose metabolism demonstrates reduced pyruvic and lactic acid production. *Chinese Medical Journal, 124 (21)*: 3499–3503.

El-Holi, M.A. and Al-Delamy, K.S. 2003. Citric acid production from whey with sugars and additives by *Aspergillus niger*. *African Journal of Biotechnology, 2 (10)*: 356–359.

Goldberg, I., Rokem, J.S. and Pines, O. 2006. Organic acids: Old metabolites, new themes. *Journal of Chemical Technology and Biotechnology, 81*: 1601–1611.

Hallberg, L., Brune, M. and Rossander, L. 1996. Effect of ascorbic acid on absorption from different types of meals. *Human Nutrition: Applied Nutrition, 40 A*: 97–113.

Haq, I., Ali, S., Qadeer, M.A. and Iqbal, J. 2004. Citric acid production by mutants of *Aspergillus niger* from cane molasses. *Bioresource Technology, 93 (2)*: 125–130.

Hofvendahl, K.B. and Hahn-Hägerdal, B. 2000. Factors affecting the fermentative lactic acid production from renewable resources. *Enzyme and Microbial Technology, 26:* 87–107.

Kamzolova, S.V., Shishkanova, N.V., Morgunov, I.G. and Finogenova, T.V. 2003. Oxygen requirements for growth and citric acid production of *Yarrowia lipolytica*. *FEMS Yeast Research, 3*: 217–222.

Kapoor, K.K., Chaudhary, K. and Tauro, P. 1983. *Prescott and Dunn's Industrial Microbiology*, 4th edition. Reed, G. ed. MacMillan Publishers, UK, p. 709.

Kareem, S.O., Akpan, I. and Alebiowu, O.O. 2010. Production of citric acid by *Aspergillus niger* using pineapple waste. *Malaysian Journal of Microbiology, 6 (2)*: 161–165.

Kareem, S.O., Oladipupo, I.O., Omemu, A.M. and Babajide, J.M. 2013. Production of citric acid by *Aspergillus niger* immobilized in *Detarium microcarpum* matrix. *Malaysian Journal of Microbiology, 9 (2)*: 161–165.

Kautola, H. and Linko, Y.Y. 1989. Fumaric acid production from xylose by immobilized *Rhizopus arrhizus* cells. *Applied Microbiology and Biotechnology, 31*: 448–452.

Kenealy, W., Zaady, E., Dupreez, J.C., Stieglitz, B. and Goldberg, I. 1986. Biochemical aspects of fumaric acid accumulation by *Rhizopus arrhizus*. *Applied and Environmental Microbiology, 52*: 128–133.

Kiel, H., Gurin, R. and Henis, Y. 1981. Citric acid fermentation by *Aspergillus niger* on low sugar concentration and cotton waste. *Applied Environmental Microbiology, 42*: 1–4.

Li, Y., Chen, J. and Lun, S.Y. 2001. Biotechnological production of pyruvic acid. *Applied Microbiology and Biotechnology*, *57*: 451–459.
Magnuson, J.K. and Lasure, L. 2004. Organic acid production by filamentous fungi. In: *Advances in Fungal Biotechnology for Industry, Agriculture and Medicine*, Lange, J. and Lange, L. eds. Kluwer Academic Publisher, Netherlands.
Mantzouridou, F., Roukas, T. and Kotzekidou, P. 2002. Effect of the aeration rate and agitation speed on carotene production and morphology of *Blakeslea trispora* in a stirrer tank reactor. Mathematical modeling. *Biochemical Engineering Journal*, *10*: 1232–1235.
Najafpour, G.D. 2007. *Biochemical Engineering and Biotechnology*, 1st edition. Elsevier, Netherlands, pp. 170–185.
Prescott, L.M., Harley, J.P. and Klein, D.A. 2005. *Microbiology*, 6th edition. McGraw-Hill, New York, pp. 176–177.
Raimbault, M. 1997. *General and Microbiological Aspects of Solid Substrate Fermentation*. International Training Course. Solid-State fermentation. Curitiba-Parana, Brazil.
Ramachandran, S., Fontanille, P., Pandey, A. and Larroche, C. 2006. Gluconic acid: A review. *Food Technology and Biotechnology*, *44 (2)*: 185–195.
Roukas, T. 2005. Citric acid and gluconic acid production from fig by *Aspergillus niger* using solid-state fermentation. *Journal of Industrial Microbiology and Biotechnology*, *25*: 298–304.
Sankpal, N.V. and Kulkarni, B.D. 2002. Optimization of fermentation conditions for gluconic acid production using *Aspergillus niger* immobilized on cellulose microfibrils. *Process Biochemistry*, *37*: 1343–1350.
Sauer, M., Porro, D., Mattanovich, D. and Brandwardi, P. 2007. Microbial production of organic acids: Expanding the markets. *Trends in Biotechnology*, *26 (2)*: 100–108.
Schmidt, K. and Moser, U. 1985. Vitamin C—A modulator of host defense mechanisms. *International Journal for Vitamin and Nutrition Research Supplement*, *27*: 363–379.
Stanko, R.T., Reynolds, H.R., Hoyson, R., Janosky, J.E. and Wolf, R. 1994. Pyruvate supplementation of a low-cholesterol, low-fat diet: Effects on plasma lipid concentrations and body composition in hyperlipidemic patients. *American Journal of Clinical Nutrition*, *59*: 423–427.
Tay, W. and Yang, S.T. 2002. Production of L(+)-lactic acid from glucose and starch by immobilized cells of *Rhizopus oryzae* in a rotating fibrous bed bioreactor. *Biotechnology and Bioengineering*, *80*: 1–12.
VickRoy, T.B. 2000. Lactic acid. In: *Comprehensive Biotechnology*, Vol. 3, Moo-Young, M. ed. Pergamon Press, New York, pp. 761–776.
Wee, Y.J., Kim, J.N. and Ryu, H.W. 2006. Biotechnological production of lactic acid and its recent applications. *Food Technology and Biotechnology 44 (2)*: 163–172.
Yigitolu, M. 1992. Biotechnology. *Journal of Islamic Academy*, *5 (2)*: 100–106.

chapter nine

Microbial amino acids production

Zafar Alam Mahmood

Contents

9.1 Introduction

Amino acids are the constructive components (subunits) of proteins and peptides and are thus regarded as the building blocks of life. Amino acids contain a high percentage of nitrogen (~16%), which distinguishes them from fats and carbohydrates (Mahmood 2010). Amino acids are classified into two categories: essential amino acids, which are not synthesized in the body; and nonessential amino acids, which are synthesized by the body (Table 9.1). Amino acids have great demand and application in view of their importance in the food, feed, personal care, and pharmaceutical industries as nutrients, additives, rejuvenators, and drugs. Their roles are highly prominent and specific in different applications as

Table 9.1 List of Essential and Nonessential Amino Acids

S. no	Essential amino acids	S. no	Nonessential amino acids
01	Arginine	01	Alanine
02	Histidine	02	Asparagine
03	Isoleucine	03	Aspartate
04	Leucine	04	Cysteine
05	Lysine	05	Glutamate
06	Methionine	06	Glutamine
07	Phenylalanine	07	Glycine
08	Threonine	08	Proline
09	Tryptophan	09	Serine
10	Valine	10	Tyrosine

well as in therapeutic potential. During the last two decades, the global demand for some essential amino acids, for example, lysine and methionine, has tremendously increased because of their extensive use in the feed, food, and pharmaceutical industries. The addition of these amino acids in animal feed not only optimizes the growth of animals but also improves the quality and quantity of meat. The availability of such animal products has certainly helped a lot in overcoming the deficiency of essential amino acids, which is quite prominent in the primary foodstuffs of underdeveloped and overpopulated areas of the globe. The demand for nonessential amino acids, for example, glutamic acid, has also significantly increased in recent years due to its extensive application in the food industry. The global demand is further expected to increase in the next decades, requiring researchers to focus more on developing advanced manufacturing techniques utilizing biotechnological tools to meet the challenging demand of various amino acids.

The aim of this chapter is to present comprehensive information on some of the abovementioned amino acids, such as lysine, methionine, tryptophan, and glutamic acid (Figure 9.1) with respect to their historical backgrounds, new challenges along with the

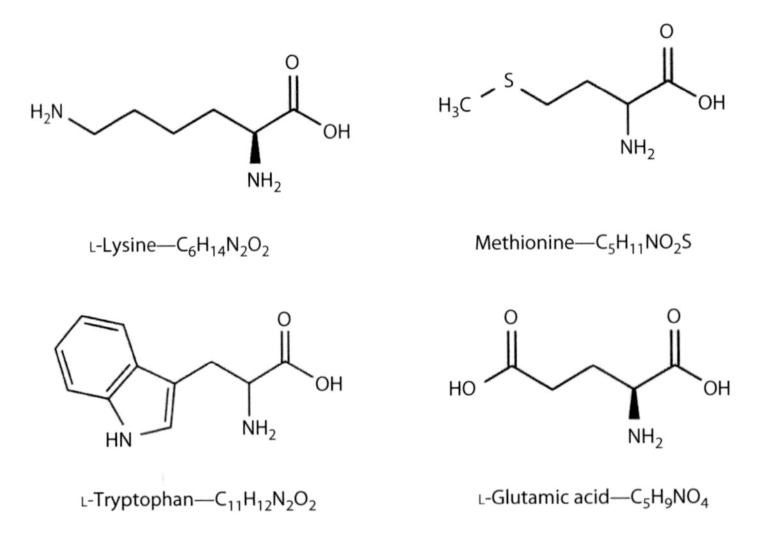

Figure 9.1 Structure and molecular formula of lysine, methionine, tryptophan, and glutamic acid.

utilization of biotechnological approaches and tools for increasing the production capability of bacterial strains, their industrial applications, and therapeutic roles.

The utilization of microorganisms for the development of commercial commodities is supposed to have initiated in primordial times. This has resulted in the production of fermented beverages and food stuffs but without the true concept of fermentation. It was Pasteur who introduced the word *fermentation* for the first time and negated the theory of spontaneous generation. Subsequent studies on these lines have been performed by a large number of researchers using bacteria and filamentous fungi. However, the scientific interpretation of fermentation started only when the perception and understanding of microbial physiology and metabolic processes were fully understood (Mahmood 2010).

The biotransformation of organic molecules into useful products provided a new concept in the history of fermentation and, as a result, a number of metabolites such as antibiotics and amino acids were identified. The continued studies on the subject resulted in the discovery of an immense number of pharmaceutical and biochemical products. At the same time, advancements in fermentation engineering provided great support in the production of these useful products on a commercial scale. With advanced knowledge of microbial physiology, molecular biology, and genetic engineering, the fermentation industry has grown considerably in terms of introducing some novel products, which have great industrial and medical applications. These advancements have provided further significant knowledge about these microorganisms, their isolation and modification, and their growth conditions along with the possibility to improve the productivity of the desired product through process optimization or by the addition or subtraction of a particular nutrient.

Initial progress in terms of mutating the strains and utilization of microbial physiology to develop suitable media composition for enhanced yield of desired products was the beginning of modern biotechnology. The evaluation of new biological commodities that could be used as pharmaceuticals, nutraceuticals, or cosmeceuticals has produced phenomenal progress—with industrial microbiology on one side and fermentation engineering on the other. The intensified industrial exploitation of biological practices advocates that biotechnology will be a major growing industry in the near future and that this will alter the lives and prosperity of people all over the world.

Over the last 50 years, the knowledge of microbial physiology and molecular biology has increased substantially in terms of amino acid production and its utilization in human nutrition as well as in domestic animals. The diversified application of amino acids as a raw material in the food, pharmaceutical, and cosmetic industries has played a significant role in boosting the research activities in this particular field. Older methods, such as extraction of amino acids from natural sources or production through chemical synthesis, have largely been replaced by biotechnological processes such as production through fermentation or by enzymatic catalysis. Organized research on the microbial production of amino acids apparently started during the late 1940s, and by the end of the 1950s, a number of amino acids were being produced using microbial sources. The best example is of direct L-glutamic acid fermentation using *Corynebacterium glutamicum* (Figure 9.2) in Japan by Kyowa Hakka Kogyo Company (Kinoshita et al. 1957a).

This progress had a great economic effect in the field of amino acid production and, as a consequence, L-lysine was successfully produced and commercialized using a mutant strain of *C. glutamicum* (Kinoshita et al. 1958). This has resulted in a new concept of fermentative production of amino acids with the introduction of a series of artificially derived auxotrophic or regulatory mutants for the production of other amino acids as well, such as methionine, tryptophan, and others. During the last three decades, a large number of

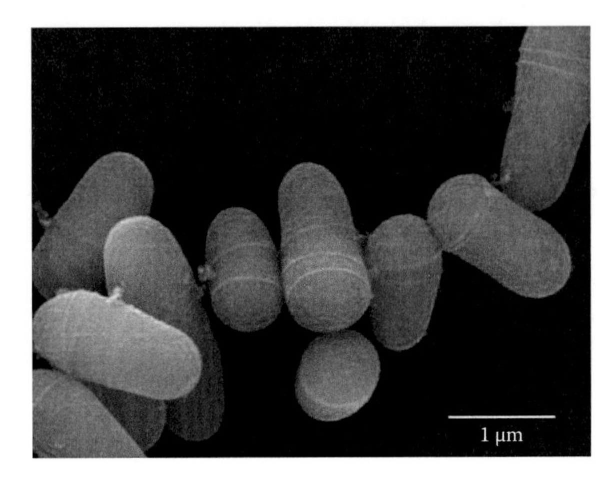

Figure 9.2 Electron micrograph of *C. glutamicum* ATCC 13032. (Adapted from Wittmann, C., and Becker, J. The L-lysine story: from metabolic pathways to industrial production. *Microbiology Monographs* DOI 10.1007/7171_2006_089/Published online: February 24, 2007.)

mutant strains of *C. glutamicum* have been developed to produce and meet the market demand for various amino acids. Biotechnology has certainly played a major role in the fermentative production of various amino acids.

With the development of new applications for amino acids in the feed, food, and pharmaceutical industries, enormous improvements have been made in production technology during the latter half of the twentieth century. Submerged fermentation technology plays a key role in this progress, and fermentative production of amino acids currently represent the leading products of biotechnology in terms of both volume and value. The field is very competitive and the cost of production is a major factor, which can certainly support the manufacturers to stay in the market with profitable business. For cost-effective fermentative production, many biotechnological methodologies have been applied to establish high productivity and improvement in product recovery processes. The role of genetic engineering relating to amino acid-producing strains is now being used for the development of biosynthetic and transport capacity of organisms for enhanced production. In addition, the rapid progress in genome analysis is expected to revolutionize microbial strain improvement techniques. With these advancements, the global production and consumption of amino acids is expected to increase more in the near future and thus, will have great effect on amino acid-based industries in terms of investment. Nowadays, the major producers of amino acids are based in Japan, the United States, South Korea, China, and Europe. However, some other Asian countries, such as Indonesia, Malaysia, Thailand, and Vietnam are also entering the amino acid business. The major manufacturing companies include Ajinomoto Group, Archer Daniels Midland Company, Cheil Jedang, COFCO Biochemical (Anhui) Co. Ltd., Daesang Corporation, Evonik, Global Bio-Chem Technology, Novus International Inc., Royal DSM N.V., Sekisui Medical Co. Ltd., Shandong Zhengda Linghua Biotechnology Ltd. Co., Toronto Research Chemicals Inc., Vedan, and VitaLys I/S. The global market for amino acids has been forecast to reach US$11.6 billion by the year 2015. The demand for various amino acids is expected to increase in the production of animal feed, health foods, dietary supplement products, artificial sweeteners, and cosmetics in the subsequent years (Jose 2011; Wippler 2011). The world's consumption pattern in 2011 for lysine, methionine, and tryptophan is shown in Figure 9.3, which is self-explanatory

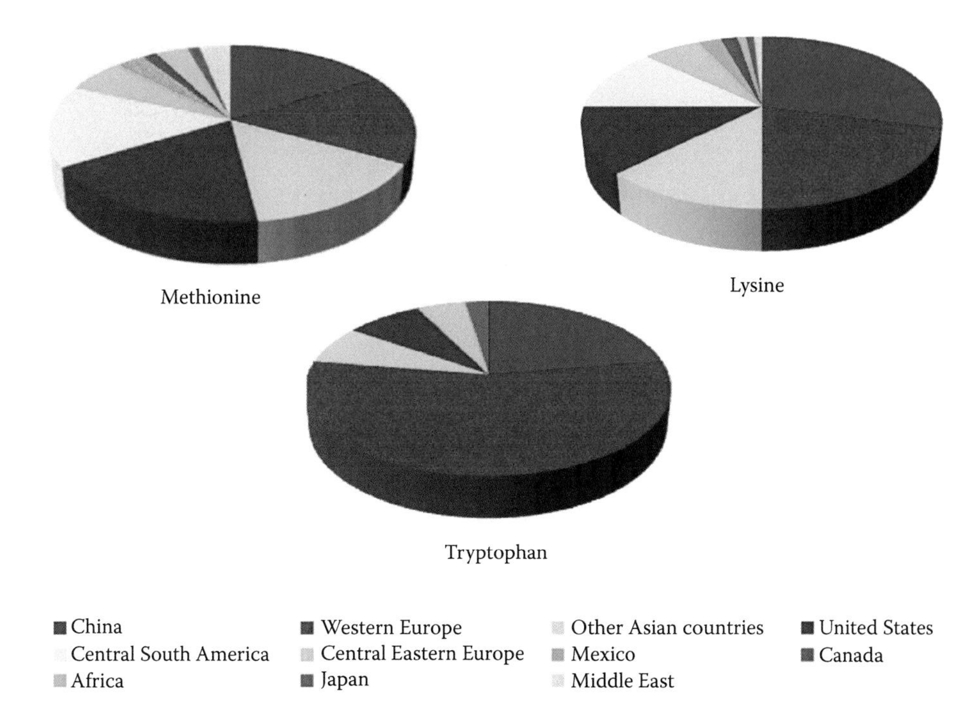

Figure 9.3 World's consumption of methionine, lysine, and tryptophan during 2011.

and provides a comprehensive idea about the consumption of these amino acids in different regions of the world (IHS Chemical 2012).

9.2 L-Glutamic acid

9.2.1 Historical background and new challenges

From an industrial application or commercial point of view, L-glutamic acid is one of the most important amino acids. The carboxylate ions and salts of glutamic acid are known as glutamates. Its initial discovery is quite old and was first reported in 1866 by a German scientist, Karl Heinrich Leopold Ritthausen, in wheat gluten treated with sulfuric acid (Plimmer and Hopkins 1912). In 1907, the first successful commercial production of glutamic acid was performed at Tokyo Imperial University by Kikunae Ikeda, who developed and patented a method to produce crystalline salt of glutamic acid, monosodium glutamate (MSG). Subsequently, MSG was patented by Ajinomoto Corporation of Japan in 1909. The turning point in the history of glutamic acid production was the discovery of *Micrococcus glutamicus* (later identified as *C. glutamicum* in 1957), which produces 30 g/L of L-glutamic acid in glucose medium through fermentation (Kinoshita et al. 1957b). The interest in large-scale L-glutamic acid fermentation was primarily developed due to increased demand for MSG, which was and still is used as a flavor-enhancing agent. Glutamic acid and its salt MSG represent the largest product segment within the amino acid market. In 2007, the world product volume was indicated as 1.6 million tons (Demain 2007); in 2009, the figure reached 2.0 million tons (Sano 2009), and in 2012, the figure was reported to be 2.2 million tons (Abdenacer et al. 2012).

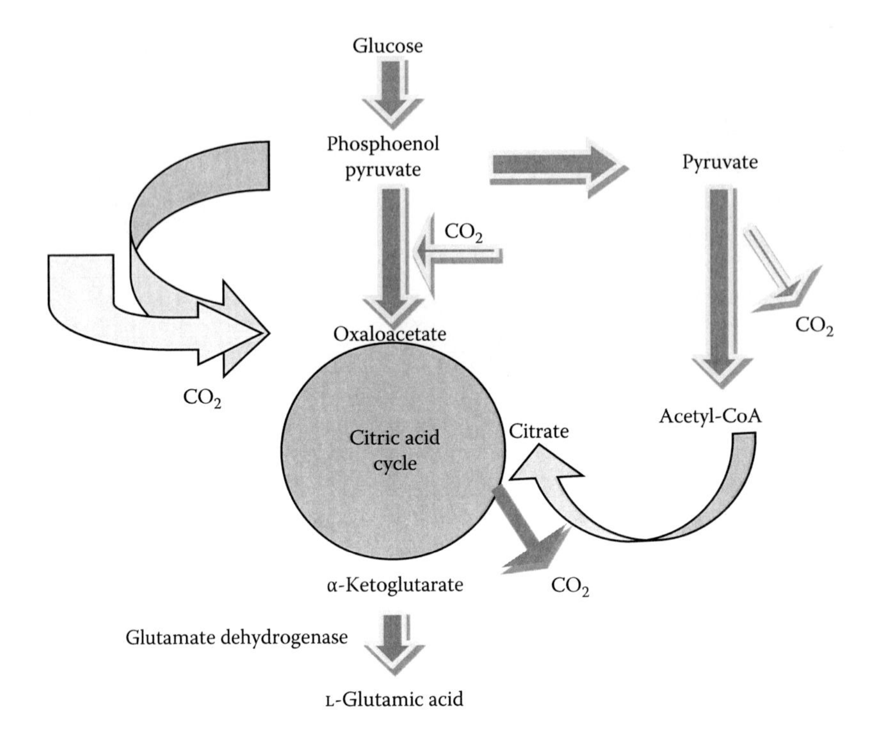

Figure 9.4 Biosynthesis of L-glutamic acid in *C. glutamicum* using glucose as a carbon source.

The market demand is continuously increasing and is creating a challenge for the manufacturers, either to increase the capacity of existing plants or to utilize advanced biotechnological tools to increase the production both in terms of strain improvement and modification in fermentation medium. The biosynthesis of L-glutamic acid using *C. glutamicum* involves relatively simple pathways of metabolism. Glucose is mainly metabolized through glycolytic pathways. Pyruvate and phosphoenol pyruvate play important roles in the production of oxaloacetate and then α-ketoglutarate. The α-ketoglutarate is converted to L-glutamic acid in the presence of enzyme glutamate dehydrogenase (Figure 9.4). The phosphorylation stage is quite important because the genes controlling this conversion process are located adjacent to the genes of cell wall synthesis and cell division. Therefore, any disorder in these mechanisms, cell wall synthesis or cell division, might inhibit the activity of α-ketoglutarate dehydrogenase activity. As a result, the conversion of α-ketoglutarate to glutamic acid might be affected (Eggeling and Sahm 2011).

9.2.2 *Biotechnological approach and production*

A great deal of biotechnological tools have been used by various researchers with an understanding of microbial physiology to modify the strains for increased production of glutamic acid, especially utilizing *C. glutamicum* (Ikeda and Takeno 2013). The progresses, with respect to the capabilities of the different strains of *C. glutamicum* to produce glutamic acid, have been reported from time to time over the last few decades. This was done having comprehensive knowledge of the biology, biosynthetic pathways, mutation, nutritional requirements, and putative mechanosensitive channels associated with this organism. Microbial production of L-glutamic acid has been extensively studied by a large number of research investigators using different strains of *Coryneform* species and the process with specific strains have been patented. The most popular *Coryneform* species include *C. glutamicum, Corynebacterium lilium, Corynebacterium herculis, Brevibacterium flavum, Brevibacterium lactofermentum, Brevibacterium divarticum, Brevibacterium ammoniagenes, Brevibacterium thiogenetalis, Brevibacterium saccharoliticum,* and *Brevibacterium roseum* (Kinoshita 1999). Other glutamic acid-producing organisms include *Escherichia coli, Bacillus megaterium, Bacillus circulans, Bacillus cereus,* and *Sarcina lutea.* Industrially, glutamic acid is usually manufactured by batch/fed-batch submerged fermentation processes using genetically modified strains of *Corynebacterium* or *Brevibacterium.* Commercial-scale production is carried out in large fermenters equipped with all the facilities to control different parameters and optimize production such as provision of cooling, measurement of dissolved oxygen and pH, and others. A typical example of the submerged fermentation process on a laboratory scale for the production of L-glutamic acid utilizing a mutant strain of *C. glutamicum* has been summarized below.

Organisms are usually maintained on Hottinger slant agar containing beef trypsin digestion, sodium chloride, dipotassium phosphate, and agar at low temperature. After subculturing, a 24-h-old culture is used to inoculate 25 mL of sterile seed medium in 250 mL Erlenmeyer flasks in triplicate. The following seed medium composition can be used: glucose (8%), NH_4Cl (0.5%), corn steep liquor (0.3%), K_2HPO_4 (0.5%), KH_2PO_4 (0.5%), $MgSO_4 \cdot 7H_2O$ (0.03%), $CaCO_3$ (1.0%), and deionized water to make 100%. The pH of the medium can be adjusted to 7.2 using NaOH. The inoculated flasks are grown in an orbital shaker incubator maintained at 30°C and 230 rpm for 15 h. The entire contents of the one flask is then transferred to a 2.0 L capacity Eyla Mini Jar fermenter with 500 mL of sterile nutrient medium containing molasses (20%), KH_2PO_4 (0.5%), KH_2PO_4 (0.5%), $MgSO_4 \cdot 7H_2O$ (0.3%), urea (0.8%), $CaCO_3$ (1.0%), and deionized water to make 100%. In most cases, the optimum pH of the medium was recorded as 7.0. The fermentation is usually initiated with continuous agitation and aeration for 48 h at 30°C. At the end of the incubation period, the culture broth is separated from the cells by centrifugation at 10,000 rpm for 5 min at 4°C and the supernatant is diluted 50-fold with 7% (*v/v*) of glacial acetic acid. The diluted sample is then further centrifuged at 10,000 rpm for 5 min at 4°C. The supernatant is then finally filtered through a nylon membrane of 0.22 μm pore size and then collected for further analysis and crystallization (Khan et al. 2013a,b; Pasha et al. 2011).

Since the discovery of the original strain of *C. glutamicum,* a large number of strains, which are capable of producing significant quantities of glutamic acid, have been developed through the process of genetic engineering. Yields have now exceeded 100 g/L and even more with modified strains compared with original wild strains, which produced only 10 g/L. Most original glutamic acid-producing strains of *C. glutamicum* were observed as biotin auxotrophs. Thus, growing such strains in a medium deficient in biotin was found to "trigger" glutamic acid production. The biotin-deficient medium has the property of altering the cell membrane of the organism because of suboptimal fatty acid biosynthesis and is thus responsible for increased production. In view of this hypothesis, it is believed that some other physicochemical factors such as fermentation at higher temperatures, the addition of surfactants (e.g., Tween 40) or antibiotics (e.g., penicillin) in the culture medium can also trigger the excretion of glutamic acid. These factors have been linked to a triggering mechanism resulting in a decrease or repression of the enzyme α-ketoglutarate dehydrogenase. Thus, finally causing redistribution of metabolites at the branch point in the tricarboxylic acid (TCA) cycle leading from α-ketoglutarate to succinyl-CoA or glutamate (Figure 9.4). However, this mechanism has limitations, therefore, a further increase or overproduction of glutamate can be achieved through optimization of the metabolic flux by tuning the glutamate dehydrogenase activity (Asakura et al. 2007). L-Glutamic acid production was also achieved through immobilized cells of *C. glutamicum*. Using more than 93 g/L, good yields were achieved through batch fermentation but with low productivity (3.8 g/L/h), whereas 73 g/L of glutamic acid was recovered through continuous fermentation with high productivity of approximately 29.1 g/L/h (Amin and Al-Talhi 2007). A number of other studies have also been published, documenting the role of immobilized cells in the production of L-glutamic acid. It has been reported that immobilization of microbial cells in biological processes can occur either as a natural phenomenon or through artificial processes. Although attached cells in their natural habitat exhibit significant growth, the artificially immobilized cells also allow restructure of growth. The authors reported obtaining 48.5 g/L of L-glutamic acid through the immobilized cells of a mutant strain of *C. glutamicum* (Pasha et al. 2011; Prasad et al. 2009). The use of immobilized cells for the production of L-glutamic acid and some other amino acids is progressing rapidly and is expected to have some more valuable and useful data for commercial exploitation in the near future.

9.2.3 *Industrial application and therapeutic role*

The greatest application of glutamic acid and its salt is in the food industry as a flavor enhancer. A considerable quantity of free glutamic acid, or its salt MSG, is present in different food products, for example, in different kinds of cheeses and soy sauces. In addition, because glutamate is a key compound in the cellular metabolism, it therefore serves as a unique brain fuel and performs some other important functions such as detoxification of ammonia, as a hepatoprotective agent (glutamine), to aid in peptic ulcer healing, and others (Chaitow 1985; Zareian et al. 2012).

One of the leading roles of glutamic acid in pharmaceuticals is that of a neurotransmitter. L-Glutamic acid is widely distributed in the central nervous system (CNS) and is reported to act as a neurotransmitter. Furthermore, it may have a stimulating effect on the metabolism of the cerebral cortex thus improving mental performance and memory function. The concentration of glutamic acid in the CNS is higher than any other commonly recognized neurotransmitter. In addition, the biochemical and electrophysiological data also suggest that glutamic acid or its analogue acts as an excitatory neurotransmitter of CNS. There is a high concentration of *N*-methyl-D-aspartate (NMDA) receptor in the hippocampus (Bloom 2001). The blockage of NMDA receptors can greatly affect the memory and overall mental performance of an individual. Glutamic acid and aspartic acid have the capability to combine with NMDA receptors thus increasing cation conductance, depolarizing the cell membrane, and deblocking the NMDA receptors. The ultimate results of these biochemical features lead to improvements in memory function and mental performance (McEntee and Crook 1993).

9.3 *L-Lysine*

9.3.1 *Historical background and new challenges*

L-Lysine is one of the leading and most exploited amino acid among the essential amino acids list. It has vast applications and role in human and animal nutrition. The systematic research to explore the possibility of amino acid production through microorganisms started during the late 1940s, and during the later part of the 1950s, some fruitful results were reported (Mitchell and Houlahan 1948; Windsor 1951). In continuation, a number of research investigators published their findings that L-lysine can be synthesized from α-aminoadipic acid by yeast and *Neurospora* mold, or from diaminopimelic acid (DAP) by *E. coli* (Davis 1952; Dewey and Work 1952; Mitchell and Houlahan 1948; Windsor 1951). However, microbial production of L-lysine through the decarboxylation of DAP by Chas Pfizer and Company Inc. in the United States was regarded as the first commercial scale production (Casida and Baldwin 1956). A modified process (Figure 9.5) utilizing a single organism, which produced DAP and converted it to L-lysine, was then simultaneously discovered and reported (Kita et al. 1958).

The introduction of L-glutamic acid fermentation using *C. glutamicum* had a great economic effect in the field of L-lysine fermentation. Further research in this direction resulted in the development of an efficient mutant of *C. glutamicum* for the commercial production of L-lysine (Kinoshita et al. 1957b, 1958). This modified method was also successfully used to produce some other amino acids (Mahmood 2010). Despite the commercial challenges, the production of L-lysine using mutant strains brought about a new concept in the fermentative production of amino acids. With the emergence of new biotechnological approaches and tools, more focus was given to improving the strains and taking artificially derived auxotrophic and regulatory mutants that were resistant to feedback inhibition. Using these genetically modified organisms, increased production of L-lysine was achieved by controlling the biosynthetic pathways. The biosynthetic block most conducive for effective L-lysine accumulation in auxotrophic mutants was observed to be those requiring homoserine or threonine plus methionine for the growth of organisms (Figures 9.6 and 9.7).

Modifications in the biosynthetic pathway extended one of the biggest advantages by producing some other amino acids along with L-lysine using *C. glutamicum*, *B. flavum*, and *B. lactofermentum*.

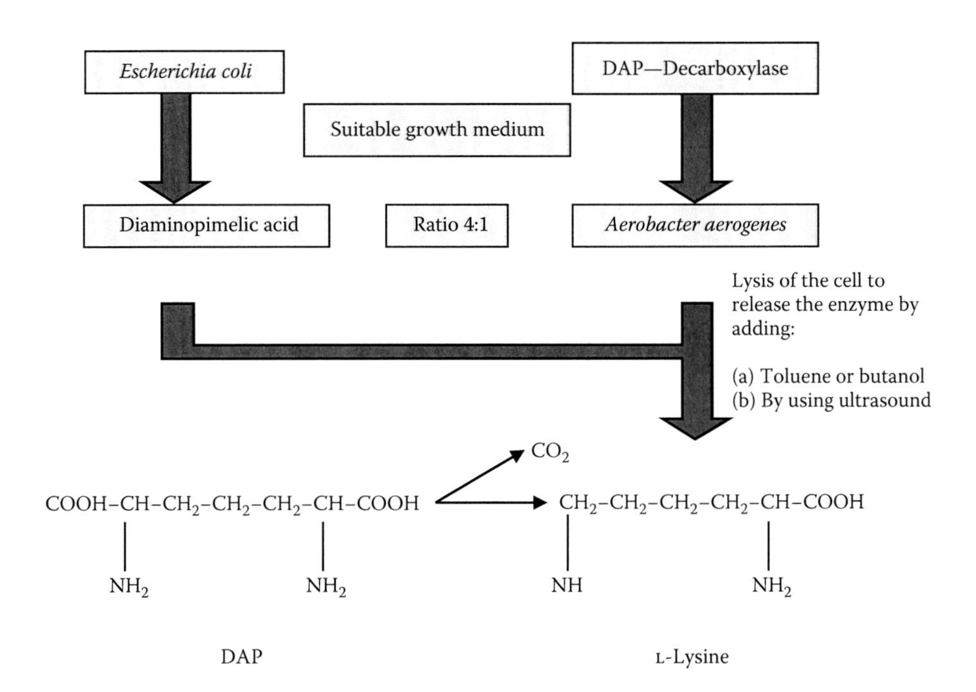

Figure 9.5 Production of L-lysine through the conversion of its immediate precursor DAP (two-step process).

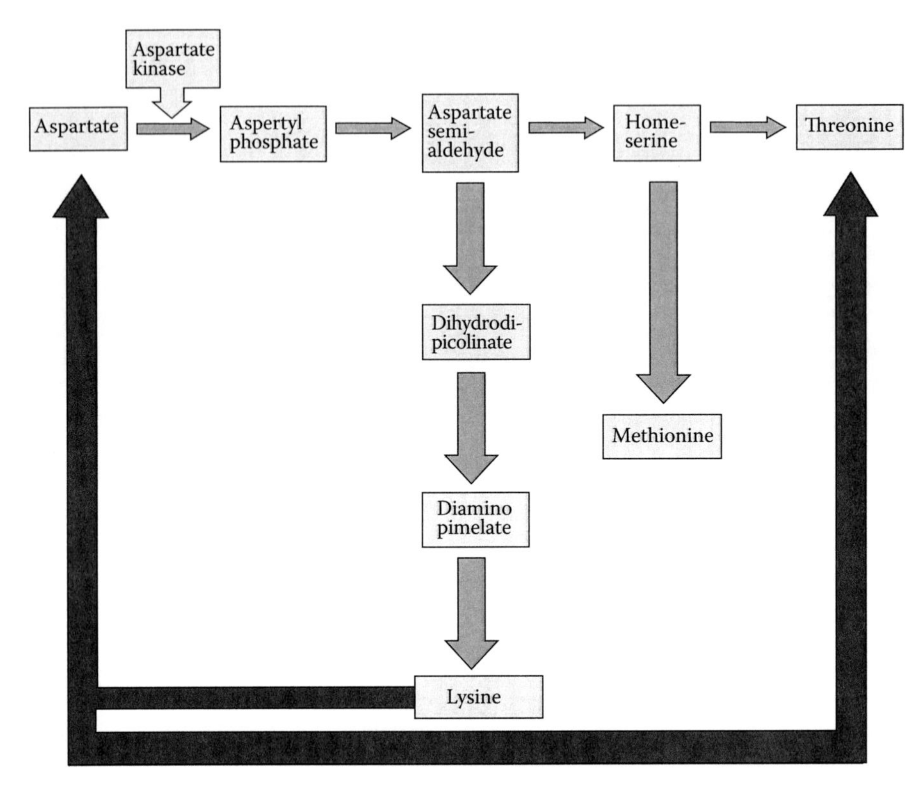

Figure 9.6 Regulatory pathways of L-lysine biosynthetic pathways in *C. glutamicum*.

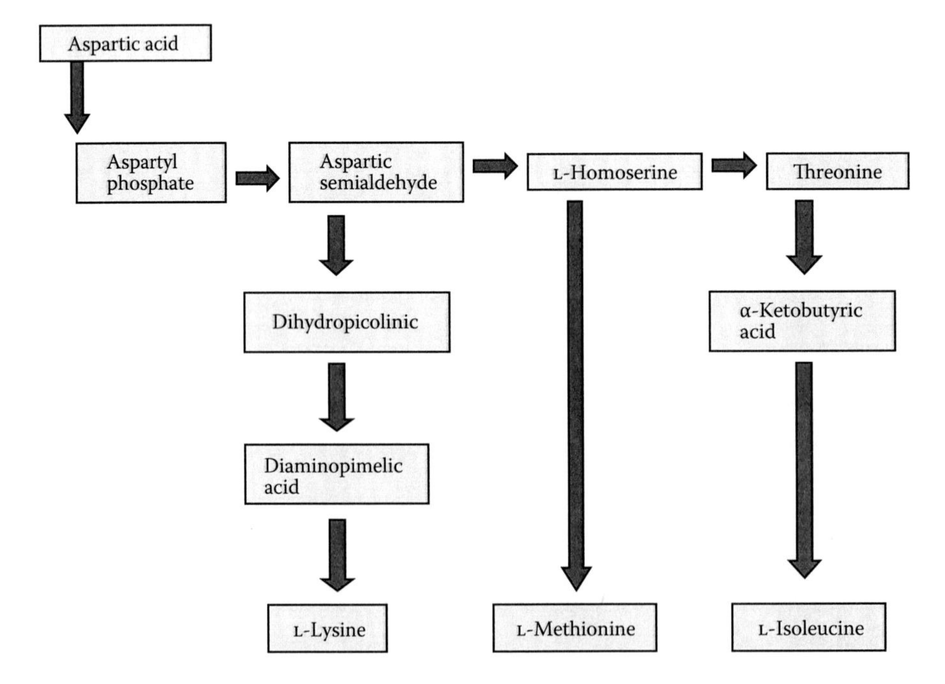

Figure 9.7 Production of L-lysine by the auxotrophic mutants (one-step process).

The isolation of auxotrophs can be carried out by using penicillin selection techniques followed by ultraviolet radiation (Adelberg and Myers 1953; Nakayama and Kinoshita 1961). The identification of growth factors (such as vitamins and amino acids) of the auxotrophs can be made auxanographically (Pontecorvo 1949). Further advancements in the production of L-lysine were achieved by introducing analogue-resistant *C. glutamicum* and *B. flavum*. An L-lysine analogue, *S*-(2-aminoethyl)-L-cysteine (AEC), is the best example of developing such analogue-resistant strains. The development of an AEC-resistant strain is primarily based on the treatment of a suitable homoserine auxotroph in a suitably complete culture medium containing meat extract, peptone, yeast extract, sodium chloride, and agar. The medium is supplemented with *N*-methyl-*N*-nitro-*N*-nitrosoguanidine in phosphate buffer. The treatment time is short (~30 min) and at a very low temperature (0°C). This is followed by the treatment of cells with AEC after washing with saline in a minimal medium supplemented with glucose as a carbon source, inorganic salts, and amino acids, for example, L-threonine and DL-methionine. It is now evident that auxotrophy and resistance to feedback inhibition, when genetically combined into a single strain (i.e., mutants resistant to AEC), results in increased production of L-lysine. The overproduction of L-lysine might be due to the starvation of threonine, which decreases the feedback inhibition of aspartate kinase, the first key enzyme in L-lysine biosynthesis (Figure 9.6). The genetic induction of an auxotropic or regulatory mutant that can overproduce L-lysine is of course extremely important. However, at the same time, process optimization (especially the medium composition and the addition of certain growth factors and culture conditions) is also indispensable for the commercial production of L-lysine.

The worldwide increased demand for L-lysine has certainly created a challenging situation for researchers to further improve the existing strains or develop new genetically modified organisms capable of producing exceptional quantities of L-lysine. The major producers of L-lysine in the world are Ajinomoto (with manufacturing facilities in Japan, France, Italy, United States, Brazil, China, and Thailand), Archer Daniels Midlands (United States), Changchun Dacheng (China), Cheil Jedang (with facilities in China, Indonesia, and Brazil), Global Biochem Technology (China), and COFCO Biochemical (China). In terms of tonnage, these major manufacturers along with other companies produced 1,755,300 tons of L-lysine in 2011 and the forecast for 2013 is around 2,376,300 tons, which is expected to reach 2,518,000 tons, corresponding to around US$5.9 billion by the end of 2018. China is the biggest producer and will continue to dominate as the world's largest L-lysine-producing country accounting for approximately 65% of the world lysine production capacity (IHS Chemical 2012; Wippler 2011).

Based on the demand for L-lysine, which is still increasing, meeting the production demand with cost-effective raw materials and recovery processes is quite challenging and will play a major role in its success. In the fermentative production of L-lysine, all raw materials are natural or biologically available substances. No harmful by-products have been found in L-lysine fermentation. On the contrary, useful substances remain in the spent-broth from which many by-products could be recovered. The spent-broth still contains various useful substances including organic and inorganic nitrogen compounds, phosphorus compounds, and potassium salts, which could be used as animal feed additives or fertilizers. A recent process that involves the spray-drying of the whole fermented broth at the end of the incubation period has also been reported to produce feed-grade L-lysine hydrochloride very successfully (Eggeling and Sahm 2011). This will certainly reduce the cost of production of feed-grade L-lysine as well as the disposal of waste material.

9.3.2 *Biotechnological approach and production*

Despite tremendous advancements in the production of L-lysine, researchers are still trying to develop a cost-effective fermentation process using new biotechnological methodologies. Also, biotechnology companies are continuously seeking novel research developments and trying to use multifaceted management models and business approaches toward achieving market leadership in the field of L-lysine and other amino acids production. Along with highly productive strains, the use of cost-effective raw materials and recovery processes has now become highly essential for different manufacturers to compete and stay in the market. Genetically modified strains, which have been introduced for the production of L-lysine, can produce as much as approximately 170 g/L. With the use of metabolic engineering approaches, more advanced and targeted developments can be made to further improve the L-lysine biosynthetic pathway and its secretion. Control over these dynamics will certainly warrant effective precursor delivery and ensure the energy requirement of the cells. It has been indicated that along with the genetic engineering of central metabolic pathways, biosynthetic pathways, and transport systems, energy metabolism and osmoregulation are also gaining interest among researchers as new targets to be engineered for the overproduction of L-lysine. Furthermore, strains other than *C. glutamicum,* such as *Bacillus methanolicus* (methylotropic bacteria) are also gaining interest for the production of L-lysine using methanol and lignocellulose as an alternate raw material (Brautaset and Ellingsen 2011).

With the progress in molecular biology and genetic engineering, some focus has been driven to ensure an optimal carbon and energy flow within the central metabolism (CCM) of bacterial cells to achieve optimized metabolic production. A specific example is that of *C. glutamicum* in which the CCM involves glycolysis, the pentose phosphate pathway, and the TCA along with the anaplerotic and gluconeogenic reactions. It is expected that the in-depth knowledge of CCM along with the specific enzyme activities of the pathways will certainly provide an opportunity to create cell factories ideal for the production of not only L-lysine but also some other industrially important metabolites as well (Papagianni 2012).

In the industrial production of L-lysine, right from the selection of bacterial strains, the media composition, process optimization, and recovery of the end product have a great significance on the overall yield and economics of the process. A balance between carbon and nitrogen sources along with the nutritional requirements of the strain and the addition of certain precursors should be taken into consideration. In addition, a sufficient supply of oxygen to satisfy the cell's oxygen requirement is highly essential for increased yield of L-lysine. On an industrial scale, submerged fermentation is highly effective and is the most widely used method worldwide. The major raw materials such as starch, glucose, molasses, and others used in L-lysine fermentation are quite common and are used in other fermentation industries as well. The most commonly used organism for L-lysine fermentation is *C. glutamicum.* Therefore, to understand its production, a specific example has been mentioned below using a genetically modified strain of *C. glutamicum.*

Organisms are usually maintained at a low temperature, around 6°C to 8°C, on nutrient agar slants on a monthly transfer schedule. A 24-h culture, produced by subculturing from stock culture, is usually used for seed, followed by fermentation. A synthetic or natural medium that has the ability to fulfill the requirements of the particular strain can be used. In commercial-scale starches, molasses and glucose are mostly used as the carbon source. Care must be taken to create a balance between carbon and nitrogen sources such as corn steep liquor, soybean cake acid hydrolysate, yeast extract, peptone, and the like, along with the addition of inorganic salts such as KH_2PO_4, K_2HPO_4, $MgSO_4{\cdot}7H_2O$,

$FeSO_4 \cdot 7H_2O$, $ZnSO_4 \cdot 7H_2O$, $MnSO_4 \cdot 7H_2O$, $(NH_4)_2SO_4$, and others. Other necessary nutrients are biotin and vitamin B_1. In some cases, the production of L-lysine can be enhanced by using the leucine fermentation liquor (0.2%–15%) in the production medium. In most cases, the optimum pH of the medium has been recorded as 7.2 and temperature at 30°C. The seed stage cultivation requires around 24 h, whereas the fermentation stage is complete by approximately 96 h. After this, harvesting is done and the product, L-lysine, is recovered using some suitable and economical method. More details on the subject can be obtained through the case study available at the end of the chapter.

9.3.3 Industrial application and therapeutic role

L-Lysine has one of most diversified applications and roles in feed, food, and pharmaceutical industries. It is the most important amino acid for monogastric animals and is regarded as the first limiting amino acid for pigs and the second limiting amino acid for poultry. Therefore, it is an important additive to animal feed for optimizing the growth of pigs and chickens. The metabolic fate of L-lysine is highlighted in Figure 9.8.

In the food industry, L-lysine is used in a number of dietary or nutritional supplements that are popularly used by athletes, weight lifters, bodybuilders, and even some individuals to boost their energy level and protect their muscles from deterioration. Nutritional supplements with a high quantity of L-lysine are available in various dosage forms such as syrups, film-coated tablets, capsules, and powder for instant drinks. L-Lysine is required by the body to synthesize L-carnitine, which is a substance required for the conversion of fatty acids into energy. L-Lysine also helps in calcium absorption and collagen formation, which are important for muscle and bone health. It also supports or acts as a precursor in the synthesis of enzymes, antibodies, and some hormones as well.

Additionally, L-lysine is also recommended for the treatment of some viral infections, for example, herpes simplex, cold sores, shingles, and human papillomavirus infections such as genital warts and genital herpes. It has also been reportedly used in the management of migraines and in some other types of pain and inflammation. Its deficiency may cause severe health problems, growth and development problems, formation of kidney stone, low thyroid hormone production, asthma, and chronic viral infection. Symptoms include nausea, fatigue, dizziness, anemia, loss of appetite and energy, inability to concentrate, irritability, bloodshot eyes, hair loss, growth retardation (especially in children), and reproductive disorders. The reported therapeutic dosage of L-lysine for the maintenance of an antiherpes effect is 500 to 1500 mg, and up to 3000 mg in active stages in a divided dosage with a low-arginine diet. The human body's daily requirement for L-lysine is 103 mg/day for infants, 1600 mg/day for children, 800 mg/day for adult males, and 500 mg/day for adult females (Balch and Balch 1990; Chaitow 1985; Mahmood 1996, 2010).

9.4 Methionine

9.4.1 Historical background and new challenges

The discovery of glutamic acid-producing bacterium, *C. glutamicum*, eventually led to fermentation processes for producing various amino acids. However, during the early phase, not much attention was given to the fermentative production of methionine. There is a severe dearth of scientific publications relating to methionine production by microorganisms. Until 1967, very few references were available on the subject. In this period, a leucine-requiring strain of *Ustilago maydis* was reported to accumulate 6 g/L of methionine on

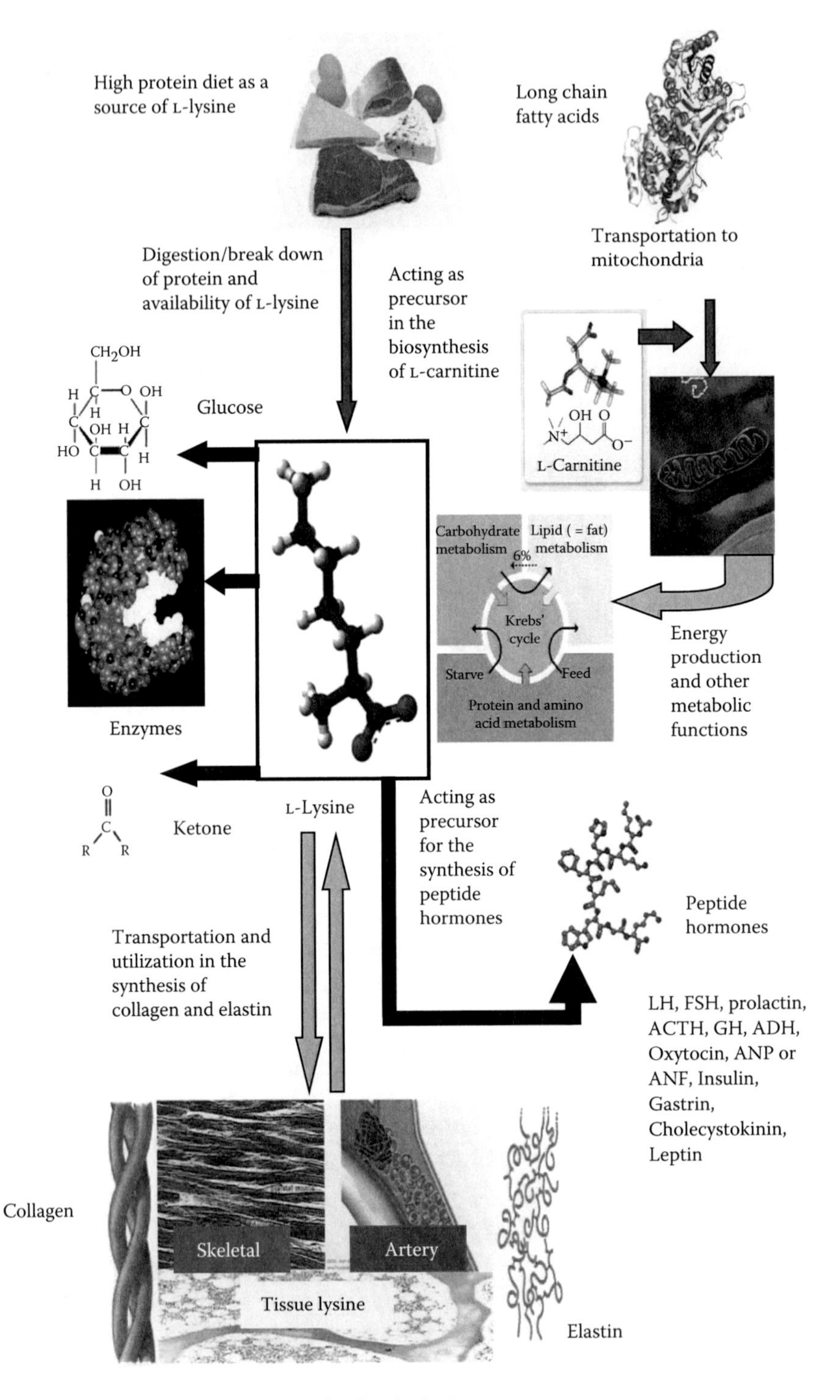

Figure 9.8 Metabolic fate: application and role of L-lysine.

a synthetic medium, and a strain of *Pseudomonas* sp. G-132-13 accumulated 13.2 g/L of methionine (Dulaney 1967).

During 1990 to early 2001, a number of constructive contributions were made by various researchers relating to the commercial production of methionine (Odunfa et al. 2001; Pham et al. 1992; Umerie et al. 2000). However, the fermentative process still needs more elaborate studies for its reproducibility and better yield before being implemented on a large scale. Currently, methionine is produced either by chemical synthesis or by enzymatic hydrolysis of proteins. Both processes are quite expensive. Chemical synthesis delivers a mixture of dextro and levo rotatory methionine, whereas enzymatic hydrolysis of proteins produces a complex mixture from which methionine needs to be separated through an expensive recovery process. Methionine isomers produced through chemical reactions can be resolved with the help of fungal enzyme aminoacylase using continuous-flow immobilized enzyme bioreactors (Tosa et al. 1967). Nevertheless, the chemical manufacturing process is still not suitable because of the hazardous nature of chemicals such as acrolein, methyl mercaotan, and ammonia, which are used during different manufacturing stages (Fong et al. 1981).

It is really challenging for the researchers as well as the manufacturers to overcome the problems associated with chemical synthesis. They need to find an optimum solution for these problems and look into the introduction of fermentative production using genetically modified organisms on a commercial scale. In bacteria, like other essential amino acids, methionine is synthesized from oxaloacetate-derived aspartate. Along with L-lysine, methionine is also a dominant amino acid used in animal feed. However, it is almost half of the L-lysine market volume. The global methionine market for animal feed has been reported as 850,000 tons in 2011, and is worth US$2.85 billion (Roquette 2012). Two big companies, Cheil Jedang (South Korea) and Arkema SA (France), are establishing a huge plant in Malaysia to manufacture biomethionine. The operation is expected to start in the latter part of 2013 and will produce 80,000 tons per year (Saidak 2012).

9.4.2 *Biotechnological approach and production*

Biotechnological tools have been applied for the production of methionine through auxotrophic, regulatory, or auxotrophic regulatory mutants. Auxotrophic mutants are generally less feasible for methionine (branched pathway amino acids) and therefore, auxotrophic regulatory are most suitable, because methionine itself will not inhibit or repress its own production. Microorganisms such as *Brevibacterium heali* can accumulate higher quantities (up to 25 g/L) of methionine (Mondal et al. 1994a,b). Regulation of methionine biosynthesis in *Corynebacterium* and *Brevibacterium* is shown in Figure 9.9.

Regulatory mutants are also used for producing methionine (Kumar et al. 2003; Kumar and Gomes 2005). Recently, a new approach to screening for methionine-producing bacteria was reported using methionine auxotrophs of *E. coli* (Ozulu et al. 2012), whereas another study relating to the production of L-methionine using *B. cereus* isolated from soil samples was also reported (Dike and Ekwealor 2012). The regulation of methionine biosynthesis in *E. coli* is shown in Figure 9.10.

In *Corynebacterium* and *Brevibacterium*, the biosynthesis mechanism for methionine production is quite simple (Figure 9.9) as compared with the biosynthesis mechanism in *E. coli* (Figure 9.10) in which all enzymes are inhibited or repressed by the end product.

Most researchers reported approximately 10% glucose or 5% maltose as the carbon source supplemented with inorganic salts and different concentrations of biotin and vitamin B_1 for the fermentative production of L-methionine (Banik and Majumdar 1975; Kase

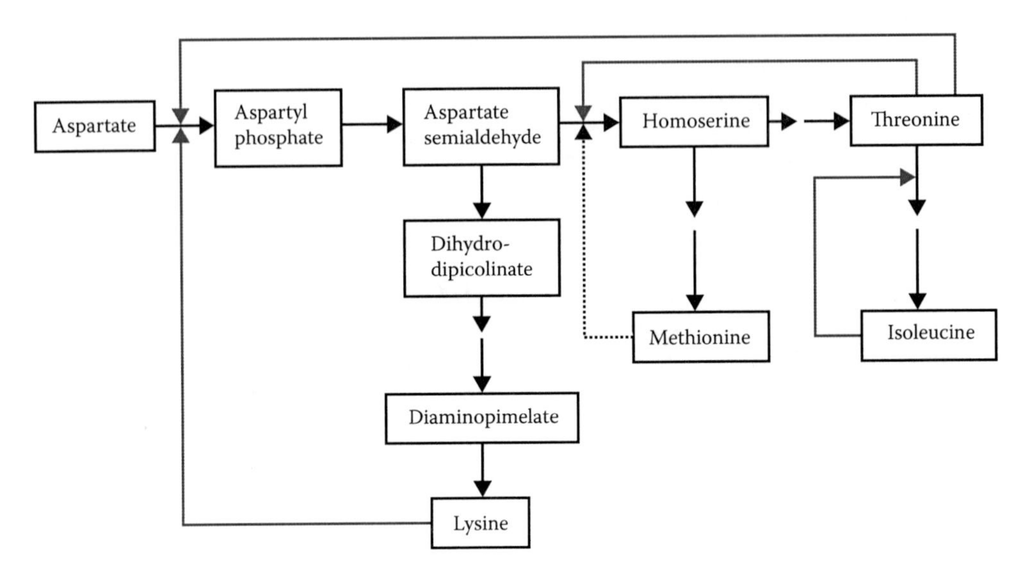

Figure 9.9 Regulation of methionine biosynthesis in *C. glutamicum*. The blue arrows show inhibition and dotted arrows represent repression.

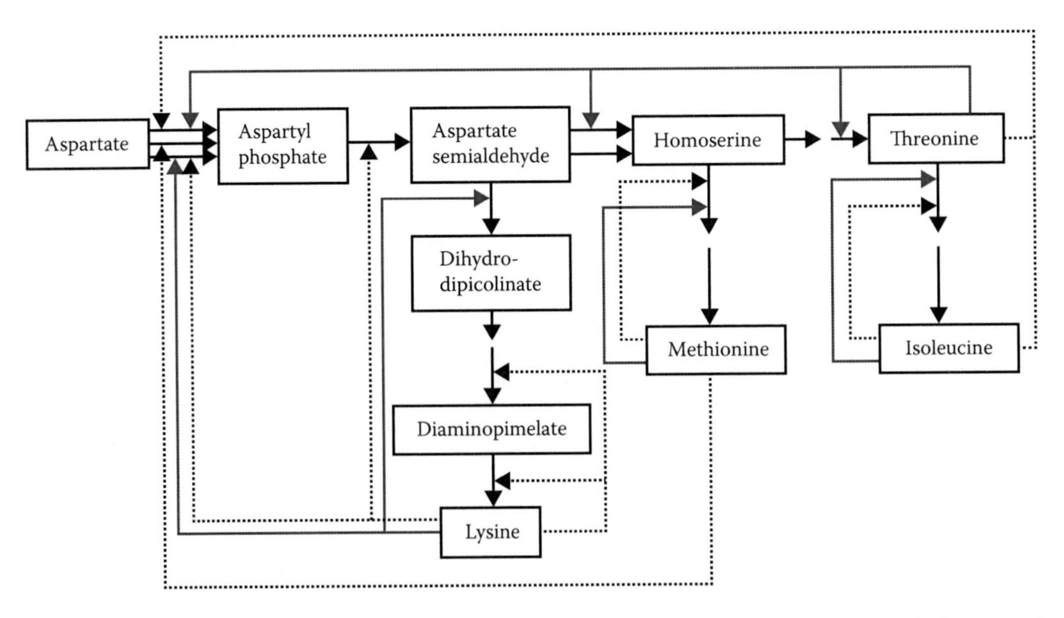

Figure 9.10 Regulation of methionine biosynthesis in *E. coli*. The blue arrows show inhibition and dotted arrows show repression.

and Nakayama 1975). Recovery of the product is based on treatment with ion exchange resins. Crystalline methionine can be obtained by concentrating under vacuum, treating with absolute alcohol, and drying at a low temperature of approximately 4°C for 24 h (Kumar and Gomes 2005).

9.4.3 *Industrial application and therapeutic role*

The application of methionine is mostly reported in some pharmaceutical and livestock formulations. Because it is an essential amino acid, it is therefore required in the diets of humans and in livestock. Methionine is an excellent natural lipotrophic agent that processes and eliminates fats from the liver and acts as a natural detoxifying agent, removing heavy metals from the body and excess histamine from the brain. In addition, it has antioxidant properties as well and thus protects the body against free radicals (Chaitow 1985). Plant proteins are frequently deficient in methionine and, consequently, an exclusively vegetarian diet may fail to meet nutritional requirements. Methionine deficiency has been linked to the development of various diseases and complications. These include toxemia, childhood rheumatic fever, muscle paralysis, hair loss, depression, schizophrenia, Parkinson's liver deterioration, and impaired growth (Chaitow 1985). These symptoms appear due to deficiencies that can be overcome by incorporating methionine in the diet (Parcell 2002). Methionine is extensively used in the poultry and livestock industries (Funfstuck et al. 1997; Neuvonen et al. 1985).

9.5 *Tryptophan*

9.5.1 *Historical background and new challenges*

Until 1967, no scientific reports were available relating to the microbial direct production of tryptophan. During this period, more attention was given by researchers looking into the possibility of tryptophan production through the conversion of anthranilic acid, coupling of indole and serine, transamination of 3-indolepyruvic acid, and conversion of β-indolyllactic acid (Dulaney 1967). The chemical method to produce tryptophan was the first method used for industrial-scale manufacturing (Sidransky 2001). This was followed by production through enzymatic reaction and fermentation (Potera 1991). A variety of microorganisms such as *Bacillus subtilis, Pseudomonas aeruginosa,* and others containing tryptophan synthase (TSase; EC 4.2.1.20) or tryptophanase (TPase; EC 4.1.pp.1) can be used for enzymatic reaction. *E. coli* contains both tryptophan synthase and tryptophanase (Austin and Esmond 1965; Hamilton et al. 1985). With the introduction of efficient strains of *Corynebacterium* and *E. coli,* now tryptophan is largely produced by fermentation. The market for tryptophan is not big. The 2012 figure indicates the production of 8,400 tons, and the main companies involved are Ajinomoto, Cheil Jedang, and Evonik. However, China is expected to dominate the market (Dong 2013; Gunderson and Koehler 2012).

9.5.2 *Biotechnological approach and production*

The initial commercial production method used was based on chemical synthesis. However, by the end of 1980, other methods such as production through enzymatic reaction and fermentation were also included in view of the increased demand of tryptophan, especially by the feed and pharmaceutical industries. Thus, continuous attempts are now being made, especially utilizing the biotechnological approach to meet the desired production demand. Among different options, metabolic engineering for increased L-tryptophan production has been given special attention. The results were encouraging when *E. coli* and *Corynebacterium* were used. The tryptophan biosynthetic pathway in *E. coli* is shown in Figure 9.11.

Overproducing strains of *E. coli* and *C. glutamicum* containing overexpressed genes have been developed, which can produce 40.2 g/L of tryptophan on suitable medium containing glucose as a carbon source in 40 h (Shen et al. 2012). A genetically engineered strain of *E. coli* was also reported to produce approximately 10.15 g/L of L-tryptophan in 48 h (Gu et al. 2012). It is believed that further research using more advanced biotechnological tools will be able to increase the productivity of tryptophan.

To understand the phenomenon of tryptophan fermentation, a brief method has been highlighted, which reflects the use of a genetically modified strain of *C. glutamicum* that is capable of producing tryptophan. Organisms are usually maintained on a monthly transfer schedule on nutrient agar slants supplemented with 0.5% sodium chloride. A seed medium containing glucose (20.0), peptone (5.0), yeast extract (1.0), sodium chloride (0.5), KH_2PO_4 (0.5), K_2HPO_4 (1.5), $MgSO_4$ (0.5), NH_4SO_4 (5.0), and calcium carbonate, (5.0 g/L) can be used. Optimum pH was observed at 6.8. Cultivation can be done in Erlenmeyer flasks of 250 mL capacity containing 30 mL of the above sterilized medium. Inoculation can be done with a loopful of culture grown on nutrient agar slant for 24 h. The flask should be shaken at 200 rpm at 30°C for 24 h in an orbital shaker-incubator to prepare the seed culture.

Fermentation medium may be prepared from molasses (30%), corn-steep liquor (0.7%), KH_2PO_4 (0.05%), K_2HPO_4 (0.15%), $MgSO_4{\cdot}7H_2O$ (0.025%), $(NH_4)_2SO_4$ (1.5%), and calcium carbonate (1%). A mixture of the following ingredients should be added in grams per liter: vitamin B_1 (1000.0), biotin (50.0), L-phenylalanine (200.0), and L-tyrosine (175.0). The pH of the seed and fermentation media remain unchanged, that is, adjusted to 6.8, whereas 1.0 mL of 20% silicon RD in deionized water is added as antifoam. For the laboratory scale, mini jar fermenters of 2 L capacity containing 1 L of the medium can be used. Serialization can be affected by autoclaving for 15 min at 15 psi. After cooling, the fermenter is inoculated

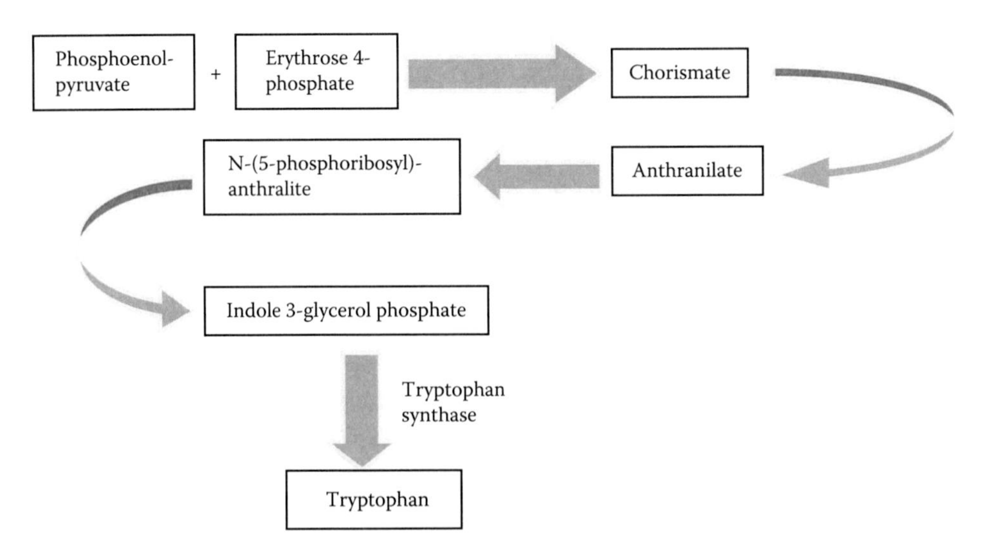

Figure 9.11 Tryptophan biosynthetic pathway in *E. coli*.

by one seed flask and stirred at 400 rpm with an air flow of 1.0 vvm (air volume/liquid volume per minute) at 30°C. The fermenters can be harvested after 72 h. Product recovery is usually done using ultracentrifugation at around 10,000 rpm, followed by treatment with cation exchange resin and decolorization with activated carbon. After further centrifugation, the mixture can be subjected to drying under a vacuum dryer. The product yield usually ranges between 10 and 25 g/L.

9.5.3 Industrial application and therapeutic role

Tryptophan has a wide range of applications in the feed and pharmaceutical industries. As an essential amino acid with a unique indole side chain, which indicates its use as a precursor for a number of neurotransmitters in the brain, for example, serotonin, melatonin, and niacin associated with appetite, sleep, mood, and pain perception. Its application in the chemical synthesis of some antidepressant drugs and in the treatment of schizophrenia is quite prominent (Chaitow 1985; Porter et al. 2005; Van der Heijden et al. 2005).

9.6 Case study: production of L-lysine on glucose and molasses medium supplemented with fish meal under submerged condition in mini jar fermenter

The importance of L-lysine has been highlighted by many researchers based on its application in the food, feed, and pharmaceuticals industries. The deficiency of L-lysine can greatly affect the growth and development of bone in children; in adults, L-lysine helps calcium absorption and maintains nitrogen balance. Therefore, it may also help prevent bone loss associated with osteoporosis. Also, it plays a role in enhancing the defense mechanisms of the body, especially acting against cold sores and herpesviruses, and as an aid in antibodies, hormones, enzymes, and collagen formation; its benefits in the repair of tissues also cannot be ignored (Mahmood 2010).

The demand for L-lysine is thus continuously increasing, asking investigators to meet the challenge by increasing the production capacity using high-yield bacterial strains and the best-possible economical source of raw material, especially carbon and nitrogen for its production. This case study is also part of this program. In the current experiment, L-lysine production was evaluated with two different types of media using a homoserine auxotrophic mutant of *C. glutamicum* FRL no. 61989, which is also resistant to AEC. The auxotrophic and regulatory mutants are good options for increased L-lysine production. Isolation of the auxotrophic mutant is based on treatment of the wild strain of *C. glutamicum* with ultraviolet radiation followed by a penicillin selection method, whereas AEC-resistant strains were developed by treating the homoserine auxotrophic mutant with *N*-methyl-*N*-nitro-*N*-nitrosoguanidine, followed by growing the cells in a medium supplemented with AEC. Colonies that appeared on the surface of the agar plate during 2 to 7 days of incubation were picked up as AEC-resistant mutants (Mahmood 1996).

9.6.1 Cultivation of C. glutamicum *FRL no. 61989*

The organism was maintained on a monthly transfer schedule on nutrient agar slants. A seed medium of following composition in grams per liter was used: glucose (30.0), peptone (10.0), yeast extract (5.0), sodium chloride (3.0), KH_2PO_4 (0.5), K_2HPO_4 (1.5), $MgSO_4$ (0.5), NH_4SO_4 (0.5), sodium acetate (0.1), and 25.5 μg/L of biotin. The pH was adjusted to 7.2. Erlenmeyer flasks of 250 mL capacity containing 30 mL of the sterilized medium were inoculated with a loopful of culture grown on nutrient agar slant for 24 h. The flasks were shaken at 200 rpm at 30°C for 24 h in an orbital shaker-incubator to prepare the seed culture.

Fermentation media I and II with the following composition in grams per liter were used for comparative studies in terms of yield and cost affectivity. Medium I contains glucose (110.0), yeast extract (5.0), peptone (10.0), and meat extract (5.0). Medium II contains molasses (400.0), fish meal (25.0), and corn steep liquor (5.0). Both media were supplemented with the following ingredients in grams per liter: KH_2PO_4 (0.5), K_2HPO_4 (1.5), $MgSO_4{\cdot}7H_2O$ (0.5), $(NH_4)_2SO_4$ (15.0), calcium carbonate (10.0), $MnSO_4{\cdot}4H_2O$ (0.01), ferrous sulfate $4H_2O$ (0.01), sodium acetate (2.0), and vitamin B_1 (0.001). The concentration of biotin was 125.0 and 25.0 μg/L in media I and II, respectively. The pH was adjusted to 7.2 for both media, and 1.0 mL of 20% silicon RD in deionized water was added as antifoam.

One liter of media I and II was transferred separately to 2.0 L capacity mini jar fermenters and serialized by autoclaving for 15 min at 15 psi. The respective fermenters containing media I and II were inoculated with the contents of one seed flask and stirred at 400 rpm with an air flow of 1.0 vvm at 30°C. The fermenters were harvested after 96 h.

9.6.2 *Isolation, extraction, and purification of L-lysine*

After the incubation period, the temperature of the fermenters was increased to 60°C for 15 min and the liquid culture was transferred to separate 3.0 L capacity Erlenmeyer flasks. Broths were centrifuged at 10,000 rpm using a high-speed centrifuge for 20 min to separate cells and other precipitated materials, and then subjected to ion exchange method for separation and purification. The supernatants from both fermenters were adjusted to pH 2.0 using 6N HCl and passed through a separate column 30 cm in length and 6 cm in diameter, packed with 250 g of Amberlite IR-120, a strong cation exchange resin. The rate of flow of supernatant through the column was fixed at approximately 100 drops/min to adsorb the L-lysine. After this, 250 mL of deionized water was passed twice through both columns and eluted with 500 mL of dilute ammonia. The whole liquid collected from each column was then passed through Amberlite IRC-50, a weak cation exchange resin. The column size, quantity of resin, and rate of flow of liquid were kept the same. Finally, columns were washed twice with a 250 mL quantity of deionized water and the solution was adjusted to pH 5.5 and treated with 10 g/L of activated carbon for 6 h. The slurry containing carbon and lysine was filtered under vacuum and concentrated to approximately 30% under reduced pressure and left overnight at 10°C to develop crystals of L-lysine. The material was filtered and, after washing with cold deionized water, the crystalline material was dried in a vacuum dryer.

The purity of the material was checked by performing thin-layer chromatography in a mixture of *n*-propanol and strong ammonia solution (67:33), using ninhydrin as a spray reagent as well as by comparing the infrared absorption spectrum with standard L-lysine. The purity was recorded as 98.9% and 98.2%, respectively. An assay for L-lysine was also performed and the endpoint was determined potentiometrically. The average of three experiments produced 67.80 and 65.20 g/L of L-lysine monohydrochloride by glucose and molasses medium with a corresponding yield ($Y_{p/s}$) of 0.6116 gg^{-1} (61.6% conversion efficiency) and 0.592 gg^{-1} (59.2% conversion efficiency), respectively.

9.6.3 *Effect of media composition and fermentation parameters on L-lysine production*

A complex microbiological process is observed to be involved in the fermentation of L-lysine, which is greatly influenced by several biochemical and physical parameters. Therefore, in-depth knowledge of bacterial physiology and biochemistry is essential to achieve high yield of the product by choosing the most suitable carbon and nitrogen source along with the fermentation parameters optimization. In the present case study, after several sideline studies, the nutritional requirement and fermentation parameters were established for a strain of *C. glutamicum* FRL no. 61989 and accordingly applied for a final comparative study to observe the production capacity on glucose and molasses medium. The strain FRL no. 61989 was developed by treating a homoserine auxotroph with AEC.

During industrial production of L-lysine, the choice of raw material, especially the carbon and nitrogen source, largely depends on economic consideration. Therefore, it is essential to determine the optimum level of carbon and nitrogen source depending on the nutritional characteristics of the production culture. In the present study, a method that utilizes shake flask experiments with medium containing different concentrations of sugar supplemented with a sufficient quantity of nitrogen source to satisfy the demand imposed by increasing the carbon level was used. An analysis of the amount of L-lysine produced and the amount of sugar remaining after the process provides the data necessary to calculate the percentage yield of product. The sugar concentration, which provides the maximum percentage yield of L-lysine, can be treated as the optimum concentration of sugar to further proceed (Daoust 1976). In addition, the overall method used for the extraction and purification of the product including the fermentation time, yield, and treatment of waste material should also be taken into consideration. Similarly, the optimization of

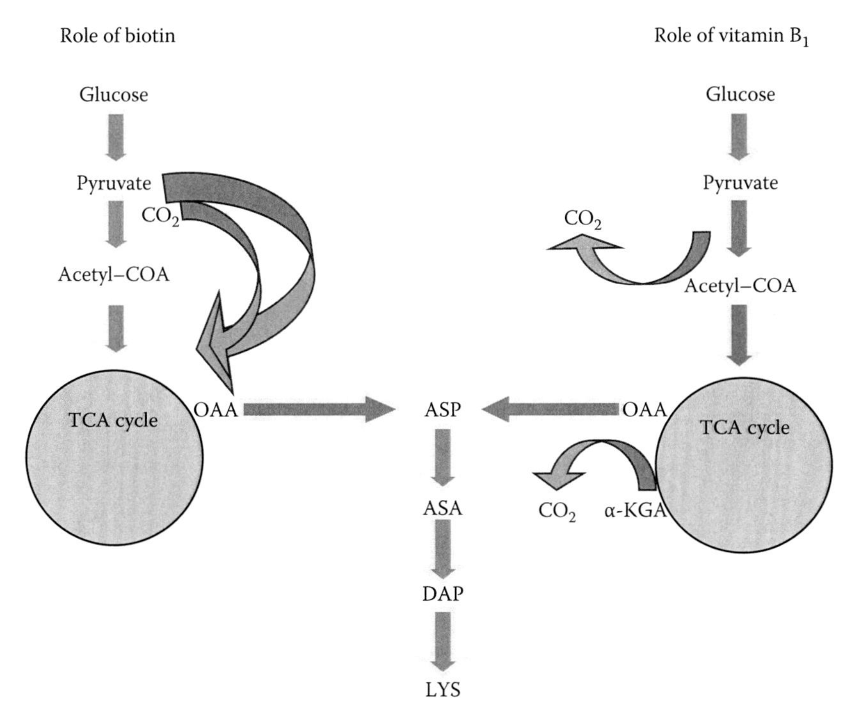

Figure 9.12 Effect of biotin and vitamin B_1 on L-lysine production.

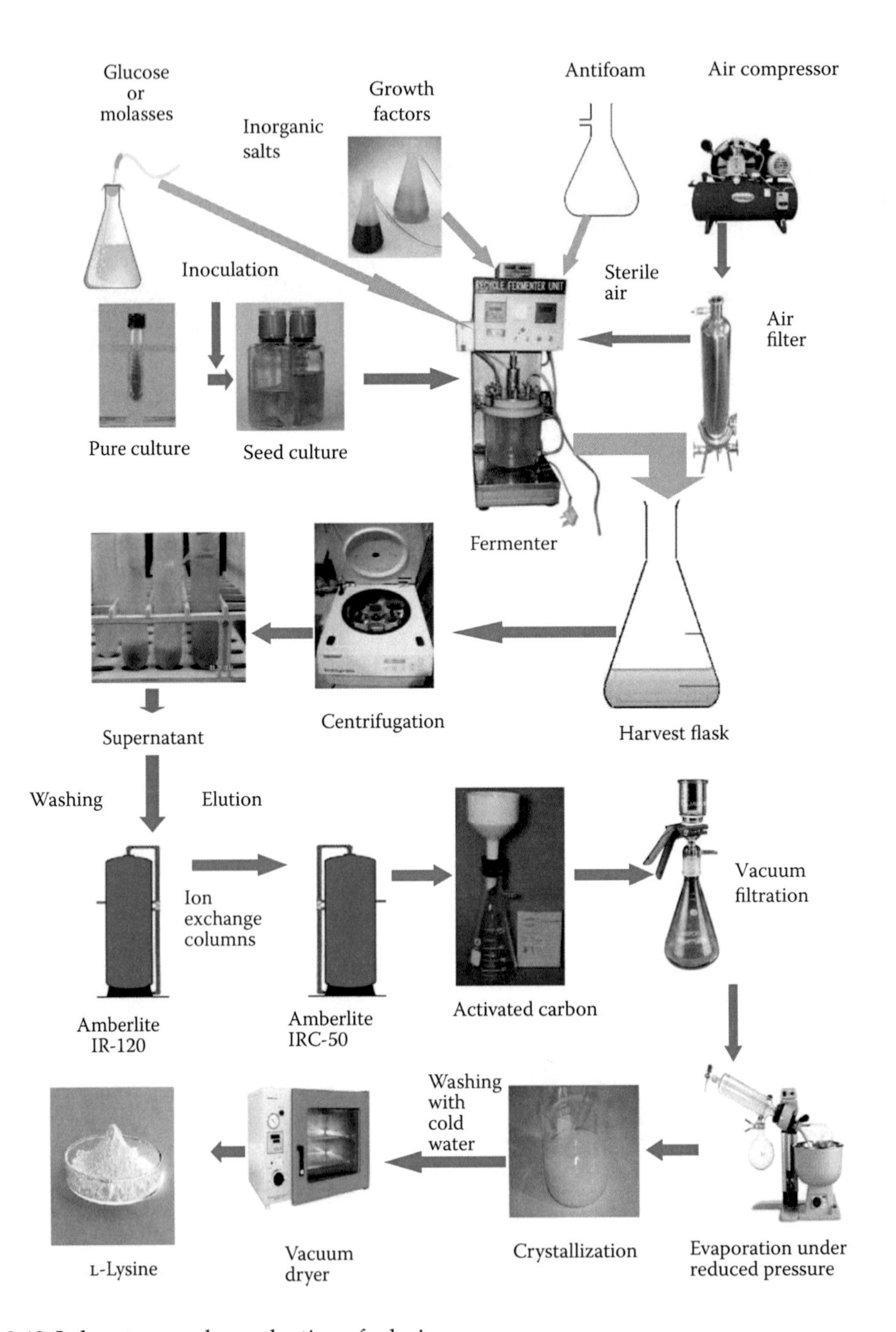

Figure 9.13 Laboratory-scale production of L-lysine.

culture conditions in relation to oxygen supply and carbon dioxide removal have played a key role in the scale-up and economic production of L-lysine. These factors determined both the rate of cell growth and product formation.

Because of the present case study, our finding is that the yield of L-lysine is more or less similar on glucose and molasses medium. Therefore, comparing the cost of glucose and molasses, the composition of medium II justified the use of molasses as a carbon source for the production of L-lysine. Furthermore, because molasses contain a sufficient quantity of biotin, the cost of the product can therefore be further reduced by decreasing the concentration of biotin in the molasses medium. Similarly, the use of fish meal as a nitrogen source was also justified in view of the high cost of soybean protein acid hydrolysate, yeast extract, peptone, and meat extract. Hence, a substantially less expensive L-lysine fermentation can be carried out using such reasonably low-cost materials.

The concentrations of biotin and vitamin B_1 were established as 125 μg/L and 1.0 mg/L, respectively. The roles of biotin and vitamin B_1 in amino acid fermentation are quite established. Biotin is reported to be involved in the oxidation of glucose and in the synthesis of proteins as well as in cell permeability (Tosaka et al. 1979), and as a carbon dioxide carrier of covalent bonds (Moss and Lane 1971). Furthermore, enzymes containing biotin provide aspartic acids, an important intermediate that leads to the synthesis of L-lysine (Tosaka and Takinami 1978). Vitamin B_1 is reported to take part in the oxidative decarboxylation of pyruvate and α-ketoglutaric acid in the bacterial system, leading to increased formation of acetyl-CoA and oxaloacetic acid, thus taking part in the increased formation of L-lysine (Koike and Reed 1960; Mahmood 1996). The effect of biotin and vitamin B_1 is highlighted in Figure 9.12. Also, the overall production mechanism of L-lysine on the laboratory-scale is presented in Figure 9.13.

9.7 Conclusion

The microbial production of amino acids has gained significant attention in recent years because of its extensive application in food, feed, and pharmaceutical industries. Therefore, evaluation of the topic in view of recent biotechnological advancements is fully justified for increased production to meet the market demand. Some very specific amino acids such as glutamic acid, lysine, tryptophan, and methionine were selected for review in this chapter, and which are of considerable importance because of the role they play in our daily lives. It is expected that the selection of cost-effective raw materials and the utilization of a biotechnological approach will certainly bring a revolution with the introduction of new microbial strains capable of producing increased quantities of the desired amino acids. Also, an in-depth knowledge of submerged fermentation for the production of amino acids will have a significant contribution. With this approach, the chapter has been written to provide comprehensive information on the production, application, and market dimensions covering four important amino acids for both students and researchers interested or working on such topics. It is expected that the information cited will help in analyzing and understanding the dynamics of the topic and to overcome the challenges expected during process design and optimization for amino acid fermentation.

Acknowledgments

I am extremely thankful to Martti Hedman, Vice President, Corporate Development, Colorcon and Jacques Michaud, Director EMEA, Colorcon Limited—England for their encouragement and support. My special thanks to IHS Chemical for providing marketing information.

References

Abdenacer, M., Kahina, B.I., Aicha, N., Nabil, N., Jean-Louis, G. and Joseph, B. 2012. Sequential optimization approach for enhanced production of glutamic acid from *Corynebacterium glutamicum* 2262 using date juice. *Biotechnol. Bioprocess Eng.*, **Vol. 17**: pp. 795–803.

Adelberg, E.A. and Myers, J.W. 1953. Modification of the penicillin technique for the selection of auxotrophic bacteria. *J. Bacteriol.*, ***Vol. 65****(3)*: pp. 348–353.

Amin, G.A. and Al-Talhi, A. 2007. Production of L-glutamic acid by immobilized cell reactor of the bacterium *Corynebacterium glutamicum* entrapped into carrageenan gel beads. *World Appl. Sci. J.*, ***Vol. 2****(1)*: pp. 62–67.

Asakura, Y., Kimura, E., Usuda, Y., Kawahara, Y., Matsui, K., Osumi, T. and Nakamatsu, T. 2007. Altered metabolic flux due to deletion of *odhA* causes L-glutamate overproduction in *Corynebacterium glutamicum*. *Appl. Environ. Microbiol.*, ***Vol. 73****(4)*: pp. 1308–1319.

Austin, N.W. and Esmond, E.S. 1965. Formation and interrelation ships of tryptophanase and tryptophan synthetases in *E. coli*. *J. Bactriol.*, ***Vol. 89****(2)*: pp. 355–363.

Balch, J.F. and Balch, P.A. 1990. *Prescription for Nutritional Healing*. Avery Publishing Group Inc., New York: pp. 27–31.

Banik, A.K. and Majumdar, S.K. 1975. Effect of minerals on production of methionine by *Micrococcus glutamicus*. *Indian J. Exp. Biol.*, ***Vol. 13***: pp. 510–520.

Bloom, F.E. 2001. Neurotransmission and the central nervous system. *In*: *Goodman and Gilman's, The Pharmacological Basis of Therapeutics*. Hardman, J.G., Limbird, L.E. and Gilman, A.G. (editors). McGraw-Hill, New York: pp. 293–335.

Brautaset, T. and Ellingsen, T.E. 2011. Lysine: Industrial uses and production. *In*: *Comprehensive Biotechnology*. Moo-Young, M. (editor-in-chief). Elsevier, Amsterdam, The Netherlands: pp. 541–554.

Casida, L.E. and Baldwin, N.Y. 1956. Preparation of diaminopimelic acid and lysine. *U.S. Patent 2,771,396*: pp. 1–3.

Chaitow, L. 1985. *Amino Acids in Therapy*. Thorsons Publishers Limited, Wellingborough, Northamptonshire: pp. 43–106.

Daoust, D.R. 1976. Microbial synthesis of amino acids. *In: Industrial Microbiology*. Miller, B.M. and Litsky, W. (editors). McGraw-Hill Book Company, New York: pp. 106–127.

Davis, B.D. 1952. Biosynthesis interrelations of lysine, diaminopimelic acid, and threonine in mutants of *E. coli*. *Nature*, ***Vol. 169***: pp. 534–536.

Demain, A.L. 2007. The business of biotechnology. *Ind. Biotechnol.*, ***Vol. 3****(3)*: pp. 269–283.

Dewey, D.L. and Work, E. 1952. Diaminopimelic acid and lysine. *Nature*, ***Vol. 169***: pp. 533–534.

Dike, K.S. and Ekwealor, I.A. 2012. Production of L-methionine by *Bacillus cereus* isolated from different soil ecovars in Owerri, South East Nigeria. *European J. Exp. Biol.*, ***Vol. 2****(2)*: pp. 311–314.

Dong, Z. 2013. 2012 Market research report on global and China tryptophan industry. Available from: http://www.qyresearch.com/yw/shop/html/?472.html.

Dulaney, E.L. 1967. Microbial production of amino acids. *In*: *Microbial Technology*. Peppler, H.J. (editor). Reinhold Publishing Corporation, New York: pp. 308–343.

Eggeling, L. and Sahm, H. 2011. Amino acid production. *In*: *Comprehensive Biotechnology*. Moo-Young, M. (editor-in-chief). Elsevier, Amsterdam, The Netherlands: pp. 625–649.

Fong, C.V., Goldgraben, G.R., Konz, J., Walker, P. and Zank, N.S. 1981. Condensation process for DL-methionine production. *In*: *Organic Chemicals Manufacturing Hazards*. Goldfrab, A.S. (editor). Ann Arbor Science Publisher, Ann Arbor: pp. 115–194.

Funfstuck, R., Straube, E., Schildbach, O. and Tietz, U. 1997. Prevention of reinfection by L-methionine in patients with recurrent urinary tract infection. *Med. Klin.*, ***Vol. 92***: pp. 574–581.

Gu, P., Yang, F., Kang, J., Wang, Q. and Qi, Q. 2012. One-step of tryptophan attenuator inactivation and promoter swapping to improve the production of L-tryptophan in *Escherichia coli*. *Microb. Cell Fact.*, ***Vol. 11***: pp. 30–36.

Gunderson, A. and Koehler, D. 2012. Not enough tryptophan in the whole, wide world. Future tryptophan outlook, *In*: Feedinfo News Service. Available from: http://swineperformance.vitaplus.com/2012/10/not-enough-tryptophan-in-the-whole-wide-world/.

Hamilton, B.K., Hasiao, H., Swann, W.E., Andersen, D.M. and Delente, J.J. 1985. Manufacture of amino acids with bioreactors. *Trends Biotechnol.*, ***Vol. 3***: pp. 64–68.

IHS Chemical. 2012. Amino acids—Chemical insight and forecasting (HIS Chemical). Available from: http://www.ihs.com/products/chemical/planning/ceh/major-amino-acids.aspx.

Ikeda, M. and Takeno, S. 2013. Amino acid production by *Corynebacterium glutamicum*. *Microbiol. Monogr.*, ***Vol. 23***: pp. 107–147.

Jose, S. 2011. Amino acids—A global strategic business report. Available from: http://www.prweb.com/releases/2011/2/prweb8151116.htm.

Kase, H. and Nakayama, K. 1975. L-methionine production by methionine analog-resistant mutants of *Corynebacterium glutamicum*. *Agric. Biol. Chem.*, **Vol. 39**: pp. 153–160.

Khan, N.S., Sing, R.P. and Prasad, B. 2013a. Studies on substrate inhibition in the microbial production of L-glutamic acid. *Int. J. Eng. Res. Technol.*, **Vol. 2**(1): pp. 1–7.

Khan, N.S., Sing, R.P. and Prasad, B. 2013b. Modelling the fermentative production of L-glutamic acid production by Corynebacterium glutamicum in a batch bioreactor. *Int. J. Eng. Res. Technol.*, **Vol. 5**(1): pp. 192–199.

Kinoshita, S. 1999. Taxonomic position of glutamic acid producing bacteria. *In*: *Encyclopaedia of Bioprocess Technology. Fermentation, Biocatalysis and Bioseparation*. Flickinger, M.C. and Drew, S.W. (editors). John Wiley & Sons, New York: pp. 1330–1336.

Kinoshita, S., Nakayama, K. and Kitada, S. 1958. L-lysine production using microbial auxotroph. *J. Gen. Appl. Microbiol.*, **Vol. 4**(2): pp. 128–129.

Kinoshita, S., Udaka, S. and Shimono, M. 1957a. Studies on amino acid fermentation: Part—I. Production of L-glutamic acid by various microorganisms. *J. Gen. Appl. Microbiol.*, **Vol. 3**(3): pp. 193–205.

Kinoshita, S., Tanaka, T., Udaka, S. and Akita, S. 1957b. Glutamic acid fermentation. *Proc. Symp. Enzyme Chem.*, **Vol. 2**: pp. 464–468.

Kita, D.A., Heights, J. and Huang, H.T. 1958. Fermentation process for the production of L-lysine. *U.S. Patent 2,841,532*: pp. 1–2.

Koike, M. and Reed, L.J. 1960. A-keto acid dehydrogenation complex. II. The role of protein bound lipoic acid and flavin adenine dinucleotide. *J. Biol. Chem.*, **Vol. 235**: pp. 1931–1938.

Kumar, D., Bisaria, V.S., Sreekrishan, T.R. and Gomes, J. 2003. Production of methionine by multi-analogues resistant mutant of *Corynebacterium lilium*. *Process Biochem.*, **Vol. 38**: pp. 1165–1171.

Kumar, D. and Gomes, J. 2005. Methionine production by fermentation. *Biotechnol. Adv.*, **Vol. 23**: pp. 41–61.

Mahmood, Z.A. 1996. Production of L-lysine through fermentation. Ph.D thesis, Department of Pharmaceutics, University of Karachi, Karachi-Pakistan: pp. 1–192.

Mahmood, Z.A. 2010. *L-lysine: Production through Fermentation*. VDM Verlag, Saarbruken, Germany: pp. 1–166.

McEntee, W.J. and Crook, T.H. 1993. Glutamate: Its role in learning, memory, and the aging brain. *Psychopharmacology*, **Vol. 111**(4): pp. 391–401.

Mitchell, H.K. and Houlahan, M.B. 1948. An intermediate in the biosynthesis of lysine in *Neurospora*. *J. Biol. Chem.*, **Vol. 174**: pp. 883–887.

Mondal, S. and Chatterjee, S.P. 1994a. Enhancement of methionine production by methionine analogue resistant mutants of *Brevibacterium heali*. *Acta Biotechnol.*, **Vol. 14**: pp. 199–204.

Mondal, S., Das, Y.B. and Chatterjee, S.P. 1994b. L-methionine production by double auxotrophic mutants of an ethionine resistant strain of *Brevibacterium heali*. *Acta Biotechnol.*, **Vol. 14**: pp. 61–66.

Moss, J. and Lane, M.D. 1971. The biotin dependent enzymes. In: *Advances in Enzymology*, **Vol. 35**. Miester, A. (editor). John Wiley & Sons Inc, New York: pp. 321–332.

Nakayama, K. and Kinoshita, S. 1961. Studies on lysine fermentation II. α,ε-diaminopimelic acid and its decarboxylase in lysine producing strain and parent strain. *J. Gen. Appl. Microbiol.*, **Vol. 7**(3): pp. 155–160.

Neuvonen, P.J., Tokola, O., Toivonen, M.L. and Simell, O. 1985. Methionine in paracetamol tablets, a tool to reduce paracetamol toxicity. *Int. J. Clin. Pharmacol. Ther. Toxicol.*, **Vol. 23**: pp. 497–500.

Odunfa, S.A., Adeniran, S.A., Teniola, O.D. and Nordstorm, J. 2001. Evaluation of lysine and methionine production in some lactobacilli and yeast from Ogi. *Int. J. Food Microbiol.*, **Vol. 63**: pp. 159–163.

Ozulu, U.S., Nwanah, O.U., Ekwealor, C.C., Dike, S.K., Nwikpo, C.L. and Ekwealor, I.A. 2012. A new approach to screening for methionine producing bacteria. *Br Microbiol. Res. J.*, **Vol. 2**(1): pp. 36–39.

Papagianni, M. 2012. Recent advances in engineering the central carbon metabolism of industrially important bacteria. *Microb. Cell Fact.*, **Vol. 11**: pp. 50–63.

Parcell, S. 2002. Sulfur in human nutrition and application in medicine. *Atern. Med. Rev.*, **Vol. 7**: pp. 22–44.

Pasha, S.Y., Ali, M.N., Tabassum, H. and Mohammad, M.K. 2011. Comparative studies on production of glutamic acid using wild type, mutants, immobilized cells and immobilized mutants of *Corynebacterium glutamicum*. *Int. J. Eng. Sci. Technol.*, **Vol. 3**(5): pp. 3941–3949.
Pham, C.B., Galvez, C.F. and Padolina, W.G. 1992. Methionine fermentation by batch fermentation from various carbohydrates. *Asian Food J.*, **Vol. 7**: pp. 34–37.
Plimmer, R.H.A. and Hopkins, F.G. 1912. *The Chemical Composition of the Proteins. Monographs on Biochemistry. Part 1. Analysis* (2nd ed). Longmans, London: pp. 114–124.
Pontecorvo, G. 1949. Auxanographic technique in biochemical genetics. *J. Gen. Microbiol.*, **Vol. 3**(1): pp. 122–126.
Porter, R.J., Mulder, R.T., Joyce, P.R. and Luty, S.E. 2005. Tryptophan and tyrosine availability and response to antidepressant treatment in major to antidepressant treatment in major depression. *J. Affect. Disord.*, **Vol. 86**(2–3): pp. 129–134.
Potera, C. 1991. Researchers identify tryptophan contaminant. *Gen. Eng. News*, **Vol. 11**(2): pp. 1–24.
Prasad, M.P., Gupta, N., Gaudani, H., Gupta, M., Gupta, G., Krishna, V.K., Trivedi, S. and Londhe, M. 2009. Production of glutamic acid using whole and immobilised cells of Corynebacterium glutamicum. *Int. J. Microbiol. Res.*, **Vol. 1**(1): pp. 8–13.
Roquette. 2012. Metabolic explorer and Roquette communicate on L-Methionine. Available from: http://www.roquette.com/news-2012-health-nutrition-food-paper/l-methionine-metabolic-explorer-and-roquette-pave-the-way-to-industrialisation/.
Saidak, T. 2012. Arkema, CJ to build bio-methionine, thiochemicals complex in Malaysia. Available from: http://www.chemanageronline.com/en/news-opinions/headlines/arkema-cj-build-bio methioninethiochemicals-complex-malaysia.
Sano, C. 2009. History of glutamate production. *Am. J. Clin. Nutr.*, **Vol. 90**: pp. 728S–732S.
Shen, T., Liu, Q., Xie, X., Xu, Q. and Chen, N. 2012. Improved production of tryptophan in genetically engineered E. coli with TktA and PpsA overexpression. *J. Biomed. Biotechnol.*, **Vol. 2012**: pp. 1–8.
Sidransky, H. 2001. *Tryptophan: Biochemical and Health Implications*. CRS Press, Boca Raton, FL: pp. 1–280.
Tosa, T., Mori, T., Fuse, N. and Chibata, I. 1967. Studies on continuous enzyme reactions for preparation of a DEAE-sephadex-aminoacylase column and continuous optical resolution of acyl-DL-amino acids. *Biotechnol. Bioeng.*, **Vol. 9**: pp. 603–608.
Tosaka, O., Hirakawa, H. and Takinami, K. 1979. Effect of biotin levels on L-lysine formation in *Brevibacterium lactofermentum*. *Agric. Biol. Chem.*, **Vol. 43**(3): pp. 491–495.
Tosaka, O. and Takinami, K. 1978. Pathway and regulation of lysine biosynthesis in *Brevibacterium lactofermentum*. *Agric. Biol. Chem.*, **Vol. 42**(1): pp. 95–100.
Umerie, S.C., Ekwealor, I.A. and Nawabo, I.O. 2000. Lysine production from various carbohydrate and seed meals. *Bioresour. Technol.*, **Vol. 75**: pp. 249–252.
Van der Heijden, F.M., Fekkes, D., Tuinier, S., Sijben, A.E., Kahan, R.S. and Verhoeven, W.M. 2005. Amino acids in schizophrenia: Evidence for lower tryptophan availability during treatment with atypical antipsychotics. *J. Neural Transm.*, **Vol. 112**(4): pp. 577–585.
Windsor, E. 1951. α-Aminoadipic acid as a precursor to lysine in *Neurospora*. *J. Biol. Chem.*, **Vol. 192**: pp. 607–609.
Wippler, B. 2011. Asia in the driver's seat of the global lysine market. Available from: http://renewablechemicals.agra-net.com/2011/10/asia-in-the-driver%E2%80%99s-seat-of-the-global-lysine-market/.
Zareian, M., Ebrahimpour, A., Bakar, F.A., Mohamed, A.K.S., Forghani, B., Safuan, M., Kadir, A. and Saari, N. 2012. A glutamic acid producing lactic acid bacteria isolated from Malaysian fermented foods. *Int. J. Mol. Sci.*, **Vol. 13**: pp. 5482–5497.

chapter ten

Probiotics

The possible alternative to disease chemotherapy

Adel M. Mahasneh and Muna M. Abbas

Contents

10.1 Introduction

"Let food be thy medicine and medicine be thy food," as Hippocrates said, is the principle of today (Suvarna and Boby 2005). Probiotics are one of the functional foods that link diet and health. The term *probiotic* is derived from the combined Greek/Latin word "pro" and "bios," meaning for life (Chuayana et al. 2003; Gupta and Garg 2009). Probiotics are living microorganisms that, administered in adequate amounts, confer a health benefit to the host (Ayeni et al. 2011; de Los Reyes-Gavilán et al. 2011; Gupta and Garg 2009; Vijayendra and Gupta 2012).

The concept of probiotics was first postulated in 1908 by the Russian scientist and Nobel Prize winner Elie Metchnikoff who suggested that the long life of Bulgarian peasants resulted from their consumption of fermented milk products (Mercenier et al. 2002; Sonal et al. 2008; Tannock 2003). Metchnikoff thought that when the fermented milk products were consumed, the fermenting *Lactobacillus* positively influenced the microflora of the gut, decreasing toxic microbial activities there (Chuayana et al. 2003; Mercenier et al. 2002; Ouwehand et al. 2011).

In 1965, the term "probiotic" was first used by Lilly and Stillwell for describing "substances secreted by one organism which stimulates the growth of another" (Gupta and Garg 2009; Saraf et al. 2010).

The concept of probiotics has been expanded to include bacteria from intestinal origin, those that are expected to beneficially affect the host by improving the intestinal microbial balance and hence are selected as probiotics besides the bacteria isolated from fermented dairy products (Ishibashi and Yamazaki 2001; Moreira et al. 2005; Zeng et al. 2010).

Lactic acid bacteria (LAB) are the most common type of microorganisms used as probiotics (Sonal et al. 2008). Strains of the genera *Lactobacillus* and *Bifidobacterium* are the most widely used and commonly studied probiotic bacteria (Ayeni et al. 2011; Yateem et al. 2008). Lactobacilli are Gram-positive, nonspore-forming rods, catalase-negative, and usually nonmotile and do not reduce nitrate (Bernardeau et al. 2008; Gill and Prasad 2008; Singh and Sharma 2009). They are either homofermentative or heterofermentative, aerotolerant or anaerobic, aciduric or acidophilic, and have complex nutritional requirements (Bernardeau et al. 2006). Lactobacilli are found in a variety of habitats, including human and animal mucosal membranes, on material of plant origin, in sewage, and in fermented milk products and spoiled food (Chiang and Pan 2012). The species of *Lactobacillus* most frequently used as probiotics are *Lactobacillus acidophilus, Lactobacillus casei, Lactobacillus lactis, Lactobacillus helviticus, Lactobacillus plantarum, Lactobacillus bulgaricus, Lactobacillus rhamnosus, Lactobacillus johnsonii, Lactobacillus fermentum, Lactobacillus brevis,* and *Lactobacillus salivarius* (Ishibashi and Yamazaki 2001; Sonal et al. 2008). *Bifidobacterium* genera include species of *Bifidobacterium animalis, Bifidobacterium bifidum, Bifidobacterium longum, Bifidobacterium lactis, Bifidobacterium infantis,* and *Bifidobacterium breve,* which are used as probiotics (Bhadoria and Mahapatra 2011). However, species belonging to the genera *Enterococcus* and *Streptococcus* are also considered as probiotic microorganisms (Ljungh and Wadström 2006; Saraf et al. 2010). Other bacteria that are nonlactic acid producers and considered as probiotics with less potential are isolates from the genus *Propionibacterium* and the genus *Bacillus* such as *Bacillus lausii, Bacillus pumilus, Bacillus coagulans, Bacillus subtilis,* and *Bacillus cereus* var. *vietnami* (Mahasneh and Abbas 2010; Marianelli et al. 2010; Saraf et al. 2010; Vinderola and Reinheimer 2003). *Escherichia coli* strain Nissle 1917 is also one of the nonlactic acid-producing bacteria used as a probiotic (Sonal et al. 2008). The yeasts *Saccharomyces boulardii, Saccharomyces cerevisiae,* and *Kluyveromyces marxianus* have been used as probiotics (Gupta and Garg 2009; Kurugol and Koturoglu 2005; Ljungh and Wadström 2006; Maccaferri et al. 2012). In addition, the filamentous fungi *Aspergillus oryzae* has been used as a probiotic (Parvez et al. 2006). The microorganisms used in probiotic preparations should be chosen according to several criteria, such as being nonpathogenic and nontoxic (Parvez et al. 2006; Suvarna and Boby 2005; Vijayendra and Gupta 2012). Probiotic preparations should be "Generally Recognized as Safe" (GRAS; Gupta and Garg 2009). Even if some of the microorganisms used in probiotics preparations are known to be GRAS or have the "Qualified Presumption of Safety" (QPS) status, the working groups of the WHO/FAO recommend proof that a given probiotic strain is safe (Ayeni et al. 2011). Moreover, some probiotics' safety have not been fully studied and understood scientifically. More information is especially needed on how safe they are for infants, children, the elderly, and those with compromised immune systems (Mahasneh and Abbas 2010). On the other hand, human origin is thought to be important for host-specific interactions by the probiotics, although many of them are not of human origin (Dunne et al. 2001; Ouwehand et al. 2002). Probiotics should also produce a beneficial effect on the host by changing the composition of the normal intestinal microflora from a potentially harmful composition toward a microflora that would be beneficial for the host (Ouwehand et al. 2002, 2011; Saraf et al. 2010; Suvarna and Boby 2005). Another important criteria that should be included is the ability to survive transit through the gastrointestinal tract (GIT), this means having the ability to resist the low pH of the stomach, bile, and pancreatic enzymes (Crittenden et al. 2005; Liong and Shah 2005; Ouwehand et al. 2011; Saito 2004; Sonal et al. 2008). Potential probiotics need to have technological properties so that they can be cultured on a large scale, have an acceptable shelf life, and in case of application in fermented products, to be of good taste by the production of flavoring compounds and other metabolites that will provide a product with the organoleptic properties desired by the consumer (Ouwehand et al. 2002; Parvez et al. 2006; Saraf et al. 2010; Sonal et al. 2008; Vijayendra and Gupta 2012). Effective properties also include viability during storage and use, although the requirement of viability has been questioned (Izquierdo et al. 2008; Sonal et al. 2008), because nonviable probiotics have been found to have a longer shelf life than the viable ones and are easier to store and transport, and exhibit less interaction with other compounds of food products during storage (Ouwehand et al. 2002; Tanaka et al. 2011).

10.2 History of probiotics

The history of good health associated with living microorganisms in food, mainly LAB, is as old as civilization. In a Persian version of the Old Testament (Genesis 18:8), it states that "Abraham owed his longevity to the consumption of sour milk" (Schrezenmeir and de Vrese 2001). In 76 BC, the Roman historian Plinius recommended the administration of fermented milk products for gastroenteritis treatment (Reddy et al. 2010).

The Ukrainian-born biologist and Nobel Prize winner Elie Metchnikoff was perhaps the first researcher to propose that fermented dairy products have beneficial properties (Azizpour et al. 2009). In 1894, he showed that cholera could be prevented by the presence of antagonistic organisms in the intestine (Reddy et al. 2010). In 1906, Henry Tissier, a French pediatrician, observed that children with diarrhea had in their stools a low number of bacteria compared with that in healthy children (Oyetayo and Oyetayo 2005). He suggested that these "bifid" bacteria could be administered to patients with diarrhea to help restore a healthy gut flora.

In 1908, while Metchnikoff was working at the Pasteur Institute in Paris, after discovering *L. bulgaricus*, he developed a theory that LAB present in Bulgarian yoghurt and in the GIT of Bulgarian farmers could, by preventing putrefaction, prolong life. This was based on his observation that Bulgarians lived longer than other people (Soccol et al. 2010). He devoted the last decade of his life to the study of lactic acid-producing bacteria as a means of increasing human longevity. The concept of probiotics was then coined and a new field of microbiology was opened and developed to what we know now as probiotic science as an offshoot of microbial biotechnology.

The term probiotic was first used by Lilly and Stillwell in 1965 to describe "substances secreted by one organism, which stimulates the growth of another" (Gupta and Garg 2009).

In 1971, Sperti applied the term to tissue extracts that stimulate microbial growth (Schrezenmeir and de Vrese 2001). In 1974, Parker was the first to use the term probiotic in the sense that it is used today: "Organisms and substances which contribute to intestinal microbial balance" (Fioramonti et al. 2003). Fuller (1989) broadened the definition of probiotics by including their effect on animals, in which he described probiotics to be "a live microbial feed supplement which beneficially affects the host animal by improving its intestinal microbial balance" (Goel et al. 2006).

10.3 Functional properties of probiotics

10.3.1 Tolerance to gastric transit and bile salt

To secure functionality at best, the probiotic bacteria should overcome the many barriers of the host body. These barriers start with the oral cavity enzymes such as lysozyme, and the extremely low pH of the stomach, which may be as low as 1.5 (Bhadoria and Mahapatra 2011; Singhal et al. 2010). Bile and pancreatin that bacteria face in the small intestine are also among the main barriers that probiotics must overcome to arrive active and alive at the target action sites, the small and large intestine (Ayeni et al. 2011; Both et al. 2010). To entertain and to obtain the maximum efficiency of any probiotic bacterial strain, numbers should be no less than 10^6 to 10^8 CFU/g of intestinal contents (Charteris et al. 1998; de Los Reyes-Gavilan et al. 2011).

Probiotics should be able to tolerate the bile salt in the GIT to function properly (Klayraung et al. 2008). Bile, which is a yellow to green aqueous solution of organic and inorganic compounds, plays a major role in the emulsification and solubilization of lipids (Begley et al. 2005). The physiological concentration of the bile in the small intestine lies between 0.2% and 2.0% (Klayraung et al. 2008). Bile has the ability to affect phospholipids and proteins of cell membranes and disrupt cellular homeostasis. So, the ability of bacteria to tolerate bile is considered to be essential for their survival and subsequent colonization of the GIT (Begley et al. 2005; Mourad and Nour-Eddine 2006a,b; WHO 2006). Chang and coworkers (2010) isolated a potential LAB from kimchi that was resistant to biological barriers (acid and bile salts), whereas Klayraung and coworkers (2008) obtained acid-resistant and bile-resistant lactobacilli from traditional Thai foods. Ten strains of the LAB that have been isolated from Kung-Som, naturally fermented shrimp, showed a survival rate of more than 50% in simulated gastric juice and only four of them showed a survival rate of more than 50% in simulated small intestinal juices (Hwanhlem et al. 2010). In studying traditional Mongolian dairy products, 126 strains of the 543 LAB were assumed to be tolerant to bile acid, and 114 of the 126 were tolerant to low pH (Takeda et al. 2011). Resistance of potential probiotics to the stresses of the human GIT and their beneficial effects vary greatly between strains and species depending on the source of isolation (Both et al. 2010; Bhadoria and Mahapatra 2011; de Los Reyes-Gavilan et al. 2011). For example, eight different strains of *B. longum* showed a great variation in their resistance to GIT juices, but none of them showed viability or ability to grow after exposure to gastric juices at pH 1.5 (Izquierdo et al. 2008).

10.3.2 Cholesterol reduction

Hypercholesterolemia is a risk factor for cardiovascular diseases (Zeng et al. 2010). Liong and Shah (2005) reported that for each 1 mmol above the normal cholesterol level, the risk for coronary heart disease is approximately 35% higher, and coronary death is 45% higher. Several studies have indicated that some *Lactobacillus* species such as *L. acidophilus* could reduce total cholesterol and low-density lipoprotein (LDL) cholesterol (Gilliland et al. 1985; Sanders 2000; Vijayendra and Gupta 2012). *In vitro* studies indicated the presence of a number of possible mechanisms that have been observed for their cholesterol-lowering action of probiotic bacteria (Pereira and Gibson 2002). One of these mechanisms is cholesterol assimilation by probiotics (Margolles et al. 2003). Others include enzymatic deconjugation of bile salts by bile salt hydrolases (Zeng et al. 2010). Furthermore, incorporation of cholesterol into the membrane of cells is not excluded and finally binding of cholesterol to the bacterial cell wall (Liong and Shah 2005; Pereira and Gibson 2002).

Zeng and coworkers (2010) isolated *Lactobacillus buchneri* from pickled juice with a high cholesterol-reducing rate of 43.95%. Major portions of cholesterol were removed by cholesterol assimilation and the rest were coprecipitated with deconjugated bile salt and adsorbed to the cell surface. In addition to *L. buchneri*, *L. fermentum*, some strains of *L. acidophilus*, *L. plantarum*, *Lactobacillus paracasei* subsp. *paracasei*, and *Lactobacillus reuteri* have been found to reduce serum cholesterol level (Cardona et al. 2000; Chiang and Pan 2012; Gilliland et al. 1985; Gilliland and Walker 1990; Pereira and Gibson 2002; Zeng et al. 2010).

Chiang and Pan (2012) reported that milk fermented by *L. paracasei* subsp. *paracasei* NTU101, *L. plantarum* NTU102, and *L. acidophilus* BCRC17010 significantly decreased the levels of serum and liver total cholesterol in Syrian hamsters when fed a high-cholesterol diet. A similar trend was observed using Sprague-Dawley rat models after supplementation with *L. plantarum* 9-41-A and *L. fermentum* M1–16 (Jiang et al. 2010; Xie et al. 2011).

10.3.3 Antimicrobial activity

Potential probiotic bacteria should present their antimicrobial actions against pathogens that are found as food-borne microorganisms that might cause infections or intoxication in the GIT (Klayraung et al. 2008). Primary metabolites of LAB such as lactic acid and acetic acid, ethanol and carbon dioxide, hydrogen peroxide and diacetyl are the main causes of their inhibitory action against Gram-positive and Gram-negative bacteria (Coeuret et al. 2004; Hernández et al. 2005; Kazemipoor et al. 2012; Khay et al. 2011; Klayraung et al. 2008; Mahasneh and Abbas 2010).

Four *Lactobacillus* isolates have been found to inhibit the growth of strains of *Listeria monocytogenes*, *Staphylococcus aureus*, *Salmonella* spp., and *E. coli* at 15°C more than at 37°C—possibly by producing inhibitory compounds such as organic acids, hydrogen peroxide, and bacteriocins (Coeuret et al. 2004). Kazemipoor and coworkers (2012) have isolated LAB from home-made fermented vegetables, and they exhibited a remarkable inhibitory effect against both Gram-positive and Gram-negative pathogens due to the accumulation of many primary metabolites.

The inhibitory activities of LAB against Gram-positive pathogens are thought to be due to the different bacteriocins produced (Klayraung et al. 2008). Bacteriocins produced by Gram-positive bacteria have a bactericidal or bacteriostatic effect on other species and genera, but activity is usually limited to other genera (Corr et al. 2007). Bacteriocins have attracted great interest in the food industry due to their potential application in food preservation and in providing curative properties to these foods (Khay et al. 2011; Tiwari and Srivastava 2008). The bacteriocins of LAB are arranged into four classes in which classes I and II are the most thoroughly studied (Klaenhammer 1993). Class I bacteriocins, called lantibiotics, are small peptides that contain posttransitionally modified amino acid residues such as lanthionine (Batdorj et al. 2006); class II bacteriocins are small, heat stable, nonlanthionine-containing peptides that were recently organized into subgroups; class III bacteriocins are large, heat-labile nonlantibiotics; and class IV bacteriocins are complex bacteriocins containing chemical moieties such as lipid and carbohydrate (Khay et al. 2011; Savadogo et al. 2006).

The lantibiotic, nisin, which is produced by different *Lactococcus lactis* spp., is the most thoroughly studied bacteriocin and the only bacteriocin that is applied as an additive in food worldwide (Khay et al. 2011; Savadogo et al. 2006). *L. acidophilus* produces acidophilin, lactocidin, and acidolin (Vilà et al. 2010). They have demonstrated an *in vitro* inhibitory activity against *Bacillus*, *Klebsiella*, *Pseudomonas*, *Proteus*, *Salmonella*, *Shigella*, *Staphylococcus*, and *Vibrio* species and enteropathogenic *E. coli* (Brown 2011; Vilà et al. 2010). Various bacteriocins produced by *L. plantarum* strains have been described such as plantaricin A, B, C, and others and glycocin F (Stepper et al. 2011; Tiwari and Srivastava 2008). Strains of *Enterococcus faecium* and *Enterococcus faecalis* are known to produce bacteriocins called enterocins, and generally belong to class II such as enterocin A, L50A, L50B and P, as well as lactococcin A from *L. lactis* ssp. *cremoris* (Batdorj et al. 2006). Moreover, the production of two synergistic enterocins and three enterocins from several *Enterococcus* spp. have been reported (Basanta et al. 2008; Casaus et al. 1997; Ghrairi et al. 2007).

Tiwari and Srivastava (2008) found that bacteriocin from a rhizospheric isolate of LAB identified as *L. plantarum* strain LR/14, inhibited not only related strains but also other Gram-positive and Gram-negative bacteria such as *S. aureus*, *L. monocytogenes*, and urogenic *E. coli*. However, the inhibitory effect of nisin, bacteriocin produced by *L. paracasei*, thermophilin produced by *Streptococcus thermophillus*, and some enterocins on Gram-negative bacteria, through their synergism with other antimicrobials, have gained increased interest (Kabuki et al. 2007; Khay et al. 2011; Li et al. 2005; Tolinački et al. 2010). Ennahar and coworkers (1998) found that enterocin 81 produced by *E. faecium*, and isolated from cheese, exhibited a narrow spectrum against *Listeria* spp. including *L. monocytogenes*. Batdorj and coworkers (2006) characterized two bacteriocins, enterocin A5-11A and enterocin A5-11B, produced by *Enterococcus durans* isolated from Mongolian airag, a traditional fermented mare's milk, which exhibited an antibacterial effect on a broad spectrum of bacterial species. Khay and coworkers (2011) studied the inhibitory activity of bacteriocin-like compounds produced by LAB such as *L. lactis*, *Lactococcus cremoris*, *E. faecium*, and *E. durans* isolated from Moroccan dromedary milk and they found that these isolates exhibited a strong inhibitory activity toward pathogenic bacteria and other closely related species. Wouters and coworkers (2002) investigated the production of bacteriocins and bacteriocin-like compounds by the wild lactococci in raw milk. They found nisin, diplococcin, lactococcin, and some unidentified bacteriocin-like compounds.

Lactobacillus gasseri, *L. rhamnosus*, and *E. faecium* are considered among the potential probiotic bacteria that have been isolated from human milk (Martín et al. 2004). Nisin-producing *L. lactis* ssp. *lactis* have been isolated from human milk during the early lactation period, and its antagonistic activity was determined against *Micrococcus luteus* (Beasley and Saris 2004).

The antagonistic activity of *Lactobacillus* and *Bifidobacterium* strains against various entero- and urinary pathogens have been determined. Lactobacilli strains *L. paracasei* 8700:2, *L. plantarum* 299v, and *L. fermentum* ME-3 were the most effective against *Salmonella enterica* spp. (Hütt et al. 2006).

Todorov and coworkers (2011) studied bacteriocins that were produced by *Lactobacillus curvatus* ET06, ET30, and ET31; *Lactobacillus delbrueckii* ET32; *L. fermentum* ET35; *Pediococcus acidilactici* ET34; and *E. faecium* ET05, ET12, and ET88 isolated from smoked salmon. Testing their antagonistic activity against several strains of *L. monocytogenes* of food origin and belonging to different serological groups, showed the ability of bacteriocins to suppress the growth of all tested *L. monocytogenes*. On the other hand, *L. salivarius* UCC 118 of human origin produces a potent broad spectrum class II bacteriocin, Abp 118, which is active against *L. monocytogenes* (Dunne et al. 2001). Corr and coworkers (2007) proved the ability of *L. salivarius* UCC 118 bacteriocin to protect mice against infection with *L. monocytogenes*. Mutants of *L. salivarius* UCC 118 that are unable to produce the Abp118 bacteriocin failed to protect mice against infection with two strains of *L. monocytogenes* (Corr et al. 2007; Sherman et al. 2009). Kaktcham and coworkers (2012) reported that the inhibitory spectrum of the bacteriocins produced by *Lactobacillus* species that have been isolated from Cameroonian traditional fermented foods such as *L. rhamnosus*, *L. fermentum*, *L. plantarum*, and *Lactobacillus coprophilus* was impressive and included both Gram-positive and Gram-negative bacteria, several of which are classified by the World Health Organization as especially dangerous such as *S. enterica* subsp. *enterica*, *B. cereus*, *Streptococcus mutans*, and *Pseudomonas aeruginosa* as well as multidrug-resistant (MDR) strains of *S. aureus* and *E. coli*. Recently, Kazemipoor and coworkers (2012) have isolated LAB from home-made fermented vegetables and their isolates showed remarkable inhibitory effect against some Gram-positive and Gram-negative pathogenic bacteria.

10.3.4 *Antibiotic susceptibility*

Bosch and coworkers (2011) studied two *L. plantarum* strains that have been isolated from the feces of healthy children to test their response to different types of antibiotics such as ampicillin, tetracycline, chloramphenicol, clindamycin, erythromycin, and gentamycin. They found that both strains were susceptible to all the antibiotics tested. Similarly, *L. plantarum* strains isolated from fermented olives were susceptible to most antibiotics tested including ciprofloxacin, rifampicin, clindamycin, vancomycin, chloramphenicol, ampicillin, and penicillin and showed resistance to others, for example, *L. plantarum* strains OL2, OL7, OL9, OL12, and OL15 showed resistance to kanamycin; OL12 and OL40 showed resistance to cefoxitine; and OL12, OL15, and OL40 showed resistance to oxacillin (Mourad and Nour-Eddine 2006b). Ortu and coworkers (2007) reported that all *Lactobacillus* strains that related to species of *L. paracasei*, *L. plantarum*, and *L. reuteri* and which were isolated from milk and Gioddu, traditional Sardinian fermented milk, were resistant to vancomycin. This resistance to vancomycin is commonly found in the genus *Lactobacillus* (Klayraung et al. 2008).

L. plantarum strains that were isolated from fermented vegetables showed variable behavior toward penicillin and tetracycline (Karasu et al. 2010). Variations in resistance of *Lactobacillus* to both of these antibiotics were also reported (Temmerman et al. 2003). Klayraung and coworkers (2008) isolated three strains of *L. fermentum* from Thai traditional food and found that they were sensitive to chloramphenicol, quinipristin, erythromycin, kanamycin, linezolid, rifampicin, streptomycin, and tetracycline but resistant to ciprofloxacin and vancomycin. Todorov and coworkers (2011) found an array of antibiotics that inhibited the growth of the probiotic strains of *L. curvatus*, *L. fermentum*, *L. delbrueckii*, and *E. faecium* with varying degrees. In this same study, amikacin, metronidazole, oxacillin, vancomycin, and nalidixic acid showed prominent activity against all probiotic bacteria tested. Lavanya and coworkers (2011) tested the response of different isolates of LAB originating from fermented milk to eight different types of antibiotics including penicillin, ampicillin, vancomycin, rifampicin, trimethoprim, kanamycin, streptomycin, and bacitracin and they found that almost all the strains tested were resistant to penicillin and only 10% were susceptible to ampicillin (β-lactam antibiotics) whereas Zhou and coworkers (2000) reported that some *Lactobacillus* and *Bifidobacterium* strains were susceptible to β-lactam antibiotics. Other studies showed that 20% of the tested strains were resistant to kanamycin and streptomycin, 70% were resistant to rifampicin, 20% were resistant to trimethoprim, and only 6% were resistant to bacitracin (Lavanya et al. 2011). Nawaz and coworkers (2011) isolated *L. fermentum* species and *L. rhamnosus* from breast-fed healthy babies in Pakistan. They were susceptible to penicillin, erythromycin, clindamycin, gentamycin, tetracycline, chloramphenicol, and streptomycin except *L. fermentum* NWS09 and NWS14, which were resistant to erythromycin and tetracycline, respectively.

10.3.5 *Adherence to intestinal epithelial cells*

The ileum is the preferential site of colonization for intestinal lactobacilli, although they have been isolated from all portions of the human GIT (Charteris et al. 1998). To prevent the intestinal epithelium from pathogenic bacterial invasion by *Lactobacillus* strains, the latter should resist biological barriers such as gastric acid and bile salt, and then they should adhere to the host intestinal epithelium (Tsai et al. 2005). This may confer a competitive advantage and is important for bacterial maintenance in the human GIT (Chiu et al. 2008; Tsai et al. 2005). This property provides an interaction with mucosal surfaces, facilitating contact with gut-associated lymphoid tissue, mediating local and systemic immune effects, and allowing the competitive exclusion of pathogenic bacteria (Bosch et al. 2011; Chiu et al. 2008). Lactobacilli have been frequently observed to bind to epithelial cells and dissected tissue samples of the alimentary canal from humans and animals, to human vaginal epithelial cells, to intestinal mucus, to cultured human carcinomal intestinal cell lines, and to the components of the extracellular matrix (Izquierdo et al. 2008; Osset et al. 2001; Styriak et al. 2003; Todoroki et al. 2001; Yuki et al. 2000).

The role of exopolysaccharides and the involvement of lipoteichoic acids in the adherence of lactobacilli to intestinal and genital epithelia have been reported (Boris et al. 1998; Granato et al. 1999; Izquierdo et al. 2008). Furthermore, the reduced adhesiveness of lactobacilli treated with proteinases has led to the hypothesis that proteinaceous molecules mediate the adhesion of lactobacilli in the host intestine (Greene and Klaenhammer 1994; Izquierdo et al. 2008).

Few adhesions of lactobacilli have been identified and characterized at the molecular level. These include the collagen binding CnBP of *L. reuteri*, the collagen and laminin binding CbsA of *L. crispatus*, and fibronectin binding SlpA of *L. brevis* (Roos et al. 1996; Antikainen et al. 2002; Hynönen et al. 2002). Furthermore, a 29-kDa surface protein of *L. fermentum* that binds to pig small intestinal mucus and gastric mucins has been identified (Rojas et al. 2002). In preliminary studies, *L. acidophilus* expressed a 15-kDa protein that binds to fibronectin and 45 and 58 kDa proteins that bind to collagen type I (Lorca et al. 2002).

In vitro data have demonstrated that *B. longum* possessed high adherent activity to Caco-2 cell monolayers and suppressed the adherence of *P. aeruginosa* to the monolayers (Matsumoto et al. 2008). Izquierdo and coworkers (2008) showed the highest adhesion of *B. longum* human isolate to mucin compared with the commercial probiotic strains of *B. longum*. This adhesion was reduced by lysozyme, indicating the involvement of bacterial cell wall components in the adhesion process. Adhesion ability of different *Lactobacillus* strains to Caco-2 cells and MIM/PPK is variable depending on the strain and the cell line tested (Ortu et al. 2007). *L. plantarum*, which was isolated from pickled vegetables, was able to adhere to the mouse intestinal epithelial cells, whereas *Pediococcus pentosaceus* could not (Chiu et al. 2008). Bosch and coworkers (2011) revealed the high adhesion capacity of two *L. plantarum* strains isolated from the feces of healthy children to those of *L. rhamnosus* and *L. reuteri* commercial strains.

Tsai and coworkers (2005) found that *Lactobacillus* strains from animal origins were able to adhere strongly to human Int-407 and Caco-2 epithelial cells in addition to poultry and swine intestinal epithelium and to the columnar epithelial cells isolated from the BALB/c mouse intestine.

Ayeni and coworkers (2011) evaluated the adhesion of two *Weissella confusa* and two *L. paracasei* isolates to the epithelial intestinal cell lines Caco-2, HT-29, and HT-29-MTX, as well as to the epithelial vaginal cell line HeLa. Adhesion of the four selected isolates to Caco-2 and HT-29 cell monolayers was poor when compared with the positive control *L. rhamnosus* GG. Adhesion percentages were better when HT-29-MTX was used as the matrix. *L. paracasei* showed similar or better adhesion than the positive control to HeLa cell monolayers, whereas the adhesion of the *Weissella* to this cell line was lower. Additionally, the ability of the four isolates to competitively inhibit the adhesion of *E. coli* to HT-29 or HeLa cell monolayers indicated that *L. paracasei* favored the adhesion of *E. coli* to the vaginal HeLa cell line, whereas the other three strains increased the *E. coli* adherence to the intestinal cell line HT-29 (Ayeni et al. 2011).

Maccaferri and coworkers (2012) demonstrated the highly adhesive ability of the probiotic yeast *K. marxianus* B0399 to Caco-2 cells and its modulation of the immune response. Tahamtan and coworkers (2011) studied the effectiveness of probiotic strains *B. lactis*, *B. animalis*, *B. longum*, and *B. bifidum* against the cytopathic effect of *E. coli* O157:H7 infection using a Vero cell line model. The bifidobacterial strains significantly reduced the adhesion of *E. coli* O157:H7 to the Vero cells and they inhibited the Shiga-like toxins (Stx-1 and Stx-2) activity produced by *E. coli* O157:H7 through neutralization of their cytotoxicity.

10.4 Mechanisms of action of probiotics

Probiotic bacterial strains have different mechanisms of action including enhancement of epithelial barrier integrity by increasing the secretion of mucus and by triggering inflammation in enterocytes of the intestines (Boirivant and Strober 2007; Boyle et al. 2006; Brown 2011). Chichlowski and coworkers (2007) found an increase in the number of goblet cells on chicken intestinal villi as a response to probiotics *L. casei*, *L. acidophilus*, *Bifidobacterium thermophilum*, and *E. faecium* mixture treatment, and suggested that metabolites produced during bacterial fermentation may play a role in the growth and maturation of goblet cells. Ohland and MacNaughton (2010) reported that several *Lactobacillus* species increased mucin expression in the human intestinal cell lines Caco-2 (MUC2) and HT29 (MUC2 and MUC3), thus blocking pathogenic *E. coli* invasion and adherence.

Probiotics also have the ability to enhance the intestinal barrier function through modulation of the epithelial tight junction and cytoskeleton architecture in the intestinal mucosa (Ng et al. 2009). Resta-Lenert and Barrett (2003) found that probiotic bacteria *Streptococcus thermophilus* and *L. acidophilus* enhanced the activation of tight junction proteins, therefore avoiding the development of a leaky intestine in HT29/cl.19A and Caco-2 cell lines exposed to enteroinvasive *E. coli* (EIEC 029:NM). Furthermore, live *S. thermophilus* and *L. acidophilus* alone increased transepithelial resistance accompanied by maintenance of cytoskeletal structures such as actin, zonula occludens-1, actinin, and occludin (Resta-Lenert and Barrett 2003; Ng et al. 2009). Mennigen and Bruewer (2009) reported that *L. acidophilus* increased the expression of occludin in the gut mucosa of animals with cecal ligation and perforation, leading to a reduced bacterial translocation. Similarly, Mennigen and coworkers (2009) found that VSL#3, a probiotic cocktail containing eight different bacterial strains protected the epithelial barrier by maintaining tight junction proteins, such as occludin, zonula occludens-1, and claudin-1, claudin-3, claudin-4, and claudin-5 expression and preventing epithelial apoptosis in a murine model of colitis.

Mechanisms of action also include probiotic competition with potential pathogens for nutrients and energy sources, thus preventing such pathogens from growth and proliferation in the gut. Probiotics compete with pathogens and toxins for adherence to the intestinal epithelium (Bosch et al. 2011; Brown 2011; Corcionivoschi et al. 2010).

The production of antimicrobial substances by probiotics have been reported to inhibit the growth of pathogenic bacteria such as antimicrobial effect of *Lactobacillus* species on

Helicobacter pylori infection of the gastric mucosa (Bosch et al. 2011; Boirivant and Strober 2007). In the colon, probiotics are capable of producing short-chain fatty acids, such as acetic acid, propionic acid, and butyric acid, which lower the colonic pH, preventing overgrowth of pathogenic organisms (Vandenbergh 1993; Goel et al. 2006; Chichlowski et al. 2007).

Autoinducers are chemical signaling molecules that allow bacteria to communicate with each other and with the surrounding environment, a phenomenon called quorum sensing (Vilà et al. 2010; Sherman et al. 2009). Probiotics are thought to be able to degrade the autoinducers of pathogenic bacteria through enzymatic secretion or production of autoinducer antagonists (Brown 2011). *L. acidophilus* secretes factors that affect the quorum sensing signaling of *E. coli* O157:H7 by reducing the secretion of autoinducer 2 molecules (Medellina-Pena et al. 2007; Sherman et al. 2009). Villamil and coworkers (2012) observed that viable *L. acidophilus* and its extracellular products were able to inhibit violacein production, which is mediated by quorum sensing activity in *Chromobacterium violaceum*. Medina-Martínez and coworkers (2007) described the degradation of *N*-acyl-L-homoserine lactones (AHL) autoinducers by probiotic bacteria *B. cereus*.

Furthermore, probiotics have the ability to modulate the immune system against pathogens via antibody production and activation of lymphocytes (Ng et al. 2009). Roos and coworkers (2010) reported improved humoral immune response against *E. coli* and bovine herpes vaccines in lambs through feeding with probiotic *B. cereus* and *S. boulardii* strains.

Corcionivoschi and coworkers (2010) found that administrating *L. rhamnosus* HN001 and *B. lactis* HN019 stimulates the activity of cytotoxic lymphocytes that had a direct role in preventing the development of malignant tumors. Talib and Mahasneh (2012) reported that a combination of *B. longum* and some plant extracts were able to more efficiently target solid mammary gland tumors in mice.

Boirivant and Strober (2007) reported that *Lactobacillus* and *Bifidobacterium* probiotics induced macrophages to express more nitric oxide and inflammatory cytokines to increase antiviral activity. *In vitro* preincubation of epithelial cell line HT-29 with *L. salivarius* and *B. infantis* was able to down-regulate the proinflammatory response induced in epithelial cells by *S. typhimurium* (O'Hara et al. 2006).

10.5 Oriental sources of probiotic strains

With the increased data about benefits of probiotics for human health and other applications and because most isolated and patented strains were mostly of western origin, it is greatly inviting to try and isolate such probiotics from the untapped sources exemplified by a wide variety of different fermented foods from the orient, hoping to find new and novel potential probiotic microorganisms (Figueroa-González et al. 2011; Mahasneh and Abbas 2010). For example, keshik, which is a Jordanian traditional fermented food made-up of parboiled dried wheat and buttermilk, is of interest (Tamime and O'Connor 1995). The product is similar to tarhana, Turkish traditional fermented food, which proved to be a rich source of probiotic LAB (Erdoğrul and Erbilir 2006; Sengun et al. 2009).

Among other potential foods is the fermented eggplant, locally called makdoos, which is made up of baby aubergines stuffed with ground walnut, garlic, parsley, and fermented in olive oil. Jameed, which is solar-dried curd of sheep or goat, naturally fermented milk prepared and used traditionally by Jordanian Beduins, the old desert dwellers, is another unique source of probiotic bacteria (Mahasneh and Abbas 2010).

Fermented vegetables such as olives, kimchi in Korea, and khalpi in India are main sources of probiotics in addition to fermented cucumber, carrot, green beans, and cabbage (Karasu et al. 2010; Tamang 2001; Chang et al. 2010; Mourad and Nour-Eddine 2006a; Kazemipoor et al. 2012).

In Thailand there are famous traditional fermented products that are highly rich in probiotic bacteria as Kung-Som (traditional fermented shrimp), nham (traditional salt-fermented ground meat with garlic, pepper, salt, and cooked rice) and miang (fermented tea leaves; Hwanhlem et al. 2010; Klayraung et al. 2008). In the western highlands region of Cameroon, Kossam (fermented cow milk) and Sha'a (a maize-based beverage) are two traditional fermented foods that are natural reservoirs of LAB with high antimicrobial activity (Kaktcham et al. 2012). Reddy and coworkers (2007) studied the ability to isolate a potent probiotic LAB from kanjika, an Indian probiotic product, which is a lactic acid-fermented rice product and is prepared from raw material of plant origin and is devoid of dairy products.

Probiotic potential LAB were isolated from different Mongolian dairy products such as airag, a fermented mare's milk beverage; tarag, which is made from cow, yak, goat, or camel milk; and cow milk aaruul (Batdorj et al. 2006; Watanabe et al. 2008; Takeda et al. 2011).

LAB have been detected in various fermented products of camel's milk. These products vary according to the method of processing (Hassan et al. 2008). Shubat is camel's sour milk from Kazakhstan and China (Serikbayeva et al. 2005; Rahman et al. 2009). Gariss is a traditional Sudanese fermented camel milk that is widely consumed by the pastoralist communities living in arid and semiarid regions of Sudan (Abdelgadir et al. 2008; Ashmaig et al. 2009). Lehban is the fermented product from camel's milk in Syria and Egypt, whereas kefir is Caucasian fermented camel's milk (Hassan et al. 2008). Other studies on shmen or semma, a traditional butter made from camel's milk in the Sahara, Algeria, indicated the presence of different LAB genera (Mourad and Nour-Eddine 2006a,b). LAB strains have been found in suusac, a Kenyan traditional fermented camel milk product (Lore et al. 2005). Yaqoob and Nawaz (2007) reported that the most common products made from camel milk are dahi (yoghurt), lassi (sour milk), and kurth (cheese) in north-eastern Baluchistan. All these products were main sources of LAB (Harun-ur-Rashid et al. 2007; Pawan and Bhatia 2007).

Some unique probiotic LAB strains *L. plantarum*, *L. pentosus*, and *L. lactis* ssp. *lactis* have been isolated from unpasteurized natural camel's milk with superior probiotic characteristics (Yateem et al. 2008). Similarly, LAB have been isolated from camel's raw milk in Morocco with the predominance of *Lactobacillus* and *Lactococcus* genera (Khedid et al. 2009). Benkerroum and coworkers (2003) also isolated different genera of LAB including *Pediococcus*, *Streptococcus*, *Lactococcus*, and *Leuconostoc* from raw camel's milk.

10.6 Specific uses of probiotics

Probiotics have been used for the prevention and treatment of a diverse range of disorders. Diarrhea cases are of the most common adverse events associated with bacterial infections and misuse of antibiotics (Saraf et al. 2010). McFarland and coworkers (1994) found that oral administration of the yeast *S. boulardii* significantly reduced the occurrence of diarrhea after administration of antibiotics. Other probiotics like *L. rhamnosus* GG, *L. acidophilus*, and *E. faecium* also seemed to be effective in preventing antibiotic-associated diarrhea (Saraf et al. 2010; Quigley 2011; Ouwehand et al. 2002).

Viral diarrhea, such as that induced by rotavirus, has been treated successfully by probiotics such as *L. casei* Shirota and *L. rhamnosus* GG (Quigley 2011; Ouwehand et al. 2002). A recent study demonstrated the effectiveness of *L. reuteri* in the treatment of rotaviral infection in a neonatal mouse gastroenteritis model (Preidis et al. 2012). *L. rhamnosus* strains 573L/1, 573L/2, and 573L/3 have reduced the duration of rotavirus diarrhea in children 2 months to 6 years of age (Szymański et al. 2006).

Traveler's diarrhea, which is a common health problem among travelers (Gupta and Garg 2009) and is caused by the ingestion of contaminated foods and water with pathogens such as *E. coli*, *Salmonella* species, and some protozoa like *Entamoeba histolytica* and *Giardia lambalia* (Lannitti and Palmieri 2010). Goel and coworkers (2006) reported that *S. boulardii*, *E. faecium*, and *L. rhamnosus* probiotics for the treatment of traveler's diarrhea were limited, whereas McFarland (2007) reported that *S. boulardii* and a mixture of *L. acidophilus and B. bifidum* had significant efficacy. Some studies revealed fewer or shorter episodes of traveler's diarrhea in subjects consuming the probiotic whereas others found no such effect (de Vrese and Marteau 2007).

Among other gastrointestinal disorders that have been treated with probiotics are inflammatory bowel diseases (IBD). Yan and coworkers (2011) described how a certain protein derived from *L. rhamnosus* GG reduced epithelial apoptosis and suppressed inflammation in mouse models of IBD.

VSL#3 probiotic cocktail was effective in the induction of remission among patients with ulcerative colitis, which is another form of IBD other than Crohn's disease (Bibiloni et al. 2005; Quigley 2011). This same cocktail alleviated symptoms in individuals with irritable bowel syndrome (IBS) such as decreasing flatulence, pain, and bloating (Quigley 2011; Yan and Polk 2011). Clarke and coworkers (2012) reported that *L. rhamnosus* GG variably reduced the frequency and severity of pain in children.

Saulnier and coworkers (2013) reviewed and discussed the role of gut microbiota and the implications for probiotics and prebiotics on epithelial cell function, gastrointestinal mobility, visceral sensitivity, perception, and behavior as well as their interaction with the central nervous system. Their remarks suggested the potential role of gut microbiota communities on brain functions such as psychological disorders, particularly in the field of psychological comorbidities associated with bowel disorders.

The gastrointestinal tract is a site of early HIV replication and CD4+ cell destruction, and HIV-associated enteropathy emerges as a critical outcome of the transition from HIV infection to AIDS (Yan and Polk 2011). *L. reuteri* and *L. rhamnosus* were found to resolve diarrhea and improve CD4+ cell number in patients infected with HIV (Anukam et al. 2008). Narwal (2011) reported that some *Lactobacillus* strains that were taken from the saliva of volunteers had produced proteins capable of binding mannose, a particular type of sugar found on the HIV envelope. This binding enables the *Lactobacillus* to stick to the mouth and digestive tract mucosal lining, thus initiating colonization. One strain secreted abundant mannose-binding protein particles into its surroundings, neutralizing HIV by binding to its sugar coating. They also observed that immune cells trapped by lactobacilli formed a clamp. This configuration would immobilize any immune cells harboring HIV and prevent them from infecting other cells.

H. pylori is a major cause of chronic gastritis and peptic ulcer and a risk factor for gastric malignancies (Gupta and Garg 2009; Medeiros and Pereira 2013). Cruchet and coworkers (2003) have shown that *L. johnsonii* La1 may interfere with *H. pylori* colonization in asymptomatic children and may be an effective alternative to modulate *H. pylori* infection. Furthermore, Hütt and coworkers (2006) found that both *L. paracasei* 8700:2

and *L. plantarum* 299v were antagonistic against *H. pylori,* which colonize the stomach mucosa.

Probiotics have been investigated for their role in the prevention of colon cancer (Saraf et al. 2010). Diets, especially those high in meat and fat or low in fiber, have a role in changing the composition of the intestinal microflora, which increases the fecal enzyme activity that converts procarcinogens into carcinogens (Ouwehand et al. 2002). Suvarna and Boby (2005) reported that *L. acidophilus* obviously reduced both the number and size of colon tumors induced by a carcinogen, 1,2-dimethyl hydrazine (DMH). Dietary administration of *B. longum* in laboratory animals significantly inhibited the incidences of colon adenocarcinomas and colon tumor multiplicity. Liong (2008) reported the effect of fermented milks on colon cancer cell growth. Milks were fermented with individual strains of *L. helveticus, L. acidophilus, Bifidobacterium,* or a mix of *S. thermophilus* and *L. delbrueckii* ssp. *bulgaricus* and when HT-29 cells were subsequently added into the fermented milk, 10% to 50% of these cells showed a decrease in growth. Goel and coworkers (2006) observed that consumption of *L. casei* would delay the recurrence of bladder tumors.

In dental practice, probiotics significantly lowered the rates of dental caries in children who were given *L. rhamnosus* GG (Goldin and Gorbach 2008). Saraf and coworkers (2010) reported the inhibitory effect of *L. rhamnosus* GG and *Bifidobacterium* isolates on salivary counts of *S. mutans* and yeast. Other studies showed the ability of *Bifidobacterium* to reduce gingival and periodontal inflammation (Reddy et al. 2010). Krasse and coworkers (2006) observed that *L. reuteri* was effective in reducing both gingivitis and plaque in patients with moderate to severe gingivitis.

Probiotic therapy has also been explored in nongastrointestinal disorders. There is ample evidence suggesting the use of probiotics to treat and prevent urinary tract infection (Velraeds et al. 1998). Reid and coworkers (2001) have shown that *L. rhamnosus* and *L. fermentum* inserted into the vagina can colonize and compete against uropathogens and reduce the risk of urinary tract infections. Petricevic and Witt (2008) found that vaginal flora can be restored after antibiotic treatment of bacterial vaginosis by exogenously applying *L. casei rhamnosus.*

Many studies have examined the efficacy of probiotics in the prevention and treatment of allergic disorders (Gill and Guarner 2004). Probiotics could have a good effect in infants at high risk of atopy and those having allergies to cow's milk (Viljanen et al. 2005), atopic eczema, and dermatitis (Sleator and Hill 2008; Prescott et al. 2005). This is linked to reports associating neonatal gut microecology with the development of atopic disease (Ouwehand et al. 2002). Therefore, it seems that intervention with probiotics before or immediately after birth is more effective in inducing immunological tolerance while the immune system is still immature, compared with older children with fully mature immune systems (Ouwehand 2007).

Moreover, LAB have antiobesity effects (Chiang and Pan 2012). Kadooka and coworkers (2010) showed that *L. gasseri* had lowering effects on abdominal adiposity, body weight, and other measures such as body mass index and waist and hip circumference. Takemura and coworkers (2010) proposed that *L. plantarum* no. 14 reduced adipocyte size in mice fed a high-fat diet.

Several studies reported the ability of probiotics to relieve constipation, prevent necrotizing enterocolitis, eliminate nasal pathogens, lower cholesterol and hypertension, and alleviate lactose intolerance. Furthermore, probiotics have been proposed to relieve the symptoms of anxiety (Azizpour et al. 2009; Culligan et al. 2009; Popova et al. 2012; Goel et al. 2006; Suvarna and Boby 2005; Gupta and Garg 2009).

10.7 *Safety of probiotics*

Many different species and strains and preparations have been used as probiotics for the last decade with growing interest for the development of new and more active strains, and the question for their safety has increased with time (Quigley 2011; Marteau 2001). The criteria in place for the use of microorganisms as probiotics assures their safe use. However, more investigations are needed on how safe probiotics are for certain patients such as severely immunocompromised people, neonates, the elderly, and hospitalized patients, although ample studies support the safety of particular probiotic strains in particular high-risk populations (Sanders et al. 2010; Boyle et al. 2006).

Probiotics side effects, if they exist, would probably be mild digestive disturbances including gas and bloating (Mahasneh and Abbas 2010). Theoretically, probiotics may be responsible for more sophisticated and serious side effects (Marteau 2001). *Lactobacillus* species are a rare if any well-recognized cause of local or systemic infections including septicemia and endocarditis (Boyle et al. 2006). Many cases of fungemia have been reported in humans treated with the probiotic *S. boulardii* (Hannequin et al. 2000). This probably means that the risk of infection is not nil but is extremely low (Marteau 2001).

Among other probable side effects are the unhealthy metabolic activities of probiotics such as intestinal mucus layer degradation in which the accumulation of probiotics along the GIT might cause gastrointestinal disturbances including intestinal inflammation (Soccol et al. 2010; Ooi and Liong 2010).

Production of bile hydrolase and accumulation of deconjugated bile could be transformed into detrimental cytotoxic secondary bile acids, which could increase the risk of gastrointestinal diseases (Ooi and Liong 2010). Probiotics may also cause too much stimulation of the immune system (Boyle et al. 2006).

Another possible risk of using probiotics is transferring antibiotic resistance genes between microorganisms especially for more pathogenic bacteria in the intestinal microbiota (Lannitti and Palmieri 2010; Boyle et al. 2006). For example, *L. reuteri* and *L. plantarum* have been found to carry antibiotic resistance genes (Yan and Polk 2011). On the other hand, vancomycin-resistant genes of many *Lactobacillus* strains are chromosomally encoded and not inducible or transferable, which would increase the safety level of probiotic use (Marteau 2001; Boyle et al. 2006).

10.8 *Case study: Characterization of probiotic lactobacilli isolates from fermented camel milk*

In the quest for the isolation of probiotic lactobacilli from exotic sources, spontaneously fermented camel milk was used as one of the least studied oriental foods. The ongoing project uses well-documented materials and methods. Prospective probiotic candidates were isolated, identified, and tested to find out if they meet the criteria for probiotic use. Figure 10.1 shows the guiding protocol to isolate such probiotics as presented by Fontana and coworkers (2013), which could be modified to enhance any stage. In this section, our findings about antibacterial activity and the cholesterol-reducing effects of probiotic lactobacilli isolates from fermented camel milk will be described.

10.8.1 *Detection of antibacterial activity of the bacterial isolates*

For the detection of antagonistic activities of the isolates, an agar spot procedure was used. The antibacterial activity of the selected *Lactobacillus* isolates was determined by the agar

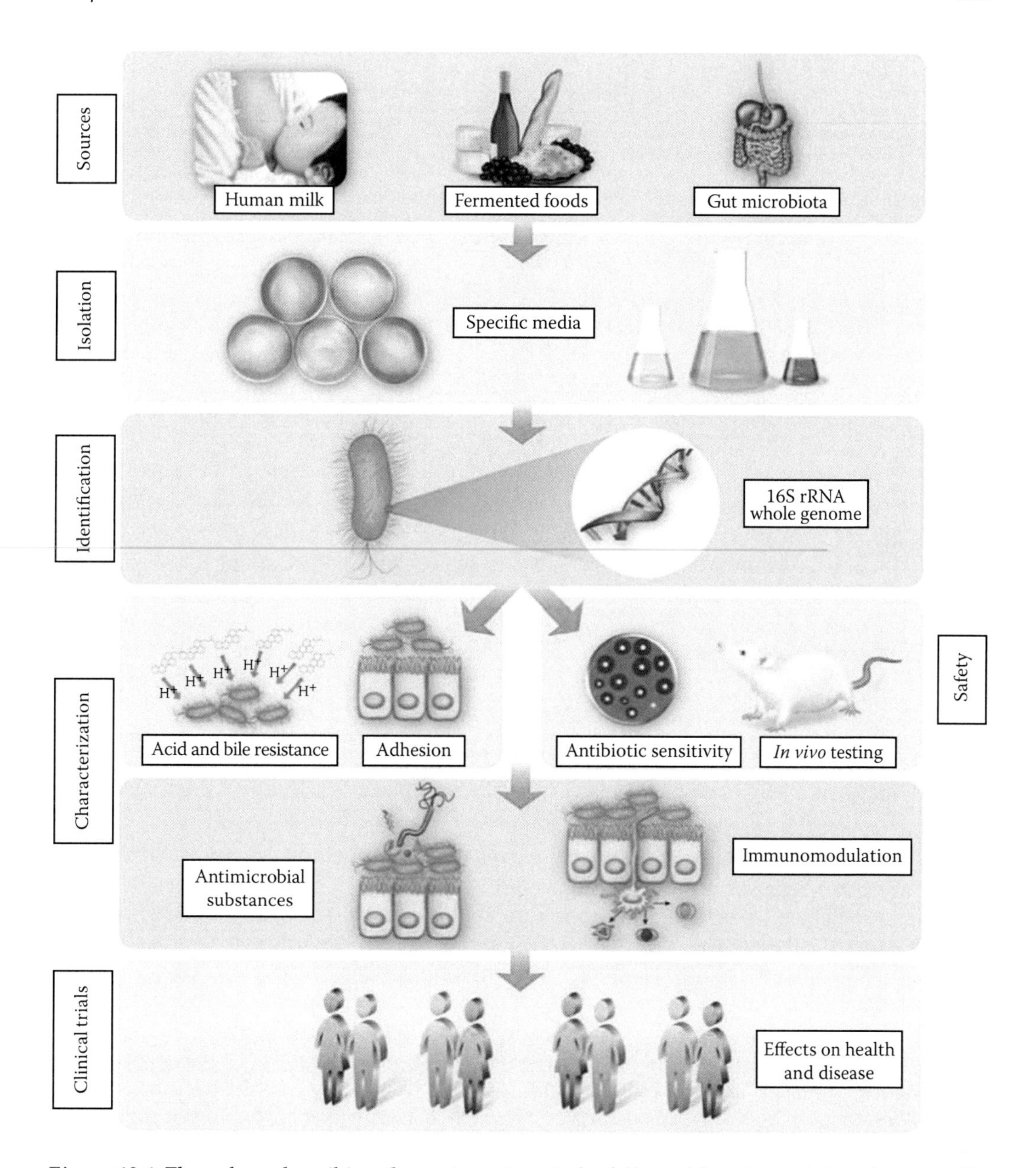

Figure 10.1 Flow chart describing the various steps to be followed for a bacterial strain to qualify as a novel probiotic. rRNA, ribosomal RNA. (From Fontana L. et al., *British Journal of Nutrition* 109 (2013): S35–S50.)

spot test described by Schillinger and Lücke (1989) with some modifications as follows: 5 mL of each overnight culture of *Lactobacillus* isolate were spotted onto the surface of MRS agar plates containing 0.2% glucose and then incubated under anaerobic conditions at 37°C for 48 h. An overnight culture of four indicator strains (*E. coli* ATCC 25922, *S. typhimurium* ATCC 14028, *B. cereus* toxigenic strain TS, and methicillin-resistant *S. aureus* [MRSA] clinical isolate) were grown in nutrient broth and were adjusted to 0.5 McFarland and then were diluted 1:10 using nutrient broth to reach 10^7 CFU/mL. Aliquots of 0.25 mL were

Table 10.1 Antibacterial Activity (Inhibition Zones Diameter, in Millimeters) of the Selected *Lactobacillus* Species

	B. cereus	MRSA	*E. coli*	*S. typhimurium*
M2	28.5 ± 0.7	54.5 ± 2.1	28.5 ± 0.7	32.5 ± 0.7
M5	36.0 ± 1.4	36.5 ± 0.7	34.0 ± 4.2	30.0 ± 2.8
M10	54.0 ± 1.4	44.0 ± 1.4	39.0 ± 1.4	28.5 ± 3.5
M15	42.5 ± 4.9	48.5 ± 3.5	36.5 ± 0.7	32.0 ± 2.8
M29	41.0 ± 0.0	42.0 ± 1.4	36.5 ± 2.1	29.5 ± 6.3

Note: Low activity, <32 mm; moderate activity, 33–39 mm; and significant activity, >39 mm. Inhibition zone (mm) of indicator strains, mean ± S.D, $n = 2$.

inoculated into 7 mL of soft MRS containing 0.2% glucose and 0.7% agar. Inoculated soft MRS agar was immediately poured in duplicate over the MRS plate on which the tested *Lactobacillus* isolate was grown. The plates were incubated aerobically at 37°C for 24 h.

The antibacterial activity was detected by measuring the inhibition zones around the *Lactobacillus* bacterial spots. Inhibition was recorded as positive if the diameter of the zone around the colonies of the producer was 2 mm or larger (Mami et al. 2008).

Microbial growth inhibition by different fermented camel milk *Lactobacillus* isolates: M2, *L. fermentum;* M5, *L. brevis;* M10, *L. plantarum;* M15, *L. paracasei* ssp. *paracasei;* and M29, *L. rhamnosus,* would produce inhibition zones. The inhibitory activity of the five selected isolates varied depending on the *Lactobacillus* species tested and type of host microorganisms, which included both Gram-positive and Gram-negative bacteria.

It can be seen from Table 10.1 that *Lactobacillus* isolates showed varying degrees of activity against four human pathogenic challenge indicators: two Gram-positive bacteria *B. cereus* toxigenic strain, TS and MRSA and two Gram-negative bacteria *E. coli* ATCC 25922 and *Salmonella typhimurium* ATCC 14028.

The highest activity was recorded for the isolate M2 (*L. fermentum*) against MRSA with an inhibition zone of 54.5 mm followed by M10 (*L. plantarum*) against *B. cereus* with an inhibition zone of 54.0 mm. The least inhibition zones were observed for isolate M2 (*L. fermentum*) against *B. cereus* and *E. coli* with inhibition zones of 28.5 mm as well as by M10 (*L. plantarum*) against *S. typhimurium* with an inhibition zone of 28.5 mm.

10.8.2 Cholesterol-lowering effect

To screen for *Lactobacillus* with cholesterol-lowering effects, one fresh colony from each *Lactobacillus* isolate was inoculated into 5 mL MRS broths separately and incubated anaerobically for 24 h at 37°C. Then, they were inoculated in 1% MRS-THIO broth with 0.1 g/L filter-sterilized water-soluble cholesterol (polyoxyethanyl cholesteryl sebacate; Sigma-Aldrich) under anaerobic conditions at 37°C for 24 h. After the incubation period, cells were centrifuged at 12,000 × *g* at 4°C for 10 min, and the remaining cholesterol concentration in the broth was determined by using *o*-phthalaldehyde modified colorimetric method as described by Rudel and Morris (1973). One milliliter of the supernatant (broth containing the remaining cholesterol) aliquot was added with 1 mL of KOH 50% (w/v) and 2 mL of absolute ethanol, vortexed for 1 min, followed by heating at 37°C for 15 min. After cooling, 2 mL of distilled water and 5 mL of hexane were added and vortexed for 1 min.

The hexane layer of 2.5 mL was transferred into a glass tube. The hexane was evaporated from each tube at 60°C under the flow of nitrogen gas. The residue was immediately dissolved in 2 mL of *o*-phthalaldehyde reagent. The reagent contained 0.5 mg of

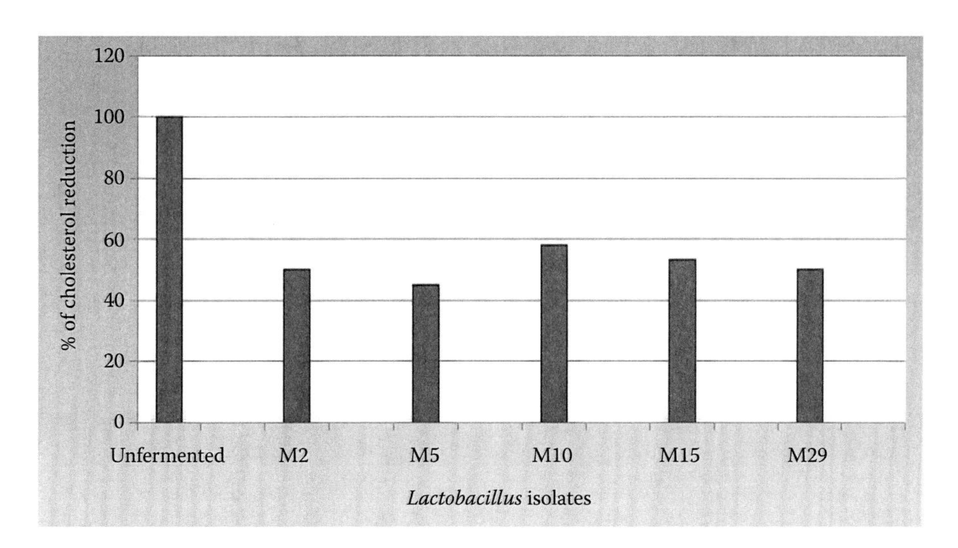

Figure 10.2 Cholesterol-reducing rate of *Lactobacillus* isolates. The first column represents the unfermented MRS-THIO broth.

o-phthalaldehyde per milliliter of glacial acetic acid. After complete mixing, the tubes were allowed to stand at room temperature for 10 min, and then 0.5 mL of concentrated sulfuric acid was added and the mixture was vortexed for 1 min. After standing at room temperature for an additional 10 min, the absorbance was read at 550 nm against a reagent blank. The removal rate of every strain was computed by the following formula:

$$\text{The cholesterol reducing rate} = [(A_0 - A)/A_0] \times 100\%$$

where A_0, absorbance of the unfermented broth; A, absorbance of the broth fermented for 24 h.

All *Lactobacillus* isolates (M2, M5, M10, M15, and M29) showed high ability of cholesterol reduction with no significant variation in cholesterol-reducing ability among the isolates. M10 (*L. plantarum*) had the highest cholesterol-reducing rate of 58.0% followed by M15 (*L. paracasei* ssp. *paracasei* 1) with a cholesterol reducing rate of 53.2%. M5 (*L. brevis*) had the lowest cholesterol-reducing rate of 45.0% (Figure 10.2).

10.9 Conclusions

Probiotics have shown their ability in promoting a wide variety of desired physiological effects for better human health. Traditional fermented foods are among the most promising sources for the isolation of unique probiotic bacteria, especially of the lactic acid group. These candidate probiotic microorganisms must meet the approved international criteria to include resistance to low pH, gastric and intestinal enzymes, and bile toxicity. The ability to adhere to epithelial cells and antimicrobial substances is also as important as the safety concerns. To substantiate the role of probiotics in disease therapy, more *in vivo* studies are needed and furthermore the need for specific designer probiotics should be thoroughly investigated and developed.

References

Abdelgadir W., Nielsen D.S., Hamad S. and Jakobsen M. "A traditional Sudanese fermented camel's milk product, Gariss, as a habitat of *Streptococcus infantarius* subsp. *infantarius*." *International Journal of Food Microbiology* 127 (2008): 215–219.

Antikainen J., Anton L., Sillanpää J. and Korhonen T.K. "Domains in the S-layer protein CbsA of *Lactobacillus crispatus* involved in adherence to collagens, laminin and lipoteichoic acids and in self-assembly." *Molecular Microbiology* 46 (2002): 381–394.

Anukam K.C., Osazuwa E.O., Osadolor H.B., Bruce A.W. and Reid G. "Yogurt containing probiotic *Lactobacillus rhamnosus* GR-1 and *L. reuteri* RC-14 helps resolve moderate diarrhea and increases CD4 count in HIV/AIDS patients." *Journal of Clinical Gastroenterology* 42 (2008): 239–243.

Ashmaig A., Hasan A. and El Gaali E. "Identification of lactic acid bacteria isolated from traditional Sudanese fermented camel's milk (*Gariss*)." *African Journal of Microbiology Research* 3 (2009): 451–457.

Ayeni F.A., Sánchez B., Adeniyi B.A., de Los Reyes-Gavilán C.G., Margolles A. and Ruas-Madiedo P. "Evaluation of the functional potential of *Weissella* and *Lactobacillus* isolates obtained from Nigerian traditional fermented foods and cow's intestine." *International Journal of Food Microbiology* 147 (2011): 97–104.

Azizpour K., Bahrambeygi S., Mahmoodpour S. and Azizpour A. "History and basic of probiotics." *Research Journal of Biological Sciences* 4 (2009): 409–426.

Basanta A., Sánchez J., Gómez-Sala B., Herranz C., Hernández P.E. and Cintas L.M. "Antimicrobial activity of *Enterococcus faecium* L50, a strain producing enterocins L50 (L50A and L50B), P and Q, against beer-spoilage lactic acid bacteria in broth, wort (hopped and unhopped), and alcoholic and non-alcoholic lager beers." *International Journal of Food Microbiology* 125 (2008): 293–307.

Batdorj B., Dalgalarrondo M., Choiset Y. et al. "Purification and characterization of two bacteriocins produced by lactic acid bacteria isolated from Mongolian airag." *Journal of Applied Microbiology* 101 (2006): 837–848.

Beasley S.S. and Saris P.E.J. "Nisin-producing *Lactococcus lactis* strains isolated from human milk." *Applied and Environmental Microbiology* 70 (2004): 5051–5053.

Begley M., Gahan C.G.M. and Hill C. "The interaction between bacteria and bile." *FEMS Microbiology Reviews* 29 (2005): 625–651.

Benkerroum N., Boughdadi A., Bennani N. and Hidane K. "Microbiological quality assessment of Moroccan camel's milk and identification of predominating lactic acid bacteria." *World Journal of Microbiology and Biotechnology* 19 (2003): 645–648.

Bernardeau M., Guéguen M. and Vernoux J.P. "Beneficial lactobacilli in food and feed: Long-term use, biodiversity and proposals for specific and realistic safety assessments." *FEMS Microbiology Review* 30 (2006): 487–513.

Bernardeau M., Vernoux J.P., Dubernet S.H. and Guguen M. "Safety assessment of dairy micro-organisms: The *Lactobacillus* genus." *International Journal of Food Microbiology* 126 (2008): 278–285.

Bhadoria P.B.S. and Mahapatra S.C. "Prospects, technological aspects and limitations of probiotics—A worldwide review." *European Journal of Food Research and Review* 1 (2011): 23–42.

Bibiloni R., Fedorak R.N., Tannock G.W. et al. "VSL#3 probiotic-mixture induces remission in patients with active ulcerative colitis." *American Journal of Gastroenterology* 100 (2005): 1539–1546.

Boirivant M. and Strober W. "The mechanism of action of probiotics." *Current Opinion of Gastroenterology* 23 (2007): 679–692.

Boris S., Suárez J.E., Vázquez F. and Barbés C. "Adherence of human vaginal lactobacilli to vaginal epithelial cells and interaction with uropathogens." *Infection and Immunity* 66 (1998): 1985–1989.

Bosch M., Rodriguez M., Garcia F., Fernández E., Fuentes M.C. and Cuñé J. "Probiotic properties of *Lactobacillus plantarum* CECT 7315 and CECT 7316 isolated from faeces of healthy children." *Letters in Applied Microbiology* 54 (2011): 240–246.

Both E., György É., Kibédi-Szabó C.Z. et al. "Acid and bile tolerance, adhesion to epithelial cells of probiotic microorganisms." *University Politehnica of Bucharest Science Bulletin* 72 (2010): 37–44.

Boyle R.J., Robins-Browne R.M. and Tang M.L.K. "Probiotic use in clinical practice: What are the risks?" *American Journal of Clinical Nutrition* 83 (2006): 1256–1264.

Brown M. "Modes of action of probiotics: Recent developments." *Journal of Animal and Veterinary Advances* 10 (2011): 1895–1900.

Cardona M.E., Vanay V.V., Midtvedt T. and Norin K.E. "Probiotics in gnotobiotic mice conversion of cholesterol to coprostanol *in vitro* and *in vivo* and bile acid deconjugation *in vitro*." *Microbial Ecology in Health and Disease* 12 (2000): 219–224.

Casaus P., Nilsen T., Cintas L.M., Nes I.F., Hernández P.E. and Holo H. "Enterocin B, a new bacteriocin from *Enterococcus faecium* T13 which can act synergistically with enterocin A." *Microbiology* 143 (1997): 2287–2294.

Chang J.-H., Shim Y.Y. and Chee K.M. "Probiotic characteristics of lactic acid bacteria isolated from kimchi." *Journal of Applied Microbiology* 109 (2010): 220–230.

Charteris W.P., Kelly P.M., Morelli L. and Collins J.K. "Development and application of an in vitro methodology to determine the transit tolerance of potentially probiotic *Lactobacillus* and *Bifidobacterium* species in the upper human gastrointestinal tract." *Journal of Applied Microbiology* 84 (1998): 759–768.

Chiang S.S. and Pan T.M. "Beneficial effects of *Lactobacillus paracasei* subsp. *paracasei* NTU 101 and its fermented products." *Journal of Applied Microbiology and Biotechnology* 93 (2012): 903–916.

Chichlowski M., Croom J., McBride B.W., Havenstein G.B. and Koci M.D. "Metabolic and physiological impact of probiotics or direct-fed-microbials on poultry: A brief review of current knowledge." *International Journal of Poultry Science* 6 (2007): 694–704.

Chiu H.-H., Tsai C.-C., Hsih H.-Y. and Tsen H.-Y. "Screening from pickled vegetables the potential probiotic strains of lactic acid bacteria able to inhibit the *Salmonella* invasion in mice." *Journal of Applied Microbiology* 104 (2008): 605–612.

Chuayana Jr. E.L., Ponce C.V., Rivera M.R.B. and Cabrera E.C. "Antimicrobial activity of probiotics from milk products." *The Philippine Journal of Microbiology and Infectious Diseases* 32 (2003): 71–74.

Clarke G., Cryan J.F., Dinan T.G. and Quigley E.M. "Probiotics for the treatment of irritable bowel syndrome." *Alimentary Pharmacology and Therapeutics* 35 (2012): 403–413.

Coeuret V., Guguen M. and Vernoux J.P. "*In vitro* screening of potential probiotic activities of selected lactobacilli isolated from unpasteurized milk products for incorporation into soft cheese." *Journal of Dairy Research* 71 (2004): 451–460.

Corcionivoschi N., Drinceanu D., Stef L., Luca I., Julean C. and Mingyart O. "Probiotics—Identification and ways of action." *Innovative Romanian Food Biotechnology* 6 (2010): 1–11.

Corr S.C., Li Y., Riedel C.U., O'Toole P.W., Hill C. and Gahan C.G.M. "Bacteriocin production as a mechanism for the antiinfective activity of *Lactobacillus salivarius* UCC118." *Proceedings of the National Academy of Sciences* 104 (2007): 7617–7621.

Crittenden R., Bird A.R., Gopal P., Henriksson A., Lee Y.K. and Playne M.J. "Probiotic research in Australia, New Zealand and the Asia-Pacific Region." *Current Pharmaceutical Design* 11 (2005): 37–53.

Cruchet S., Obregon M.C., Salazar G., Diaz E. and Gotteland M. "Effect of the ingestion of a dietary product containing *Lactobacillus johnsonii* La1 on *Helicobacter pylori* colonization in children." *Nutrition* 19 (2003): 16–21.

Culligan E.P., Hill C. and Sleator R.D. "Probiotics and gastrointestinal disease: Successes, problems and future prospects." *Gut Pathogens* 1 (2009): 19. doi:10.1186/1757-4749-1-19.

de Los Reyes-Gavilán C.G., Suárez A., Fernández-García M., Margolles A., Gueimonde M. and Ruas-Madiedo P. "Adhesion of bile-adapted *Bifidobacterium* strains to the HT29-MTX cell line is modified after sequential gastrointestinal challenge simulated *in vitro* using human gastric and duodenal juices." *Research in Microbiology* 162 (2011): 514–519.

de Vrese M. and Marteau P.R. "Probiotics and prebiotics: Effects on diarrhea." *The Journal of Nutrition* 137 (2007): 803S–811S.

Dunne C., O'Mahony L., Murphy L. et al. "*In vitro* selection criteria for probiotic bacteria of human origin: Correlation with *in vivo* findings." *American Journal of Clinical Nutrition* 73 (2001): 386S–392S.

Ennahar S., Aoude-Werner D., Assobhei O. and Hasselman C. "Antilisterial activity of enterocin 81, a bacteriocin produced by *Enterococcus faecium* WHE 81 isolated from cheese." *Journal of Applied Microbiology* 85 (1998): 521–526.

Erdoğrul Ö. and Erbilir F. "Isolation and characterization of *Lactobacillus bulgaricus* and *Lactobacillus casei* from Various Foods." *Turkish Journal of Biology* 30 (2006): 39–44.

Figueroa-González I., Cruz-Guerrero A. and Quijano G. "The benefits of probiotics on human health." *Microbial and Biochemical Technology* S1 (2011): 003. doi:10.4172/1948-5948.S1-003.
Fioramonti J., Theodorou V. and Bueno L. "Probiotics: What are they? What are their effects on gut physiology?" *Best Practice and Research Clinical Gastroenterology* 17 (2003): 711–724.
Fontana L., Bermudez-Brito M., Plaza-Diaz J., Muñoz-Quezada S. and Gil A. "Sources, isolation, characterization and evaluation of probiotics." *British Journal of Nutrition* 109 (2013): S35–S50.
Fuller R. "Probiotics in man and animals." *Journal of Applied Bacteriology* 66 (1989): 365–378.
Ghrairi T., Frere J., Berjeaud J.M. and Manai M. "Purification and characterization of bacteriocins produced by *Enterococcus faecium* from Tunisian rigouta cheese." *Food Control* 19 (2007): 162–169.
Gill H.S. and Guarner F. "Probiotics and human health: A clinical perspective." *Postgraduate Medical Journal* 80 (2004): 516–526.
Gill H. and Prasad J. "Probiotics, immunomodulation, and health benefits." *Advances in Experimental Medicine and Biology* 606 (2008): 423–454.
Gilliland S.E., Nelson C.R. and Maxwell C. "Assimilation of cholesterol by *Lactobacillus acidophilus*." *Applied and Environmental Microbiology* 49 (1985): 377–381.
Gilliland S.E. and Walker D.K. "Factors to consider when selecting a culture of *L. acidophilus* as a dietary adjunct to produce a hypercholesterolemic effect in humans." *Journal of Dairy Science* 73 (1990): 905–909.
Goel A.K., Dilbaghi N., Kamboj D.V. and Singh L. "Probiotics: Microbial therapy for health modulation." *Defense Science Journal* 56 (2006): 513–529.
Goldin B.R. and Gorbach S.L. "Clinical indications for probiotics: An overview." *Clinical Infectious Disease* 46 (2008): S96–S100.
Granato D., Perotti F., Masserey I. et al. "Cell surface-associated lipoteichoic acid acts as an adhesion factor for attachment of *Lactobacillus johnsonii* La1 to human enterocyte-Like Caco-2 Cells." *Applied Environmental Microbiology* 65 (1999): 1071–1077.
Greene J.D. and Klaenhammer T.R. "Factors involved in adherence of lactobacilli to human Caco-2 cells." *Applied Environmental Microbiology* 60 (1994): 4487–4494.
Gupta V. and Garg R. "Probiotics." *Indian Journal of Medical Microbiology* 27 (2009): 202–209.
Hannequin C., Kauffmann-Lacroix C., Jobert A. et al. "Possible role of catheters in *Saccharomyces boulardii* fungemia." *European Journal of Clinical Microbiology and Infectious Diseases* 19 (2000): 16–20.
Harun-ur-Rashid M.D., Togo Æ.K., Ueda Æ. and Miyamoto Æ. "Identification and characterization of dominant lactic acid bacteria isolated from traditional fermented milk Dahi in Bangladesh." *World Journal of Microbiology and Biotechnology* 23 (2007): 125–133.
Hassan R.A., El Zubeir I.E.M. and Babiker S.A. "Chemical and microbial measurements of fermented camel milk "Gariss" from transhumance and nomadic herds in Sudan." *Australian Journal of Basic and Applied Sciences* 2 (2008): 800–804.
Hernández D., Cardell E. and Zárate V. "Antimicrobial activity of lactic acid bacteria isolated from Tenerife cheese: Initial characterization of plantaricin TF711, a bacteriocin-like substance produced by *Lactobacillus plantarum* TF711." *Journal of Applied Microbiology* 99 (2005): 77–84.
Hütt P., Shchepetova J., Loivukene K., Kullisaar T. and Mikelsaar M. "Antagonistic activity of probiotic lactobacilli and bifidobacteria against entero- and uropathogens." *Journal of Applied Microbiology* 100 (2006): 1324–1332.
Hwanhlem N., Wattanasakphuban N., Riebroy S., Benjakul S., Kittikun A. and Maneerat S. "Probiotic lactic acid bacteria from Kung-Som: Isolation, screening, inhibition of pathogenic bacteria." *International Journal of Food Science and Technology* 45 (2010): 594–601.
Hynönen U., Westerlund-Wikström B., Palva A. and Korhonen T.K. "Identification by flagellum display of an epithelial cell- and fibronectin-binding function in the SlpA surface protein of *Lactobacillus brevis*." *Journal of Bacteriology* 184 (2002): 3360–3367.
Ishibashi N. and Yamazaki S. "Probiotics and safety." *American Journal of Clinical Nutrition* 73 (2001): 465S–470S.
Izquierdo E., Medina M., Ennahar S., Marchioni E. and Sanz Y. "Resistance to simulated gastrointestinal conditions and adhesion to mucus as probiotic criteria for *Bifidobacterium Longum* strains." *Current Microbiology* 56 (2008): 613–618.
Jiang J., Hang X., Zhang M., Liu X., Li D. and Yang H. "Diversity of bile salt hydrolase activities in different Lactobacilli toward human bile salts." *Annals of Microbiology* 60 (2010): 81–88.

Kabuki T., Uenishi H., Watanabe M., Seto Y. and Nakajima H. "Characterization of a bacteriocin, Thermophilin 1277, produced by *Streptococcus thermophilus* SBT1277." *Journal of Applied Microbiology* 102 (2007): 971–980.

Kadooka Y., Sato M. and Imaizumi K. "Regulation of abdominal adiposity by probiotics (*Lactobacillus gasseri* SBT2055) in adults with obese tendencies in a randomized controlled trial." *European Journal of Clinical Nutrition* 64 (2010): 636–643.

Kaktcham P.M., Zambou N.F., Tchouanguep F.M., Soda M. and Choudhary M.I. "Antimicrobial and safety properties of Lactobacilli isolated from two Cameroonian traditional fermented foods." *Scientia Pharmaceutica* 80 (2012): 189–203.

Karasu N., Şimşek Ö. and Çon A.H. "Technological and probiotic characteristics of *Lactobacillus plantarum* strains isolated from traditionally produced fermented vegetables." *Annals of Microbiology* 60 (2010): 227–234.

Kazemipoor M., Radzi C.W.J.W.M., Begum K. and Yaze I. "Screening of antibacterial activity of lactic acid bacteria isolated from fermented vegetables against food borne pathogens." *Archives Des Sciences* 65 (2012). arXiv:1206.6366.

Khay O., Idaomar M., Castro L.M.P., Bernárdez P.F., Senhaji N.S. and Abrini J. "Antimicrobial activities of the bacteriocin-like substances produced by lactic acid bacteria from Moroccan dromedary milk." *African Journal of Biotechnology* 10 (2011): 10447–10455.

Khedid K., Faid M., Mokhtari A., Soulaymani A. and Zinedine A. "Characterization of lactic acid bacteria isolated from the one humped camel milk produced in Morocco." *Microbiological Research* 164 (2009): 81–91.

Klaenhammer T.R. "Genetics of bacteriocins produced by lactic acid bacteria." *FEMS Microbiology Reviews* 12 (1993): 39–85.

Klayraung S., Viernstein H., Sirithunyalug J. and Okonogi S. "Probiotic properties of lactobacilli isolated from Thai traditional food." *Scientia Pharmaceutica* 76 (2008): 485–503.

Krasse P., Carlsson B., Dahl C., Paulsson A., Nilsson A. and Sinkiewicz G. "Decreased gum bleeding and reduced gingivitis by the probiotic *Lactobacillus reuteri*." *Swedish Dental Journal* 30 (2006): 55–60.

Kurugol Z. and Koturoglu G. "Effects of *Saccharomyces boulardii* in children with acute diarrhea." *Acta Paediatrica* 94 (2005): 44–47.

Lannitti T. and Palmieri B. "Therapeutical use of probiotic formulations in clinical practice." *Clinical Nutrition* 29 (2010): 701–725.

Lavanya B., Sowmiya S., Balaji S. and Muthuvelan B. "Screening and characterization of lactic acid bacteria from fermented milk." *British Journal of Dairy Sciences* 2 (2011): 5–10.

Li T., Tao J. and Hong F. "Study on the inhibition effect of nisin." *The Journal of American Science* 1 (2005): 33–37.

Liong M.T. "Roles of probiotics and prebiotics in colon cancer prevention: Postulated mechanisms and *in-vivo* evidence." *International Journal of Molecular Sciences* 9 (2008): 854–863.

Liong M.T. and Shah N.P. "Acid and bile tolerance and cholesterol removal ability of Lactobacilli strains." *Journal of Dairy Science* 88 (2005): 55–66.

Ljungh A. and Wadström T. "Lactic acid bacteria as probiotics." *Current Issues in Intestinal Microbiology* 7 (2006): 73–89.

Lorca G., Torino M.I., de Valdez G.F. and Ljungh Å. "Lactobacilli express cell surface proteins which mediate binding of immobilized collagen and fibronectin." *FEMS Microbiology Letters* 206 (2002): 31–37.

Lore T.A., Mbugua S.K. and Wangoh J. "Enumeration and identification of microflora in *suusac*, a Kenyan traditional fermented camel milk product." *LWT—Food Science and Technology* 38 (2005): 125–130.

Maccaferri S., Klinder A., Brigidi P., Cavina P. and Costabile A. "Potential probiotic *Kluyveromyces marxianus* B0399 modulates the immune response in Caco-2 cells and peripheral blood mononuclear cells and impacts the human gut microbiota in an in vitro colonic model system." *Applied and Environmental Microbiology* 78 (2012): 956–964.

Mahasneh A.M. and Abbas M.M. "Probiotics and traditional fermented foods: The eternal connection." *Jordan Journal of Biological Sciences* 3 (2010): 133–140.

Mami A., Henni J.-E. and Kihal M. "Antimicrobial activity of *Lactobacillus* species isolated from Algerian raw goat's milk against *Staphylococcus aureus*." *World Journal of Dairy and Food Sciences* 3 (2008): 39–49.

Margolles A., García L., Sánchez B., Gueimonde M. and de los Reyes-Gavilán C.G. "Characterization of *Bifidobacterium* strain with acquired resistance to cholate—A preliminary study." *International Journal of Food Microbiology* 82 (2003): 191–198.

Marianelli C., Cifani N. and Pasquali P. "Evaluation of antimicrobial activity of probiotic bacteria against *Salmonella enterica* subsp. *enterica* serovar typhimurium 1344 in a common medium under different environmental conditions." *Research in Microbiology* 161 (2010): 673–680.

Marteau P. "Safety aspects of probiotic products." *Scandinavian Journal of Nutrition/Näringsforskning* 45 (2001): 22–24.

Martín R., Langa S., Reviriego C. et al. "The commensal microflora of human milk: New prospective for food bacteriotherapy and probiotics." *Trends in Food Science and Technology* 15 (2004): 121–127.

Matsumoto T., Ishikawa H., Tateda K., Yaeshima T., Ishibashi N. and Yamaguchi K. "Oral administration of *Bifidobacterium longum* prevents gut-derived *Pseudomonas aeruginosa* sepsis in mice." *Journal of Applied Microbiology* 104 (2008): 672–680.

McFarland L.V. "Meta-analysis of probiotics for the prevention of traveler's diarrhea." *Travel Medicine and Infectious Disease* 5 (2007): 97–105.

McFarland L.V., Surawicz C.M., Greenberg R.N. et al. "A randomized placebo-controlled trial of *Saccharomyces boulardii* in combination with standard antibiotics for *Clostridium difficile* disease." *The Journal of the American Medical Association* 271 (1994): 1913–1918.

Medeiros J.A. and Pereira M.I. "The use of probiotics in *Helicobacter pylori* eradication therapy." *Journal of Clinical Gastroenterology* 47 (2013): 1–5.

Medellina-Pena M.J., Wang H., Johnson R. and Griffhs M.W. "Probiotics affect virulence-related gene expression in *Escherichia coli* O157:H7." *Applied and Environmental Microbiology* 73 (2007): 4259–4267.

Medina-Martínez M.S., Uyttendaele M., Rajkovic A., Nadal P. and Debevere J. "Degradation of N-acyl-L-homoserine lactones by *Bacillus cereus* in culture media and pork extract." *Applied and Environmental Microbiology* 73 (2007): 2329–2332.

Mennigen R. and Bruewer M. "Effect of probiotics on intestinal barrier function." *Annals of the New York Academy of Sciences* 1165 (2009): 183–189.

Mennigen R., Nolte K., Rijcken E. et al. "Probiotic mixture VSL#3 protects the epithelial barrier by maintaining tight junction protein expression and preventing apoptosis in a murine model of colitis." *American Journal of Physiology—Gastrointestinal and Liver Physiology* 296 (2009): G1140–G1149.

Mercenier A., Pavan S. and Pot P. "Probiotics as biotherapeutic agents: Present knowledge and future prospects." *Current Pharmaceutical Design* 8 (2002): 99–110.

Moreira J.L.S., Mota M.R., Horta M.F. et al. "Identification to the species level of *Lactobacillus* isolated in probiotic prospecting studies of human, animal or food origin by 16S-23S rRNA restriction profiling." *BMC Microbiology* 5 (2005): 15. doi:10.1186/1471-2180-5-15.

Mourad K. and Nour-Eddine K. "*In vitro* preselection criteria for probiotic *Lactobacillus plantarum* strains of fermented olives origin." *International Journal of Probiotics and Prebiotics* 1 (2006a): 27–32.

Mourad K. and Nour-Eddine K. "Physicochemical and microbiological study of "*shmen*", a traditional butter made from camel milk in the Sahara (Algeria): Isolation and identification of lactic acid bacteria and yeasts." *Grasas Y Aceites* 57 (2006b): 198–204.

Narwal A. "Probiotics in dentistry—A review." *Journal of Nutrition and Food Sciences* 1 (2011): 114. doi:10.4172/2155-9600.1000114.

Nawaz M., Wang J., Zhou A., Ma C., Wu X. and Xu J. "Screening and characterization of new potentially probiotic lactobacilli from breast-fed healthy babies in Pakistan." *African Journal of Microbiology Research* 5 (2011): 1428–1436.

Ng S.C., Hart A.L., Kamm M.A., Stagg A.G. and Knight S.C. "Mechanisms of action of probiotics: Recent advances." *Inflammatory Bowel Disease* 15 (2009): 300–310.

O'Hara A.M., O'Regan P., Fanning A. et al. "Functional modulation of human intestinal epithelial cell responses by *Bifidobacterium infantis* and *Lactobacillus salivarius*." *Immunology* 118 (2006): 202–215.

Ohland C.L. and MacNaughton W.K. "Probiotic bacteria and intestinal epithelial barrier function." *American Journal of Physiology—Gastrointestinal and Liver Physiology* 298 (2010): G807–G819.

Ooi L.G. and Liong M.T. "Cholesterol-lowering effects of probiotics and prebiotics: A review of in vivo and in vitro findings." *International Journal of Molecular Sciences* 11 (2010): 2499–2522.
Ortu S., Felis G.E., Marzotto M. et al. "Identification and functional characterization of *Lactobacillus* strains isolated from milk and Gioddu, a traditional Sardinian fermented milk." *International Dairy Journal* 17 (2007): 1312–1320.
Osset J., Bartolomé R.M., García E. and Andreu A. "Assessment of the capacity of *Lactobacillus* to inhibit the growth of uropathogens and block their adhesion to vaginal epithelial cell." *Journal of Infected Diseases* 183 (2001): 485–491.
Ouwehand A.C. "Antiallergic effects of probiotics." *Journal of Nutrition* 137 (2007): 794S–797S.
Ouwehand A.C., Salminen S. and Isolauri E. "Probiotics: An overview of beneficial effects." *Antonie van Leeuwenhoek* 82 (2002): 279–289.
Ouwehand A.C., Svendsen L.S. and Leyer G. "Probiotics: From strain to product." *Probiotic and Health Claims*, Wiley-Blackwell, Oxford, 2011. doi:10.1002/9781444329384.ch3.
Oyetayo V.O. and Oyetayo F.L. "Potential of probiotics as biotherapeutic agents targeting the innate immune system." *African Journal of Biotechnology* 4 (2005): 123–127.
Parvez S., Malik K.A., Kang S. and Kim H.-Y. "Probiotics and their fermented food products are beneficial for health." *Journal of Applied Microbiology* 100 (2006): 1171–1185.
Pawan R. and Bhatia A. "Systemic immunomodulation and hypocholesteraemia by dietary probiotics: A clinical study." *Journal of Clinical and Diagnostic Research* 1 (2007): 467–475.
Pereira D.I.A. and Gibson G.R. "Cholesterol assimilation by lactic acid bacteria and Bifidobacteria isolated from the human gut." *Applied and Environmental Microbiology* 68 (2002): 4689–4693.
Petricevic L. and Witt A. "The role of *Lactobacillus casei rhamnosus* Lcr35 in restoring the normal vaginal flora after antibiotic treatment of bacterial vaginosis." *BJOG: An International Journal of Obstetrics and Gynaecology* 115 (2008): 1369–1374.
Popova M., Molimard P., Courau S. et al. "Beneficial effects of probiotics in upper respiratory tract infections and their mechanical actions to antagonize pathogens." *Journal of Applied Microbiology* 113 (2012): 1305–1318.
Preidis G.A., Saulnier D.M., Blutt S.E. et al. "Host response to probiotics determined by nutritional status of rotavirus-infected neonatal mice." *Journal of Pediatric Gastroenterology and Nutrition* 55 (2012): 299–307.
Prescott S.L., Dunstan J.A., Hale J. et al. "Clinical effects of probiotics are associated with increased interferon-γ responses in very young children with atopic dermatitis." *Clinical and Experimental Allergy* 35 (2005): 1557–1564.
Quigley E.M.M. "Gut microbiota and the role of probiotics in therapy." *Current Opinion in Pharmacology* 11 (2011): 593–603.
Rahman N., Xiaohong C., Meiqin F. and Mingsheng D. "Characterization of the dominant microflora in naturally fermented camel milk shubat." *World Journal of Microbiology and Biotechnology* 25 (2009): 1941–1946.
Reddy J.J., Sampathkumar N. and Aradhya C. "Probiotics in dentistry: Review of the current status." *Revista de Clínica e Pesquisa Odontológica* 6 (2010): 261–267.
Reddy K.B.P.K., Raghavendra P., Kumar B.G., Misra M.C. and Prapulla S.G. "Screening of probiotic properties of lactic acid bacteria isolated from Kanjika, an ayruvedic lactic acid fermented product: An in-vitro evaluation." *The Journal of General and Applied Microbiology* 53 (2007): 207–213.
Reid G., Bruce A.W., Fraser N., Heinemann C., Owen J. and Henning B. "Oral probiotics can resolve urogenital infections." *FEMS Immunology and Medical Microbiology* 30 (2001): 49–52.
Resta-Lenert S. and Barrett K.E. "Live probiotics protect intestinal epithelial cells from the effects of infection with enteroinvasive *Escherichia coli* (EIEC)." *Gut* 52 (2003): 988–997.
Rojas M., Ascencio F. and Conway P.L. "Purification and characterization of a surface protein from *Lactobacillus fermentum* 104R that binds to porcine small intestinal mucus and gastric mucin." *Applied Environmental Microbiology* 68 (2002): 2330–2336.
Roos S., Aleljung P., Robert N. et al. "A collagen binding protein from *Lactobacillus reuteri* is part of an ABC transporter system?" *FEMS Microbiology Letters* 144 (1996): 33–38.
Roos T.B., Tabeleão V.C., Dümmer L.A. et al. "Effect of *Bacillus cereus* var. Toyoi and *Saccharomyces boulardii* on the immune response of sheep to vaccines." *Food and Agricultural Immunology* 21 (2010): 113–118.

Rudel L.L. and Morris M.D. "Determination of cholesterol using o-phthalaldehyde." *The Journal of Lipid Research* 14 (1973): 364–366.

Saito T. "Selection of useful probiotic lactic acid bacteria from the *Lactobacillus acidophilus* group and their applications to functional foods." *Animal Science Journal* 75 (2004): 1–13.

Sanders M.E. "Considerations for use of probiotic bacteria to modulate human health." *Journal of Nutrition* 130 (2000): 384S–390S.

Sanders M.E., Akkermans L.M., Haller D. et al. "Safety assessment of probiotics for human use." *Gut Microbes* 1 (2010): 164–185.

Saraf K., Shashikanth M.C., Priya T., Sultana N. and Chaitanya N. "Probiotics—Do they have a role in medicine and dentistry?" *Journal of the Association of Physicians of India* 58 (2010): 488–492.

Saulnier D.M., Ringel Y., Heyman M.B. et al. "The intestinal microbiome, probiotics and prebiotics in neurogastroenterology." *Landes Bioscience Journals: Gut Microbes* 4 (2013): 17–27.

Savadogo A., Ouattara C.A.T., Bassole I.H.N. and Traore S.A. "Bacteriocins and lactic acid bacteria—A minireview." *African Journal of Biotechnology* 5 (2006): 678–683.

Schillinger U. and Lücke F.-K. "Antibacterial activity of *Lactobacillus sakei* isolated from meat." *Applied and Environmental Microbiology* 55 (1989): 1901–1906.

Schrezenmeir J. and de Vrese M. "Probiotics, prebiotics, and synbiotics-approaching a definition." *American Journal of Clinical Nutrition* 73 (2001): 361S–364S.

Sengun I.Y., Neilsen D.S., Karapinar M. and Jakobsen M. "Identification of lactic acid bacteria isolated from Tarhana, a traditional Turkish fermented food." *International Journal of Food Microbiology* 135 (2009): 105–111.

Serikbayeva A., Konuspayeva G., Faye B., Loiseau G. and Narmuratova M. "Probiotic properties of a sour-milk product: Shubat from the camel milk." *Desertification Combat and Food Safety: The Added Value of Camel Producers*, Amsterdam, 187–191, 2005.

Sherman P.M., Ossa J.C. and Johnson-Henry K. "Unraveling mechanisms of action of probiotics." *Nutrition in Clinical Practice* 24 (2009): 10–14.

Singh G.P. and Sharma R.R. "Dominating species of Lactobacilli and Leuconostocs present among the lactic acid bacteria of milk of different cattles." *Asian Journal of Experimental Sciences* 23 (2009): 173–179.

Singhal K., Joshi H. and Chaudhary B.L. "Bile and acid tolerance ability of probiotic *Lactobacillus* strains." *Journal of Global Pharma Technology* 2 (2010): 17–25.

Sleator R.D. and Hill C. "New frontiers in probiotic research." *Letters in Applied Microbiology* 46 (2008): 143–147.

Soccol C.R., Vandenberghe L.P.S., Spier M.R. et al. "The potential of probiotics: A review." *Food Technology and Biotechnology* 48 (2010): 413–434.

Sonal S.M., Suja A., Lima T.B. and Aneesh T. "Probiotics: Friendly microbes for better health." *The Internet Journal of Nutrition and Wellness* 6 (2008). doi:10.5580/1041.

Stepper J., Shastri S., Loo T. et al. "Cysteine S-glycosylation, a new post-translational modification found in glycopeptide bacteriocins." *FEBS Letters* 585 (2011): 645–650.

Styriak I., Nemcová R., Chang Y.-H. and Ljungh Å. "Binding of extracellular matrix molecules by probiotic bacteria." *Letter of Applied Microbiology* 37 (2003): 329–333.

Suvarna V.C. and Boby V.U. "Probiotics in human health: A current assessment." *Current Science* 88 (2005): 1744–1748.

Szymański H., Chmielarczyk A., Strus M. et al. "Colonization of the gastrointestinal tract by probiotic *L. rhamnosus* strains in acute diarrhea in children." *Digestive and Liver Disease* 38 (2006): S274–S276.

Tahamtan Y., Kargar M., Namdar N., Rahimian A., Hayati M. and Namavari M.M. "Probiotic inhibits the cytopathic effect induced by *Escherichia coli* O157:H7 in Vero cell line model." *Letters in Applied Microbiology* 52 (2011): 527–531.

Takeda S., Yamasaki K., Takeshita M. et al. "The investigation of probiotic potential of lactic acid bacteria isolated from traditional Mongolian dairy products." *Animal Science Journal* 82 (2011): 571–579.

Takemura N., Okubo T. and Sonoyama K. "*Lactobacillus plantarum* strain no. 14 reduces adipocyte size in mice fed high-fat diet." *Experimental Biology and Medicine* 235 (2010): 849–856.

Talib W.H. and Mahasneh A.M. "Combination of *Ononishirta* and *Bifidobacterium longum* decreases syngeneic mouse mammary tumor burden and enhances immune response." *Journal of Cancer Research and Therapeutics* 8 (2012): 417–423.
Tamang J.P. "Food culture in the Eastern Himalayas." *Journal of Himalayan Research and Cultural Foundation* 5 (2001): 107–118.
Tamime A.Y. and O'Connor T.P. "Kishk-A dried fermented milk/cereal mixture." *International Dairy Journal* 5 (1995): 109–128.
Tanaka A., Seki M., Yamahira S. et al. "*Lactobacillus pentosus* strain b240 suppresses pneumonia induced by *Streptococcus pneumonia* in mice." *Letters in Applied Microbiology* 53 (2011): 35–43.
Tannock G.W. "Probiotics: Time for a dose of realism." *Current Issues in Intestinal Microbiology* 4 (2003): 33–42.
Temmerman R., Pot B., Huys G. and Swings J. "Identification and susceptibility of bacterial isolates probiotic products." *International Journal of Food Microbiology* 81 (2003): 1–10.
Tiwari S.K. and Srivastava S. "Characterization of a bacteriocin from *Lactobacillus plantarum* strain LR/14." *Food Biotechnology* 22 (2008): 247–261.
Todoroki K., Mukai T., Soto S. and Toba T. "Inhibition of adhesion of food-borne pathogens to Caco-2 cells by *Lactobacillus* strains." *Journal of Applied Microbiology* 91 (2001): 1–6.
Todorov S.D., Furtado D.N., Saad S.M., Tome E. and Franco B.D. "Potential beneficial properties of bacteriocin-producing lactic acid bacteria isolated from smoked salmon." *Journal of Applied Microbiology* 110 (2011): 971–986.
Tolinački M.M., Kojić M., Lozo J., Terzić-Vidojević A., Topisirović L. and Fira D. "Characterization of the bacteriocin-producing strain *Lactobacillus paracasei* subsp. *paracasei* BGUB 9." *Archives of Biological Sciences* 62 (2010): 889–899.
Tsai C.-C., Hsih H.-Y., Chiu H.-H. et al. "Antagonistic activity against *Salmonella* infection *in vitro* and *in vivo* for two *Lactobacillus* strains from swine and poultry." *International Journal of Food Microbiology* 102 (2005): 185–194.
Vandenbergh P.A. "Lactic acid bacteria, their metabolic products and interference with microbial growth." *FEMS Microbiology Review* 12 (1993): 221–238.
Velraeds M.M.C., van der Belt-Gritter B., van der Mei H.C., Reid G. and Busscher H.J. "Interference in initial adhesion of uropathogenic bacteria and yeasts silicone rubber by a *Lactobacillus acidophilus* biosurfactant." *Journal of Medical Microbiology* 47 (1998): 1081–1085.
Vijayendra S.V.N. and Gupta R.C. "Assessment of probiotic and sensory properties of dahi and yoghurt prepared using bulk freeze-dried cultures in buffalo milk." *Annals of Microbiology* 62 (2012): 939–947.
Vilà B., Esteve-Garcia E. and Brufau J. "Probiotic microorganisms: 100 years of innovation and efficacy. Modes of action." *World's Poultry Science Journal* 66 (2010): 369–380.
Viljanen M., Savilahti E. and Haahtela T. "Probiotics in the treatment of atopic eczema/dermatitis syndrome in infants: A double-blind placebo-controlled trial." *Allergy* 60 (2005): 494–500.
Villamil L., Reyes C. and Martínez-Silva M.A. "*In vivo* and *in vitro* assessment of *Lactobacillus acidophilus* as probiotic for tilapia (*Oreochromis niloticus*, Perciformes: Cichlidae) culture improvement." *Aquaculture Research* 1 (2012): 10. doi:10.1111/are.12051.
Vinderola C.G. and Reinheimer J.A. "Lactic acid starter and probiotic bacteria: A comparative "*in vitro*" study of probiotic characteristics and biological barrier resistance." *Food Research International* 36 (2003): 895–904.
Watanabe K., Fujimoto J., Sasamoto M., Dugersuren J., Tumursuh T. and Demberel S. "Diversity of lactic acid bacteria and yeasts in Airag an Tarag traditional fermented milk products of Mongolia." *World Journal of Microbiology and Biotechnology* 24 (2008): 1313–1325.
WHO/FAO (World Health Organization/Food and Agriculture organization). 2006. "Probiotics in foods. Health and nutritional properties and guidelines for evaluation." FAO Food and Nutrition Paper 85 (ISBN-92-5-105513-0).
Wouters J.T.M., Ayad E.H.E., Hugenholtz J. and Smit G. "Microbes from raw milk for fermented dairy products." *International Dairy Journal* 12 (2002): 91–109.
Xie N., Cui Y., Yin Y.-N., Zhao X., Yang J.-W. and Wang Z.-G. "Effects of two *Lactobacillus* strains on lipid metabolism and intestinal microflora in rats fed a high-cholesterol diet." *BMC Complementary and Alternative Medicine* 11 (2011): 53–63.

Yan F., Cao H. and Cover T.L. "Colon-specific delivery of a probiotic-derived soluble protein ameliorates intestinal inflammation in mice through an EGFR-dependent mechanism." *American Society for Clinical Investigation* 121 (2011): 2242–2253.

Yan F. and Polk D.B. "Probiotics: Progress toward novel therapies for intestinal diseases." *Current Opinion of Gastroenterology* 26 (2011): 95–101.

Yaqoob M. and Nawaz H. "Potential of Pakistani camel for dairy and other uses." *Animal Science Journal* 78 (2007): 467–475.

Yateem A., Balba M.T., Al-Surrayai T., Al-Mutairi B. and Al-Daher R. "Isolation of lactic acid bacteria with probiotic potential from camel milk." *International Journal of Dairy Science* 3 (2008): 194–199.

Yuki N., Shimazaki T. and Kushiro A. "Colonization of the stratified squamous epithelium of the nonsecreting area of horse stomach by lactobacilli." *Applied Environmental Microbiology* 66 (2000): 5030–5034.

Zeng X.Q., Pan D.D. and Guo Y.X. "The probiotic properties of *Lactobacillus buchneri* P2." *Journal of Applied Microbiology* 108 (2010): 2059–2066.

Zhou J.S., Gopal P.K. and Gill H.S. "Potential probiotic lactic acid bacteria *Lactobacillus rhamnosus* (HN001), *Lactobacillus acidophilus* (HN017) and *Bifidobacterium lactis* (HN019) do not degrade gastric mucin *in vitro*." *International Journal of Food Microbiology* 63 (2000): 81–89.

chapter eleven

Microbial healthcare products

Zafar Alam Mahmood and Saad Bin Zafar Mahmood

Contents

11.1 Introduction

Microbial healthcare products have now gained huge application in both primary and secondary human healthcare systems irrespective of whether people are living in underdeveloped or developed countries. In the primary healthcare system, microbial products can play a role in immunization, prevention of locally endemic diseases, treatment of common diseases, and others, whereas their use and application is fully warranted in secondary healthcare systems for more organized treatment in hospitals for complicated infectious diseases and in chemotherapy. The vast majority of multinational companies are marketing a number of products of microbial origin and are generating huge revenues out of these products. The introduction of new products manufactured through cost-effective raw materials for both the primary and secondary healthcare sectors is still open and desirable both on the global and regional level. Healthcare products for human application can further be grouped under biopharmaceuticals, cosmeceuticals, and nutraceuticals, with prominent properties of prophylactic and therapeutic applications. Microorganisms have been utilized by human beings for centuries in view of their biochemical features and for the production of healthcare products. This can still be regarded as a very early phase of biotechnology, with the best possible economic use of microorganisms in food and allied industries. However, as knowledge of microbial physiology grew, the scope of utilizing microorganisms to develop healthcare products increased tremendously and their utilization was also expanded for the development of pharmaceuticals and, later on, by the personal and healthcare product manufacturing industries. The discovery of pathways for the development or biosynthesis of secondary metabolites such as antibiotics, alkaloids, flavonoids, antitumor agents, cholesterol-lowering drugs, and others, certainly provided a better understanding and platform for their role in the human healthcare system. A significant number of various products belonging to these diverse groups are now available commercially both for treatment and prophylaxis.

With the advancements and understanding in the fields of microbial physiology and molecular biology, followed by the extensive role of genetics, ultimately resulting in the isolation of DNA, has completely changed the concept and role of microorganisms in the production of healthcare products. Although the ultimate outcome of these findings resulted in the production of a host of products, the first one which came into the picture was "insulin," developed through genetic engineering. The in-depth studies involved in resolving the structure of DNA provided a great hope and toll to scientists to use advanced biotechnological techniques for the removal of DNA from one region of the genome and transport the DNA in a controlled way to another region of the same DNA, or DNA in a completely different organism for more economic gain and benefits by expanding the role and type of products in substantial quantities. In the current scenario, the use of microorganisms as biotechnological agents and the technology of profit have not only continued but have exclusively increased. Indeed, the biotechnology sector, as it is recognized today, is already a multibillion dollar sector worldwide. With these objectives, some of the major microbial healthcare products have been taken into consideration to study and to provide reasonable information to highlight their significance in the human healthcare system.

11.2 Antibiotics

11.2.1 Historical background and new challenges

Microbial production of secondary metabolites and especially those with therapeutic importance such as "antibiotics" have been given significant attention worldwide in view of their role in the management of infectious disease. It has been nearly a century since the first therapeutic compound "penicillin," from the fungal species, *Penicillium notatum*, followed by *Penicillium chrysogenum* (Figure 11.1), was discovered by Alexander Fleming in 1928. The actual name penicillin was given on March 7, 1929 and reported to be highly inhibitory effect against *Staphylococcus aureus*, a Gram-positive bacteria. Before this, Fleming also had the honor of discovering the antibacterial enzyme "lysozyme" in 1923. In 1924, actinomycetin was discovered, but was only used in the production of vaccines.

Despite the fact that the term antibiotics was used and noted in the literature before the discovery of penicillin (Burkholder 1952), it was not until 1942 when Selman Waksman suggested a proper definition of the term antibiotics as "a chemical substance obtained from certain microorganisms, capable of inhibiting the growth or destroying other microorganism in small concentrations." The term *low-molecular weight organic natural products* has been used in some recent literature in place of chemical substance, as originally used by Waksman (Sanchez and Demain 2011). By the end of 1940, Fleming was able to test the effect of penicillin (benzylpenicillin or penicillin G) on some other Gram-positive organisms responsible for causing scarlet fever, pneumonia, meningitis, and diphtheria and one Gram-negative organism causing gonorrhea, but no prominent place or identification was given to penicillin.

In 1941, Howard Florey and Emst Boris Chain completely revolutionized and changed the concept of infectious disease therapy with the successful commercial production of penicillin by surface culture techniques using thousands and thousands of glass bottles (Figure 11.2); this period was called the "Golden Age of Antimicrobial Therapy." This was followed by production through submerged fermentation. In 1945, Fleming, Florey, and Chain shared the Nobel Prize in Physiology or Medicine. Based on his discovery, Sir Alexander Fleming (after his death in 1955) was named one of the 100 top most famous important people of the twentieth century in 1999. Further success in the field was achieved with the introduction of the first oral penicillin (penicillin V) in 1952 by Hans Margreiter and Ernst Brandl (Greenwood 2008). The structures of penicillin G and penicillin V are shown in Figures 11.3 and 11.4, respectively.

After penicillin, subsequent research in the field of secondary metabolites, especially from fungi and actinomycetes (filamentous bacteria), have led to the discovery of a series of antibiotics for human and veterinary use, such as cephalosporins, tetracyclines,

Figure 11.1 *P. notatum* colony.

Figure 11.2 Glass bottle used for the production of penicillin by surface culture technique.

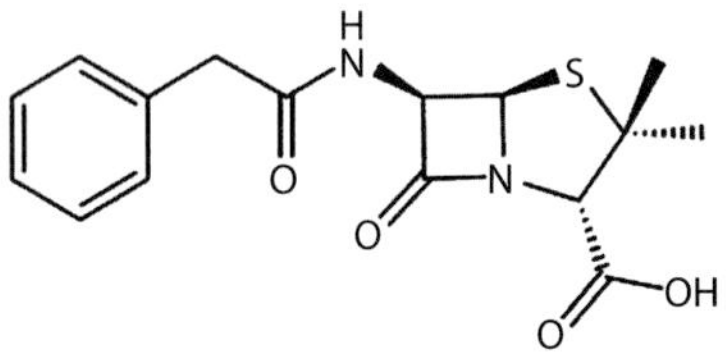

Figure 11.3 Benzylpenicillin (penicillin G).

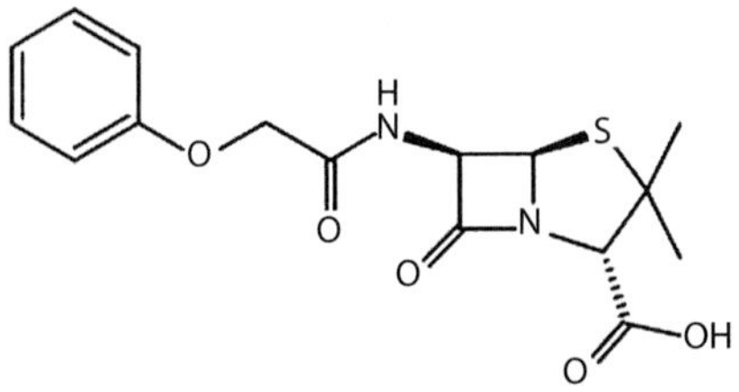

Figure 11.4 Phenoxymethylpenicillin (penicillin V).

aminoglycosides, chloramphenicol, macrolides, and glycopeptides. The actinomycetes are the most extensively studied organism and are capable of producing a large number of antibiotics (~75% out of the total), whereas its genus, *Streptomyces* has the largest contribution (~75%) and the specific species, *Streptomyces hygroscopicus* and *Streptomyces griseus*, have been observed to produce approximately 200 and 40 different antibiotics, respectively. The contribution of nonfilamentous bacteria is small, approximately 12% and that of filamentous fungi is approximately 20% out of total antibiotics reported thus far (Sanchez and Demain 2011). The discovery of tyrothricin by Dubos in 1939 has, to some extent, supported infectious disease therapy; however, the susceptible organisms are still Gram-positive. The introduction of streptomycin in 1944 certainly extended the antimicrobial spectrum as it was capable of inhibiting some Gram-negative organisms as well as *Mycobacterium* (Figure 11.5). The concept of broad-spectrum antibiotics was introduced for the first time after the

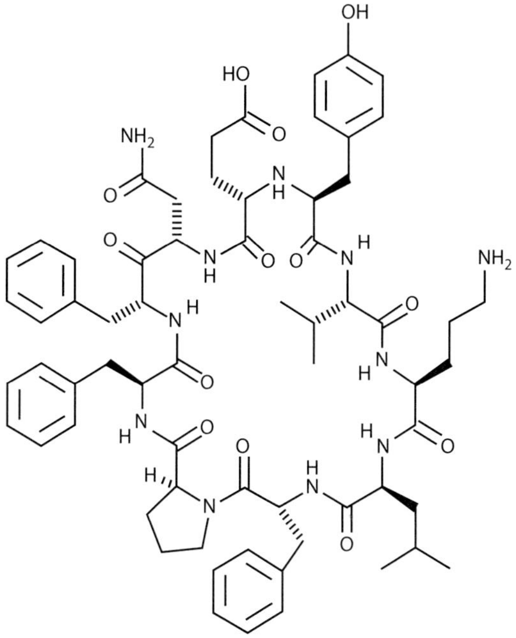

Figure 11.5 Tyrothricin.

discovery of chloramphenicol in 1947 and chlortetracycline in 1948, as these were quite effective against both Gram-positive and Gram-negative bacteria along with *Rickettsia*, large viruses, and certain protozoa. This was followed by the discovery of other members of the tetracyclines (Figures 11.6 and 11.7).

The subsequent discoveries of various antibiotics in about 25 years of duration after the commercial production of penicillin in 1941, such as "aminoglycoside" and "tetracycline,"

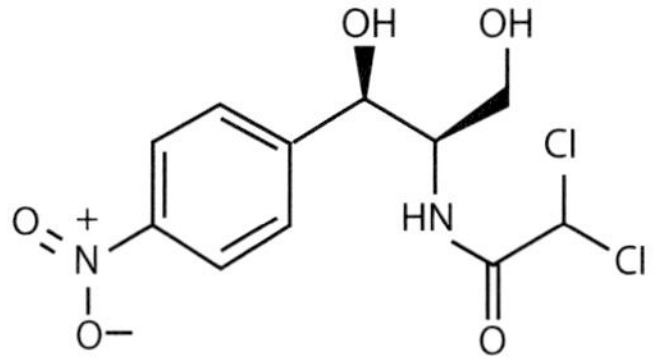

Figure 11.6 Chloramphenicol.

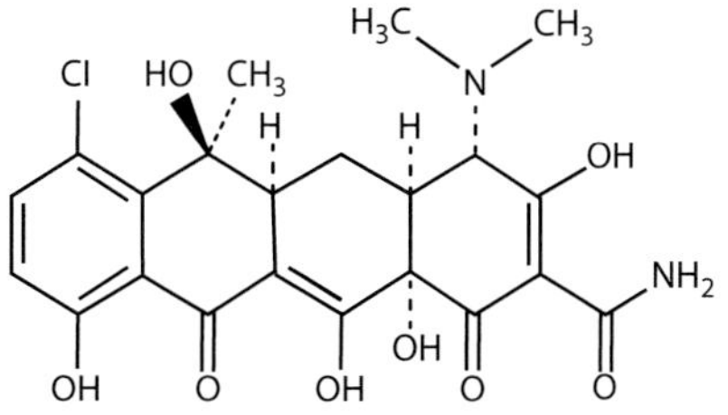

Figure 11.7 Chlortetracycline.

have had a great effect in the therapy of infectious disease because the spectrum of activity has significantly increased. The natural aminoglycosides (Figure 11.8) are bactericidal in action and act by inhibition of protein synthesis, for example, streptomycin, neomycin, kanamycin, gentamicin, and tobramycin are active against both Gram-positive and Gram-negative bacteria, but predominantly act on Gram-negative bacteria.

Because of nephrotoxicity and ototoxicity, significant precautions are required during therapy with aminoglycosides, which restricts their use to clinical applications when used alone, although it is still very commonly used along with other antibiotics, when used synergistically, by adjusting the dose. The introduction of semisynthetic aminoglycosides, such as netilmicin in 1976 (Figure 11.9) and amikacin in 1972 (Figure 11.10), further expanded the spectrum of activity; however, toxicity still remains the main factor for consideration.

Approximately 150 aminoglycoside antibiotics have thus far been discovered from various species of *Micromonospora*, the most well-known of these is gentamicin produced by *M. echinospora*. A series of other antibiotics (Figure 11.11) were discovered between 1950 and 1965, for example, nystatin (1951), erythromycin (1953), vancomycin (1955), novobiocin (1956), rifamycin (1959), and lincomycin (1962).

On the other hand, although tetracyclines (Figure 11.12) produced better results against both Gram-positive and Gram-negative, the spectrum of activity was further increased

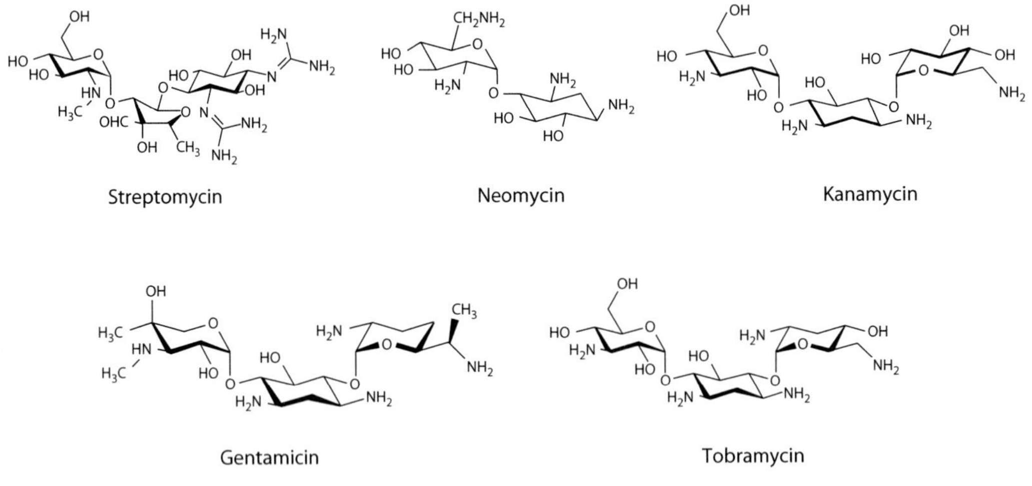

Streptomycin Neomycin Kanamycin

Gentamicin Tobramycin

Figure 11.8 Natural aminoglycosides.

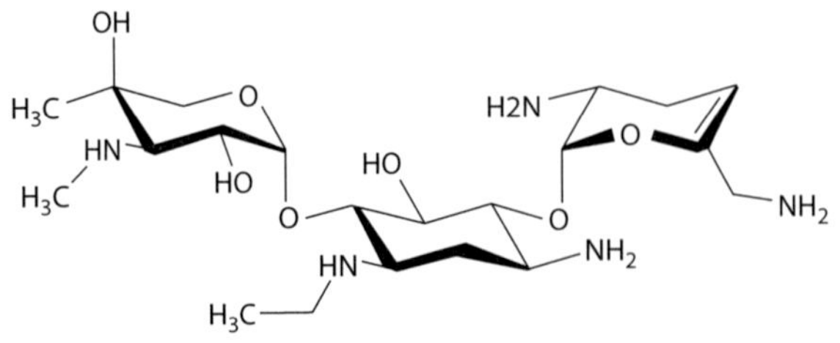

Figure 11.9 Netilmicin.

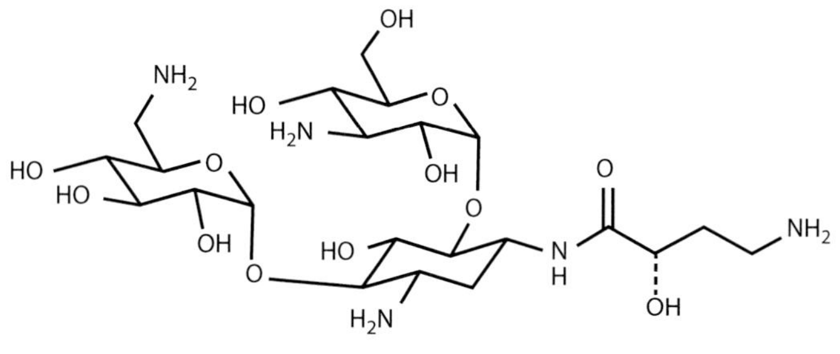

Figure 11.10 Amikacin.

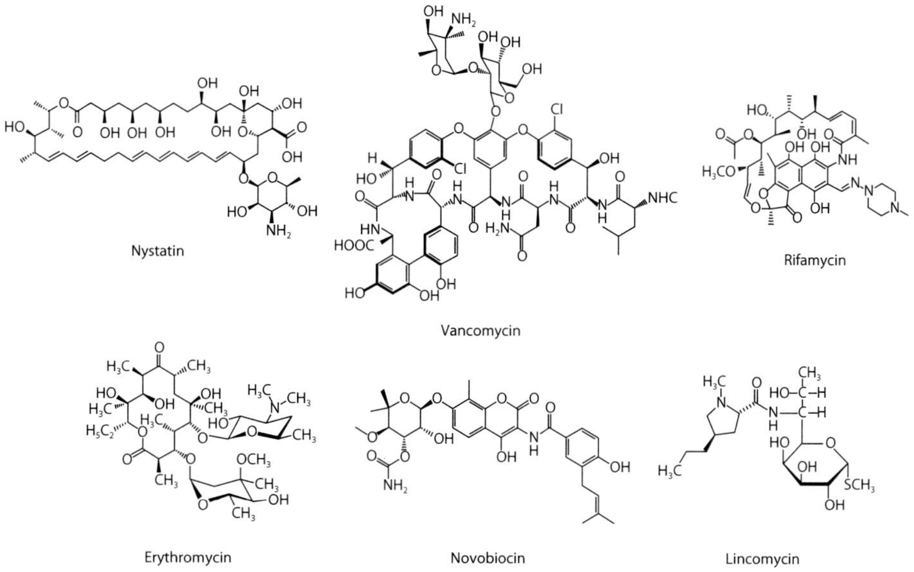

Nystatin

Vancomycin

Rifamycin

Erythromycin

Novobiocin

Lincomycin

Figure 11.11 Antibiotics discovered between 1950 and 1965.

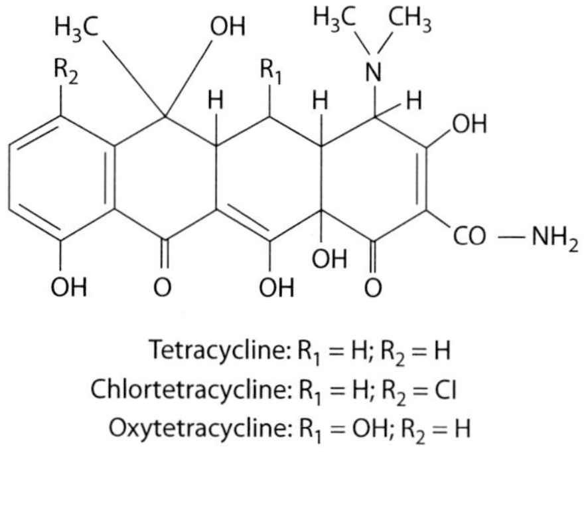

Tetracycline: $R_1 = H$; $R_2 = H$

Chlortetracycline: $R_1 = H$; $R_2 = Cl$

Oxytetracycline: $R_1 = OH$; $R_2 = H$

Figure 11.12 Tetracyclines.

with the introduction of semisynthetic tetracyclines such as doxycycline, minocycline, and tigecycline (Figures 11.13 through 11.15).

Macrolides, for example, erythromycin and its semisynthetic forms (Figure 11.16), have also found tremendous application, especially against respiratory tract infections. The macrolides contain a large lactone ring with multiple keto and hydroxyl groups, linked to one or more sugars. The modified macrolides derived from erythromycin, such as clarithromycin, azithromycin, roxithromycin, and the ketolides, for example, telithromycin, have been successfully used in clinical practice. These modified forms are active against penicillin and erythromycin-resistant *Streptococcus pneumonia, Haemophilus influenzae,* group A *Streptococci, Legionella* spp., *Chlamydia* spp., and *Mycoplasma pneumoniae,* and can be used both parenterally and orally.

The greatest moment in the history of antibiotics, which once again revolutionized therapy, was the introduction of semisynthetic penicillins and cephalosporins, classified under β-lactam antibiotics. The discovery of 6-aminopenicillinic acid (6-APA) followed by 7-aminocephalosporanic acid (7-ACA) by the action of enzyme penicillin amidase on penicillin G, and cephalosporin acylase on cephalosporin C, respectively, has lead to the production of a series of semisynthetic penicillins and cephalosporins with an extended spectrum of activity. Because the semisynthetic penicillins have shown remarkable

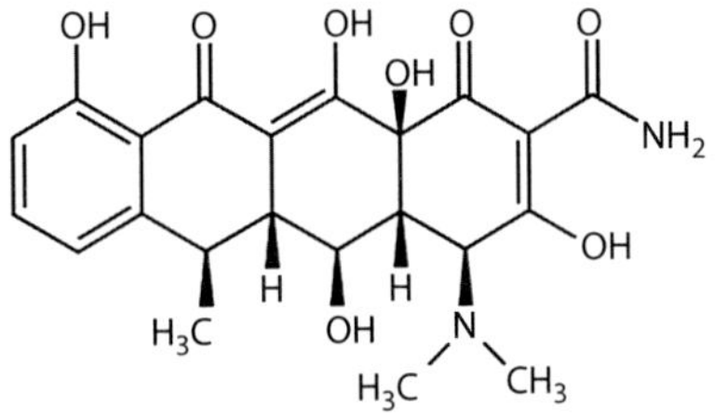

Figure 11.13 Doxycycline.

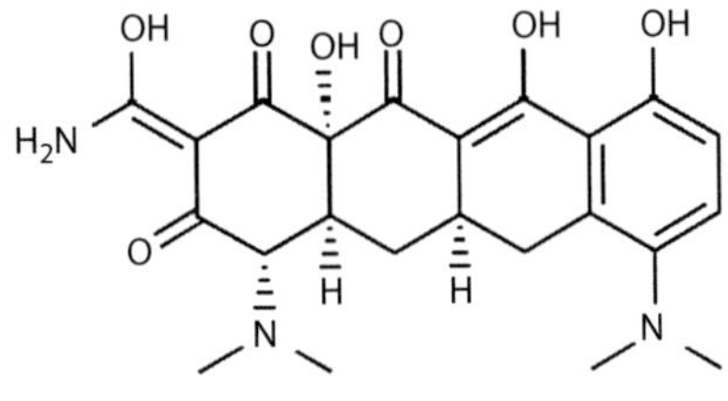

Figure 11.14 Minocycline.

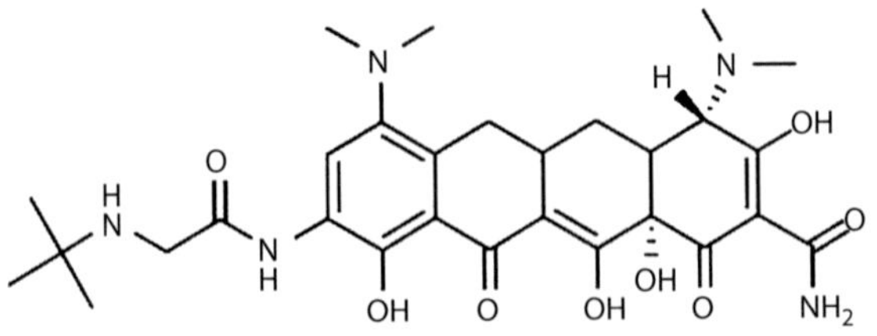

Figure 11.15 Tigecycline.

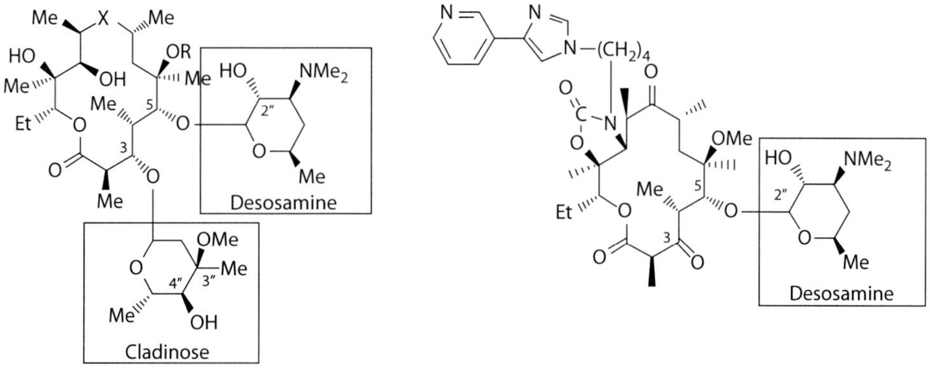

1 R=H, X = CO Erythromycin
2 R=Me, X = CO Clarithromycin
3 R=H, X=N(Me)CH_2 Azithromycin

4 Telithromycin

Figure 11.16 Structures of modified macrolides.

therapeutic activity, 6-APA and 7-ACA have naturally become very important pharmaceutical raw materials. Nowadays, a considerable number of semisynthetic penicillins and cephalosporins, all derived from 6-APA and 7-ACA (Figures 11.17 and 11.18), are in widespread clinical use, whereas a numbers of other novel structures are under clinical investigation.

The β-lactams are the most important class of antibiotics in terms of clinical use and safety profile. Members of the β-lactams constitute a major part of the antibiotic market today and include the penicillins, cephalosporins, monobactams, and carbapenems. Natural carbapenems such as thienamycin, which is resistant to β-lactamase, are made by *Streptomyces cattleya, Erwinia carotovora* subsp. *caratovora, Serratia* sp., and *Photorhabdus luminescens*; however, the commercial carbapenems, such as imipenem, meropenem, and ertapenem are manufactured synthetically (Sanchez and Demain 2011).

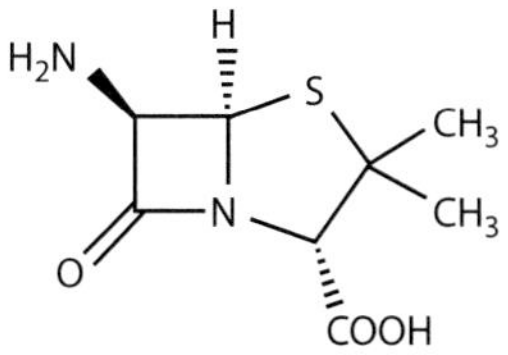

Figure 11.17 6-Aminopenicillinic acid.

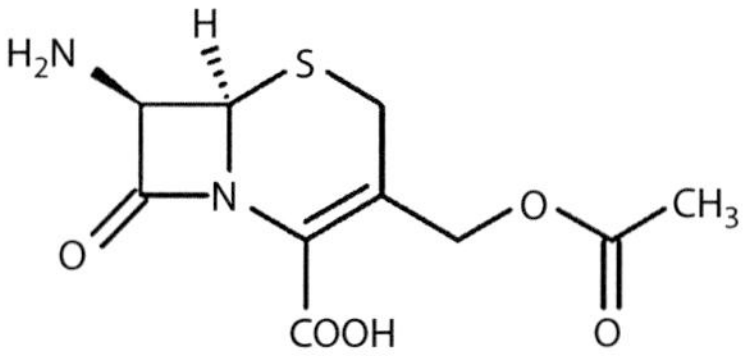

Figure 11.18 7-Aminocephalosporanic acid.

Other antibiotic groups include glycopeptides such as vancomycin, balhimycin, and chloroeremomycin (Figures 11.19 through 11.21) and lipopeptides such as daptomycin (Figure 11.22). Vancomycin has been the molecule of choice to treat infections caused by resistant organisms and is considered as the "antibiotic of last resort." However, lipoglycopeptides (e.g., teicoplanin; Figure 11.23) and oxazolidinones (e.g., linezolid; Figure 11.24) were later found to inhibit vancomycin-resistant bacteria. This was followed by

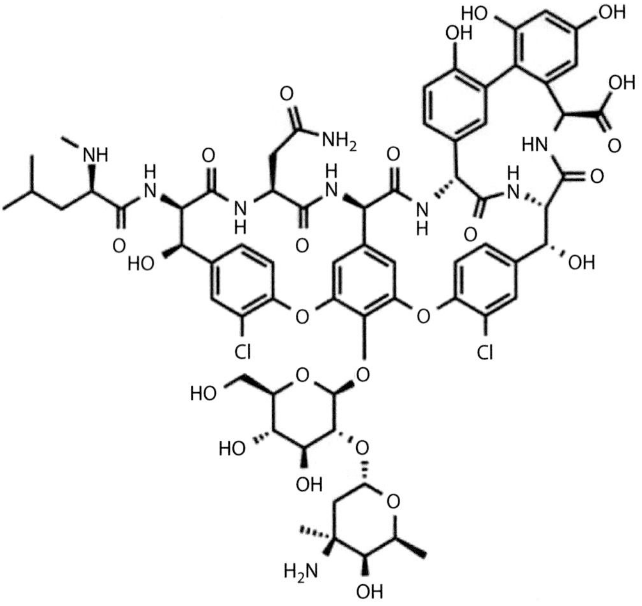

Figure 11.19 Vancomycin.

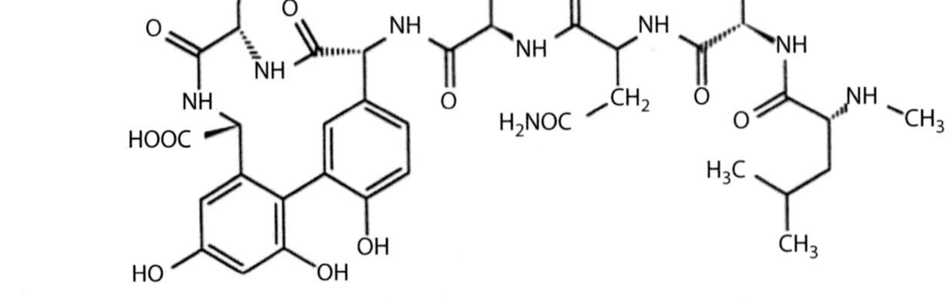

Figure 11.20 Balhimycin.

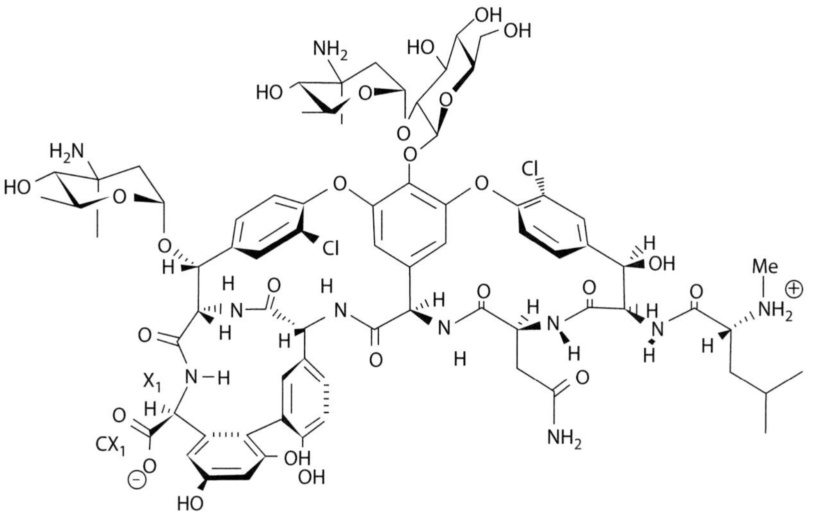

Figure 11.21 Chloroeremomycin.

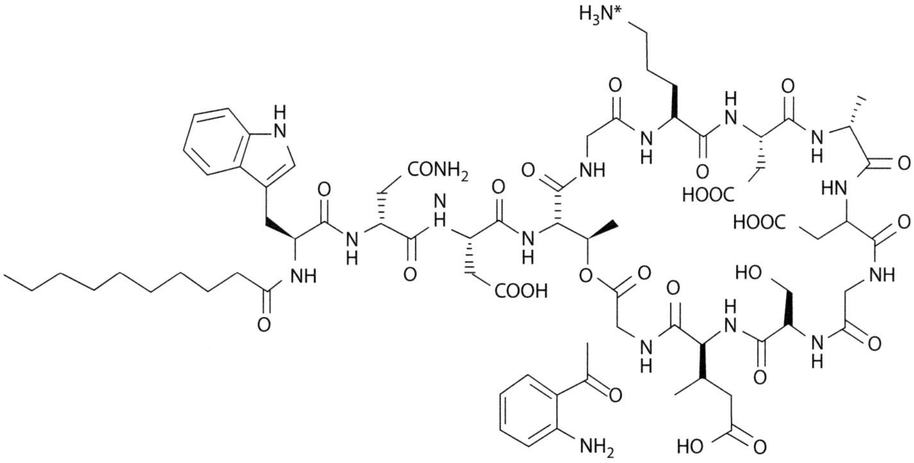

Figure 11.22 Daptomycin.

the discovery of streptogramins such as virginiamycin (Figure 11.25) and pristinamycin (Figure 11.26), as well as the derivatives of pristinamycin, for example, quinupristin/dalfopristin, used in combination under the brand name of Synercid (Figure 11.27), with the hope and findings that it will treat even Gram-positive bacteria resistant to vancomycin and teicoplanin.

The world market for antibacterial drugs (covering natural, semisynthetic, and synthetic) was valued at more than US$40 billion in 2012, having grown 1% since 2011. In both years, the leading submarket was cephalosporins, accounting for more than a quarter of the antibacterial drug revenue. In 2012, the United States was the leading national market for antibacterial drugs, with a market share of more than 25%. However,

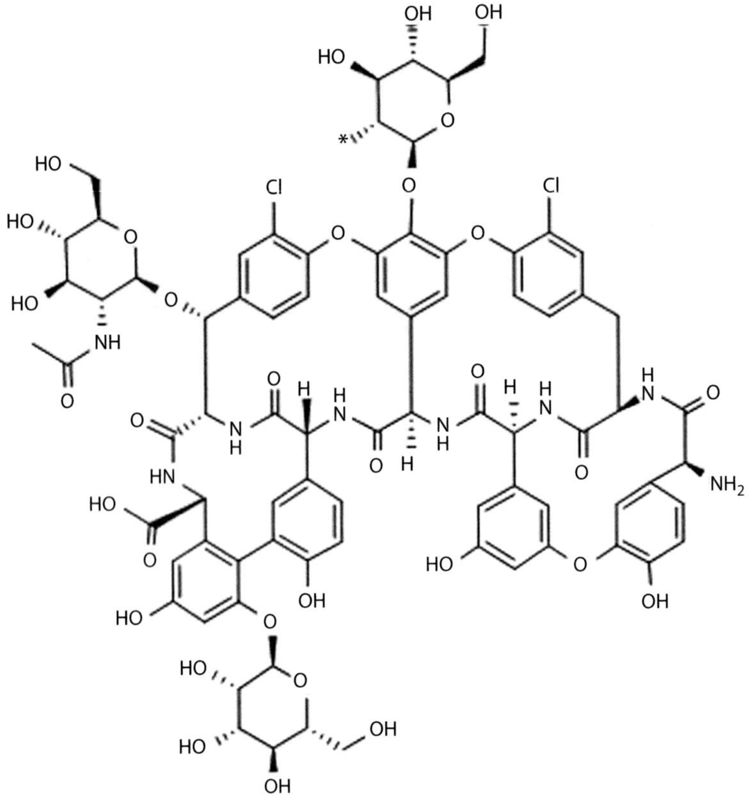

Figure 11.23 Teicoplanin.

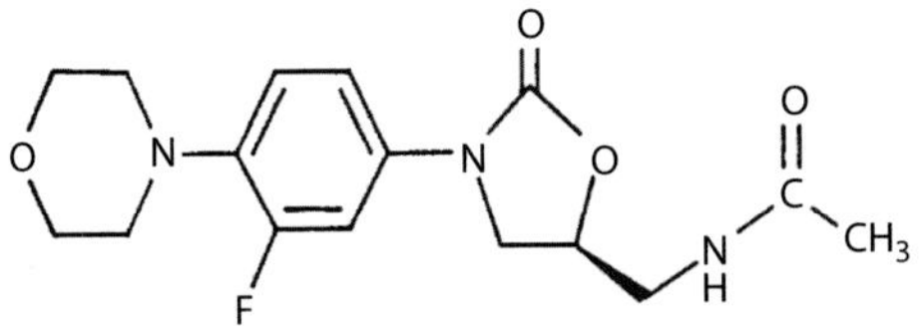

Figure 11.24 Linezolid.

emerging markets also account for a significant proportion of the global market. China's market share in 2012 was more than 15%. The leading antibacterial drug in 2012 was reportedly Zyvox (linezolid, oxazolidinone antibacterial from Pfizer), with a revenue of US$1.35 billion. The top 10 antibacterial drugs had a combined revenue of US$6.92 billion in 2012. There were approximately 10 new antibacterial drugs in phase III trials in 2013. Most of these could be launched in major markets by 2015 (Visiongain 2013). However, the pipeline for antibacterial drugs is much leaner than for other most important sectors of the pharmaceutical industry, despite unmet treatment needs existing for many bacterial infections.

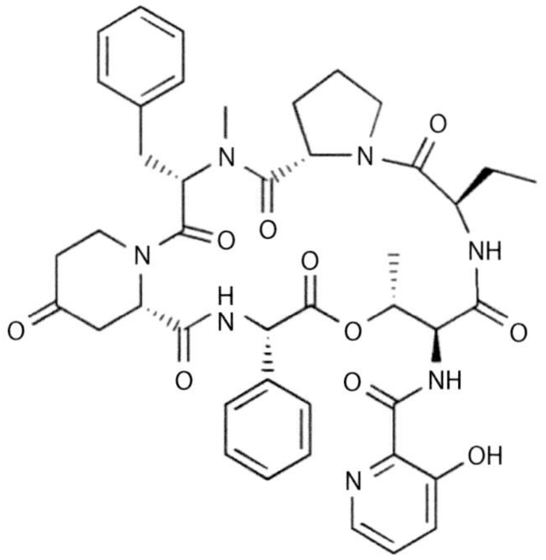

Figure 11.25 Virginiamycin.

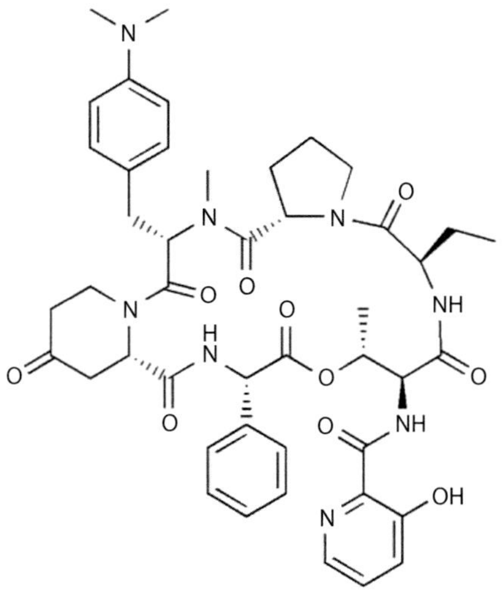

Figure 11.26 Pristinamycin.

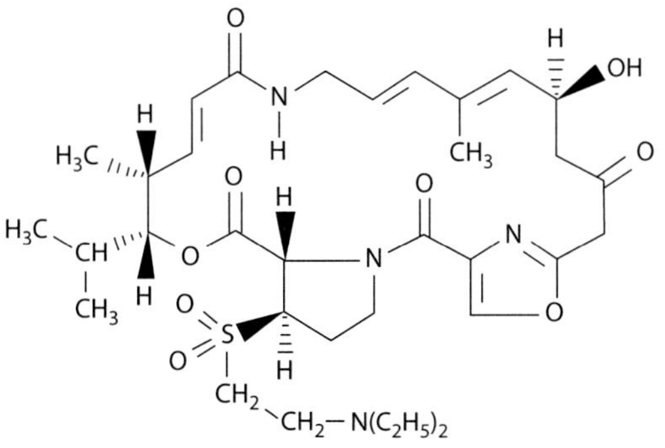

Figure 11.27 Synercid.

11.2.2 Production and pharmaceutical presentation

The production of antibiotics is mainly based on the knowledge of microbiological and biochemical methods, which distinguish them from similar compound (antimicrobials), obtained through chemical synthesis. Most manufacturing processes have been patented by big multinational companies; however, the outline of these methods is fairly well known. The most feasible method for the production of antibiotics is through fermentation, which certainly has now reached tremendous advancement in terms of process optimization and extraction, and purification of the desired product. Classic fermentation processes include propagation (cultivation of microorganisms), isolation/extraction, purification, and bulk packaging.

Propagation starts with the preparation of inoculum from stock culture usually maintained on a suitable medium based on the type of organism used. This is followed by seed culture and then growth in large fermenters (Figure 11.28) with 100,000 to 150,000 L capacity containing suitable sterile liquid growth medium. The fermentation medium should be sufficient to provide carbon and nitrogen sources (such as molasses or soy meal for the growth of microorganisms and ammonium salts for the regulation of the metabolic cycle). The medium is supplemented with trace elements/inorganic salts (such as phosphorus, sulfur, magnesium, zinc, iron, and copper) and precursor according to the strain's requirements. Aeration, agitation, pH, and temperature are controlled through automatic/computerized systems and the optimum level of nutrients is maintained at all times. This is closely monitored and adjusted as and when necessary.

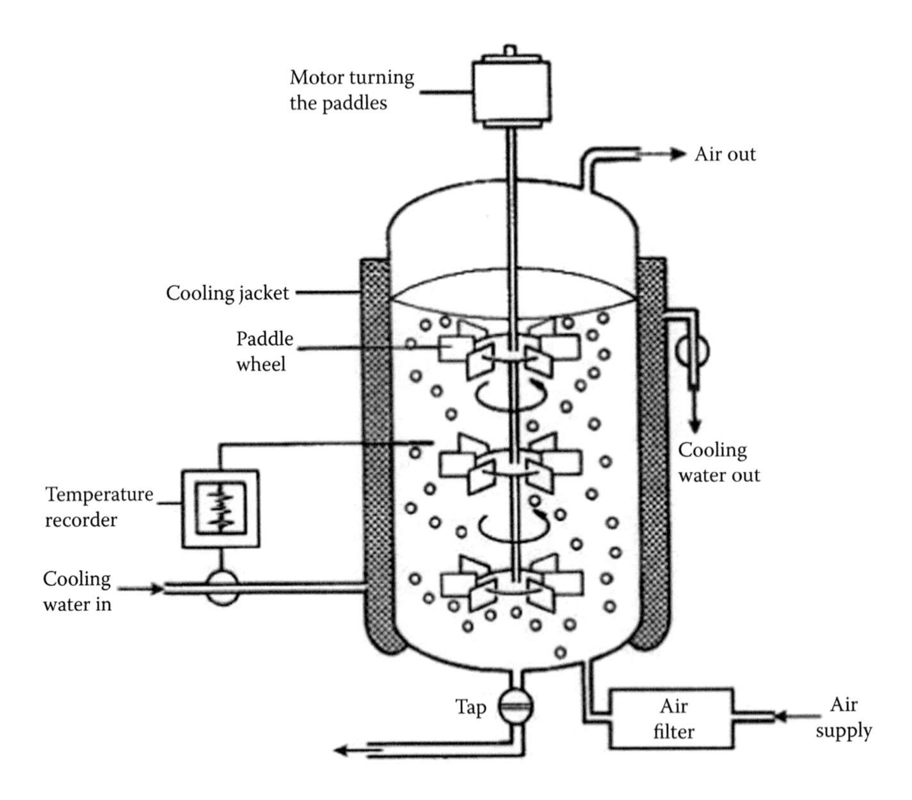

Figure 11.28 Structure of a modern fermenter used for submerged fermentation.

To prevent foaming during fermentation, a suitable concentration of antifoaming agents, such as octadecanol and silicones, are used. In most cases, the composition of media and the process conditions are slightly different for both seed and fermentation stages based on the objectives of the two stages. Process development in antibiotic fermentation is highly critical and thus needs comprehensive attention and consideration. At the end of the fermentation period, the fermenter is harvested and the contents are subjected to isolation, extraction, and purification of the product. During the entire manufacturing process, maintenance of sterile conditions is highly crucial because a slight contamination of antibiotic-resistant microorganisms or with phages will lead to complete loss of the production batch.

Pharmaceutical presentation of antibiotics is mainly based on its uses and application. Because antibiotics are used in many different forms, each of which enforces fairly different kinds of manufacturing requirements and packaging material. For bacterial infections on the skin surface, an antibiotic may be applied as an ointment, cream, or lotion packed in suitable collapsible containers. For eye and ear applications, the antibiotic is dispensed in sterile containers and the product is known as eye/ear drops. In addition, for some antibiotics, eye ointments are also available in sterile containers. If the infection is internal, the antibiotic can be swallowed in the form of immediate release or extended release film-coated tablet formulations or injected directly into the body through the intramuscular or intravenous route dispensed in sterile containers (vials or ampoules). In these cases, the antibiotic is delivered throughout the body by absorption into the bloodstream.

11.2.3 Biotechnological approach

Soon after the discovery and commercial production of penicillin, scientists started working on the genetic aspects of antibiotic formation, primarily because of the increased demand for penicillin to treat the victims of World War II. The initial studies were based on the development of high-yielding mutant strains using both physical and chemical treatment of the wild strains. The process of mutation is a classic approach in which strain improvement is usually observed stepwise after several rounds of mutagenization are required before the identification and selection of a good and efficient strain. Some great success was recorded with mutation, not only with antibiotics but for other secondary metabolites as well. As a result, penicillin, productivity significantly increased and a number of multinational companies started working to develop their own mutant strains, not only for the production of penicillin, but also to discover other antibiotics as well. It has been reported that mutation may increase yield by approximately 100-fold to 1000-fold, and indeed it is the chief factor in antibiotic production. By the end of 1944, sufficient amounts of penicillin were produced with the help of mutant strains and, within 10 years' time, approximately 20 metric tons of penicillin was being produced every month (Porter 1976).

With the discovery of the complex mechanisms responsible for the regulation of antibiotics in various microbial strains, the classic mutation processes were further linked with genetic engineering techniques. This has not only shortened the strains' improvement and development time but has been able to reduce the unnecessary involvement of human resources during the mutant screening program. With the further advancements of new biotechnological approaches and tools, genetic engineering alone is quite sufficient to modify the strain, provided the biosynthesis mechanism is fully known for the desired antibiotic. Even if the biological mechanism is not fully known, protoplast techniques along with random mutagenization can be employed for strain improvement.

11.2.4 Therapeutic role

In the history of medicine, antibiotics have had great therapeutic applications and a great role in the management of various infectious diseases. However, with the introduction of new antibiotics as well as new synthetic antimicrobials, significant revisions have been made in the selection of the drug of choice for treatment (since the initial discoveries). The choice or selection is usually made on the basis of clinical diagnosis and culture sensitivity. The mechanisms by which the antibiotics acts on greatly helps in assessing the therapeutic role of various antibiotics. Broad-spectrum (or extended spectrum) antibiotics can be used to treat a wide range of infections, whereas narrow-spectrum antibiotics have limited applications as these are only effective against a few types of organisms. Some antibiotics also act against anaerobic bacteria and are of significant importance. Along with the therapeutic role, some antibiotics have prophylactic roles as well and are also used to prevent infection, especially before abdominal and orthopedic surgeries. The therapeutic role of some antibiotics is also dependent on drug–drug and drug–food interactions.

Most antibiotics produce prompt effect against infection depending on the site and type of application/dosage form. The onset of action depends on the peak plasma concentration, which is very short in case of the intravenous route, and takes time, maybe hours, if delivered orally. The most important part of antibiotic therapy is to complete the whole course of medication to prevent the infection from relapsing. With incomplete therapy or misuse of drugs, both have significant effects in developing resistant strains, which are quite difficult to treat. In addition, antibiotic overuse is also an important factor that contributes toward the growing number of bacterial infections which are becoming resistant to antibacterial medications. According to the European Centre for Disease Prevention and Control (ECDE), antibiotic resistance is becoming a serious public health threat worldwide. The ECDE statement issued on November 19, 2012 highlighted that approximately 25,000 people die each year in the European Union from antibiotic-resistant bacterial infections.

11.3 Antitumor agents

Exploration of new and effective antitumor drugs perhaps remains one of the most important topics in recent years throughout the world. This is primarily because the number of new incidences of cancer have significantly increased and is approaching more than six million every year (Chandra 2012). The initial antitumor drug development was mostly based on the discovery of molecules that have significant cytotoxic or cytostatic activity on tumor cells. However, during the last two decades, a large number of chemical compounds (with some degree of success) have been developed through chemical synthesis, or isolated from natural sources (such as plants, animals, and microorganisms) after having a better understanding of the complexities of tumor biology. The secondary metabolites produced by plants, microorganisms, and marine organisms have been extensively investigated for such products. Therefore, there is hardly any doubt in highlighting the importance of secondary metabolites in the discovery of antitumor drugs produced from natural sources. However, the great biodiversity of microorganisms represents an excellent possibility of producing secondary metabolites, which can be used as antitumor drugs. Along with these dynamic sources, microorganisms, especially bacteria, have also been explored to deliver antitumor drugs. In-depth studies on bacterial metabolites, with the advanced knowledge in molecular biology and genetic engineering, have widened the scope of antitumor drugs from bacteria and other microorganisms. The antitumor drugs, as reported during recent decades, from microbial sources have proven quite effective and less toxic for chemotherapy (Cragg et al. 2009; Ma and Wang 2009; Ravelo 2004).

The role of antitumor drugs from microbial sources has significantly increased and are now very much dominating as chemotherapeutic agents. This is mainly because of the resistance to conventional antitumor therapies, especially in patients with advanced solid tumors. Based on these, different forms of bacteria (e.g., live, attenuated, nonpathogenic, or genetically modified) have now begun to appear as promising sources of antitumor drugs. In recent years, the pigments produced by some bacterial species such as *Serratia* have been extensively investigated to develop antitumor drugs. A natural red bioactive pigment, prodigiosin (Figure 11.29), produced by *Serratia marcescens* (Figure 11.30) has demonstrated potential as an antitumor drug by exerting immunosuppressive activity *in vitro* and apoptotic effect *in vivo*. More research on prodigiosin from other bacterial species, such as *Vibrio psychroerythrus, Hahella chejuensis,* and *Streptomyces coelicolor* is under investigation to establish the production system, fermentation strategies, purification, and identification processes. Certainly, the role and application of biotechnology will play a significant role in establishing these processes.

The global market for antitumor drugs is increasing very rapidly. The leading pharmaceutical companies involved in antitumor drug business are Roche, Novartis, AstraZeneca, Eli Lilly, Celgene and some other pharmaceutical companies. Based on 2011 business

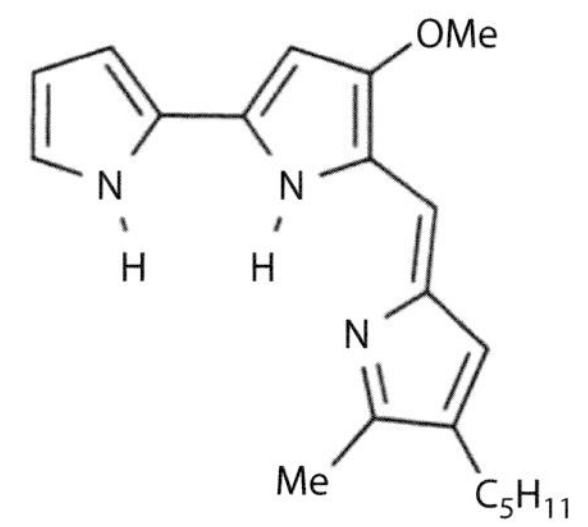

Figure 11.29 Prodigiosin.

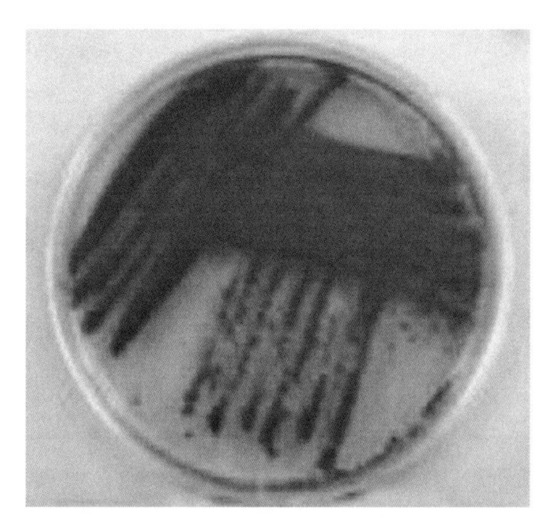

Figure 11.30 *S. marcescens* culture.

figures for antitumor drugs (US$68.8 billion), the proposed figure for 2012 was estimated at US$75 billion and further forecasted to 139.3 billion by the end of 2022. The top 25 highest revenue-generating drugs dominate the world market, generating 83% of revenues in 2010. Avastin, Rituxan, Herceptin, and Glivec/Gleevec are a few examples out of the 25 leading drugs in the anticancer treatment market. The overall revenues for anticancer treatment are expected to increase strongly from 2012 to 2022. In addition, individual drugs are expected to benefit too. With the emerging advancement in pharmaceutical biotechnology and the ever-increasing demand in antitumor drugs because of the increase in cancer incidence and prevalence, both in developed and developing countries, the future of this particular field holds great promise. Research and development (R&D) pipelines in oncology are quite strong, and a number of new drug molecules are under critical investigation and some are waiting for Food and Drug Administration approval. Thus, high revenues are likely to be generated from many drugs—small-molecule products and biological agents (biologicals) by the end of 2022. The top 25 highest revenue-generating drugs dominated the world market, generating 83% of revenues in 2010.

11.3.1 *Historical background and new challenges*

The true scientific approach to treating cancer was observed during the 1950s, after the discovery of vincristine and vinblastine (Figures 11.31 and 11.32) as alkaloids isolated from

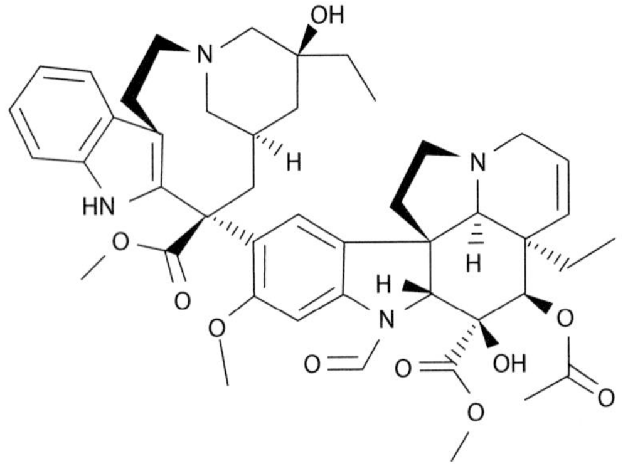

Figure 11.31 Vincristine.

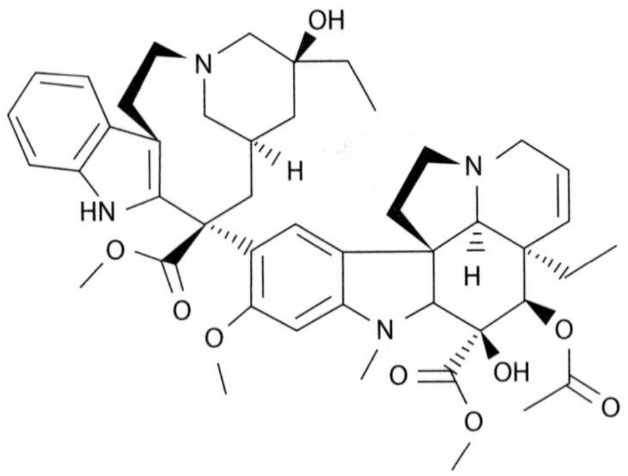

Figure 11.32 Vinblastine.

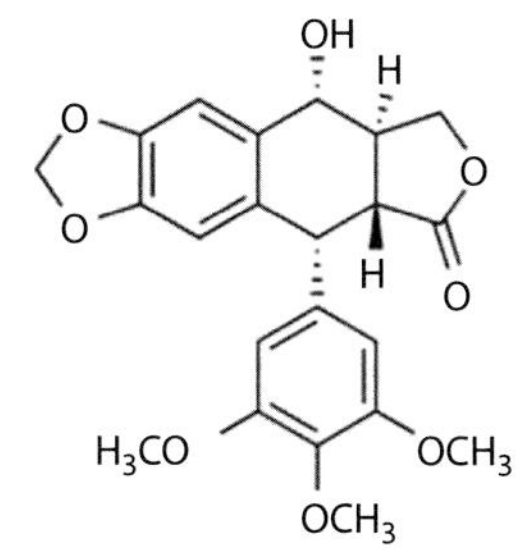

Figure 11.33 Podophyllotoxin.

Catharantus roseus. Vincristine was introduced successfully against non-Hodgkin's lymphoma, whereas vinblastine was introduced for Hodgkin's lymphoma, along with breast, testicular, and small lung cancer. Another nonalkaloidal drug, podophyllotoxin (Figure 11.33), isolated from the rhizomes of *Podophyllum peltatum* and *Podophyllum hexandrum*, was introduced for the treatment of genital warts caused by human papillomavirus (Harwell and Schrecker 1952).

By the end of 1990, after 40 years of struggle, scientists were able to identify a significant number of drugs suitable for treating different types of cancer belonging to different groups and chemical structures. Out of several classifications, the most acceptable was the one that included: DNA-damaging agents, antimetabolites, natural products and their analogues, hormonally directed agents, and biological response modifiers (American Medical Association 1995). A majority of the antitumor drugs (~60%) are reported to be of natural origin, that is, plants and microorganisms. Among the microorganisms, doxorubicin, dactinomicines, mitomycin, bleomycin, borrelidin, β-glucan, becatecarin, Eco-4601, NPI-0052, romidepsin, texol, spicamycine, amrubicin HCl, tartrolon D, elsamirucini, and prodigiosin are quite important. It is now well-documented that antitumor drugs from microbial sources have some reasonable advantages in terms of the diversity of the compounds and manipulation in production. One of the great challenges for researchers is to utilize bacteria to enhance the antigenicity of tumor cells using immunotherapeutic strategies, an emerging and effective approach in cancer therapy. Some work in this direction is underway using attenuated *Salmonella typhimurium*, which has demonstrated successful invasion of melanoma cells. *Salmonella* can infect the malignant cells both *in vitro* and *in vivo*, thus triggering the immune response (Patyar et al. 2010).

11.3.2 Production and pharmaceutical presentation

Submerged fermentation is the method of choice for the production of microbial antitumor drugs. Primarily, the process can be divided into a number of steps such as maintenance of microorganisms, preparation of inoculum, propagation of organism, isolation/extraction, purification, and bulk packaging. All these steps are quite specific and depend on the type of microorganisms planned for use in the production of a drug. To understand the basic phenomenon, an example of prodigiosin produced by a strain of *S. marcescens* is described below.

The maintenance of a culture is usually done at low temperature using suitable maintenance media such as nutrient agar. For subculturing, nutrient broth or peptone glycerol broth can be used. The inoculum is usually prepared using shaken flask conditions for about 24 h. The inoculum is aseptically transferred to a seed vessel containing nutrient broth. The pH of the medium is maintained between 3 and 7, temperature at 28°C, and agitation at 200 rpm. After 12 h, the entire contents of the seed vessel is transferred to a fermenter containing nutrient broth (peptone 10 g/L, sodium chloride 5 g/L, yeast extract 3 g/L + 0.5% glucose or maltose) supplemented with glucose or maltose. Sesame seed medium is also equally effective in producing prodigiosin without adding glucose or maltose. Small amounts of thiamine and ferric acid can enhance the productivity of the product. The pH of the medium, fermentation temperature, and agitation rate is kept similar with that of the seed vessel. A temperature higher than 37°C and lower than 20°C can greatly inhibit the product of prodigiosin.

After 36 h of fermentation, the pigment can be isolated, extracted, and purified. The broth is filtered and then centrifuged at 10,000 rpm for 15 min using an ultracentrifuge. The supernatant is extracted with ethyl acetate and the pigment is extracted with acetone followed by ethyl acetate to remove the impurities. The two fractions after mixing are subjected to evaporation. Silica column of 80 to 100 mesh size can be used for separation of the noncolored impurities from the pigment. The dried powder at different concentrations was used for plotting the standard graph versus absorbance at 535 nm. The purified sample showing a single peak absorbance at 535 nm in the UV spectrophotometer was further analyzed for determination of molecular weight using a mass spectrophotometer.

Antitumor drugs are available in different dosage forms based on their chemical nature and the types of application required for therapy, such as topical, oral, subcutaneous, intramuscular, intravenous, intrathecal, intraperitoneal, intralesional, intrahepatic artery, intracavitary, intravesical, and isolated perfusion. Chemotherapy is mostly delivered through the intravenous route, although a number of drugs such as melphalan, busulfan, and capecitabine can be administered orally. In some special cases, such as melanoma, the isolated limb perfusion or isolated infusion of chemotherapy into the liver or the lung have been used. The main function and objectives of these approaches is to supply the required dose of antitumor drugs, which is usually very high at the site of the tumor, and protect from significant systemic damage. Infection risk is common during antitumor therapy. Therefore, to lower the risk of infection during continuous, frequent, or prolonged intravenous chemotherapy, various systems, such as the Hickman line, the Port-a-Cath, and the PICC line are commonly inserted (following surgery) into the vasculature to maintain access. These systems have reasonably lower infection risk and are much less prone to phlebitis or extravasation, and eliminate the demand for repeated insertion of peripheral cannulae.

11.3.3 Biotechnological approach

The future role of biotechnology in developing and understanding antitumor drugs looks quite promising. Recent advancements and approaches have significantly increased the possibility of exploring genetically modified microorganisms and the use of gene-directed enzymes, especially bacteria, to deliver potent drugs for the selective destruction of tumors. Microbial toxins for the destruction of tumors along with the introduction of cancer vaccines are leading projects as well as objectives of various research institution and top multinational groups of companies, presently engaged and dominating the antitumor and vaccine market. For more targeted tumor therapies, some recent biotechnological approaches, such as conjugation of toxins with monoclonal antibodies or some polymeric carriers have also been suggested and are under investigation. Tumor-amplified protein expression therapy using genetically modified organisms, especially bacteria as a vector or vehicle for preferentially delivering drugs to solid tumors (e.g., brain tumors) is also under study and development. In addition, newer biotechnological techniques can also be applied during submerged fermentation of the organism to enhance the production yield of the desired product. One of the leading approaches is the immobilization of microorganisms in a special carrier to enhance the proliferation and productivity of the desired product.

11.3.4 Therapeutic role

The antitumor drugs available today have diverse modes of action and thus vary in their effects on different types of cancer and normal cells. A single "cure" for cancer has proved subtle because of the large number of different types of cancer cells identified thus far. Furthermore, the biochemical differences between normal and cancerous cells is also one of the key factors in selective treatment and drug design. Therefore, the therapeutic role and application of many antitumor drugs, especially the traditional chemotherapeutic agents, are limited due to their toxicity to normal, rapidly growing cells in bone marrow, digestive tract, and hair follicles. The ultimate results/side effects of chemotherapy appear as myelosuppression (decreased production of blood cells, hence also immunosuppression), mucositis (inflammation of the lining of the digestive tract), and alopecia (hair loss). Some newer antitumor drugs such as monoclonal antibodies are not extensively cytotoxic, but rather focus proteins that are uncharacteristically expressed in cancer cells and are indicated as essential for their growth. The mechanism of these antitumor drugs is stated as targeted therapy, which is quite distinct from classic or traditional chemotherapeutic agents and are also used alone or in combination with traditional chemotherapeutic agents in antineoplastic treatment regimens.

11.4 Ergot alkaloids

Fungal metabolites, and especially the alkaloids, have been focused on and studied with great importance and interest in view of their significant biological activities. It symbolizes a cluster of fascinating and multifaceted chemical compounds containing at least one nitrogen atom in a ring structure in the molecule. These chemical compounds are produced as metabolites by the living organisms, including fungi in different biotopes. Fungal alkaloids have significant biopharmaceutical importance as some are used as drugs. The best example that can be given here is that of "ergot alkaloids," produced particularly by *Claviceps purpurea*, which is usually regarded as a parasite of cereals and grasses. Industrial production started in 1918 with ergotamine tartrate, and Sandoz (now Novartis) was the first company to start marketing it in 1921 and dominated the market until the 1950, when competitors began to emerge. Other reputable companies producing ergot alkaloids include: Boehringer Ingelheim (Germany), Galena (Czech Republic), Gedeon Richter (Hungary), Lek (Slovenia), Poli (Italy), Eli Lilly (USA), and Farmitalia (Italy). The 2006 figures indicated world production of ergot alkaloid between 5.0 and 8.0 metric tons, of which approximately 60% resulted from fermentation and the rest (40%) from field cultivation of a hybrid of wheat and rye (Schiff 2006). However, in 2010, the total production of these alkaloids significantly increased to about 20.0 metric tons, and the production ratio between field cultivation and fermentation became equal (50% each). Chemically, the identification of ergot alkaloids is based on the presence of a tetracyclic ergoline ring system and is derived from an amino acid, L-tryptophan (Figure 11.34). By nature, ergot alkaloids are indole compounds and signify the biggest group of nitrogenous fungal metabolites.

More than 80 different ergot alkaloids have been reported in the literature and *Claviceps* species were noted to contribute the highest number, more than 70 alkaloids.

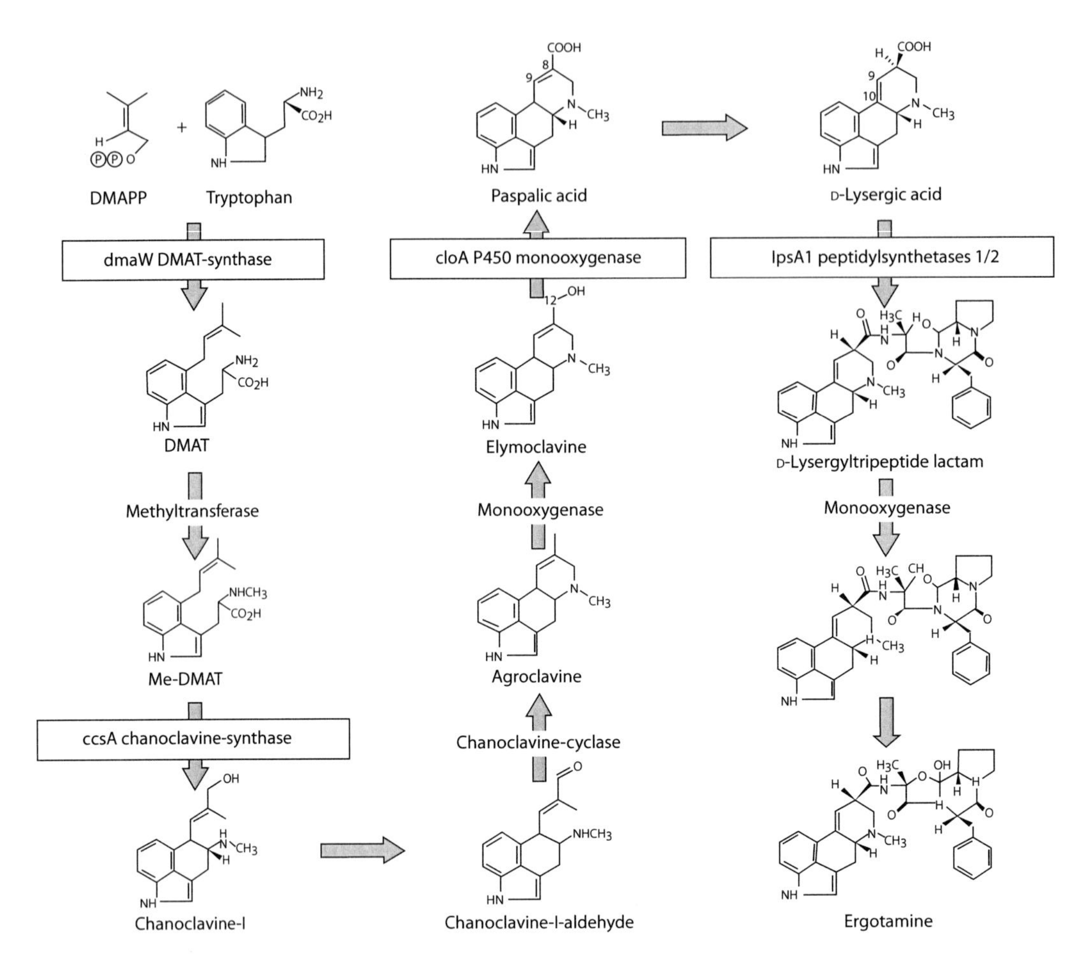

Figure 11.34 Biosynthesis of ergot alkaloids from tryptophan.

From a medicinal point of view, ergot alkaloids can be divided into two main groups, that is, the water-soluble amino alcohol derivatives and the water-insoluble peptide derivatives. The water-soluble amino alcohol contains 20% of the total alkaloid mixture, whereas the water-insoluble peptide derivatives constitute the remaining 80% of the alkaloids. However, based on their chemical nature, it will be more realistic to divide ergot alkaloids into three major groups: clavine alkaloids, D-lysergic acid and its derivatives, and ergopeptines (Hulvova et al. 2013). The clavine alkaloids (Figure 11.35) are substituted 6,8-dimethylergolines, which include approximately 35 different members, such as the ergolines, chanoclavines, agroclavines, and others, but none of the members of this group are of medicinal importance.

Lysergic acid and its derivatives (lysergic acid-derived amides) are significantly active pharmacologically. Ergonovine and its semisynthetic derivatives, such as methylergonovine and methysergide are quite important and have applications in medicine. The naturally occurring ergopeptines, such as ergotamine is the only example that is used medicinally. However, useful semisynthetic derivatives of peptide alkaloids (Figure 11.36) include: brominated ergocrytine (e.g., dihydroergotamine) and bromocriptine and dihydrogenated ergot alkaloids (e.g., ergoloid).

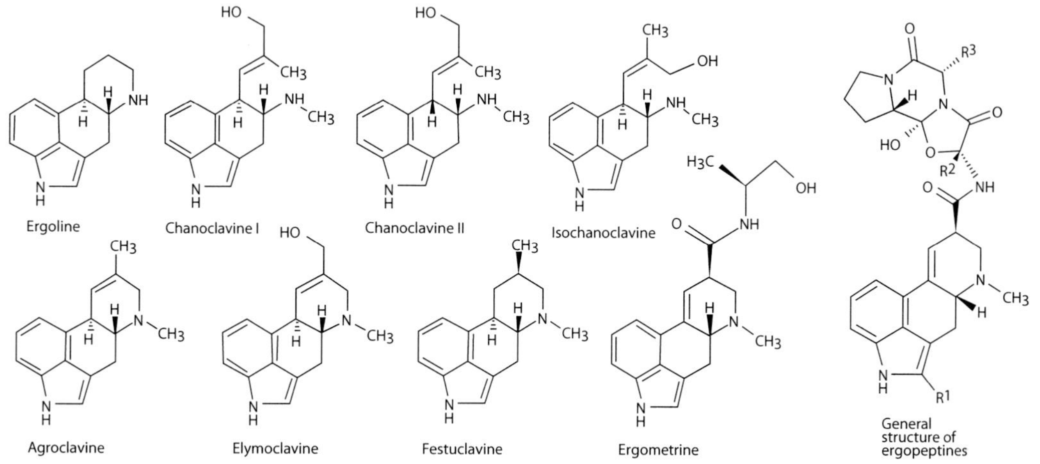

Figure 11.35 Structures of clavine alkaloids.

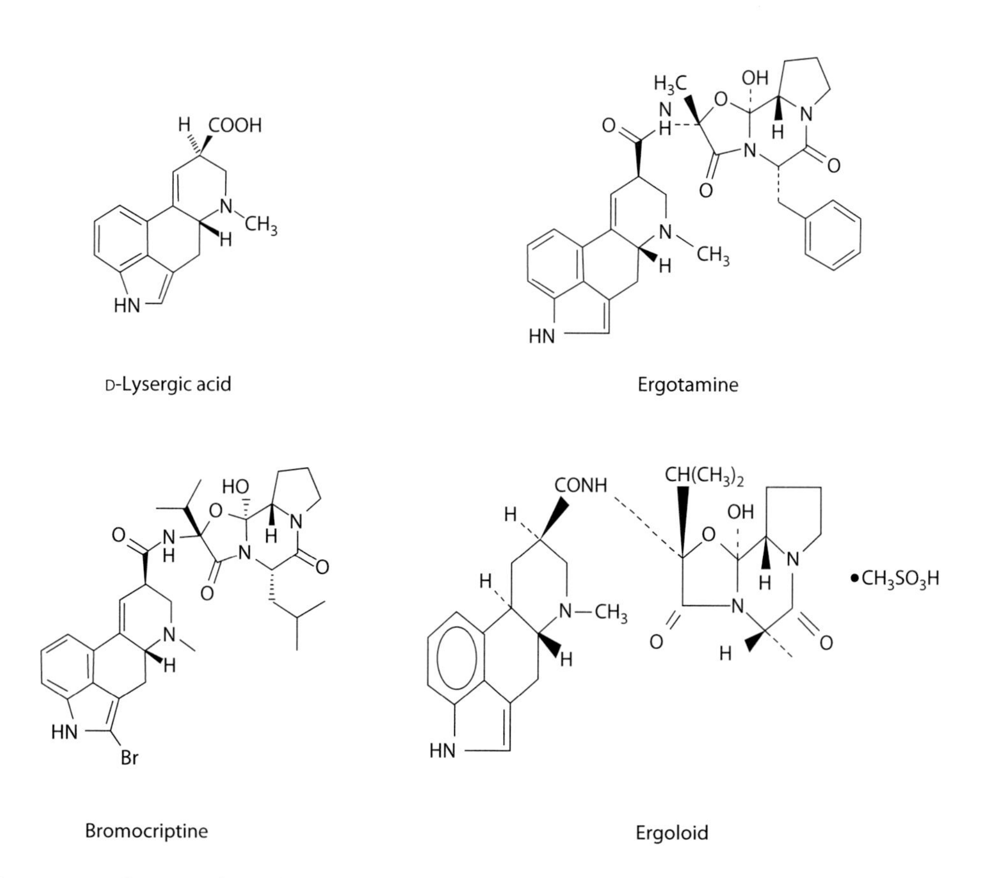

Figure 11.36 Semisynthetic derivatives of peptide alkaloids.

11.4.1 Historical background and new challenges

The first microbial alkaloids were identified and reviewed in *C. purpurea*. Collectively, these are known as ergot alkaloids and can be isolated from the sclerotia produced after infection of the ovaries of the plant by *Claviceps* ascospores or conidia. Apart from *C. purpurea*, ergot alkaloids have also been identified in other fungal genera, such as *Aspergillus* and *Penicillium* (Rehacek 1984). The number of alkaloids found in fungi and other microorganisms is not large but it is believed that it would increase rapidly if more extensive research was carried out. In the case of ergot, it was observed that no comprehensive attempt was made until 1960 to obtain these alkaloids on a commercial scale for its proper utilization in medicine. In 1960, a number of studies were published that claimed to produce around 1,000 μg/mL of ergot alkaloids (lysergic acid hydroxyethylamide) through submerged fermentation using *Claviceps paspali*.

Continued research on the subject all over the world lead to the discovery of more important alkaloids such as lysergol, lysergine, and lysergene from the saprophytic culture of ergot fungi in 1961. This was the beginning of applying submerged fermentation technology for the production of ergot alkaloids. Finally, a group of scientists reported on the successful fermentative production of lysergic acid in 1964 using *C. paspali*, followed by ergotamine from *C. purpurea* in 1966 to 1967 (Amici et al. 1966, 1967; Kobel et al. 1964; Tonolo 1966). Further research in these directions, contributed to the fermentative production of ergocryptine and ergotamine, ergocornine, ergosine, and ergocristine with high yield using *C. purpurea* (Amici et al. 1969).

11.4.2 Production and pharmaceutical presentation

The production of ergot alkaloids can be done through artificial parasitic cultivation on rye (traditional field cultivation) or by using fermentation, surface culture, or submerged fermentation techniques. Depending on the resources in different regions of the globe, these methods are still used for the production of alkaloids. In the traditional field cultivation process, a conidial suspension is sprayed onto field-cultivated rye and is observed for the development of various stages. The genus *Claviceps* has 36 members that can infect approximately 600 different species of monocotyledonous plants (Hulvova et al. 2013). Once the mycelium (mat-like body) of the fungus develops in the ovaries of the host plant; it eventually turns into a hard pink or purple body, the sclerotium, or ergot, which resembles a grain of rye in shape. The maximum alkaloidal content is usually observed after approximately 20 days. The sclerotium contains alkaloids that are subjected to extraction with a toluene/ethanol solvent mixture to obtain the primary extract. Further purification of the alkaloids is done by liquid–liquid extraction and a semipurified toluene extract is obtained. For further purification, partial evaporation of the toluene extract is done to obtain a crystalline product. A number of other solvents, such as methanol and chloroform, have also been used for extraction. The isolation and extraction of ergot alkaloids with efficient processes using nontoxic and environmentally friendly solvents are still desirable. According to some studies, approximately 10 to 20 kg of ergot alkaloids can be recovered from 1 to 2 metric tons of sclerotia. The climatic conditions can have great effect on the yield of the product (Tudzynski et al. 2001).

For surface culture, the *Claviceps* are grown on suitable artificial medium containing a carbon source (such as sorbitol, mannitol, or sucrose) supplemented with some inorganic salts, and maintained at a required temperature in flat glass bottles or using sterile plastic bags.

Submerged fermentation is now widely used in different regions to produce some ergot alkaloids through the production of paspalic acid, which is ultimately converted to D-lysergic acid through the process to some other semisynthetic alkaloid derivatives. Process parameters, media composition, and the addition of certain nutrients and precursors are key factors in the fermentative production of ergot alkaloids and have the capacity to affect overall yield of alkaloids. The cultures can be maintained on potato dextrose agar slants at low temperature. After extensive studies, the optimum pH for the production of alkaloids in saprophytic culture was reported to lie between pH 5 and 6. For surface culture technique, a medium containing yeast extract, glucose, mannitol, and peptone could be used. Strains of *C. purpurea* grown on a peptone-mannitol medium can produce high alkaloid content (fivefold increase), if concentrations were increased from 5% to 25%. The effect of various macroelements and microelements, along with carbon and nitrogen sources for the production of ergot alkaloids have been studied by various research investigators to optimize the production of alkaloids. The concentration of phosphate in the medium plays the most important role because it is essential for the production of secondary metabolites. The exhaustion of phosphate in the medium will affect the yield of alkaloids. KH_2PO_4 added to the medium at a concentration of 1000 mg/L has been observed to produce the maximum yield of fermented broth, whereas the addition of glycols and Tween 80 has also been reported to increase the alkaloid yield (Mahmood 1983).

The pharmaceutical presentations of ergot alkaloids are highly significant in terms of achieving the desired effects. The regulatory issues and legislation are somewhat different in various countries; therefore, the availability of the finished dosage form and their applications differ from region to region. Various dosage forms, such as film-coated tablets, sublingual tablets, capsules, drops, and powders are available for oral administration. Injectable preparations, suppositories, and nasal sprays are also available. Ergoloid mesylate is available in 0.5, 1.0, 2.0, and 4.5 mg tablets, as well as 1 mg capsules, 1 mg/mL drops, and in 0.3 mg/mL injections in many countries. These dosage forms are recommended for very selective applications, for example, to increase cerebral metabolism and blood flow and in Alzheimer's disease, in which ergoloid may lower the scores on some cognitive and behavioral rating scales. Application in dementia needs more controlled studies to determine the risk–benefit profile. Ergotamine tartrate (Ergomar) is available as 2 mg sublingual tablets for the treatment of severe migraine headache and cluster headache. Ergotamine tartrate is also available in combination with caffeine (Cafergot) containing 1 mg ergotamine + 100 mg caffeine for the treatment of migraine attacks with or without aura. Another combination formulation containing 2 mg of ergotamine tartrate, 50 mg of cyclizine hydrochloride, 100 mg of caffeine hydrate (Migril) as well as 2 mg of ergotamine + 100 mg of caffeine as suppositories (Cafergot suppository) are also available for the same indication. Methylergonovine maleate in 0.2 mg tablets and 0.2 mg/mL injection (Methergine) are available for migraine. Dihydroergotamine is available as a 4 mg/mL nasal spray, and in 1 mg/mL injections. Bromocriptine mesylate (Parlodel) is available in 2.5 mg tablets and 5 mg capsules.

11.4.3 Biotechnological approach

Despite the valuable information and material published on ergot alkaloids, less attention has been focused on the use of biotechnological tools for further improvements in yield through submerged fermentation by optimizing the process condition and manipulating the media composition. In fact, the field of fungal alkaloids is demanding that researchers pay more attention and explore new molecules to combat the new challenges emerging in the healthcare system. During the early phase of biotechnology, ergot alkaloids were reported as direct gene products. Thus, more emphasis was given to enzymatic transformation for their synthesis via small precursor molecules. However, the enzymes linked with the relevant structural and regulatory genes were not mapped, and although the isolation and cloning of the structural genes had been proposed, it was considered a difficult task. The possibility of gene transfer via protoplast fusion remained one of the leading topics during the early phase of *Claviceps* genetic engineering (Keller et al. 1980; Maier et al. 1980). Continuation of research in the field lead to the discovery of plasmids associated with mitochondria in a wild strain of *C. purpurea* (Tudzynski et al. 1983).

Later, the construction of a plasmid that replicates automatically, the development of a transformation system with drug resistance markers, and the development of genes controlling the regulation of the entire pathway were proposed (Rehacek 1984). In 1995, the first pathway gene cluster, *dmaW*, was reportedly cloned from *C. fusiformis* SD58, which encodes the determinant step in ergot alkaloid biosynthesis (Tsai et al. 1995), this was followed by the identification of the orthologue in a *C. purpurea* strain P1 (Tudzynski et al. 1999). Further research on this cluster has been extended and, thus far, more than 68.5 kb of the cluster has sequences containing 14 genes and is coordinately induced under ergot alkaloid production (Schardl et al. 2006, Wallwey and Li 2011).

Thus far, most of the randomly mutated strains of *C. purpurea* were employed in the production of ergot alkaloids on an industrial scale, but with the recently sequenced ESA cluster of 14 genes, it is expected that a guided overexpression of certain genes or targeted upregulation of the entire cluster could significantly increase the yield of ergot alkaloids (Hulvova et al. 2013). The gene *dmaW* is now being identified in all *Claviceps* species but, at the same time, *dmaW* has also been reported to pass through duplication and losses (Liu et al. 2009). Therefore, the possibility of loss of mutation and deletions of the *dmaW* genes associated with *C. purpurea* does exist, which may affect productivity and can be validated on a pseudogene of *dmaW* that has already been detected in an unspecified *C. purpurea* strain. Gene transfer via protoplast fusion, which was proposed in the early 1980s, has successfully been used to develop a *C. purpurea* mutant strain overexpressing genes *easC* and *easG* under a constitutive promoter of glycerol-3-phosphate dehydrogenase from *Aspergillus nidulans*; however, no significant increase in the production of ergot alkaloids was noted (Hulvova et al. 2010). Genetic manipulation of the proteins engaged in the signaling cascade and ultimately influencing biosynthesis has also been proposed as a possible way to increase the productivity of ergot alkaloids (Lorenz et al. 2009; Mey et al. 2002). Most recently, genome sequencing of some *Claviceps* species has also been reported and some work is also under progress in the Czech Republic where *Claviceps africana* is being sequenced for a possible increase in alkaloid productivity (Hulvova et al. 2013).

11.4.4 *Therapeutic role*

Ergot alkaloids have a long history of benefits and are observed to play a key detriment to humans in different civilizations and, ultimately, are used primarily as an aid in childbirth and in the treatment of some neurological and cardiovascular disorders. In view of the prominent biological activity and well-established pharmacological activity, ergot alkaloids are of considerable importance in pharmaceuticals and were thus introduced into the first edition of the United States Pharmacopeia in 1820 and into the London Pharmacopeia in 1836. The alkaloid possesses adrenoblocking, antiserotonin, and dopaminomimetic effects because of the structure's similarities with receptors, such as adrenaline, serotonin, and dopamine, respectively. The best therapeutic application of ergot alkaloids is in the management of migraine headache and hypertension (e.g., ergotamine and dihydroergotamine are responsible for vasodilatation of arteriovenous anastomoses and can affect α_1 and α_2 adrenergic receptors), Parkinsonism and hyperprolactinemia (bromocriptine derived from ergocryptine) to stimulate the contraction of uterus postpartum, mastopathy, and as a sedative. As an anti-Parkinson agent, the ergot alkaloids are not prescribed as first-line therapeutic agent due to the risk of fibrotic reaction. Some ergot alkaloids such as agroclavine and elymoclavine possess antibiotic activity as well, but are not used in clinical practice.

11.5 Vaccines

Among the biologicals, one of the leading and most extensively studied products is vaccines. These are given to generate immune responses against a number of diseases. Although the process of immunization against certain infections has been observed and noted to be a centuries-old phenomenon, it was only given a proper status and recognition with the introduction of modern immunological processes and techniques. With the advancement in molecular biology and microbial physiology, the classic pragmatic approach for vaccines has now completely changed for anticipated immune response in controlling a particular infectious disease. The product presentation is usually done in the form of a pharmaceutical suspension of live, inactivated, or fractionated microorganism that has been depicted as nonpathogenic. Despite the fact that the safety and effectiveness of vaccines are highly questioned, its high benefit-to-risk ratio strongly supports its use in providing long-lasting, humoral and cellular immunity. The memory T-cells (CTL or Th1 cells) derived from normal T-cells play a significant role in the development of long-lasting immunity. The T-cells represent an extremely advanced wing of the adaptive immune system, which is able to differentiate between pathogens and is efficient at progressing or adapting during the lifetime of an individual such that immunity is enhanced with each consecutive contact with a pathogen.

The importance of immunization and the use of relevant products can further be supported by the fact that in a decade's time, the vaccine market has grown from US$5.7 billion to approximately US$27.0 billion and is further expected to grow at the rate of more than 8% through 2018 (Bryant 2012; Palmer and Bryant 2013). The success rate or efficacy of vaccines usually depends on the prevailing body immunity conditions or level (low, moderate, or high), which is again dependent on various factors such as diabetes, use of steroids, age, ethnicity, genetic predisposition, HIV infection, and others, and the capability of the cells of the host's immune system to generate antibodies to specific antigens. Apart from this, the packaging, adjuvants, and transportation of vaccines have significant implications on effectiveness and thus need special considerations to manage and maintain a high degree standard as well. The system used for storage and distribution of vaccines in a suitable condition is termed as cold chain, which consists of a series of storage and transport links, all of which are intended to retain the vaccine at an appropriate temperature until it reaches the end user.

The European companies and US-based companies are the biggest contributors supplying vaccines all over the world. Based on 2012 revenues, the top five vaccine manufacturers (which includes Sanofi, Merck, GSK, Pfizer, and Novartis) have captured 90% of the world's total vaccine business. The major part of the growth in vaccine sales has been recorded in the adult influenza market, with the huge emphasis on flu restraint and an acceptance of adult vaccines (Global Vaccine Market 2010; Palmer and Bryant 2013). According to reports, total time of development is usually 12 to 15 years with an investment of approximately US$0.5 to 1 billion. This includes the selection of proper vaccine candidates, followed by phase I, phase II, and phase III trials. The successful completion of phase III trial data permits entry into a licensing program. Currently, approximately 145 pure vaccines and 11 combination vaccines are available for human use and the segment includes pediatrics, adolescents, adults, and the elderly (Castellani 2010).

11.5.1 Historical background and new challenges

The long and tedious history of vaccines began with the discovery of the smallpox vaccine in 1796 by Edward Jenner, a term derived from "vacca," meaning cow. Jenner used cowpox virus (vaccinia) to develop a vaccine that prevented smallpox infection. This was followed by a series of discoveries, which are highlighted below under types of vaccines along with year of discovery, the name of the discoverer, and the routes of administration. The types and classifications of vaccines usually depend on the strategies and applications, not only to reduce the risk of illness but also to induce a beneficial immune response.

11.5.1.1 Types of vaccines

Killed/inactivated vaccines—these vaccines are produced from killed or inactivated microorganisms. Because the process of inactivation involves subjecting the pathogens to rigorous lethal conditions, there can be variations in the antigenicity of these vaccines. Therefore, repeated booster doses are recommended to maintain long-term immunity. Some specific examples include: vaccines against cholera (1879, Louis Pasteur, now not in use), rabies (1885, Louis Pasteur and Emile Roux), plague (1897, Alexandre Yersin), pertussis (1926, Thorvald Madsen), influenza (1945, Thomas Francis), polio (1952, Jonas Salk), anthrax (1930s in Russia; 1950s in the UK and the US), rabies HDCV (1967, Koprowski, Wiktor, and Plotkin), rabies inactivated vaccine (1971, Koprowski, Wiktor, and Plotkin), and hepatitis A (1992, Hilleman).

Live, attenuated vaccines—these vaccines contain live microorganisms that have been weakened (attenuated) to disable their virulent properties but to retain their antigen-producing capacity. Therefore, these can no longer be capable of producing any pathological complaints. However, some special precautions are required so that the attenuated forms could not mutate back into their virulent forms, producing the same disease that they were meant to prevent. Some specific examples include: typhoid (1896, Wright), tuberculosis BCG (1921, Calmette and Guerin), yellow fever (1926, Max Theiler), polio oral (1950, Koprowski), measles (1963, Enders; 1969, Hilleman), mumps (1967, Hilleman), rubella (1969, Hilleman), MMR (1971, Hilleman), chicken pox (1974, Hilleman), cholera (1980, oral), and rotavirus (1998).

Toxoids—these vaccines are manufactured using toxins produced by a pathogenic organism. After chemical treatment, the harmful effects of the toxin are removed and the inactivated toxin is used to produce a lower level of immunological protection. Some specific examples include: tetanus (1890, Behring) and diphtheria (1913, Behring).

Subunit vaccines—these represent antigens such as proteins to the immune system without introducing whole organisms. Specific examples include: vaccine against hepatitis B with protein subunits (1981, Hilleman), the virus-like particle vaccine against human papillomavirus (2006, Frazer and Zhou), and the hemagglutinin and neuraminidase subunits of the influenza virus.

Conjugate vaccines—these contain the polysaccharide outer coats with proteins or toxins. Specific examples include: *H. influenzae* type B vaccine (1977, Hilleman), pneumococcal (1977, Hilleman), meningococcal (1978, Hilleman), and typhoid oral (1975).

11.5.1.2 Challenges

Despite the fact that the vaccine industry has been proved to be one of the fastest-growing sectors during the past decade, the vaccine business also has had great challenges, both from the development and marketing points of view. Manufacturers are trying to develop new and advanced methods, not only to produce cost-effective and safe vaccines but also at the same time looking into novel vaccine delivery systems, such as needle-free patches, edible vaccines, and nasal sprays. Currently, the US and European multinational companies are dominating the global market; however, they face a great threat with the emerging Chinese and Indian companies. With the introduction of new tools and the growing awareness for vaccine-preventable diseases, adults and the elderly are entering into focus as an appealing target population for future vaccine development. The most emerging and challenging area for big companies is to introduce new vaccines to combat malaria, dengue, tuberculosis, HIV/AIDS, and vaccines against cancer.

In addition, vaccines against diabetes, asthma, allergy, and hypertension are also under development. These vaccines are quite challenging and still need considerable work to move forward. Furthermore, it will be too early to predict any reasonable progress that has been noted in developing these specific vaccines. Some highly constructive developments have been noted to introduce peptide-based malaria vaccines capable of inducing antibodies against *Plasmodium falciparum*. Although the peptide sequence (Glu-Val-Leu-Tyr-Leu-Lys-Pro-Leu-Ala-Gly-Val-Tyr-Arg-Ser-Leu-Lys-Lys-Gln-Leu-Glu) derived from MSP-1, which is conserved in all parasite strains, but success remains questionable and more work is still required to develop malaria vaccine because *P. falciparum* is highly effective at avoiding detection by the immune system through different mechanisms (Lozano et al. 2006; Owens and Schweizer 2011; Urquiza et al. 1996).

Transportation of vaccines is a highly gray area, which has great implications for stability and effectivity. The packaging conditions and temperature throughout the transportation period play a significant role in the success of the immunization process. The World Health Organization (WHO) provides reasonable guidelines for the packaging and shipping of vaccines. Any deviation from standard recommended packaging and the temperature is likely to result in an unsuccessful immunization process (WHO 2002). A time–temperature indicator known as a vaccine vial monitor (VVM) is absolutely vital to know that the vaccine has not been exposed to excessive heat and is suitable for immunization (Figure 11.37). This has now been successfully used for different vaccines. The VVM indicates three distinct stages based on color change—"use this vaccine," "discard point," and "do not use this vaccine" (Figure 11.38).

Figure 11.37 VVM. Manufactured by Temptime Corporation, Morris Plains, NJ.

In addition, the process of immunization itself is also quite challenging for some vaccines, and in a number of countries, because of some highly specific issues such as infertility, autism, and others, after the scientific data published by Slovakian researchers wherein polysorbate 80 (a nonionic surfactant), which is used as an emulsifier/solubilizing agent in some flu vaccines, was found to damage the vaginal and uterine lining, and induced hormonal changes, ovarian deformities, and infertility when injected into newborn female rats (Gajdova et al. 1993). After this, most manufacturers of flu vaccines using polysorbate 80 began highlighting "the manufacturer cannot guarantee that your fertility will be unharmed," in their package inserts. Another example of infertility and other side effects, such as anaphylactic shock, associated with polysorbate 80 present in HPV, have recently been highlighted in Australia and New Zealand through media and Internet campaigns, which asked parents not to allow their children to take part in HPV vaccination programs. Also, there is a general belief in some African countries, Afghanistan, and in northern parts of Pakistan that oral polio vaccine may cause infertility, and thus the situation remains challenging despite the fact that polio cases are still being reported from these areas. Perhaps the greatest misunderstanding or controversy developed during 2003 to 2004 when the State Government of Kano, Northern Nigeria detected the presence of two sex hormones, estrogen and progesterone, in polio vaccine. Although the concentration detected was very low and perhaps would not have been able to inhibit fertility, this caused tremendous concern among Nigerians and it was treated as a misdeclaration from the manufacturer and they launched a campaign, saying—"nobody wants their child to be crippled by polio, and nobody wants her child to be sterile, either." Although a clarification was made on the situation by saying that because the polio vaccine is developed in a culture made from monkey kidney, which contains these two sex hormones, and because the hormones are highly water-soluble, traces are bound to be found in the vaccine (Maiyali 2013).

However, it is still not clear, if by any means, the levels of these two sex hormones increased in monkey kidney due to some pathophysiological abnormality, then what will

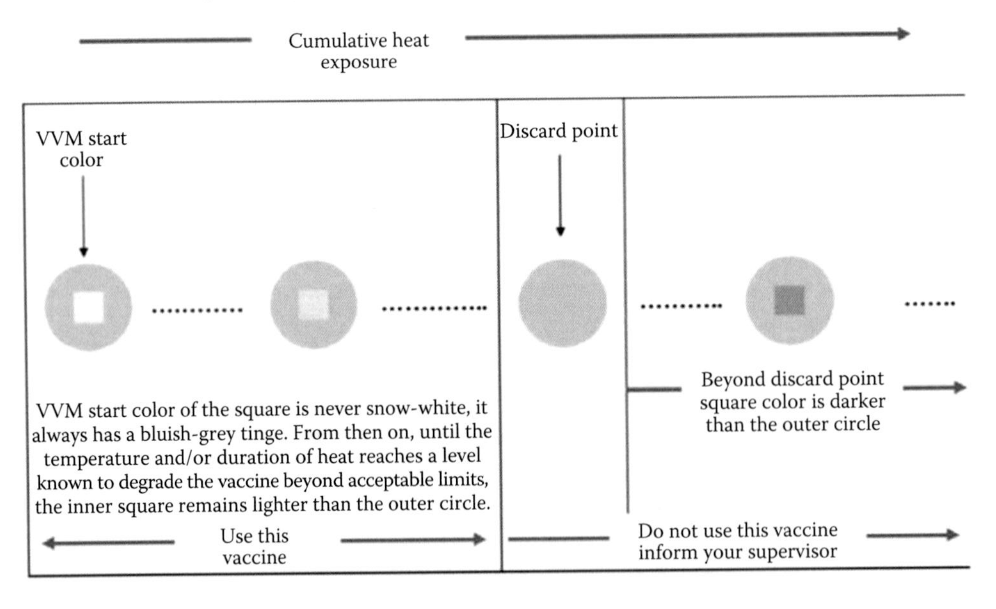

Figure 11.38 VVM showing three distinct stages based on color change—"use this vaccine," "discard point," and "do not use this vaccine." Manufactured by Temptime Corporation, Morris Plains, NJ.

be the consequences if the concentrations of these two sex hormones increased in the vaccine? Because generally, the batches of polio vaccines are released without monitoring or estimating the levels of these sex hormones.

Another challenging situation developed relating to polio vaccines when in 1992, an interview with Hilary Koprowski was published in *Rolling Stone Magazine*, mentioning oral polio vaccine (OPV) as a possible source of HIV and, in turn, the AIDS epidemic. The possible link between OPV and HIV was based on the information that chimpanzees' kidney cells were used to culture polio virus, which had been infected with simian immunodeficiency virus (SIV), and thus a vaccine produced through such a cell culture could lead to HIV infection. Hilary Koprowski was a pioneering virologist who developed the first successful oral polio vaccine (he recently died at the age of 96). Although, at that time, a number of clarifications and explanations were given at significant levels, it was really difficult and it is still challenging to regain the confidence of the African people on the subject and this certainly needs to be addressed correctly with more justifications.

There is a huge list of ingredients (thimerosal, causing autism; aluminum, a neurotoxic; 2-phenoxyethanol, affecting organs and causing CNS disorders; formaldehyde, a nephrotoxic; monosodium glutamate, causing retinal degeneration; MRC-5 and WI-38 cellular protein taken from aborted human fetuses, serious ethical objections; and oxtoxinol, a vaginal spermicide) that are used for different purposes such as preservatives, stabilizers, and adjuvants in different vaccines, which certainly needs more attention and studies so that alternatives for these toxic chemicals and objectionable ingredients can be found.

11.5.2 Production and pharmaceutical presentation

The large scale of vaccine production is generally an industry secret and most multinational companies have the patent rights to produce particular types of vaccines. However, a generalized production method with raw materials and area requirement is available to understand the basic phenomenon of vaccine production. A highly dedicated area and controlled conditions are required for vaccine manufacturing. However, sections and the modules of the manufacturing process are usually specific to either viral or bacterial vaccine production. Various manufacturing steps (such as propagation, isolation, purification, and formulation) are conducted under aseptic conditions and the procedure is strictly monitored for Good Manufacturing Practices (GMP) compliance throughout the whole process, which was designed and recommended by some international regulatory authorities.

The propagation step involves the growth and development of selected organisms in a suitable medium in a bioreactor. The nutritional requirements and fermentation conditions are maintained according to the type of microorganism used for a specific vaccine. This is followed by isolation, which involves the separation of microbial cells from the fermented media at the end of the fermentation process. The separation of microbial cells is usually done with the help of ultracentrifugation or some specific polysaccharide extraction method.

The purification step, which eliminates various impurities or materials sticking to the microbial cells in case of bacterial vaccine, is very specific to the antigens and some time may be required for chemical precipitation or fractionation, followed by ultrafiltration and chromatography. For the development of conjugated bacterial vaccines, carrier proteins may be conjugated at this stage to some polysaccharide vaccines, followed by the purification of conjugated vaccine by ultracentrifugation or chromatography. In case of viral vaccines, the purification step also involves centrifugation, followed by ultrafiltration, and chromatography. The viruses may be chemically inactivated to produce killed vaccine at this stage.

The pure cells thus obtained in the purification step are subjected to formulation, which is composed of mixing of the purified product in a suitable solution to achieve a required concentration followed by the addition of preservative to some vaccines, to warrant product stability in terms of microbial growth or contamination or to prevent cross-contamination from multidose vaccine vials. In case of bacterial vaccines, the purified products may be combined with several other antigens to develop the desired vaccine. Specific examples are of pneumococcal vaccine, which is developed by combining 23 different types of polysaccharides and diphtheria-tetanus-purtussis-*H. influenzae* type b-hepatitis B (DTP-Hib-HepB) is produced by combining bacterial vaccines with viral antigens. For manufacturing toxoids such as diphtheria and tetanus, exotoxins are used after subjecting to heat, UV light, or chemical treatment, which destroys the toxic part without altering their antigenic specificity and often boosting their immunogenicity.

Some highly specific methods are still under investigation to manufacture DNA vaccines and recombinant vector vaccines. The regulatory authorities are continuously monitoring the quality of vaccines produced and marketed worldwide. Various tests are performed by the regulatory authorities during the monitoring program, which includes safety tests (performed by injecting the product intraperitoneally into laboratory animals), identity test (based on the type and nature of vaccine), purity test (performed to check that the product meets the international standards of purity), potency testing (to check that vaccine confers protective immunity), and sterility test (performed to check that the vaccine is free from microbial contamination).

11.5.3 *Biotechnological approach*

With the introduction of recombinant DNA technology, utilizing both microorganisms and mammalian cells have been exploited enormously for the production of vaccines in recent years. However, the research and development to introduce new vaccines is still mostly moving around a few advanced countries and their respective strong multinational groups. During the last 25 years, more than two-thirds of the new vaccines were developed and introduced only by US-based companies (Douglas et al. 2008). Recent advances introduced in molecular biology and genetic engineering techniques are now available, which grants the isolation and separation of strategic genetic sequences to produce vaccines for effective immune response. A wide range of host cells, such as bacteria, yeast, chicken eggs, and even plants have been examined as potential sources for engineered vaccines in substantial quantities. Tissue culture techniques are also being investigated in genetically modified host cells to produce large quantities of antigens.

Some vaccines, such as antihuman papillomavirus vaccine and hepatitis B vaccine are now manufactured through gene techniques. A single gene, usually a surface glycoprotein of the virus, can be expressed in a foreign host by cloning. Bacteria (e.g., *Escherichia coli* and *Bacillus subtilis*) and yeasts (e.g., *Saccharomyces cerevisiae*) are commonly used as expression hosts whereas viruses or plasmids are usually used as expression vectors on a commercial scale to produce large amounts of antigens, which are used as vaccines after purification and combination with an adjuvant.

The role of nanobiotechnology in the development of vaccines is also under investigation because nanoparticles play an important role in the drug delivery system for biological therapies. No adverse reactions were noted at the site of immunization in experimental animal modules when nanobeads were used as an adjuvant, instead of alum, the most commonly used adjuvant for vaccines. Nanobeads serve a dual purpose, that is, apart from stimulating antibody production, it also produces T-cells. The size of the nanobeads (measuring 40 nm) has critical value in vaccine production. Because most viruses are of a similar size (40 nm), the nanobeads are taken up profusely by the immune system and deceived into elevated levels of T-cell production. The baculovirus expression vector system, in recent years, has also been evaluated and observed as a feasible option for large-scale production of flu vaccines. In addition, further advances in virology also provide a platform for the production of virus-like particles as vaccines against emergent disease and viral vectors for gene therapy (Roldao et al. 2011).

Some highly useful biotechnological techniques are under investigation to develop new types of vaccines, such as DNA vaccines, which use genes that code for those all-important antigens. A few such DNA vaccines against influenza and herpes are also being tested in humans. Another experimental vaccine is the recombinant vector vaccine, which uses attenuated viruses or bacterium to introduce microbial DNA to the cells of the body. A number of vaccine research institutes and multinational companies engaged in developing new vaccines are working hard on both bacterial as well as viral recombinant vector vaccines for HIV, rabies, and measles. It is expected that with the introduction of more advanced biotechnological methodologies and approaches, the vaccine industry will grow more rapidly and will benefit from exploring and developing some novel vaccines helping mankind all over the world.

11.5.4 Preventive/therapeutic role

Vaccines have both bioprophylactic and biotherapeutic roles and thus their applications have certainly generated value for mankind greater than any other biologicals or biopharmaceuticals available commercially in the pharmaceutical market. However, its established therapeutic role for the treatment of rabies and in the treatment of cancer, based on some recent preclinical and clinical studies, have delivered reasonable testimony that this distinctive therapeutic modality will lead to new theories and models in both clinical trial design and endpoints and in combination therapies. It is expected that after a successful launch, cancer vaccines will ultimately be employed for the therapy of many different types of cancer (Schlom et al. 2007).

The preventive roles of vaccines are quite predominant. According to a WHO report, every year, 3 million deaths are prevented and 750,000 children are saved from disability by vaccines (WHO 2009). The WHO has a comprehensive vaccine-preventable disease monitoring system, which highlights immunization schedules by disease covered with antigens all over the world, divided into six different regions: Africa, the Americas, Eastern Mediterranean, Europe, Southeast Asia, and the Western Pacific.

11.6 Flavonoids

The natural polyphenolic compounds with low molecular weight represent a diverse group of secondary metabolites, known as flavonoids, with distinct and pronounced biological activities. Their importance has now been explored comprehensively to ascertain their role as antimicrobials (antiviral, antibacterial, antifungal, antiprotozoal, etc.), antioxidants (responsible for beneficial effects on cardiovascular disease), anti-inflammatory, antiallergic, anticancer, antiobesity, and others. Because of the health-promoting effects, the usage and application of flavonoids and flavonoid-derived compounds have now become very popular and have drawn considerable attention all over the world to promote as nutritional supplements or as pharmaceutical products (Fowler and Koffas 2009; Shaik et al. 2006).

11.6.1 Historical background and new challenges

Historically, flavonoids were reported in plants in different forms such as free aglycones and glycosides. Bioflavonoids are responsible for the development of various colors. In

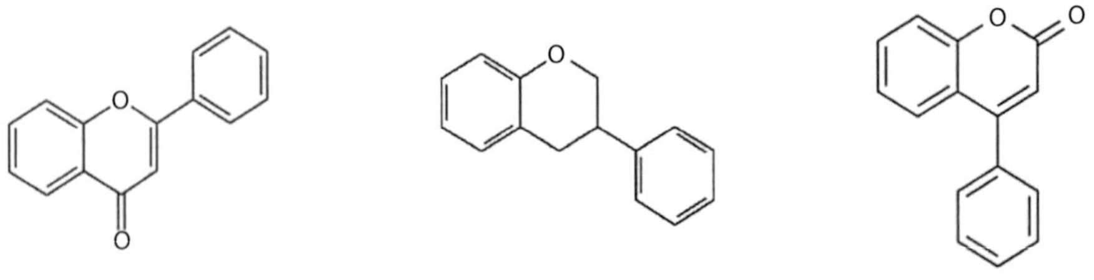

Figure 11.39 Classification of flavonoids under IUPAC nomenclature 1997.

view of their effects on the permeability of vascular capillaries, an additional name, "vitamin P" was also assigned; however, the name was later declared obsolete during the early 1950s. Based on the IUPAC (1997) nomenclature system, flavonoids are now classified under three categories: flavones, isoflavonoids, and neoflavonoids (Figure 11.39).

11.6.2 Production and pharmaceutical presentation

The biosynthesis mechanism of flavonoids in plants is quite complex and involves a series of enzymatic reactions. The major skeleton is phenylpropanoids, which are synthesized from phenylalanine after the removal of an amino group by the enzymatic action of phenylammonium lyase to produce transcinnamic acid. The hydroxylation of the aromatic ring of transcinnamic acid by the enzyme cinnamate 4-hydroxylase thus produces *p*-coumaric acid, which can then be ligated to coenzyme A by another enzyme, 4-coumaroyl-CoA ligase.

A number of flavonoid-based oral preparations are available in the international market for the management of various diseases. However, these products are largely sold as nutritional or food supplements in the US and European markets, but in China, India, and in some South Asian countries, a proper place has been given to these products as a drug in the respective/traditional systems of medicine practiced in these countries. Although in some cases, comprehensive pharmacological investigation results and even the clinical studies data are available to support their usage and application. Yet more information on treatment, health benefits, and side effects with marketed flavonoid drugs products are desired for more authentications and for U.S. Food and Drug Administration consideration for approval as a drug. Currently, flavonoids are mostly available in the form of film-coated tablets/caplets, soft gel capsules, powders in a sachet, and creams for topical application.

11.6.3 Biotechnological approach

The isolation, extraction, and purification of secondary metabolites from plants have always been posed as a difficult task in view of several factors affecting the final yield of the desired product. The seasonal and regional variations along with low concentrations of the pure compound certainly create challenges for researchers. Although, in some cases, chemical synthesis usually provides a reasonable alternative, but the intense reaction environments and disastrous chemicals required to synthesize flavonoids makes it less acceptable. Parallel to this, production through tissue culture has also been reported, but it is still difficult to organize the cell viability along with the engineering challenges affecting cultivation during large-scale production of flavonoids. Therefore, to meet the increased demand for flavonoids and to effectively utilize their therapeutic role, microbial production through genetically modified organisms is becoming progressively imperative from a recombinant manufacturing point of view. The field is interesting and open, but challenging for the biotechnologist looking for a successful commercial plan and roadmap because despite substantial developments in improving microbial strains and the yields of certain flavonoids, the overall fermentation process still needs considerable attention from researchers for more improvements in terms of process optimization and media composition.

A large number of genetically modified organisms have been used to biosynthesize various flavonoids, during which the precursor molecules play a significant role in the yield of the desired compound. *E. coli, S. cerevisiae, Streptomyces venezuelae*, and *Phellinus igniarius* have been extensively studied in the production of some important flavonoids such as pinocembrin, naringenin, apigenin, luteolin, genkwanin, eriodictyol, kaempferol, chrysin, galagin, genistein, flavanone, dihydroflavonol, flavone, flavonol, dihydrokaempferol, quercetin, and exopolysaccharides (Figure 11.40).

In view of the structural similarity of yeasts with plant cells, it has been extensively investigated as a module to observe the genes relating to the biosynthesis of flavonoids. There are some definite advantages with using yeast, which permits the posttranslational alterations of protein and endomembrane systems in eukaryotes into which plant P450 enzymes are typically embedded, together with their cytochrome P450 reductase partner.

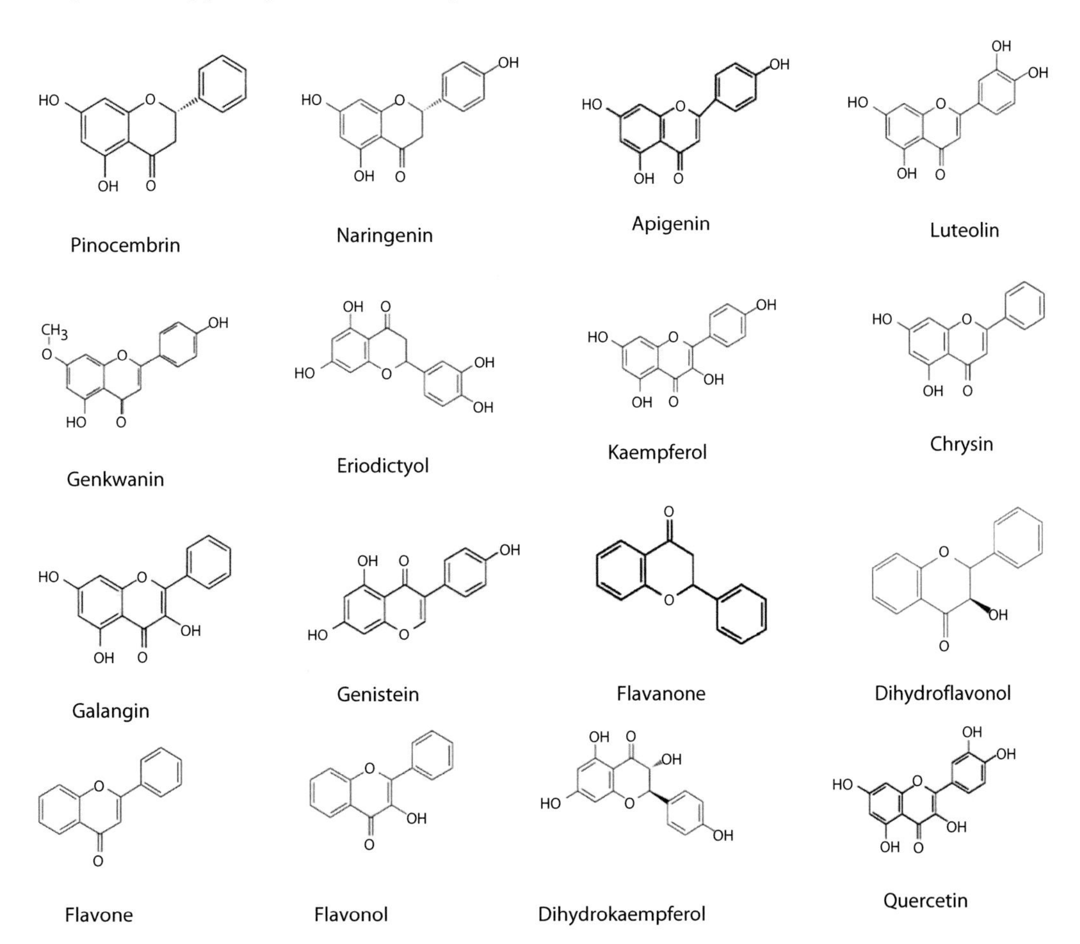

Figure 11.40 Some important flavonoids produced by microorganisms.

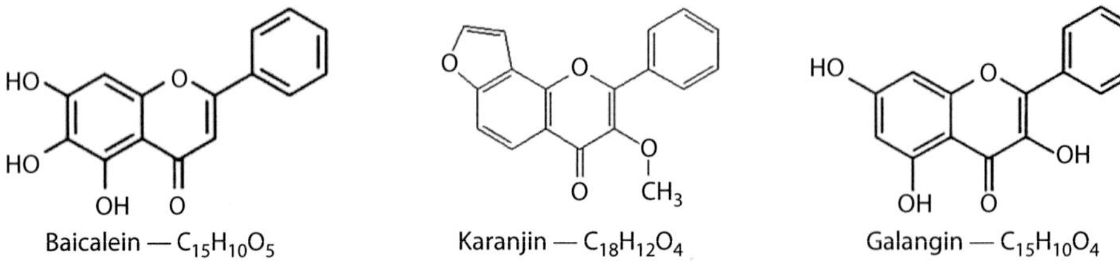

Baicalein — $C_{15}H_{10}O_5$ Karanjin — $C_{18}H_{12}O_4$ Galangin — $C_{15}H_{10}O_4$

Figure 11.41 Flavonoids with antimicrobial properties.

11.6.4 Therapeutic role

aFlavonoids have now been identified in playing a significant therapeutic role in the management of various diseases. The role of flavonoids as antimicrobials, including controlling HIV, has been investigated and reported by a number of authors. Its antioxidant property ultimately confers its beneficial effect on cardiovascular disease (including reductions in low-density lipoproteins), allergies, and cancer. The antimicrobial activity is linked with the phenolic groups of flavonoids. Baicalein, karanjin, and galangin (Figure 11.41) are the most important flavonoids identified to possess antimicrobial activity.

11.7 Case study: Optimization of media composition for the production of ergot alkaloids from a mutant strain of Botryodiplodia *sp. FRL-1955 through surface culture technique*

With the increased medical importance of ergot alkaloids and its analogues, optimization processes for increased production of ergot alkaloids have been taken into consideration by research workers worldwide. However, major emphasis was given to several species of *Claviceps, Penicillium,* and *Aspergillus* for process optimization to increase yield.

In the present study, we report the production of ergot alkaloids in a mutant strain of *Botryodiplodia* species, which belongs to the class *Ascomycetes* and lives in a saprophytic way. It normally needs injured tissue to parasitize the plant. The presence of alkaloids has been reported in *Botryodiplodia* sp. during the earlier phase of the studies. The alkaloids have also been shown to possess antibacterial activity against some common pathogens when tested through bioautographic techniques (Mahmood 1983; Mahmood et al. 1986). However, more in-depth studies are required to further explore and establish the medicinal value of *Botryodiplodia* species in which this study is part of the program.

The generation of pycnidia, in which spores are formed, is a characteristic feature of *Botryodiplodia*. Pycnidia rarely develop on artificial media or may develop after a long time. If generated, it can be seen with the naked eye as black balls the size of a pinhead. The spores are elliptical and large. Young and immature spores are colorless and unicellular, whereas mature spores are brown, distichous, and thick-walled. The fast and constantly growing mycelium of the fungus is snow-white at first, turning black within 3 to 4 weeks. This discoloring is not due to the generation of spores.

11.7.1 Materials and methods

The fungus was grown by surface culture technique using 1-L capacity Roux bottles containing 100 mL of sterile media with glucose (25 g/L), mannitol (25 g/L), and triammonium citrate (12 g/L), which is supplemented with the following minerals: $MgSO_4{\cdot}7H_2O$ (2,000 mg/L), KH_2PO_4 (250 mg/L), $ZnSO_4{\cdot}7H_2O$ (20 mg/L), $CuSO_4{\cdot}5H_2O$ (2.5 mg/L), and $MnSO_4{\cdot}H_2O$ (15 mg/L) at pH 5.6. One milliliter of spore suspension, prepared from a 5- to 7-day-old stock culture of *Botryodiplodia* sp. FRL-1955, grown on potato dextrose agar slant at 25°C, was used to inoculate each Roux bottle and were incubated at 30°C for 28 days. Selection of the media was based on several studies to optimize the production capacity of strain FRL-1955.

Upon completion of the incubation period, the mycelial cells were separated from the fermented medium by filtration using a suction pump, washed with sterile distilled water, and homogenized for extracting the alkaloids in 50 mL solvent (1:1 mixture of acetone and 2% tartaric acid) with 6 h contact time. The solids were removed by centrifugation and the extract was mixed with filtered fermented broth. The content was acidified with 2% acetic acid and treated with ethyl acetate to separate the impurities. The aqueous layer was basified with 20% NH_4OH solution up to pH 10.0 and extracted three times with chloroform. The combined chloroform was concentrated under reduced pressure at low temperature in a rotary evaporator. The chloroform extract was fractionated on thin-layer chromatography plates using chloroform–methanol (8:2). The total alkaloids content was determined using Van Urk reagent (Van Urk 1929), as modified by Smith (1930). The amount of alkaloids present was calculated from a standard curve of ergotamine.

11.7.2 Results and discussion

Approximately 1.2 mg of alkaloid per milliliter of medium was obtained in this study, which is quite encouraging. The current study revealed that a medium composed of glucose in combination with mannitol and triammonium citrate supplemented with inorganic salts is an excellent source of carbon and nitrogen for the production of ergot alkaloids. The carbon-to-nitrogen ratio of 4.16:1 was recorded as optimum after several studies and was found in accordance with a previous study using *Aspergillus fumigatus* (Narayan and Rao 1982). During the initial optimization program, starch and molasses were also employed as the sole carbon sources, but were noted to be inferior in alkaloid production capacity of *Botryodiplodia* sp. Similarly, during the optimization program, it was also observed that an increase in nitrogen source level decreases alkaloid production. The increase in nitrogen level suppresses mycelial growth and, eventually the yield of alkaloids because a major part of the alkaloids accumulate in the mycelium. Different pH values between 4 and 8 were investigated and pH 5.6 was noted as optimum. Values lower than 3.5 and more than 7.5 significantly affected the growth and thus the alkaloid content. A lower concentration of potassium phosphate was used in the present study because high concentrations were reported to inhibit the secondary metabolites (Martin 1977). In conclusion, it was observed that different sources of carbon as well as the ratio between carbon and nitrogen, along with the concentration of inorganic phosphate and pH of the medium, can greatly affect the overall yield of ergot alkaloid.

11.8 Conclusion

The microbial healthcare products classified under secondary metabolites such as antibiotics, antitumor agents, vaccines, bioactive alkaloids, and flavonoids have received considerable attention in recent years because of their importance in our daily life, both from a therapeutic and prophylactic point of view. Therefore, reviewing the topic in light of recent biotechnological advancements is fully justified. The new dimension and biotechnological approach is expected to bring a revolution with the introduction of new antibiotics with a wider spectrum of activity, highly selective antitumor drugs, more safe and effective vaccines, and better understanding of submerged fermentation for the production of bioactive compounds or drugs through genetic manipulation. Based on this concept, this chapter was written to provide recent and updated information for both students and researchers, not only to observe the dynamics of the topic but also to analyze the challenges expected during the coming decades.

Acknowledgments

We are extremely thankful to Martti Hedman, Vice President, Corporate Development, Colorcon and Jacques Michaud, Director EMEA, Colorcon Limited, England for their encouragement and support. My special thanks to Sara Peerun, Commercial Director (Visiongain) for providing marketing information.

References

American Medical Association. 1995. *Drug Evaluations Annual: Principles of Cancer Chemotherapy*. Printed by the Division of Drugs and Toxicology, Chicago, IL, pp. 2059–2093.

Amici, A.M., Minghetti, A., Scotti, T., Spalla, C. and Tognoli, L. 1966. Production of ergotamine by a strain of Claviceps purpurea (Fr.) *Tul. Experientia, 22*: 415–416.

Amici, A.M., Minghetti, A., Scotti, T., Spalla, C. and Tognoli, L. 1967. Ergotamine production in submerged culture and physiology of *Claviceps purpurea. Appl. Microbiol., 15*: 597–602.

Amici, A.M., Minghetti, A., Scotti, T., Spalla, C. and Tognoli, L. 1969. Production of peptide ergot alkaloids in submerged culture by three isolates of *Claviceps purpurea. Appl. Microbiol., 18 (3)*: 464–468.

Bryant, A. 2012. 20 Top-selling vaccines. Fierce Vaccines—Weekly vaccines industry newsletter, September 25, p. 1. Available from: http://www.fiercevaccines.com/special report/20-top-selling-vaccines/2012-09-25.

Burkholder, P.R. 1952. Cooperation and conflict among primitive organisms. *Am. Sci., 40*: 601–631.

Castellani, J.J. 2010. Biopharmaceutical research continues against infectious diseases with 395 medicines and vaccines in testing. *Medicines in Development for Infectious Diseases*. Report presented by America's Biopharmaceutical Research Companies, pp. 1–36.

Chandra, S. 2012. Endophytic fungi: Novel source of anticancer lead molecules. *Appl. Microbiol. Biotechnol., 95 (1)*: 47–59.

Cragg, G.M., Grothaus, P.G. and Newman, D.J. 2009. Impact of natural products on developing new anti-cancer agents. *Chem. Rev., 109*: 3012–3043.

Douglas, R.G., Sadoff, J. and Samant, V. 2008. The vaccine industry. In: Plotkin, S.A., Orenstein, W. and Offit, P.A. (editors) *Vaccines*. Saunders Elsevier, China, pp. 37–44.

Fowler, Z.L. and Koffas, M.A.G. 2009. Biosynthesis and biotechnological production of flavanones: Current state and perspectives. *Appl. Microbiol. Biotechnol., 83*: 799–808.

Gajdova, M., Jakubovasky, J. and Valky, J. 1993. Delayed effect of neonatal exposure to Tween 80 on female reproductive organs in rat. *Fd. Chem. Toxic., 31 (3)*: 183–190.

Greenwood, D. 2008. *Antimicrobial Drugs: Chronicle of a 20th Century Medical Triumph.* Oxford University Press, New York, pp. 120–121.

Hartwell, J.L. and Schrecker, A.W. 1951. Components of podophyllin. V. The constitution of podophyllotoxin. *J. Am. Chem. Soc., 73 (6)*: 2909–2916.

Hulvova, H., Galuszka, P., Frebortova, J. and Frebort, I. 2013. Parasitic fungus *Claviceps* as a source for biotechnological production of ergot alkaloids. *Biotechnol. Adv., 31*: 79–89.

Hulvova, H., Kubesa, V., Jaros, M. and Galuszka, P. 2010. Affecting the ergot production in *Claviceps purpurea* by genetic manipulation. IMC9—*The Biology of Fungi,* Edinburg, UK, August 1–6, pp. 4.187.

IUPAC. 1997. *Compendium of Chemical Terminology*, 2nd ed. (the "Gold Book"). Compiled by A. D. McNaught and A. Wilkinson. Blackwell Scientific Publications, Oxford.

Keller, U., Zocher, R. and Kleinkauf, H. 1980. Biosynthesis of ergotamine in protoplasts of Claviceps purpurea. *J. Gen. Microbiol., 118*: 485–494.

Kobel, H., Schreier, E. and Rutschmann, J. 1964. 6-Methyl-A8.9-ergolen-8-carbonsaure, ein neues Ergolinderivat aus Kulturen eines Stammes von Claviceps paspali Stevens et Hall. *Helv. Chim. Acta, 47*: 1052–1064.

Liu, M., Panaccione, D.G. and Schardl, C.L. 2009. Phylogenetic analyses reveal monophyletic origin of the ergot alkaloid gene *dmaW* in fungi. *Evol. Bioinform., 5*: 15–30.

Lorenz, N., Haarmann, T., Pazoutova, S., Jung, M. and Tudzynski, P. 2009. The ergot gene cluster: Functional analyses and evolutionary aspects. *Phytochemistry, 70*: 1822–1832.

Lozano, J.M., Bermudez, A. and Patarroyo, E. 2006. Peptide vaccines for Malaria. In: Kastin, A.J. (editor) *Handbook of Biologically Active Peptides*. Academic Press, London, pp. 515–526.

Ma, X. and Wang, Z. 2009. Anticancer drug discovery in the future: An evolutionary perspective. *Drug Discov. Today, 14*: 1136–1142.

Mahmood, Z.A. 1983. Studies on alkaloids and antibacterial compounds produced by some fungi. M. Pharm Thesis, Department of Pharmaceutics, Faculty of Pharmacy, University of Karachi, Karachi, Pakistan.

Mahmood, Z.A., Khan, K.H. and Shaikh, D. 1986. Antibacterial and chemical studies on alkaloids produced by *Botryodiplodia species. J. Pharm., 4 (2)*: 105–111.

Maier, W., Schumann, B., Erge, D. and Groger, D. 1980. Biosynthesis of ergot alkaloids by protoplasts of various Claviceps purpurea strains. *Biochem. Physiol. Pflanz., 175*: 815–819.

Maiyali, J.H. 2013. Global polio eradication campaign and its misconception in Northern Nigeria. Available from: http://www.freedomradionig.com/index.php/interviews/39-icetheme/editorials/436-global-polio-eradication-campaign-and-its-misconception-in-northern-nigeria.

Martin, J.F. 1977. Controlled of antibiotic synthesis by phosphate. In: Ibrahim, A.M., Dalhatu, F.M., Mohammad, A.A., Mahmud, A.B., Kiyawa, A.A.A., Abdullahi, S.A., Sani, H.R., Dalhatu, A.M., Wada, A.U.S., Mamman, A.M.K. and Ladan, A.S. (editors) *Advances in Biochemical Engineering*, Vol. 6. Springer-Verlag, Berlin, pp. 105–127.

Mey, G., Held, K., Scheffer, J., Tenberger, K.B. and Tudzynski, P. 2002. CPMK2, an SLT2-homologous mitogen-activated protein (MAP) kinase, is essential for pathogensis of Claviceps purpurea on rye: Evidence for a second conserved pathogensis-related MAP kinase cascade in phytopathogenic fungi. *Mol. Microbiol., 46*: 305–318.

Narayan, V. and Rao, K.K. 1982. Production of ergot alkaloids by *Aspergillus fumigatus. European J. Appl. Microbiol. Biotechnol., 14*: 55–58.

Owens, N.W. and Schweizer, F. 2011. Peptides and glycopeptides. In: Moo-Young, M. (editor-in-chief) *Comprehensive Biotechnology*. Elsevier, Amsterdam, Netherlands, pp. 121–138.

Palmer, E. and Bryant, A. 2013. Top 5 vaccines companies by revenue - 2012. Fierce Vaccines—Weekly vaccines industry newsletter, March 14, p. 1. Available from: http://www.fiercevaccines.com/special-reports/top-5-vaccine-companies-revenue 2012#ixzz2QAWJfPgk.

Patyar, S., Joshi, R., Prasad, D.S., Prakash, A., Medhi, B. and Das, B.K. 2010. Bacteria in cancer therapy: A novel experimental strategy. *J. Biomed. Sci., 17*: 1–9.

Porter, J.N. 1976. Antibiotics. In: Miller, B.M. and Litsky, W. (editors) *Industrial Microbiology*. McGraw-Hill Book Company, NY, pp. 60–78.

Ravelo, A.G., Estévez-Braun, A., Chávez-Orellana, H., Pérez-Sacau, E. and Mesa-Siverio, D. 2004. Recent studies on natural products as anticancer agents. *Curr. Top. Med. Chem., 4*: 241–265.

Rehacek, Z. 1984. Biotechnology of ergot alkaloids. *Trends Biotechnol., 2 (6)*: 166–172.

Roldao, A., Silva, A.C., Mellado, M.C.M., Alves, P.M. and Carrondo, M.J.T. 2011. Viruses and virus-like particles in biotechnology: Fundamentals and applications. In: Moo-Young, M. (editor-in-chief) *Comprehensive Biotechnology*. Elsevier, Amsterdam, Netherlands, pp. 625–649.

Sanchez, S. and Demain, A.L. 2011. Secondary metabolites. In: Moo-Young, M. (editor-in-chief) *Comprehensive Biotechnology*. Elsevier, Amsterdam, Netherlands, pp. 154–167.

Schardl, C.L., Panaccione, D.G., Tudzynski, P. and Geoffrey, A.C. 2006. Ergot alkaloids-biology and molecular biology. *Alkaloids Chem. Biol., 63*: 45–86.

Schiff, P.L. 2006. Teacher's topic. Ergot and its alkaloids. *Am. J. Pharm. Educ., 70 (5)*: 1–10.

Schlom, J., Gulley, J.L. and Arlen, P.M. 2007. Updates and developments in oncology: Role of vaccine therapy in cancer: Biology and practice. *Curr. Oncol., 14 (6)*: 238–245.

Shaik, Y., Castellani, M., Perrella, A., Conti, F., Salini, V., Tete, S., Madhappan, B., Vecchiet, J., De Lutiis, M. and Caraffa, A. 2006. Role of quercetin (a natural herbal compound) in allergy and inflammation. *J. Biol. Regul. Homeost. Agents, 20*: 47–52.

Smith, M.L. 1930. Quantitative colorimetric reaction for ergot alkaloids and its application in the chemical standardization of ergot preparations. *Public Health Rep., 45*: 1466–1481.

Tonolo, A. 1966. Production of peptide alkaloids in submerged culture by a strain of *Claviceps purpurea* (Fr.) Tul. *Nature, 209 (5228)*: 1134–1135.

Tsai, H.F., Wang, H., Gabler, J.C., Poulter, C.D. and Schardl, C.L. 1995. The Claviceps purpurea gene encoding dimethylallyltryptophan synthase, the committed step for ergot alkaloid biosynthesis. *Biochem. Biophys. Res. Commun., 216*: 119–125.

Tudzynski, P., Correia, T. and Keller, U. 2001. Biotechnology and genetics of ergot alkaloids. *Appl. Microbiol. Biotechnol., 57*: 593–605.

Tudzynski, P., Duvell, A. and Esser, K. 1983. Extrachromosomal genetics of *Claviceps purpurea. Curr. Gent., 7*: 145–150.

Tudzynski, P., Holter, K., Correia, T., Arntz, C., Grammel, N. and Keller, U. 1999. Evidence for an ergot alkaloid gene cluster in Claviceps purpurea. *Mol. Gen. Gent., 261*: 133–141.

Urquiza, M., Rodriguez, L.E. and Suarez, J.E. 1996. Identification of *Plasmodium falciparum* MSP-1 peptides able to bind to human red blood cells. *Parasite Immunol., 18 (10)*: 515–526.

Van Urk, H.W. 1929. A new sensitive reaction for the ergot alkaloids, ergotamine, ergotoxine and alkaloid biogenesis in *Claviceps* sp. Strain SD 58. *Indian J. Exp. Biol., 20*: 475–478.

Visiongain. 2013. Antiacterial drugs: World market prospects 2013–2023. Executive summary, pp. 26–31.

Wallwey, C. and Li, S.M. 2011. Ergot alkaloids: Structure diversity, biosynthetic gene cluster and functional proof of biosynthetic genes. *Nat. Prod. Rep., 28*: 496–510.

WHO. 2002. *Guidelines on the International Packaging and Shipping of Vaccines*. Department of Vaccines and Biological, World Health Organization, Geneva, pp. 1–14.

WHO. 2009. *Vaccine. Centers for Disease Control and Prevention*, Vol. 27S. World Health Organization, Geneva, pp. G3–G8.

chapter twelve

Microbial biomass production

Ashok K. Rathoure

Contents

12.1 Introduction

Biomass refers to microbial mass, vegetable or plant matter, and sometimes animal matter that can be converted into an energy source. Microbial biomass refers to the biomass produced by the action of microbial strains, for example, baker's yeasts, single-cell proteins (SCP), biofertilizers, and biopesticides.

The history of yeast started 9000 years ago in the production of food. Bread, wine, sake, and beer are made with the essential contribution of yeasts, especially from the species *Saccharomyces cerevisiae*. Yeasts were first used in the Caucasian and Mesopotamian regions approximately 7000 BC (Lin and Tanaka 2006). However, there was no recorded information about yeast until Louis Pasteur discovered that it could produce CO_2 and ethanol from sugar. After the discovery of fermentation by Pasteur, yeast fermentation became famous and by the end of the nineteenth century, the production of bread and beer become a common practice. Wine producers are more disinclined to change their traditional practices and only began using exogenous yeast inocula in the 1950s, especially in countries with less wine tradition. In the 1960s, yeast biomass production, which was used in European countries, contributed to the technology of producing large amounts of dry yeast (Reed and Nagodawithana 1988). Efficient and profitable processes have been developed to produce yeast biomass at the laboratory and plant scale. The standard process has been developed from previous observations and experience for highest yield by increasing biomass production and decreasing costs (Daromola and Zampraka 2008; Dien et al. 2003).

12.2 Yeast

Yeasts are single-celled eukaryotic microorganisms classified in the kingdom Fungi, and are found on and around the human body. Although some species with yeast forms may become multicellular through the formation of a string of connected budding cells known as pseudohyphae or false hyphae, as in most molds. The fungi include edible mushrooms, common baker's yeast, and molds that ripen blue cheese as well as produce antibiotics for medical and veterinary use. Many consider edible yeast and fungi to be as natural as fruits and vegetables. Yeasts were identified as living organisms after the invention of the microscope, and it is also the agent responsible for alcoholic fermentation and dough leavening. With the knowledge that yeast is a living organism, and the ability to isolate yeast strains in pure culture form, the commercial production of baker's yeast started in the early twentieth century. Since that time, bakers, scientists, and yeast manufacturers have been working to find and produce pure strains of yeast that meet the specialized needs of the baking industry (Prescott and Dunn 1959; Van Hoek et al. 1998).

More than 600 different species of yeast are known and they are widely distributed in nature. They are found in association with other microorganisms as part of normal inhabitants of soil, vegetation, marine, and other aqueous environments. Baker's yeast is used to leaven bread throughout the world. Baker's yeast is produced from a species of yeast called *S. cerevisiae*. The genus in the scientific name of baker's yeast, *Saccharomyces*, refers to saccharo meaning sugar and myces meaning fungus. The species name *cerevisiae* is derived from the name Ceres, Roman goddess of agriculture. The size of the yeast cell equals that of a human red blood cell, and the shape varies from spherical to ellipsoidal (Bekatorou et al. 2006).

The reproduction in yeast takes place by budding (vegetative), a process in which a new bud grows from the side of an existing cell wall. This bud eventually breaks away from the mother cell to form a separate daughter cell. Each yeast cell undergoes this budding process 12 to 15 times before it is no longer capable of reproducing. The yeast should grow under controlled conditions on a sugar-containing medium during commercial production; yeast cells reproduce every 2 to 3 hours under ideal growth conditions.

12.2.1 Yeast metabolism

Metabolism refers to catabolic and anabolic pathways in the cell. Catabolic pathways are oxidative processes that remove electrons from substrates or intermediates used to generate energy, whereas anabolic pathways are a reductive process for the production of new cellular material. These processes use NADP or NAD as cofactors. There is a metabolic diversity in how organisms generate and consume energy from substrates. Although all yeasts derive their chemical energy from the breakdown of organic compounds (ATP). Knowledge of the underlying regulatory mechanisms is not only valuable in understanding the general principles of regulation but also of great importance in biotechnology, if new metabolic capabilities of particular yeasts have to be exploited. Now, it is well established that most yeasts employ sugars as their main carbon and hence energy source, but there are particular yeasts that can utilize nonconventional carbon sources. With regard to nitrogen metabolism, most yeasts are capable of assimilating simple nitrogenous sources to biosynthesize amino acids and proteins. Aspects of phosphorus and sulfur metabolism as well as aspects of metabolism of other inorganic compounds have been studied in detail (Nielsen and Jewett 2007; Ostergaard et al. 2000). The metabolic pathway of *S. cerevisiae* is presented in Figure 12.1.

The yeasts reoxidize NADH to NAD in a two-step reaction from pyruvate, which is first decarboxylated by pyruvate decarboxylase and the reduction of acetaldehyde is catalyzed by alcohol dehydrogenase in alcoholic fermentation. The alternative mode of glucose oxidation is the hexose phosphate pathway, which provides pentose sugars and cytosolic NADPH for biosynthetic reactions of the cell such as the production of fatty acids, amino acids, and sugar alcohols. Another function of this pathway is the production of ribose sugars, which serve in the biosynthesis of nucleic acid precursors and nucleotide coenzymes (Mormeneo and Sentandreu 1982).

Yeasts can be categorized into several groups according to their modes of energy production, respiration, or fermentation as regulated by environmental factors. Yeasts can adapt to varying growth environments and the prevailing pathways depend on actual growth conditions, for example, glucose can be used in different ways by *S. cerevisiae* depending on the presence of oxygen and other carbon sources (Table 12.1) as well as on the yeast's mode of respiration (Table 12.2; Birhanu et al. 1982; Gunasekaran and Chandra 1999). Yeasts use disaccharides as a substrate, which can be utilized by extracellular and intracellular enzymes, as presented in Table 12.3.

Catabolite repression is caused by glucose or the products of glucose metabolism, which represses the synthesis of various respiratory and gluconeogenic enzymes, whereas catabolite inactivation results in the rapid disappearance of enzymes upon the addition of glucose. Enzyme activity is lost by dilution with cell growth in catabolite repression,

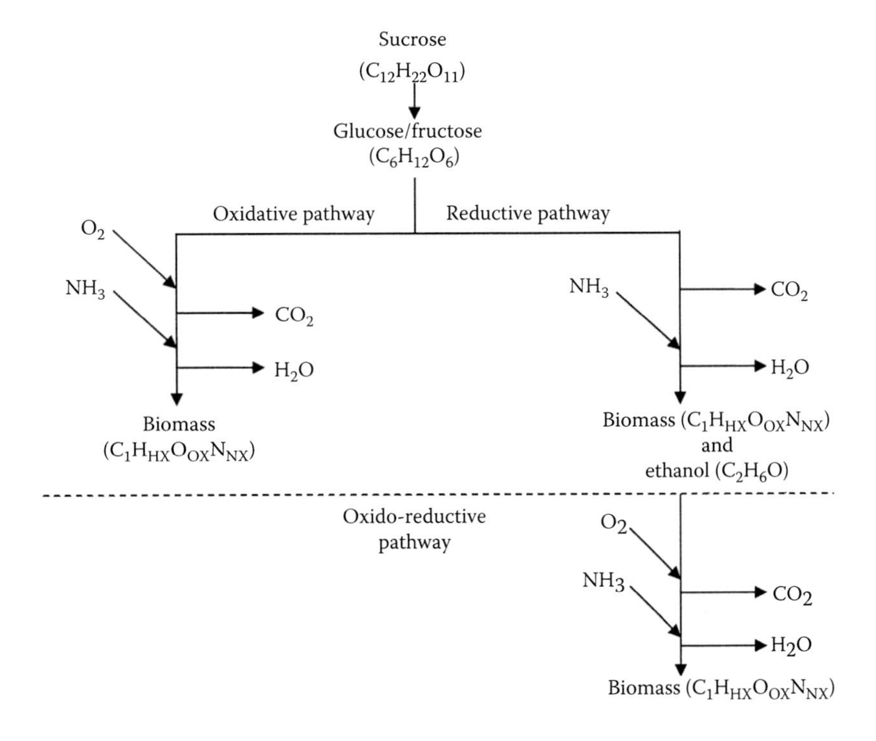

Figure 12.1 Metabolic pathways of *S. cerevisiae*. (From Sonnleitner, B., and Kappeli, O., *Biotechnology and Bioengineering* 28, 6: 927–937, 1986.)

Table 12.1 Carbon Sources for Growth of *S. cerevisiae*

Substrate	Intermediates	Enzymes	Products
Saccharose	None	Invertase	Glucose + fructose
Maltose	None	Maltase	Glucose
Melibiose	None	Melibiase	Glucose + galactose
Glucose	Glyceraldehyde-3-phosphate	Hexokinase Phosphoglucose isomerase Phosphofructokinase Fructose bisphosphate aldolase Triosephosphate isomerase Glyceraldehyde 3-phosphate dehydrogenase Phosphoglycerate kinase Phosphoglycerate mutase Enolase Pyruvate kinase	Pyruvate
Ethanol	Acetaldehayde, acetyl-CoA, oxaloacetate	Alcohol dehydrogenase	Glucose by gluconeogenesis
Lactate	Pyruvate	Lactate dehydrogenase	Glucose by gluconeogenesis
Glycerol	Glycerol-3-phosphate, dihydroxyacetonphosphate	Glycerol kinase	Glucose by gluconeogenesis

Source: Adapted from H. Feldmann. *Yeast: Molecular and Cell Biology*, 2012. Wiley-Blackwell: United Kingdom.

Table 12.2 Modes of Respiration in Yeasts

Types	Examples	Respiration	Fermentation	Anaerobic growth
Obligate respires	*Rhodotorula* sp. *Cryptococcus* sp.	Yes	No	No
Anaerobic respires	*Candida* sp. *Kluyveromyces* sp. *Pichia* sp.	Yes	Anaerobic in pregrown cells	No
Aerobic fermenters	*S. pombe*	Limited	Aerobic and anaerobic	No
Facultative aerobic fermenters	*S. cerevisiae*	Limited	Aerobic and anaerobic	Facultative
Obligate fermenters	*Torulopsis* sp.	No	Anaerobic	Yes

Source: Adapted from H. Feldmann. *Yeast: Molecular and Cell Biology*, 2012. Wiley-Blackwell: United Kingdom.

although enzymes are no longer synthesized due to gene repression by signals derived from glucose or other sugars. The nature of signals is currently not clear. Glucose repression in yeast describes a long-term regulatory adaptation to degrade glucose exclusively to ethanol and CO_2. Therefore, when *S. cerevisiae* is grown aerobically on high concentrations of glucose, fermentation will account for the bulk of glucose consumption. In batch cultures, when the levels of glucose decline, cells become gradually derepressed resulting in the induction of respiratory enzyme synthesis. This results in the oxidative consumption of ethanol, when cells enter a second phase of growth known as the diauxic shift.

Table 12.3 Disaccharides as Substrates in Yeasts

Disaccharide	Extracellular hydrolysis	Intracellular hydrolysis	Products	Organism
Maltose	–	Maltase	2 Glucose	*S. cerevisiae*
Sucrose	Invertase	–	Glucose + fructose	*S. cerevisiae*
Melibiose	α-Galactopyranosidase	–	Glucose + galactose	*S. carlsbergensis*
Lactose	–	β-Galactosidase	Glucose + galactose	*Kluyveromyces*
Cellobiose	β-Glucosidase	–	2 Glucose	*Brettanomyces*

Source: Adapted from H. Feldmann. *Yeast: Molecular and Cell Biology*, 2012. Wiley-Blackwell: United Kingdom.

Catabolite inactivation is faster than catabolite repression because glucose deactivates a limited number of key enzymes such as fructose 1,6-bisphosphatase. Inactivation occurs primarily by enzyme phosphorylation, followed by slower vascular degradation of the enzyme. The cAMP as a second messenger plays a central role in regulating catabolite repression and inactivation in *S. cerevisiae* (Cook 1958).

12.2.2 *Role of yeasts in baking*

Yeast is a key ingredient for the production of bakery products and plays an important role in CO_2 production, dough maturation, and fermentation flavor. CO_2 is generated by yeast as a result of the breakdown of fermentable sugars in the dough. The evolution of CO_2 causes an expansion of the dough trapped within the protein matrix. Dough maturation is accomplished by the chemical reactions of yeast producing alcohols and acids from the proteins of flour, and by the physical stretching of protein and CO_2, which results in the light and airy physical structure associated with yeast-leavened products. Yeast also gives bread its characteristic flavor. During dough fermentation, yeast produces many secondary metabolites such as ketones, higher alcohols, organic acids, aldehydes, and esters. Some of them escape during baking. Others react with each other and with other compounds to form new and complex flavor compounds. These reactions occur primarily in the crust and crumb of baked breads.

Yeasts can grow in the presence or absence of oxygen. Anaerobic growth is slow and inefficient, for example, in bread dough. The sugars that can sustain either fermentation or growth are used mainly to produce alcohol and CO_2. Only a small portion of the sugar is used for cell maintenance and growth. Under aerobic conditions, and in the presence of a sufficient quantity of dissolved oxygen, yeast grows by using most of the available sugar for growth and produces only negligible quantities of alcohol. If the concentration of sugar in the fermentation growth medium is greater than a very small amount, yeast can produce some alcohol with an adequate supply of oxygen or in large quantities. This problem can be solved by adding the sugar solution slowly to the yeast throughout the fermentation process. This type of fermentation is referred to as fed-batch fermentation (Sheoran et al. 1998).

12.2.3 *Yeast production*

The basic sources of carbon and energy for yeast growth are sugars. Starch cannot be used because yeast does not contain the enzymes to hydrolyze starch. Beet and cane molasses are commonly used as raw materials because the sugars present in molasses are readily fermentable. The molasses are a mixture of sucrose, fructose, and glucose. In addition, yeast also requires certain minerals, vitamins, and salts for growth. Some of these can be added to the blend of beet and cane molasses before flash sterilization whereas others are fed separately from the fermentation. Alternatively, a separate nutrient feed tank can be used to mix and deliver some of the necessary vitamins and minerals. Each of these nutrients is fed separately to the fermenter to permit better pH control of the process. The sterilized molasses are stored in a separate stainless steel tank. The sterilized molasses are referred to as mash or wort. The mash stored in this tank is used to feed sugar and other nutrients to the appropriate fermentation vessels (Atiyeh and Duvnjak 2003; Zamani et al. 2008). A variety of essential nutrients and vitamins is also required in yeast production. The nutrient and mineral requirements include nitrogen, potassium, phosphate, magnesium, and calcium, with traces of iron, zinc, copper, manganese, and molybdenum. Normally, nitrogen is supplied by adding ammonium salts, aqueous ammonia, or anhydrous ammonia to the feedstock. Phosphates and magnesium are added in the forms of phosphoric acid or phosphate salts and magnesium salts. Vitamins such as biotin, inositol, pantothenic acid, and thiamine are also required for yeast growth. Thiamine is added to the feedstock. Most other vitamins and nutrients are already present in sufficient amounts in the molasses malt (Chen and Chigar 1985).

Baker's yeast production starts with a pure culture, which serves as the inoculum for prepure culture tank. After growth, the contents are transferred to a larger pure culture fermenter in which propagation is carried out with aeration under sterile conditions. These stages should be conducted as set batch fermentations. In set batch fermentation, media and nutrients are introduced to tank before inoculation. From pure culture vessels, the grown cells are transferred to a series of progressively larger seed and semiseed fermenters. These later stages are conducted as fed-batch fermentations. During fed-batch fermentation, molasses, phosphoric acid, ammonia, and minerals are fed to yeast at a controlled rate. This rate is designed to feed enough sugar and nutrients to the yeast to maximize multiplication and prevent the production of alcohol. It is not economical to use pressurized tanks to guarantee the sterility of the large volumes of air required in these fermenters or to achieve sterile conditions during all the transfers through the many pipes, pumps, and centrifuges. Extensive cleaning of the equipment, steaming of pipes and tanks, and filtering of the air should be practiced to insure aseptic conditions. At the end of semiseed fermentation, the contents of the vessel are pumped to a series of separators that separate the yeast from spent molasses, and then the yeast is washed with cold water and pumped to a semiseed yeast storage tank where the yeast cream is held at 34°F, until it is used to inoculate the commercial fermentation tanks (Acourene et al. 2007).

The commercial fermentations should be carried out in large fermenters with working volumes of up to 200,000 L. To start the commercial fermentation, a volume of water referred to as set water, is pumped into the fermenter. At the start of the fermentation, the liquid seed yeast and additional water may occupy only about one-third to one-half the volume. Constant additions of nutrients during the course of fermentation brings the fermenter to its final volume. The rate of nutrient addition increases throughout the fermentation because more nutrients are supplied to support the growth of the increasing cell population. The number of yeast cells increases approximately fivefold to eightfold during this fermentation. Air is provided to the fermenter through a series of perforated tubes located at the bottom of the vessel. The rate of airflow is approximately one volume of air per fermenter volume per minute. A large amount of heat is generated during yeast growth, and cooling is accomplished by internal cooling coils or by pumping the fermentation liquid through an external heat exchanger. The regulation of pH, temperature, and airflow should be monitored carefully and controlled by computer systems during the entire process. Throughout fermentation, temperature should be maintained at around 30°C and pH at approximately 4.5 to 5.5. Yeast growth ranged from 120 kg in the intermediate fermenter, 420 kg in the stock fermenter, 2,500 kg in the pitch fermenter, and 15,000 to 100,000 kg in the trade fermenter (Figure 12.2; Birhanu et al. 1982; Zheng et al. 2005).

Once an optimum quantity of yeast has been grown, the yeast cells are recovered from the final fermenter. The centrifuged yeast solids should be concentrated by filter press or rotary vacuum filter. A filter press forms a filter cake containing 27% to 32% solids, whereas a rotary vacuum filter cake contains approximately 33% solids. The filter cake should be blended in mixers with small amounts of water, emulsifiers, and cutting oils to form the end product. In compressed yeast production, emulsifiers are added to give the yeast a white and creamy appearance. A small amount of oil is added to help extrude the yeast to form continuous ribbons. The ribbons are cut into pieces, wrapped, and cooled to less than 8°C. The yeast is extruded in thin ribbons and dried in a batch or continuous

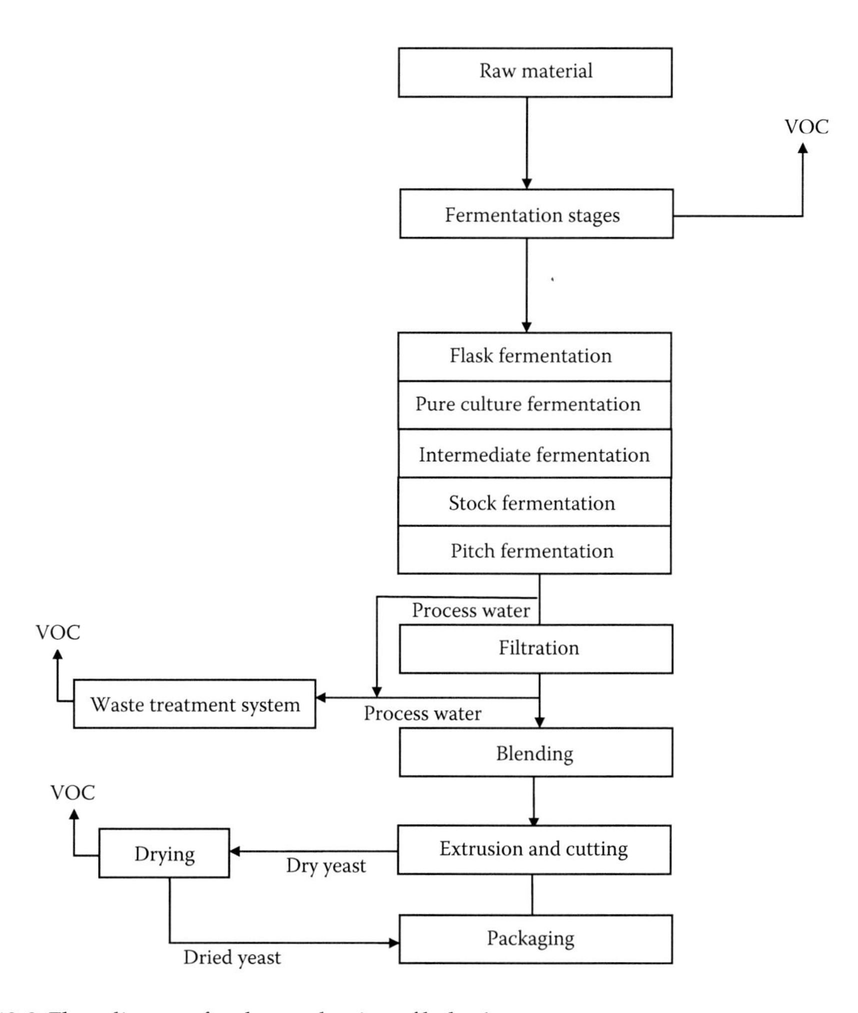

Figure 12.2 Flow diagram for the production of baker's yeast.

drying system. The yeast should be packed under vacuum or nitrogen gas before heat sealing. The shelf life at ambient temperature is 1 to 2 years (Chen and Chigar 1985).

Volatile organic compound emissions, that is, ethanol and acetaldehyde are generated as by-products of the fermentation process. Other by-products such as butanol, isopropyl alcohol, 2,3-butanediol, organic acids, and acetates are also generated during the fermentation process. Volatile by-products form as a result of either excess sugar (molasses) present in the fermenter or because of an insufficient oxygen supply. When anaerobic fermentation occurs, 2 mol of ethanol and 2 mol of CO_2 are formed from 1 mol of glucose. Under anaerobic conditions, the ethanol yield increases as the yeast yield decreases. Hence, it is essential to suppress ethanol formation in the final fermentation by incremental feeding of the molasses mixture with sufficient oxygen. The ethanol formation rate is higher in pure culture than in the final fermentation process. The earlier fermentation stages are batch fermenters, where excess sugars are present and less aeration is used during the fermentation process. These fermentations are not controlled to the degree that the final fermentations are controlled because the majority of yeast growth occurs in the final fermentation stages. The process wastewater is another potential emission from yeast production (Chen and Chigar 1985; Reed and Peppler 1973).

12.3 New researches in baker's yeast

12.3.1 Baker's yeast as natural therapy for cancer

In 2010, at a special conference on cell death mechanisms sponsored by the American Association for Cancer Research in San Diego, California, Dr. Mamdooh Ghoneum presented his findings about the role of yeasts in cancer. This conference included new complexities of cell death and cell survival, new technologies, and the clinical translational aspects necessary for the evolution of new therapeutic strategies. Dr. Ghoneum pursued a theory that cancer cells self-destruct when exposed to small quantities of yeast. Dr. Ghoneum said that in laboratory tests, cancer cells were ingested by yeast through phagocytosis, and then the cancer cells died. First, Dr. Ghoneum investigated this phenomenon *in vitro*, introducing yeast to breast, tongue, colon, and skin cancers. In later experiments on mice, a decrease in the size of the tumor mass was observed. Dr. Ghoneum's observations showed significant clearance of cancer cells from the lung and concluded that when the cancer cells eat the yeast, they die. These findings have been confirmed by similar studies at the U.S. Department of Health and Science, National Institutes of Health.

12.3.2 Model organism for somatic mutations and cancer study

Recently, new evidence has provided additional insight into cancer research in the field of epigenetics. According to this new theory, cancer is a genetic disease because most cancer cells do what they do, that is, proliferate at a rapid speed as a result of mutations in their genetic sequence. Mutations frequently occur as a normal part of cell life and most have a negligible effect on phenotype. Therefore, scientists suspect that most malignant transformations result from the accumulation of multiple mutations. Over the past 40 years, scientists have identified dozens of different cancer-causing mutations. Most of these mutations occur in genes that play some kind of role in regulating the cell cycle, or a series of events that a cell goes through as it replicates its DNA and divides. Dr. Leland H. Hartwell discovered some of these cancer-causing mutations and eventually received the Nobel Prize in 2001. Hartwell worked on the single-celled eukaryote *S. cerevisiae* for studying cancer and the cell cycle in his doctorate.

The same genes that control the cell cycle in baker's yeast and that malfunction in tumor cells exist in more or less the same capacity in human cells. Dr. Hartwell identified more than 100 genes involved in cell cycle control. These genes are known as cell division cycle (CDC) genes. Dr. Hartwell also determined the pathway of cell cycle regulation events. Many of the same genes that work to regulate cell division in yeast were also shown to work similarly in humans. Hartwell found that CDC genes regulate the cell cycle either by stimulating or by inhibiting cell division in response to the barrage of signals these cells constantly receive from their environment. However, in cancer cells, mutated genes that normally stimulate cell division only at the right time and place, the proto-oncogenes, start operating like stuck accelerators (turning into oncogenes). Meanwhile, other mutated genes that normally inhibit excessive cell division, so-called tumor suppressor genes simply stop working, similar to broken brakes. Because of their role in regulating cell division so precisely and because they are found to be mutated in tumor cells, these genes are frequently often referred to as cancer genes. Most oncogenes function as dominant mutations, that is, only one mutated copy of the gene is needed to start operating like a stuck accelerator, whereas most tumor suppressor genes function as recessive mutations, that is, two mutated copies of the gene are needed to lose its ability to apply the brakes. Despite this difference, both types of mutations, stuck accelerators and broken brakes, ultimately lead to out-of-control cell growth and reproduction. Hence, the yeast strains can be used to control the cancer cells.

12.3.3 Baker's yeast in genetics study

Baker's yeast such as *S. cerevisiae* is used as a model system for molecular research and genetics because the mechanics of replication, recombination, cell division, and metabolism are conserved between yeasts and eukaryotes including mammals. The complete genome sequence has proven to be extremely useful as a reference toward the human genome sequence and other higher eukaryotic genes. The ease of genetic manipulation of yeast allows its use for conveniently analyzing and functionally dissecting gene products from other eukaryotes. In yeast, direct transformation is unique and has become a model organism; synthetic oligonucleotides can be directly transformed into yeast allowing for easy production of altered forms of proteins. Also, yeast has the unique ability to recombine exogenous DNA with partially homologous segments directly to the specific genome location of your choice. High rates of gene conversion coupled with homologous recombination in yeast have led to the development of techniques for direct replacement of genetically engineered DNA sequences into their normal chromosome locations.

Yeasts are so easy to work with because

- It permits rapid growth in the laboratory, dispersed cells, ease of replica plating, and mutant isolation
- It has a well-defined genetic system and highly versatile DNA transformation system
- It can be handled with few safety measures; no special safety equipment required
- Remarkably inexpensive compared with higher eukaryote organisms
- It can be easily frozen and stored at –80°C for later use

Baker's yeast production practices are subject to yeast stress. The most critical phenotypes for baker's yeast improvement are those that tolerate high levels of sucrose, tolerate freeze–thawing stress, rapidly use maltose, and produce high CO_2. Dough can contain up to 30% sucrose per weight of flour, which applies harsh osmotic stress on yeast cells.

Baker's yeast strains must efficiently use maltose, which originates from flour in lean doughs. *S. cerevisiae* strains that have high sucrose tolerance and also rapidly use maltose have been developed and are used commercially. A desired characteristic for baker's yeast is the development of strains that use the disaccharide melibiose. Raffinose found in molasses is hydrolyzed by yeast invertase to fructose and melibiose. Baker's yeast does not have the ability to use melibiose because it lacks α-galactosidase. This enzyme is present in bottom-fermenting brewer's yeast strains, and this gene-encoding enzyme has been introduced into laboratory strains of baker's yeast. The resulting strains give increased biomass without alteration of growth rate in fermentations. One of the most desired properties of baker's yeast strains is a rapid fermentation rate. Free sugars in dough are sequentially consumed due to transport and catabolite repression regulatory mechanisms. Amylase found in dough releases maltose from starch, but many strains of baker's yeast use maltose. *Torulaspora delbrueckii* is highly tolerant to freezing and freeze–thawing and this yeast has been considered as a primary yeast or helper in frozen dough products.

12.4 Single-cell protein

The production of SCPs started in 1960. The term SCP was coined in 1966 by Professor Carol L. Wilson at the Massachusetts Institute of Technology (Ware 1977). SCP refers to mixed proteins from pure or mixed cultures of algae, yeasts, fungi, or bacteria used as a substitute for protein in human foods and animal feeds. Since 2500 BC, yeasts have been used in bread and beverage production. In 1781, processes for preparing highly concentrated forms of yeast were established. In 1919, a fat production process using *Endomyces vernalis* from sulfite liquor was developed, and in 1941, *Geotrichum* was introduced to improve this process. In the 1960s, researchers at British Petroleum developed the technology of SCP production by yeast from waxy *n*-paraffins, a product of oil refineries. The concept of food from oil became entirely popular in the 1970s. SCPs are primarily used as poultry and cattle feed. SCP production technologies arose as a promising way to solve the problem of a worldwide protein shortage (Haider et al. 2003; Nigam 2000).

Large-scale fermenters are required for the production of SCP and in high biomass production, oxygen transfer rate and respiration rate increases metabolic heat production, hence the need for an efficient cooling system. In continuous operation, the economics of production such as substrate cost, energy, operating costs, waste, safety, and the global market should be taken into account. The substrate cost is the largest single cost factor. Simplifying the manufacture and purification of raw material can save costs. Moreover, the manufacture of raw materials is more economical in larger plants. Factors involved in the raw materials' costs are raw material production, process capacity of the plant, and substrate yield. The energy for compressing air, cooling, sterilizing, and drying forms the next most important cost factor. Sites with cheaply available thermal, electrical, fossil, or process-derived energy are preferred. The capital-dependent costs are determined by the cost of the apparatus for the process, capacity of a plant, and its conditions (mainly the size of the plant). Small plants can be profitable only if they include simplifications of the processes and materials to a considerable degree. The greater expenditure on apparatus in processes with cheap, simple, and unpurified raw materials usually does not pay in comparison with more expensive pure substrates and simpler technology. The quality of the product is poorer for a low-value unpurified product of a rational mini process with varying compositions or one including numerous additional components than for upgraded products. Upgrading the product may consist of purification and separation into the components of the microbial biomass. Because of genetic variability, the possibility of technical control during manufacture, and the simplicity of the process, microbial biomass is more suitable for such special products compared with biomass from plants or animals.

Various bacteria, mold, yeast, and algae are used for the production of SCPs. The bacteria include *Brevibacterium, Methylophilus methylitropous, Acromobacter delvaevate, Acinetobacter calcoacenticus, Aeromonas hydrophilla, Bacillus megaterium, Bacillus subtilis, Lactobacillus* sp., *Cellulomonas* sp., *Methylomonas methylotrophus, Pseudomonas fluorescens, Rhodopseudomonas capsulata, Flavobacterium* sp., *Thermomonospora fusca*, and others. Yeasts such as *Candida utilis* (torula yeast), *Yarrowia lipolytica, Candida tropicalis, Candida novellas, Candida intermedia,* and *S. cerevisiae* are all among the various organisms that have been used for the production of SCP. Some of the algae such as *Chlorella pyrenoidosa, Chlorella sorokiana, Chondrus crispus, Scenedesmus acutus, Porphyrium* sp., and *Spirulina maxima* are used for SCP production. The filamentous fungi include *Chaetomium cellulolyticum, Fusarium graminearum, Aspergillus fumigates, Aspergillus niger, Aspergillus oryzae, Cephalosporium cichhorniae, Penicillium cyclopium, Rhizopus chinensis, Scytalidium aciduphlium, Tricoderma viridae, Tricoderma alba,* and *Paecilomyces varioti* have been employed for SCP production (Bhalla et al. 2007). The desired microorganisms should be cultured on the medium under sterile conditions. The organisms to be cultured should be nonpathogenic to plants, humans, and animals.

12.4.1 Production of SCP

Large-scale SCP production has contributed greatly to present-day advancements. The future of SCP will be heavily dependent on reducing the production costs and improving quality by fermentation, downstream processing, and improvement in the producer organisms as a result of conventional applied genetics together with recombinant DNA technologies. The process of SCP production from any microorganism or substrate would have the following basic steps:

i. Provision of a carbon source; it may need physical and chemical pretreatment
ii. Sources of nitrogen, phosphorus, and other nutrients need to support the optimal growth of selected microbial strains
iii. Prevention of contamination by maintaining aseptic conditions; the medium components may be heated or sterilized through filtration, and fermentation equipment may be sterilized
iv. The selected microbial strain should be inoculated in a pure state
v. SCP processes are highly aerobic, so adequate aeration must be provided; in addition, cooling is necessary as considerable heat is generated
vi. The microbial biomass is recovered from the medium

The production of SCP takes place in a fermentation process. This can be done using selected microbial strains cultivated on suitable raw materials directed to the growth of culture and cell mass followed by separation processes. After microbial screening, suitable production strains are obtained from samples of soil, water, air, or from swabs of inorganic or biological materials and subsequently optimized by selection, mutation, or other genetic methods. In addition, process engineering and apparatus technologies are adapted to the technical performance of the process to prepare the product for use on a large technical scale. Safety demands and environmental protection are also considered in SCP production in relation to both the process and the product. Finally, safety concerns and protection of innovation introduce legal and control aspects, that is, operating licenses, product authorizations for particular applications, and the legal protection of new processes and strains of microorganisms. The classic raw materials are substances containing monosaccharides and disaccharides because almost all microorganisms can digest glucose, other hexose and pentose sugars, and disaccharides. Other substrates for SCP include bagasse, citrus wastes, sulfite waste liquor, molasses, animal manure, whey, starch, sewage, and others (Nasseri et al. 2010).

The fermentation process requires a pure culture of the selected microbial strain, sterilized growth medium, production fermenter, cell separation, collection of cell-free supernatant, product purification, and effluent treatment. There are many methods available for concentrating the solutions such as filtration, precipitation, centrifugation, and the use of semipermeable membranes. The equipment used for these dewatering methods are expensive and thus would not be suitable for small-scale productions and operations. SCPs need to be dried to 10% moisture or they can be condensed and denatured to prevent spoilage. Therefore, an adequate technology that maintains appropriate growth conditions for a prolonged period must be implemented specifically for the purpose of obtaining high yield and productivity. Batch fermentation is inadequate for biomass production because the conditions in the reaction medium change with time. Fed-batch fermentations are better suited for the purpose of biomass production because they involve the control of carbon source supply through feeding rates. Fed-batch cultures are still in use for baker's yeast production using well-established and proven models. However, they have not been favored for the production of SCP at a large industrial scale. Prolonging a microbial culture by the continuous addition of fresh medium with the simultaneous harvesting of product has been implemented successfully in industrial fermentations destined for biomass production. The technical implications of chemostat culture are various and extremely relevant. The flow diagram for the production of SCP is illustrated in Figure 12.3.

A common problem for industrial fermentations is the appearance of a profuse foam on the headspace of the reactor. Among the various designs, deep-jet fermenters and airlift fermenters have been the most successfully applied. Airlift fermenters have had great success as the configuration of choice for continuous SCP production. The biomass from yeast fermentation processes is normally harvested by continuous centrifugation; filamentous fungi are harvested by filtration. Then, biomass is treated for RNA reduction and dried in a steam drum of spray driers. Drying is expensive, but results in a stabilized product with a shelf life of years. Solid-state fermentation (SSF) is the growth of microorganisms

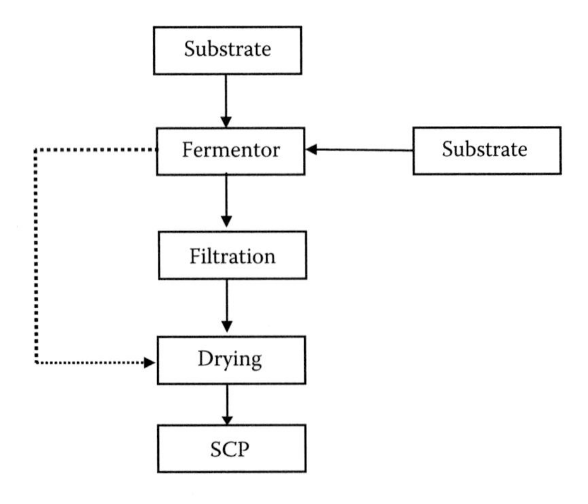

Figure 12.3 Flow chart for SCP production.

on predominantly insoluble substrate, where there is no free liquid. Generally, under combined conditions of low water activity and the presence of intractable solid substrate, fungi show luxuriant growth. Hence, proper growth of fungi in SSF gives much higher concentrations of the biomass and higher yield when compared with submerged fermentation. The advantage in SSF processes is the efficient use of waste as the substrate to produce commercially viable products. The process does not need elaborate prearrangements for media preparation. The process of SSF initially concentrated on enzyme production. Currently, there is worldwide interest for SCP production due to the dwindling conventional food resources. Large-scale production of microbial biomass has many advantages over traditional methods for producing proteins (Tipparat and Kittikun 1995).

12.4.2 SCP from waste

SCP production from waste and the effective advantages of microorganisms compared with conventional sources of protein is well known (Argyro et al. 2006). A number of agricultural and agro-industrial waste products are used for SCP production and other metabolites including orange waste, mango waste, cotton salks, kinnow-mandarin waste, barley straw, corn cops, corn straw, rice straw, onion juice, sugarcane bagasse (Nigam et al. 2000), cassava starch (Tipparat and Kittikun 1995), wheat straw (Abou Hamed 1993), banana waste (Saquido et al. 1981), capsicum powder (Zhao et al. 2010), and coconut water (Smith and Bull 1976). The use of such wastes as the sole carbon and nitrogen source for SCP production by microbial strains can be attributed to their abundance in nature and their cheap cost. The amount of agricultural and some industrial wastes used for SCP production can be locally very high and may contribute to a significant level of pollution in the water. Thus, the use of such materials in SCP processes serves two functions: the reduction in pollution and the creation of edible protein. Cellulose from agriculture and forestry sources constitutes the most abundant renewable resource in the planet as potential substrates for SCP production. Cellulose has emerged as an attractive substrate for SCP production, but in nature, it is usually found in complex form with lignin, hemicellulose, starch, and others. Therefore, if cellulose is to be used as a substrate, it must be pretreated chemically (acid hydrolysis) or enzymatically (cellulases) to make it a fermentable sugar. To use lignocellulose, pretreatment is usually necessary. Many of the pretreatment methods that have been reported vary from alkali or acid treatment, steam explosion, or x-ray radiation. To date, lignocellulosic wastes are used in mushroom production. Wood can also be cooked in a medium containing calcium sulfite with excess free sulfur dioxide. Lignin is converted to lignosulfonates and hemicellulose is hydrolyzed to monosaccharides and may be further broken down to furfurals. The amount of free sugars in the spent liquor varies with the type of procedure chosen as various cellulose fibers may be obtained with different degrees of degradation. Spent sulfite liquor has been used as a substrate for fermentation in Sweden since 1909 and later in many other parts of the world. *S. cerevisiae* is the first organism used for SCP production, although this organism is unable to metabolize pentoses that are found in considerable amounts in waste products. Later, other organisms better suited for the assimilation of all sugar monomers were chosen, that is, *C. tropicalis* and *C. utilis*.

Yeast biomass from sulfite liquor has been used as a food source during periods of war. However, in Finland, baker's yeast is produced from sulfite liquor using the Peliko process. The protein content of the fungus *P. variotii* exceeds 55% (w/w) and is officially approved as a food in Finland. The extracellular cellulases are commercially used in a cellulose-separating process. Cellulase is a complex of three enzymes including endocellulase,

1,4-β-cellobiohydrolase, and cellobiase. A number of efficient cellulase producers have been reported but *Trichoderma viride* continues to be the best known high cellulose-producing organism. *C. cellulolyticum* is another cellulolytic fungus that grows faster and forms 80% more biomass protein than *Trichoderma*. This means that *C. cellulolyticum* is suitable for SCP production, whereas *T. viride* is a hyperproducer of extracellular cellulases. The amino acid composition of *C. cellulolyticum* is generally better than *T. viride*, and is similar to alfalfa and soya meal protein. Starch is a cheaper and more amenable substrate for SCP production. Abundant amounts of carbohydrate may be obtained from rice, maize, and cereals. In tropical countries, cassava has been proposed as a good source of starch for SCP processes. The Symba process, developed in Sweden, uses starchy wastes combining two yeasts in a sequentially mixed culture; the amylase-producing *Endomycopsis fibuligira* and the fast-growing *C. utilis*. This process consists of three phases; the incoming starch waste is fed through heat exchangers and sterilized. The medium should feed to the first bioreactor where yeast grows and hydrolyzes starch. The hydrolyzed starch solution should feed to the second bioreactor where culture conditions favor the proliferation of *C. utilis*. Whey originates from the curding process in cheese production, but can now be obtained after ultrafiltration procedures for the production of spreading cheeses where the protein fraction corresponding to lactalbumins and lactoglobulins is incorporated into the casein fraction and all the proteins are in their native form, with the principal component being lactose (4%–6% w/v); although other nutrients can be found in significant amounts. Whey has been presented as an extremely suitable substrate for the production of SCP. In 1956, the French company, Fromageries Bel, produced yeast from whey using lactose-assimilating *Kluyveromyces marxianus*. Molasses is a by-product of the sugar manufacturing process. Biomass production from molasses requires supplementation with a suitable nitrogen source as well as phosphorus. Coffee-pressing wastes contain soluble carbohydrates and have a high chemical oxygen demand and soluble solid contents. In Guatemala, *Trichoderma* sp. is used to produce SCP on this substrate (Abou Hamed 1993; Prosser and Tough 1991; Rashad et al. 1990; Saquido et al. 1981).

12.4.3 *Advantages and disadvantages of SCP*

Various factors such as nutrient composition, amino acid profile, vitamin and nucleic acid content as well as palatability, allergies, and gastrointestinal effects should be considered for nutritional value. The nutritive and food values of SCP vary with the microorganisms used. The method of harvesting, drying, and processing has an effect on the nutritive value of the finished product. SCP basically comprises proteins, fats, carbohydrates, ash, water, and other elements such as phosphorus and potassium. The composition depends on the organism and substrate used for growth. Proteins not only provide a nutritional component in a food system but also perform a number of other functions. SCP has high protein and low fat content. It has good amino acid composition, which is rich in lysine but poor in sulfur-containing amino acids such as methionine and cysteine. It is a good source of vitamins, particularly B-complex, thiamine, riboflavin, and folic acid. SCP from yeast and fungi has up to 50% to 55% protein and it has high protein–carbohydrate ratio compared with forage. SCP from bacteria has more than 80% protein, although it is poor in sulfur-containing amino acids and it has high nucleic acid content (Burgents et al. 2004; Campa-Cordova et al. 2002; Mahajan and Dua 1995; Oliva-Teles and Goncalves 2001; Tovar et al. 2002).

There are many problems in the adoption of SCP on a global basis such as a high concentration of nucleic acids (6%–10%), which elevates the uric acid level in serum and results in kidney stone formation. Approximately 70% to 80% of the total nitrogen is represented by amino acids, whereas the rest occur in nucleic acids. This concentration of nucleic acids is higher than in other conventional proteins and is characteristic of all fast-growing organisms. SCP from yeasts and fungi have high nucleic acid content. The filamentous fungi show slower growth rates than yeasts and bacteria, there is a high risk of contamination, and some strains produce mycotoxins; hence, they should be well screened before consumption. Similarly, SCP from bacteria has high ribonucleic acid content, and high risk of contamination during the production process and during the recovery of the cells. All these detrimental factors affect the acceptability of SCP as a global food (Nasseri et al. 2011).

12.5 Conclusion

Microorganisms such as algae, bacteria, yeasts, and fungi are used as protein sources. Yeasts can be considered as the oldest industrial microorganism. It is believed that the early fermentation systems for alcohol production and bread making formed through the natural microbial contaminants of flour, other milled grains, and from fruits or other juices containing sugar. Such microbial flora would have included wild yeasts and lactic acid bacteria, which were associated with cultivated grains and fruits. For hundreds of years, it was traditional for bakers to obtain the yeast to leave their bread as by-products of brewing and wine making. The baker's yeast industry is a major market. New strains of baker's yeast from recombinant DNA technology, which produces CO_2 more rapidly and is more resistant to stress, or which produces proteins or metabolites that can modify the flavor of bread, dough rheology, or shelf life are now emerging. This low-value, high-volume product is produced under stringent environmental conditions to obtain the maximum product/biomass yield, which is dependent on process design, cost, and the strain of *S. cerevisiae* used. As with all biotechnological processes, research and development in this area is ongoing to create more beneficial strains and to optimize the fermentation and processing steps that make up the baker's yeast process. From many studies, it was found that yeast cells help control cancer. The same genes that control the cell cycle in baker's yeast (and that malfunction in tumor cells) exist in more or less the same capacity in human cells, and that the yeast strains can be used to control cancer cells. *S. cerevisiae* is used as a model system for molecular research because the basic cellular mechanics of replication, recombination, cell division, and metabolism are generally conserved between yeast and larger eukaryotes, including mammals. The SCP is another widely used microbial biomass, which has proteins, fats, carbohydrates, ash, water, and other elements such as phosphorus and potassium. SCP is not widely accepted as a global food due to its high nucleic acid content. Attempts to improve the acceptability of SCP products should be intensified. Further research and development will ensure the use of microbial biomass as SCP or as a supplement in the diet.

Acknowledgments

The author is thankful to Kanchan Prabha Rathoure (Mrs.) for technical suggestions and Madhav Pandey for English corrections to make the text more effective.

References

Abou Hamed, S.A.A. "Bioconversion of wheat straw by yeast into single cell protein." *Egyptian Journal of Microbiology* 28 no. 1 (1993): 1–9.

Acourene, S., Khalid, A.K., Bacha, A., Tama, M. and Taleb, B. "Optimization of baker's yeast production cultivated on musts of dates." *Journal of Applied Science Research* 3 (2007): 964–971.
Argyro, B., Costas, P. and Athanasios, A.K. "Production of food grade yeasts." *Food Technology and Biotechnology* 44 (2006): 407–415.
Atiyeh, H. and Duvnjak, Z. "Production of fructose and ethanol from cane molasses using *Saccharomyces cerevisiae* ATCC 36858." *Acta Biotechnology* 23 (2003): 37–48.
Bekatorou, A., Psarianos, C. and Koutinas, A.A. "Production of food grade yeasts." *Food Technology and Biotechnology* 44 (2006): 407–415.
Bhalla, T.C., Sharma, N.N. and Sharma, M. *Production of Metabolites, Industrial Enzymes, Amino Acids, Organic Acids, Antibiotics, Vitamins and Single Cell Proteins.* National Science Digital Library, India, 2007.
Burgents, J.E., Burnett, K.G. and Burnett, L.E. "Disease resistance of Pacific white shrimp, Litopenaeus vannamei, following the dietary administration of a yeast culture food supplement." *Aquaculture* 231 (2004): 1–8.
Campa-Cordova, A.I., Hernandez-Saavedra, N.Y., De Philippis, R. and Ascencio, F. "Generation of superoxide anion and SOD activity in haemocytes and muscle of American white shrimp (*Litopenaeus vannamei*) as a response to β-glucan and sulphated polysaccharide." *Fish Shellfish Immunology* 12 (2002): 353–366.
Chen, S.L. and Chigar, M. *Production of Baker's Yeast, Comprehensive Biotechnology*, Volume 20. Pergamon Press, New York, 1985.
Cook, A.H. *The Chemistry and Biology of Yeasts*. Academic Press Publishers, New York, 1958.
Daromola, M.O. and Zampraka, L. "Experimental study on the production of biomass by *Saccharomyces cerevisiae* in a fed batch fermenter." *African Journal of Biotechnology* 7 (2008): 1167–1114.
Dien, B.S., Cotta, M.A. and Jaffries, T.W. "Bacteria engineered for fuel ethanol production." *Journal of Applied Microbiology and Biotechnology* 63 (2003): 258–266.
Feldmann, H. Ed. *Yeast: Molecular and Cell Biology*. Wiley-Blackwell, United Kingdom, 2012.
Gashe, B.A., Girma, M. and Bisrat, A. "The role of microorganisms in fermentation and their effect on the nitrogen content of teff." *SINET: Ethiopian Journal of Science* 5 (1982): 69–76.
Gunasekaran, P. and Chandra, R.K. "Ethanol Fermentation technology." *Current Science* 77 (1999): 56–68.
Haider, M.M., El-Tajoris, N.N. and Baiu, S.H. "Single cell protein production from carob pod extracts by the yeast *Saccharomyces cerevisiae*." Thesis, Botany and Biochemistry Department of Faculties of Science and Medical University, Benghazi, Libya, 2003.
Lin, T. and Tanaka, S. "Ethanol fermentation from biomass resources: Current status and prospects." *Journal of Applied Microbiology Biotechnology* 69 (2006): 627–642.
Mahajan, A. and Dua, S. "Functional properties of repressed (*Brassica campestris* var. toria) protein isolates." *Journal of Food Science and Technology* 32 (1995): 162–165.
Mormeneo, S. and Sentandreu, R. "Regulation of invertase synthesis by glucose in *Saccharomyces cerevisiae*." *Journal of Bacteriology* 152 (1982): 14–18.
Nasseri, A.T., Rasoul-Amini, S., Morowvat, M.H. and Ghasemi, Y. "Single cell protein: Production and process." *American Journal of Food Technology* 6 no. 2 (2011): 103–116.
Nasseri, S., Rezaei Kalantary, R., Nourieh, N., Naddafi, K., Mahvi, A.H., Baradaran, N. "Influence of bioaugmentation in biodegradation of PAHs-contaminated soil in bio-slurry phase reactor." *Journal of Environmental Health Science and Engineering* 7 no. 3 (2010): 199–208.
Nielsen, J. and Jewett, C.M. "Impact of systems biology on metabolic engineering of *Saccharomyces cervisiae*." *FEMS Yeast Research* 8 (2007): 122–131.
Nigam, N.M. "Cultivation of *Candida langeronii* in sugarcane bagasse hemi cellulose hydrolysate for the production of single cell protein." *World Journal of Microbiology and Biotechnology* 16 (2000): 367–372.
Oliva-Teles, A. and Gonçalves, P. "Partial replacement of Fishmeal by brewer's yeast (*Saccaromyces cerevisae*) in diets for sea bass (*Dicentrarchus labrax*) Juveniles." *Aquaculture* 202 (2001): 269–278.
Ostergaard, S., Olsson, L. and Nielsen, J. "Metabolic engineering of *Saccharomyces cerevisiae*." *Microbiology and Molecular Biology Reviews* 64 (2000): 34–50.
Prescott, S.C. and Dunn, C.G. *Industrial Microbiology*. McGraw-Hill, New York, 1959.

Prosser, J.I. and Tough, A.J. "Growth mechanism and growth kinetics of filamentous microorganism." *Critical Reviews in Biotechnology* 10 no. 4 (1991): 253–274.
Rashad, M.M., Moharib, S.A. and Jwanny, E.W. "Yeast conversion of mango waste or methanol to SCP and other metabolites." *Biological Waste* 32 no. 4 (1990): 277.
Reed, G. and Nagodawithana, T.W. "Technology of yeast usage in winemaking." *American Journal of Enology and Viticulture* 39 no. 1 (1988): 83–90.
Reed, G. and Peppler, H. *Yeast Technology*. Avi Publishing Company, Westport, CT, 1973.
Saquido, P.M.A., Cayabyab, V.A. and Vyenco, F.R. "Bioconversion of banana waste into single cell protein." *Journal of Applied Microbiology and Biotechnology* 5 no. 3 (1981): 321–326.
Sheoran, A., Yadav, B.S., Nigam, P. and Singh, D. "Continuous ethanol production from sugar cane using a column reactor of immobilized *Saccharomyces cerevisiae* HAU-1." *Journal of Basic Microbiology* 38 (1998): 123–128.
Smith, M.E. and Bull, A.T. "Protein and other compositional analysis of *Saccharomyces fragilis* grown on coconut water waste." *Journal of Applied Bacteriology* 41 (1976): 97–107.
Sonnleitner, B. and Kappeli, O. "Growth of *Saccharomyces cerevisiae* is controlled by its limited respiratory capacity—Formulation and verification of a hypothesis." *Biotechnology and Bioengineering* 28 no. 6 (1986): 927–937.
Tipparat, H. and Kittikun, A.H. "Optimization of single cell protein production from cassava starch using *Schwanniomyces castellii*." *World Journal of Microbiology and Biotechnology* 11 (1995): 607–609.
Tovar, D., Zambonino, J., Cahu, C., Gatesoupe, F.J., Vázquez-Juárez, R. and Lésel, R. "Effect of live yeast incorporation in compound diet on digestive enzyme activity in sea bass larvae." *Aquaculture* 204 (2002): 113–123.
Van Hoek, P.V., Dijken, J.P.V. and Pronk, J.T. "Effect of specific growth rate on fermentative capacity of baker's yeast." *Applied Environmental Microbiology* 64 (1998): 4226–4233.
Ware, S.A. *Single Cell Protein and Other Food Recovery Technologies from Waste*. Municipal Environment Research Laboratory, Office of R&D. U. S. E.P.A., Cincinnati, Ohio, 1977.
Zamani, J., Pournia, P. and Seirafi, H.A. "A novel feeding method in commercial baker's yeast production." *Journal of Applied Microbiology* 105 (2008): 674–680.
Zhao, G., Zhang, W. and Zhang, G. "Production of single cell protein using waste capsicum powder produced during capsanthin extraction." *Letters of Applied Microbiology* 50 (2010): 187–191.
Zheng, T., De-Kock, S.H. and Kilian, S.G. "Anomalies in the growth kinetics of *Saccharonmyces cerevisiae* strains in Aerobic chemostat cultures." *Journal of Industrial Microbiology and Biotechnology* 47 (2005): 231–236.

chapter thirteen

Microbial biofertilizers and their pilot-scale production

Santosh Kumar Sethi, Jayanti Kumari Sahu, and Siba Prasad Adhikary

Contents

13.1 Introduction

Due to uncontrolled population growth in many developing countries, there is a heavy demand for food. Hence, the aim is to achieve sustained growth rates that are high enough to feed an enormous population without degrading the environment. Researchers are focusing on an evergreen revolution, which can be achieved through sustainable agriculture (Parr et al. 1990). However, slow poisoning of the soil due to chemical fertilizers is one of the major calamities of present-day agriculture. As a result of excessive fertilizer use, the quality of soil is continuously degraded year after year, making it difficult to sustain soil fertility. These chemicals, in addition to causing damage to the soil's health, also creates a chain of ecological and economic problems. The residual effects of these chemical fertilizers severely affect the soil microflora, especially on the nodulation of legume crops and other nitrogen-fixing free-living microorganisms that play a key role in productivity. Hence, the only alternative is to use biological nitrogen fixation technology for the maintenance of soil health and sustainable productivity in agriculture.

The goal of an evergreen revolution and sustainable agriculture to meet the needs of the present generation without endangering our natural resources is mainly dependent on soil organic matter for nutrient supply through efficient microorganisms. Microorganisms not only contribute to nitrogen fixation but also solubilize phosphate, produce growth-promoting substances, and inhibit the growth of plant phytopathogens resulting in better growth and yield of crops. The main advantages of such sustainable agriculture are ecological balance, low cost of cultivation, an increase of the soil microflora to maintain good and fertile soil, clean environment, and nutritious food without residues that are harmful to human health (Bhatnagar and Patla 1996).

The following have been given importance for sustainable agriculture:

1. Use of efficient N_2 fixing microorganisms and use of potentially indigenous (local) strains instead of exogenous ones.
2. Because chemical fertilizers reduce soil organic matter, reduce soil porosity, impede oxygen flow, reduce the water-holding capacity of soil, and obstruct natural nitrogen fixation by soil bacteria, dependence on these chemical inputs needs to be lessened.
3. Agrochemical inputs are neither economically feasible nor environmentally desirable on a long-term basis. Therefore, improvement of the crop yield by inoculation with biofertilizers such as *Rhizobium*, *Azotobacter*, *Azospirillum*, Cyanobacteria, plant growth-promoting rhizobacteria (PGPR), and so on are to be adapted as ecofriendly technologies in agriculture.

Biofertilizers are defined as preparations containing living or dormant cells of efficient strains of nitrogen-fixing, phosphate-solubilizing, or cellulolytic microorganisms used for the application of seed, soil, or composting areas with the objective of increasing the number of such microorganisms and accelerating certain microbial processes to augment the availability of nutrients in a form that can be assimilated by plants. Hence, in present-day agriculture, biofertilizers are important natural resources in integrated nutrient management and are advocated for use as supplements to chemical fertilizers for increased productivity and sustainable agriculture.

13.2 *Microorganisms used as biofertilizers*

13.2.1 Rhizobium

Rhizobium is a soil bacterium able to colonize the roots of leguminous plants and fix atmospheric nitrogen symbiotically. They are non-spore-forming rods, motile, aerobic, Gram-negative, and capable of utilizing several carbohydrates. Initially, all root nodule–forming, nitrogen-fixing bacteria were assigned to a common name, *Rhizobium*, and were classified into species based on a cross-inoculation group concept (Subba Rao 1993). Based on a polyphasic approach despite splitting the genus into five different genera, they are still commonly referred to as rhizobia. These genera are *Rhizobium* with 14 root-nodulating nitrogen-fixing species and four nonnodulating non-nitrogen fixing species, *Sinorhizobium* with nine species, *Mezorhizobium* with eight species, *Bradyrhizobium* with four species, and a single species of *Azorhizobium*. Recently, two more genera, Methylobacterium with one species under alpha-proteobacteria and *Burkholderia* under B-proteobacteria have been included in the rhizobia group. The morphology and physiology of *Rhizobium* vary from free-living conditions to the bacteroid in the nodules of legumes. The symbiotic association is spread among a wide range of host species belonging to the family Leguminosae. Out of the 748 genera and nearly 19,000 legume species, nearly 49% of the genera and 16% of the species have thus far been examined for nodulation; of which 41% of the genera and 14% of the species have been found to bear rhizobial nodules (Yadav and Mowade 2005). They are the most efficient biofertilizers according to the quantity of nitrogen they fix. It has been estimated that 40 to 250 kg N/ha/year is fixed by different legume crops due to colonization of *Rhizobium*.

13.2.2 Azotobacter

Azotobacter is one of the most extensively used plant growth-promoting microorganisms because its inoculation benefits a wide variety of crops. These are polymorphic, possess peritrichous flagella, produce polysaccharides, sensitive to acidic pH, high salts, and temperatures above 35°C, grow on a nitrogen-free medium, and fix atmospheric nitrogen. Several types of azotobacteria groups have been found in the soil and the rhizosphere such as *Azotobacter chroococcum*, *Azotobacter nigricans*, *Azotobacter paspali*, *Azotobacter armenicus*, *Azotobacter salinestris*, and *Azotobacter vinelandi*, of which *A. chroococcum* is most commonly found in Indian soils (Subba Rao 1993). Besides nitrogen fixation, they produce siderophores, antifungal substances, and plant growth regulators (Suneja et al. 1996). Production of phytohormones such as auxin, cytokinin by *A. chroococcum*, have been reported (Verma et al. 2001). The bacterium also produces an abundant amount of slime, which helps in soil aggregation.

13.2.3 Azospirillum

Azospirilla are Gram-negative, vibroid to straight rods, sometimes curved, highly pleomorphic, with an abundant accumulation of poly beta hydroxybutarate in the cytoplasm, motile, with a preference for salts and organic acids as carbon source, oxidase positive, and are capable of nitrogen fixation. *Azospirillum* is comprised of seven species. These *are Azospirillum amazonense, Azospirillum brasilense, Azospirillum doebereinnerae, Azospirillum halopraeferens, Azospirillum irakense, Azospirillum largimobile,* and *Azospirillum lipoferum*. These are isolated from the root and aboveground parts of a variety of crop plants. The organism proliferates under both anaerobic and aerobic conditions but it is preferentially microaerophilic in the presence or absence of combined nitrogen in the medium. Apart from nitrogen fixation, production of growth-promoting substances, disease resistance, and drought tolerance are the additional benefits due to *Azospirillum* inoculation (Yadav and Mowade 2005).

13.2.4 Mycorrhiza

Mycorrhizal fungi are intimately associated with plant roots, and form a symbiotic relationship. The plant provides sugars to the host and the fungi provide nutrients such as phosphorus to the plants. Mycorrhizal fungi absorb, accumulate, and transport phosphate within their hyphae and release to plant cells in the root tissue. There are three major groups of mycorrhiza: ectomycorrhiza, endomycorrhiza, and ectendomycorrhiza, of which the former two are important in agriculture and forestry. Plants inoculated with endomycorrhiza have been shown to be more resistant to root diseases. Mycorrhiza increase higher branching of plant roots facilitating contact with a wider area of the soil surface, which results in an increase of the absorbing area for water and nutrients.

13.2.5 Cyanobacteria

Cyanobacteria are also known as blue-green algae (BGA), which obtain their energy through photosynthesis. Cyanobacteria include unicellular, colonial, and filamentous species. Some filamentous forms show the ability to differentiate into several different cell types: vegetative cells or photosynthesizing cells, which form under favorable growing conditions; akinetes or the climate-resistant spores, which form when environmental conditions become harsh; and thick-walled heterocysts, which contain the enzyme nitrogenase for nitrogen fixation. Heterocyst-forming species are specialized for nitrogen fixation and are able to fix nitrogen into ammonia (NH_3) or nitrates (NO^{-3}), which can be absorbed by plants and converted into proteins. Rice plants utilize fixed nitrogen from healthy populations of cyanobacteria as well as from the aquatic fern *Azolla* and the cyanobacterium *Anabaena*, as a symbiont occurring in waterlogged fields and hence are used as biofertilizers for rice.

13.2.6 Plant growth-promoting rhizobacteria

The group of bacteria that colonize roots or rhizosphere soil and that are beneficial to crops are referred to as PGPR. The PGPR inoculants promote growth through the following mechanisms: suppression of plant disease, improved nutrient acquisition, and phytohormone production. These PGPR are principally species of *Pseudomonas* and *Bacillus*, which produce biostimulants like indole-acetic acid, cytokinins, gibberellins, and inhibitors of ethylene production. They cause crops to have greater amounts of fine roots, which have the effect of increasing the absorptive surface of plant roots for uptake of water and nutrients.

Despite promising test results when applied to various crops, biofertilizers have not received widespread application in agriculture, mainly because of the variable response of plant species to inoculation depending on the microbial strain used, the agroclimatic conditions, and cultivation practices that vary from region to region, especially in a tropical country like India with varied climatic regimes. On the other hand, good competitive ability with the indigenous flora is one of the major factors determining the success of an inoculated strain in establishing a biofertilizer. Studies into the synergistic activities and persistence of specific microbial populations in complex environments, such as the rhizosphere, need to be addressed to obtain efficient inoculants and to obtain appropriate formulations of microbial inoculants incorporating nitrogen-fixing, phosphate-solubilizing, and PGPR, which will help in promoting the use of such beneficial microbes as biofertilizers for sustainable agriculture.

13.3 Biofertilizer application methodology

Biofertilizers are applied in three different ways, that is, seed treatment, root dipping (or seedling treatment), or soil application depending on the crop for which it is used.

13.3.1 Seed treatment

Seed treatment is the most common method of application of bacterial inoculants. *Rhizobium*, *Azotobacter*, *Azospirillum*, along with phosphate-solubilizing microorganisms (PSM), and PGPR are used in this process. For small quantities of seeds up to 5 kg, the coating can be done in a plastic bag. For this purpose, the bag is filled with 2 kg or more of seeds and closed in such a way to trap as much air as possible. Then, the bag is squeezed for up to 5 min or until all the seeds are uniformly wetted. Then, the bag is opened, inflated again, and shaken gently. After ascertaining that each seed has received a uniform layer of culture coating, the bag is opened and the seeds dried under the shade before use. For large amounts of seeds, coating can be done in a bucket and inoculant mixed directly by hand.

13.3.2 Root dipping

For application of *Azospirillum*/PSM, seedlings of paddy and vegetable crops are raised in a nursery. The required quantity of *Azospirillum*/PSM biofertilizer is mixed with 5 to 10 L of water in which the seedlings are dipped for up to 30 min followed by transplantation.

13.3.3 Soil application

Almost all types of biofertilizers can be used for soil application and their recommended dose per acre varies from the biofertilizer type and the crop for which it is applied. For PSB, cow dung/farmyard manure along with rock phosphate are mixed, kept under shade overnight, and maintained at 50% moisture. The mixture is used as a soil application in rows or during the leveling of soil. This is similar with the method of application of *Azospirillum* and *Azotobacter* in addition to seed dipping; however, cyanobacteria are used for broadcasting after transplantation.

13.4 Mass production of bacterial biofertilizer

Biofertilizers are carrier-based preparations containing efficient strains of nitrogen-fixing, phosphate-solubilizing, or cellulolytic microorganisms. The organic carrier materials are more effective for the preparation of the inoculants. The carrier-based inoculants carry more microbial cells and also support the survival of cells for longer periods of time. The mass production of carrier-based bacterial biofertilizers involves the following three steps:

- Culturing of microorganisms, the process of which is dependent on the type of organism used
- Type of carrier material it's processing before mixing with a particular type of microorganism
- Mixing the carrier with the microbial culture produced, its packing and storage

An ideal carrier material is to be chosen for the production of quality biofertilizer. Peat, soil, lignite, vermiculite, charcoal, and soil are used as carrier materials. However, neutralized peat soil/lignite are found to be better carrier materials especially for *Rhizobium*, *Azotobacter*, *Azospirillum*, and PSB biofertilizer preparations. The important criteria for selection of ideal carrier material are lower cost, local availability, high organic matter content, nontoxic to microbes, water-holding capacity of more than 50%, and ease of processing during packaging. The carrier materials are crushed into a fine powder that could pass through a 212-μm IS sieve. Peat soil or lignite are acidic, thus they need to be neutralized using calcium carbonate (1:10 ratio), followed by autoclaving to eliminate contaminants before being used.

13.4.1 *Culturing of microorganisms*

Microorganisms are produced in large quantities for a preparation of biofertilizers. For this purpose, nutrient broths are prepared by inoculating the seed material in flasks containing the appropriate medium and grown in an incubator shaker at ambient temperature until a cell population of approximately 10^{10} colony-forming units per milliliter (CFU/mL) is produced. Under optimum conditions, this population level could be attained within 4 to 5 days for *Rhizobium*, 5 to 7 days for *Azospirillum*, 2 to 3 days for phosphobacteria, and 6 to 7 days for *Azotobacter*. For large-scale inoculant production, inoculum from starter culture broths are transferred to large flasks/seed tank fermentors and grown until the required cell count level is reached. During the entire process, the culture broth is checked for contamination, if any. After harvesting, the cultures are mixed with suitable carrier material and packed. After fermentation, the broth should not be stored for more than 24 h because even at 4°C storage temperature, the number of viable cells starts decreasing.

13.4.2 *Preparation of biofertilizer packs*

Biofertilizer packets are prepared by mixing the culture with carrier material after processing. First, the carrier material is spread onto a clean, dry, sterile metallic or plastic tray, and then bacterial culture is added and mixed well manually using sterile gloves or using a mechanical mixer. The culture suspension is added at 40% to 50% of the water-holding capacity of the carrier. Curing is done by spreading the inoculants on a clean floor/polyethylene sheet/by keeping it in open shallow tubs/trays with polyethylene covering for 2 to 3 days at room temperature before packaging. Using this material, biofertilizer packs of 200 g quantities are prepared and sealed. If polyethylene bags are used for packing, these should be of low density grade and the thickness should be approximately 50 to 75 μm. Each packet should contain the name of the manufacturer, product name, details of the strain used, storage instructions, batch number, the crop for which it is to be applied, mode of application, date of manufacture, date of expiry, price, full address of the manufacturer, and others. The biofertilizer packets should be stored in a cool place away from the heat or direct sunlight. The population of inoculants in the prepared packets should be checked at 15-day intervals to ascertain that they contain more than 10^9 cells/g at the time of preparation and 10^7 cells/g while using for various crops (Motsara et al. 1995).

13.5 *Pilot-scale production of* Rhizobium *and* Azotobacter *biofertilizer in a rural area of Eastern India*

For mass production of microorganisms on a pilot scale, the following equipments are required: (1) 20-L fabricated fermenter fitted with an air pressure pump, (2) horizontal rotary shaker, (3) vertical autoclave with pressure regulator, (4) incubator, (5) refrigerator, (6) inoculation chamber (with provision of UV-C), (7) distillation apparatus, (8) microscope, (9) polyethylene sealer, and (10) air-conditioning provision.

Selected region-specific bacterial strains isolated by Sethi and Adhikary (2009a,b,c) were grown in slants and transferred to a liquid broth to prepare the mother culture. The cultures were finally grown on a large scale in a fermenter for up to 5 days. The cultures were harvested in batch culture mode and then mixed with unsterile brown sterile soil/charcoal in a ratio of 1:3. Two hundred grams of shade-dried, carrier-mixed bacteria were packed in polyethylene bags, sealed, and then stored for use as biofertilizer for the desired crop. The limited scale production of bacterial biofertilizers as standardized at the extension laboratory in a rural area has been presented photographically (Figure 13.1).

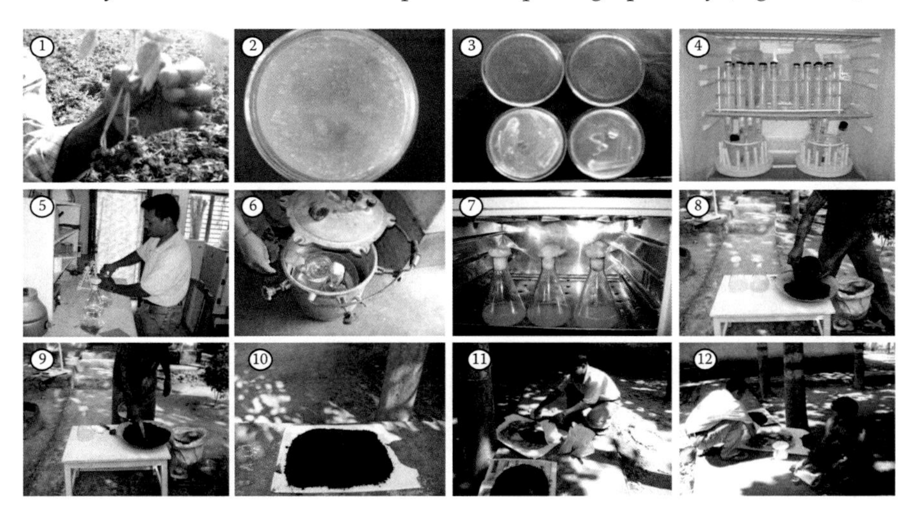

Figure 13.1 Stepwise preparation of biofertilizer from isolation of *Rhizobium* from effective nodules till packing of biofertilizer carried out at the extension laboratory in the village Maniakati under Surada block of Odisha. ① Uprooted *Vigna radiata* plant with effective nodules. ② *Rhizobium* colonies on YEMA plates. ③ Growth of *Rhizobium* on plates with YEMA and YEMA + Congo red. ④ Routine maintenance of *Rhizobium* in screw cap test tubes. ⑤ Preparation of YEMA medium. ⑥ Sterilization of medium. ⑦ Growth of *Rhizobium* strains in incubator at 28°C ± 2°C. ⑧ Five-day-old culture of *Rhizobium* and the carrier material (charcoal + brown forest soil 3:1) used for preparation of biofertilizer. ⑨ Mixing of *Rhizobium* suspension with carrier material. ⑩ Drying of prepared biofertilizer in shade. ⑪ Weighing of 200 g each of biofertilizer per packet. ⑫ Packing of biofertilizer.

The different steps of production of the biofertilizer are also given schematically in the following:

Bacterial culture slants
(stored in refrigerator at 4°C)

↓

Culture in tubes
(28°C for *Rhizobium* and 30°C for *Azotobacter* grown for 5 days)

↓

Culture in 500 mL conical flasks containing broth medium
(incubation on rotary shaker for 3–5 days)

↓

Inoculation of actively growing broth culture into fermenter
(incubation at 28°C–30°C depending on the organism for 5 days)

↓

Checking the purity of the culture
(by microscope and streaking culture methods)

↓

Bacteria harvested and mixed with carrier
(mix 5 kg of carrier grinded brown forest soil/charcoal = 1:3 for 1 L culture, and air drying)

↓

Biofertilizer prepared and stored in 200 g packs

For the production of biofertilizers, the fermenters were fabricated to make the production cost more economical. The details about the requirements for fabrication of the low-cost fermenters is given below:

1. Glass vessel (Borosil, cat. no. 1585) 10 L or 20 L capacity
2. Stainless steel fermenter head: 6-mm-thick SS plate cut into a circular piece with two extended lobes having holes for holding clamping nuts, and the setup is fitted to the mouth of two half-circular clamps around the bottle of neck. Three 6-mm tubes will be welded to the plate, of which one will have ceramic sparger for aeration
3. Membrane filter cartridges (47 mm; Hepa vent filter discs, Whatman no. 6723-5000) will be fitted to one of the outlet tubes
4. For air inlet and outlet filter, 500 mL capacity bottle (Tarson cat. no. 6-2020) fitted with glass wool will be fixed
5. One aerator (aquarium pump) will be fixed to each setup for bubbling/aeration

For limited-scale production of biofertilizer, two glass fermenters of 10 L capacity are used for each batch. For 1 L of bacterial culture, 5 kg of carrier material is required for biofertilizer preparation. Thus, 100 kg of carrier-based biofertilizer is produced from two fermenters in each batch or 500 packets (200 g each) of biofertilizer is produced. From one-time harvest in a week (on average), 500 packets of carrier-based biofertilizer can be produced. Thus, in a month, production will be 2000 packets, and in a year, it will be 24,000 packets (4800 kg from 960 L of broth). Following the protocol, the cost–benefit of commercial production of bacterial biofertilizer (*Rhizobium* as well as *Azotobacter*) developed by us is as follows.

13.5.1 *Stepwise calculation of pilot-scale production cost of* Rhizobium *and* Azotobacter *biofertilizers (according to existing cost of utilities, wages, and labor cost, etc., in eastern regions of India)*

1. Fixed cost: includes capital investment on equipments
2. Variable cost: raw material cost, carrier material, broth (i.e., chemicals), polyethylene bags for package of biofertilizer
3. Indirect cost:
   a. Salary of the staff, marketing costs (transport costs, power consumption per packet), publicity cost, marketing margin (wholesale and retail, subsidy or commission, if any, risk coverage against unsold packets)
   b. Miscellaneous expenses

Fixed cost (nonrecurring)

Equipment	Cost ($)
Glass fermenter (fabricated with aerator), ×2	3000
Autoclave (one electrically operated), ×1 (20 L capacity)	600
Glass double distillation set, 5 L/h capacity, ×1	200
Incubator, ×1	400
Compound microscope (binocular), ×1	300
Refrigerator, ×1	200
Inoculation chamber with UV, ×1 (fabricated)	100
Polyethylene sealer, ×1	100
Chemical balance, ×1	100
Total	5000

Variable cost per year (recurring)

Material	Quantity	Cost of the material ($)
Carrier material	10 quintal	80
Broth	1000 L, $0.25/L	250
Polyethylene bag	30,000, $1/1000 bags	30
Recurring expenses per year (consumables) plastic/polypropylene bottle + plastic wares + glass wares (conical flask, pipettes, test tube, measuring cylinder, beaker etc.)		100
Total cost		460

Requirements for chemicals in the production process:

For production of *Rhizobium* broth per 1000 packets require 40 L of broth, for example, Himedia chemical cost with VAT according to 2009–2010 price lists as follows:

Chemicals required	For 40 L (g)	Total cost for 40 L ($)
Mannitol	400	7.5
Yeast extract	16	1.0
NaCl	4	0.5
$MgSO_4{\cdot}7H_2O$	8	1.0
K_2HPO_4	20	1.0
Total cost		11.0

Therefore, for 40 L of broth, the cost of required chemicals = $11
For 1 L = $11 ÷ 40 = $0.275

Indirect cost

Salary and wages per annum (according to the existing wages in rural areas of eastern regions of India)	Salary component ($)
One microbiologist (skilled), $200 per month	2400
One production assistant (unskilled), $100 per month	1200
Miscellaneous expenses per year, $100	1200

(Maintenance of equipment, fuel charge, office expenses etc.), total cost = 4800
Thus, for 24,000 packets of biofertilizer, the production cost is the variable cost + indirect cost = $5260 + 10% depreciation on fixed cost $500 = $5760, for example, $6000. Therefore, per packet production cost is Rs. $6000 ÷ 24,000 packets = $0.25.

Price per packet of biofertilizer is calculated as follows:

Production cost per packet	$0.25
Power consumption (0.046 kW/packet)	$0.05
Profit margin	$0.2

So, the sale price per packet is = $0.5

Return:
Net sale value for 2000 packets per month of biofertilizer using 80 L of broth in four batches (one batch per week) is $1000 or $12,000 per annum. By subtracting the $6000 toward investment for the production, the net profit per year through this pilot-scale production of bacterial biofertilizer is $6000 or $500 per month.

Alternatively:
If the production capacity is multiplied by four times using eight fermenters (each of 10 L capacity), the cost–benefit calculation is as follows:

Fixed cost:
The cost would be $12,000 (including cost of six more fermenters) + $200 for other equipments = $14,000. Thus, from eight fermenters, the production of biofertilizer is 96,000 packets at 200 g each.

Variable cost:
If the cost of carrier material, broth, and polyethylene will be four times higher, then the total cost is $1840.

Indirect cost:
The additional manpower required with a salary component as follows:

One microbiologist, $240 per month	$2880
One production assistant, $160 per month	$1920
One marketing and sales personnel, $160 per month	$1920
One laboratory attendant cum peon, $80 per month	$960
Miscellaneous expenses	$2000
Total cost	$7680

Return:

Net sale value of 24,000 × 4 = 96,000 packets of biofertilizer per year with a price of $0.5 per packet	$48,000
Investment for production: 10% depreciation against fixed cost ($1400) + variable cost ($1840) + indirect cost ($7680)	$10,920
Gross profit per year = $48,000 – $10,920	$37,080

After deduction of the cost of advertisement in print and electronic media of about 20% from the gross profit, the net profit is = $37,080 – $7416 = $29,664 per year or, for example, $30,000 or $2500 per month.

In the above production system, unsterilized carrier (forest soil/charcoal) is used in dried form after grinding to avoid the natural population of microorganisms from competing with the inoculated bacterial broth. However, if a sterile carrier is used, the cost would be higher. From our experience, it was found that the brown forest soil available in most of the central and southern regions of Odisha state, India is free from microbial load, hence it is used as a carrier even in nonsterile conditions.

13.5.2 Quality control

The quality of the inoculants in the biofertilizer pack is one of the most important factors resulting in their success or failure and acceptance or rejection by the farmers. Basically, quality means the presence of the right type of microorganism in active form and in desired numbers. The stages requiring quality control are during the mother culture stage, during carrier selection, during broth culture stage, while mixing of broth with carrier, during packing, and during storage. Testing of the culture is usually done by taking a sample from the finished product for comparison with standard specification at the time of mixing of broth with carrier. In India, the Indian Stand Institution has developed standards for *Rhizobium* and *Azotobacter* (ISI 1986). The standard prescribed is that the inoculants shall contain a minimum 10^7 cells per gram of carrier on dry mass basis within 15 days before the expiry date marked on the packet when the inoculants are stored at 25°C–30°C (Matsara and Bhattacharyya 1994; Subba Rao 1993). If the hazards imposed by harsh storage and transport conditions under tropical climatic regime are minimized, either by decentralization of manufacturing units or by rapid transportation to farmers at the time of showing/planting, the quality of biofertilizers reaching the end users can be improved to a greater extent.

13.6 Cyanobacteria (BGA) biofertilizer production, quality control, and popularization for entrepreneurship development in rural areas of India

In India, De (1939) first suggested that BGA plays an important role in maintaining the fertility of tropical rice fields. Subsequently, Singh (1961) confirmed the role of these organisms in nitrogen fixation and emphasized the use of these organisms as biofertilizer for nitrogen fixation as well as for waste land reclamation. Then, a number of studies were carried out at different agriculture universities and Indian Council of Agricultural Research institutes to evaluate the benefit of BGA inoculation into rice crops. Venkataraman (1972) and his associates at the Indian Agricultural Research Institute (IARI), New Delhi termed the process as algalization, and through multilocation field trials, convincingly showed its beneficial effects on grain yield and nitrogen savings. Studies have shown that BGA, besides contributing nitrogen to the tune of 25 to 30 kg/ha per season, increases the productivity of rice by up to 10% to 15% (Venkataraman 1978, 1981). Even in the presence of chemical fertilizers, this biofertilizer can give supplementary effects leading to a reduction in the dose of chemical nitrogen by 25 to 30 kg. There are several other benefits that have been found due to cyanobacterial inoculation in the field. These include the addition of organic matter to the soil, mobilizing bound phosphate, improving aeration and water logging capacity of soil, and increasing soil aggregation and reclamation of saline and sodic soils (Kaushik and Subhasini 1995; Roger and Kulasooriya 1980).

13.6.1 Present status of the technology: Its strengths and weaknesses

Based on the natural ecology of BGA in the rice field ecosystem, a cost-effective and easily adaptable biofertilizer technology was developed by Venkataraman at IARI, New Delhi in 1978. This basic method involves the mass production of a mixture of cyanobacteria consisting of five nitrogen-fixing species: *Anabaena*, *Nostoc*, *Plectonema*, *Tolypothrix*, and *Aulosira* in shallow trays or cemented tanks in open air using soil as the carrier material. These organisms have been screened as better organisms in terms of growth rate and nitrogen-fixing capacity from among the several species isolated from more than 2000 soil samples collected from different parts of India through the All-India Coordinated Project on Algae at IARI.

In mass culture, for each square meter area of the cultivation tanks or trays, 2 kg of soil and 100 g of single superphosphate are added. To this, 5 mL of malathion or 25 mg of furadan (Carbofuran 3%) granules is added to prevent breeding of mosquitoes and other insects. The contents were thoroughly mixed and allowed to settle down. A BGA culture of all five species is sprinkled on the surface. These organisms multiply in the tanks exposed to direct sunlight and form a thick mat over the surface in 10 to 12 days. The contents were allowed to dry and the dried flakes are collected, packed, and used for inoculation at the rate of 10 kg/ha in the rice fields after transplantation.

Later on, this technology was further elaborated. Depending on the requirement for biofertilizers, four different production methods have been developed such as tank method, pit method (making pits of variable size with polyethylene lining at the base), field method, and the nursery-cum-field production method (cultivated in leveled fields covering large areas). All these methods have been practiced extensively since 1978; however, these have many limitations. The main limitation being essentially climatic as the production is affected by low temperatures in winter and by washing out of the organisms in the rainy season. Thus, the results of the product are highly unpredictable in terms of amount of BGA biomass. Furthermore, the open air tank culture method is easily subjected to contamination by other organisms. Thus, it was felt that although algalization technology certainly holds promise and the economics are also favorable, large-scale production and harnessing maximum benefits from BGA biofertilizers is possible only through the resolution of these major constraints during production (Adhikary 2000).

13.6.2 *Details of the appropriate technology*

13.6.2.1 *Production in polyhouse*

In this method, BGA multiplication is undertaken in roller-compacted concrete (RCC) tanks inside a polyhouse. The dimensions and number of such production units are variable depending on the turnover. The smallest recommended size of such a polyhouse is 12 × 5 × 3.5 m. Polyhouses are ideal because they provide controlled light intensity and temperature as desired and also prevent contamination from other organisms. Here, five different species of BGA are grown in equal amounts, that is, one species is grown in few adjacent tanks in a row within the polyhouse. Even one polyhouse can be exclusively used for growing one species to avoid cross-contamination of the species. A starter culture of each species is grown in 500 mL flasks containing 250 mL of BG11 medium in the laboratory for 8 to 10 days. Five of these flasks containing approximately 0.5 g of wet BGA biomass are used to inoculate 50 L of medium in a single cement tank. These were grown in the tank inside the polyhouse for about 7 days at optimum conditions. A fully grown culture containing 10 g of wet BGA biomass is collected in a container and then mixed well with 1 kg of carrier material. Then, this is dried at room temperature, preferably inside the same polyhouses, for 2 to 3 days. Different BGA species on dried carriers are then mixed in equal proportions and then packed in polyethylene packets for distribution/marketing.

13.6.2.2 *Use of alternate carrier materials*

Carrier is an important component in the biofertilizer formulation. The shelf life of the biofertilizer depends on the quality of the carrier. A few characteristics of the carrier were considered before selecting: it should not be toxic to the BGA, it should have good surface area, it should have high absorption and good buffering capacity, it should be able to carry high microbial load, and it should be easily available and cost-effective. As an alternative to soil carrier, wheat and rice straw, and coconut coir cut into small pieces have been successfully developed for use as a carrier. Individual cyanobacteria are grown separately in an alternative cheap medium in shallow tanks and the fresh harvested mass is properly mixed with a definite quantity of carrier. The BGA load is maintained at 10^6 cells/g carrier so that only 400 g of the biofertilizer is sufficient to inoculate a 1-acre rice field. Straw as a carrier has also been used in a different way in which desired organisms were grown in pulverized and sterilized straw under controlled conditions in polyalkene bags. In these methods, different BGA species are grown separately and then mixed with carriers in desired proportions that ensure an appropriate share of all the organisms used in the inoculum. Unlike soil-based inoculum, this biofertilizer is supplied in ready-to-use quality material for inoculation.

Goyal et al. (1994) have reported that "multanimetti," a montmorillonite clay, can be used as a suitable carrier for BGA biofertilizer. It fulfills most of the qualities of a good carrier. It is a powder form of carrier with small particle size, good buffering capacity, high absorptive nature, and is microbiologically inert. Using this method, cultures were grown separately under controlled conditions in polyhouses. The BGA, while in a growing state, are taken out from the tanks and mixed with clay in a ratio of 1:1 and dried. The dried material is finally powdered and packed. This biofertilizer in multanimetti carrier possesses a longer shelf life and can be used even after 2 years of storage under ambient conditions.

13.6.2.3 Quality control parameters and testing methodologies

Biofertilizers contain living or latent microbial cells and their metabolic activity after application to the field determines their efficiency. The quality of biofertilizers is the most important factor leading to the success and acceptance or failure and rejection. The quality of BGA biofertilizer can be controlled at various stages of its production and ascertaining their load in the packing material. The quality of the inoculum is assessed before packing in terms of colony-forming units, which should not be less than 10^3 CFU/g of the final product (Singh and Pabbi 1998). The packing should be done in plastic bags properly sealed to avoid moisture. The BGA biofertilizers can be stored for 2 years at ambient temperature with no appreciable loss in viability or efficiency.

The quality standard of the BGA biofertilizer is assured by the following techniques:

1. *Viable cell count*: In this method, a known quantity of BGA samples were spread on agar medium in petri dishes and incubated at 25°C to 30°C for about 10 days and the colonies that appeared were counted.
2. *Most probable number method*: BGA biofertilizer samples are diluted to known grades in culture tubes and incubated under similar conditions in a culture room.
3. *Acetylene reduction assay*: This assay quantifies the nitrogenase activity of the BGA cells in the biofertilizer packs, in which the process is carried out only in living cells. The assay is carried out using gas chromatography (GC) measuring the amount of acetylene reduced to ethylene which is an indicator of nitrogenase activity.

13.6.3 Cyanobacteria biofertilizer production on a small scale; economics of the technology and popularization for entrepreneurship development

We have isolated several cyanobacteria species from different agroclimatic zones of Odisha State in the eastern region of India and have maintained them in a germplasm collection since 1996 (Adhikary 1998). All these cyanobacteria have been documented with their strain history, for example, place of collection, habitat, mode of occurrence, and others (Nayak et al. 1996; Sahu et al. 1996; Tirkey and Adhikary 2005). Basing on their capability to tolerate several stress factors e.g., salinity, pH, pesticides, commercial inorganic fertilizers, desiccation, etc., which have been experimented in the laboratory (Rath and Adhikary 1995, 1996; Padhi et al. 1997; Das and Adhikary 1996; Sahu and Adhikary 2006). Five cyanobacterial species belonging to the genera *Trichormus, Nostoc, Cylindrospermum, Aulosira,* and *Westiellopsis,* of which some are free-floating (*Trichormus variabilis* UU 147 and *Nostoc carneum* UU25130), and others epiphytic, growing attached to rice culms (*Aulosira fritschii* UU 25118) or attached to soil (*Cylindrospermum indicum* UU131 and *Westiellopsis prolific* UU2542) were screened. They were grown year-round in bulk in cemented tanks in polyhouses (Figures 13.2 and 13.3). The performance of brown forest soil as a carrier-based inoculum containing region-specific strains (Figure

Figure 13.2 Photograph of a semipolyhouse showing four cemented tanks covered with UV-free polyethylene where BGA was grown outdoors.

Figure 13.3 Close view of growth of BGA in cemented tanks in the polyhouse.

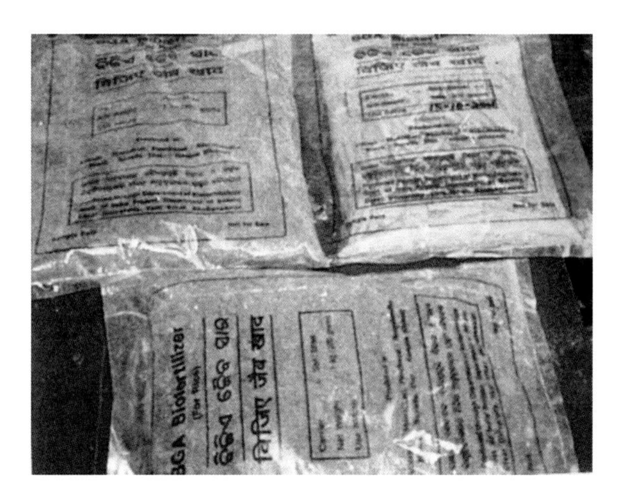

Figure 13.4 One-kilogram BGA biofertilizer packs prepared with brown forest soil as carrier used for application in rice crops.

13.4) at an application dose of 12 kg/ha was tested in the field for 3 consecutive years. The results showed an approximately 8% to 12% increase in the grain yield (Adhikary and Sahu 2000). Improvement in soil quality due to BGA inoculation was determined based on an increase in the porosity, water-holding capacity, build-up of organic carbon, and microbial population of soil. Thus, the previous methods of producing BGA biofertilizer in pits, trays, cemented tanks, or by field-cum-nursery methods were upgraded emphasizing quality biofertilizer production. Authentic region-specific strains were used for year-round production of BGA in outdoor polyhouses.

For popularization of the technology and its transfer from the laboratory to the field, a pilot-scale demonstration, production-cum-quality control of BGA biofertilizer has been set up in the village of Maniakati under Asurabandha Grampanchayat of Surada block in the Ganjam District (Odisha) since April 2000. Several training programs have been conducted for farmers, agricultural officials, and entrepreneurs including women at the project site in the village during the last 10 years. Already, a few cyanobacteria biofertilizer production units have been established by entrepreneurs in the rural areas of Odisha State in the eastern region of India with partial support through Rastriya Krshi Bikash Yojana of the State Government.

13.6.3.1 *Usefulness of the technology for providing self-employment*

Five kilograms of soil-based biofertilizer or 1 kg of straw or coir-based biofertilizer, with the requisite number of viable cyanobacteria, is sufficient to use in 1 acre of land soon after transplantation of rice. The cost per kilogram of soil-based biofertilizer for selling in India is approximately $0.5, which is easily affordable by small and marginal farmers. This technology has now been adapted as a cottage industry in rural areas of the eastern region of India, and unemployed youth and nongovernmental organizations produce this biofertilizer for income generation.

13.6.3.2 *Economics (cost–benefit of the technology adapted as cottage industry)*

Capital investment:

RCC ground floor work (12 ft. × 4 ft. × 1 ft.) for 25 tanks with brick and cement	$2000
Iron framework with UV-free polyethylene shading	$1000
Light microscope for quality control	$500
Total investment for infrastructure	$3500

Recurring expenses (including cost of raw materials for 1 year):

Superphosphate, lime, insecticide, etc., for 25 tanks	$500
Carrier material (soil/coir/straw)	$500
Bags (polybag or jute bag) for packing of biofertilizer	$500
Labor cost: $5 per day for 100 days (approximately 8 days per month)	$500
Publicity (in print, electronic media, display boards, etc.)	$500
Total investment in recurring expenses per year	$2500

Return:
Under normal conditions in the eastern region of India, 48 harvests are possible, yielding 12,000 kg of soil-based cyanobacteria biofertilizer in 25 tanks in a year

Selling price of biofertilizer of 1 kg pack	\$1
Cost of 12,000 kg biofertilizer	\$12,000
(–) Capital investment plus recurring expenses for production in a year	(–) \$6000
Net profit per year: \$12,000–\$6000	\$6000 or \$500 per month

The income would further increase by increasing the number of tanks for the production of biofertilizer. Hence, this is a viable technology for rice farming and is useful for the rural development sector.

13.7 Constraints in the commercialization of microbial biofertilizer technology

There are several constraints in biofertilizer production and its commercialization. These are physical, chemical, biological, technological, financial, and market-related issues.

i. *Temperature*—soil temperature plays a pivotal role in the survival of microorganisms. The optimum temperature of soil is between 25°C and 30°C. The major constraint in the use of biofertilizer are the summer months as the soil temperature reaches 45°C to 50°C at 5 cm depth.
ii. *Drought*—the number of microorganisms in soil declines drastically as the soil dries up. Microbial populations decrease rapidly when stored under dry conditions.
iii. *Biological constraints*—the presence of numerous parasites and predators in soil/fields pose problems in the establishment of biofertilizer. There is also the problem of competition with the native population. Some nematodes act as predators of most of these microbes in the soil.
iv. *Technical constraints*—peat has been recognized as the most suitable and standard carrier for bacterial biofertilizers. Due to high cost, few manufacturers use it. Therefore, most production units use charcoal, which is easily available locally. Mixing charcoal to brown forest soil in a 3:1 proportion as a carrier is a viable alternative due to lower costs.

13.7.1 Extension programs

Most farmers do not have sufficient and clear knowledge of the use of biofertilizers, which is necessary for deriving maximum benefit. The farmers need to be educated through effective demonstrations on the use of biofertilizers and not simply that biofertilizers are cheap fertilizers. Once the farmers are convinced, they will adapt to the use of biofertilizers, which would help in the commercialization of biofertilizers on a large scale.

13.7.2 Marketing constraints

This includes weakness in the marketing network, retail outlet, storage facility, and so on. Because biofertilizers contain living organisms, they have short shelf lives (usually

6 months; or ~2 years for cyanobacteria). Hence, there is a necessity to maintain these inoculants under proper storage conditions at the retail outlet.

13.8 Conclusion

Biofertilizer technology has an immense market potential considering the vast area of land under cultivation globally. Once appropriate production technologies for quality biofertilizer are adapted, and the farmers are convinced about their efficiency, the demand for biofertilizers will increase. Although biofertilizers are produced and also available in the market either strict quality control measures are not taken in most production units or the farmers are not apprised of the proper process of their application, including their crop specificity. As a result, in many instances, they are found ineffective as claimed. Hence, the production technology of different biofertilizers has been refined and upgraded in recent years, and quality biofertilizers are produced using inoculum from authentic sources together with quality assurance at every stage of production, and also by selecting appropriate carrier materials.

It is a primary necessity to test the biofertilizers according to their efficacy in the farmer's fields. Dissemination of knowledge for the production and quality control of biofertilizers to the production houses/industrial sectors is also important. Second, a strong network—by establishing quality control laboratories, biofertilizer-producing units at the regional level, maintaining region-specific bacterial isolates, and generation of trained manpower—is essential. Once farmers can come forward to apply different biofertilizers, being convinced on their performance in the field with the concept of "seeing is believing," it would lead to large-scale commercialization of the technology.

Acknowledgments

The authors thank the authorities of Utkal University, Bhubaneswar, Odisha, and Visva-Bharati, Santiniketan, West Bengal for providing facilities. Thanks are due to the people of Maniakati Village in the Ganjam District of Odisha for cooperating in field experiments, to the entrepreneurs for their enthusiasm to adapt the biofertilizer technologies standardized by us, and to the Department of Biotechnology, Government of India for financial assistance through societal project.

References

Adhikary, S.P. 1998. Cyanobacteria germplasm of Orissa state maintained in the Department of Botany, Utkal University. *Pl. Sci. Res.* **20**: 57–63.

Adhikary, S.P. 2000. Outdoor cultivation of cyanobacteria in polybags and selection of carrier material for using as biofertilizer. *Adv. Plant Sci.* **13**: 335–337.

Adhikary, S.P. and Sahu, J.K. 2000. Studies on the establishment and nitrogenase activity of inoculated cyanobacteria in the field and their effect on yield of rice. *Oryza* **37**: 39–43.

Bhatnagar, R.K. and Palta, R.K. 1996. *Earthworm-Vermiculture and Vermicomposting*. Kalyani Publications, Ludhiana, India, 106 p.

Das, M.K. and Adhikary, S.P. 1996. Toxicity of three pesticides to several rice field cyanobacteria. *Trop. Agric., Trinidad* **73**: 156–158.

De, P.K. 1939. The role of blue green algae in nitrogen fixation in rice fields. *Proc. Roy. Soc. Lond.* **127B**: 1–21.

Goyal, S.K., Singh, B.V., Nagpal, V. and Marwaha, T.S. 1994. An improved method for production of algal biofertilizer. *Ind. J. Agric. Sci.* **67**: 314–315.

Kaushik, B.D. and Subhasini, D. 1995. Amelioration of salt affected soils with BGA. II. Improvement of soil properties. *Proc. Natl. Sci. Acad.* **51B**: 386–389.

Motsara, M.R., Bhattacharyya, P. and Srivastava, B. 1995. *Biofertilizer Technology, Marketing and Usage*. Fertilizer Development and Consultant Organisation, New Delhi, 184 p.

Nayak, H., Sahu, J.K. and Adhikary, S.P. 1996. Blue green algae of rice fields of Orissa state. II. Growth and nitrogen fixing potential. *Phykos* **35**: 111–118.

Padhi, H., Rath, B. and Adhikary, S.P. 1997. Tolerance of nitrogen fixing cyanobacteria to NaCl. *Biol. Plant*. **40**: 262–268.

Parr, J.F., Stewart, B.A., Hornick, S.B. and Singh, R.P. 1990. Improving the sustainability of dryland farming systems: A global perspective. *Adv. Soil Sci.* **13**: 1–5.

Rath, B. and Adhikary, S.P. 1995. Toxicity of Furadan to several nitrogen fixing cyanobacteria from rice fields of coastal Orissa, India. *Trop. Agric, Trinidad* **72**: 80–84.

Rath, B. and Adhikary, S.P. 1996. Effect of pH, irradiance and population size on the toxicity of Furadon (carbofuran, 75 DB) to two different species of *Anabaena*. *Biol. Plant.* **39**: 563–570.

Roger, P.A. and Kulasooriya, S.A. 1980. *Blue-Green Algae and Rice*. The International Rice Research Institute, Manila, Philippines, 112 p.

Sahu, J.K. and Adhikary, S.P. 2006. Growth response and acetylene reduction activity of two nitrogen fixing cyanobacteria to commercial fertilizers. *Arch. Hydrobiol. Suppl. Algol. Studs.* **121**: 119–136.

Sahu, J.K., Nayak, H. and Adhikary, S.P. 1996. Blue-green algae in rice-fields of Orissa state. I. Distributional pattern in different agroclimatic zones. *Phykos* **35**: 93–110.

Sethi, S.K. and Adhikary, S.P. 2009a. Effect of region specific *Rhizobium* in combination with seaweed liquid fertilizer on vegetative growth and yield of *Arachis hypogea* and *Vigna mungo*. *Seaweed Res. Utiln.* **31**: 177–184.

Sethi, S.K. and Adhikary, S.P. 2009b. Vegetative growth and yield of *Arachis hypogea* and *Vigna radiata* in response to region specific *Rhizobium* biofertilizer treatment. *J. Pure Appl. Microbiol.* **3**: 295–300.

Sethi, S.K. and Adhikary, S.P. 2009c. Efficacy of region specific *Azotobacter* strain on vegetative growth and yield of *Solanum melongena, Lycopersicon esculentum* and *Capsicum annum*. *J. Pure Appl. Microbiol.* **3**: 331–336.

Singh, P.K. and Pabbi, S. 1998. *Mass Production of Blue Green Algal Biofertilizer*. Venus Printers and Publishers, New Delhi, 67 p.

Singh, R.N. 1961. *Role of Blue Green Algae in Nitrogen Economy of Indian Agriculture, ICAR Monograph on Algae*. ICAR, New Delhi, 128 p.

Subba Rao, N.S. 1993. *Biofertilizers in Agriculture and Forestry*. Oxford and IBH Publishers, New Delhi, 242 p.

Suneja, S., Narula, N., Anand, R.C. and Lakshminarayana, K. 1996. Relation of *Azotobacter chroococcum* siderophores with nitrogen fixation. *Folia Microbiol.* **4**: 154–158.

Tirkey, J. and Adhikary, S.P. 2005. Cyanobacteria in the biological soil crusts of India. *Curr. Sci.* **89**: 515–521.

Venkataraman, G.S. 1972. *Algal Biofertilizer and Rice Cultivation*. Today and Tomarrows Printers and Publishers, Faridabad, India, 98 p.

Venkataraman, G.S. 1978. *Algal Biofertilizer for Rice, an Information Bulletin*. Indian Agricultural Research Institute, New Delhi, 16 p.

Venkataraman, G.S. 1981. Blue green algae for rice production—A manual for its production. *FAO Soil Bull.* **46**: 102 p.

Verma, A., Kukreja, K., Pathak, D.V., Suneja, S. and Narula, N. 2001. In vitro production of plant growth regulators (PGRs) by *Azotobacter chroococcum*. *Ind. J. Microbiol.* **41**: 305–307.

Yadav, A.K. and Mowade, S.M. 2005. *Handbook of Microbial Technology*. Regional Centre for Organic Farming, Nagpur, India, 236 p.

chapter fourteen

Bacterial biocontrol agents

Rikita Gupta and Jyoti Vakhlu

Contents

14.1 Introduction

There is an increasing pressure on agriculture for meeting the ever-increasing food demand of the world population. The world population is expected to grow by 2.3 billion people between 2009 and 2050, further increasing the demand (http://www.fao.org). However, growth in agriculture is not corresponding to the increasing food demand. This is attributed to crop loss due to biotic as well as abiotic factors. The biotic factors contributing to crop loss include pests, pathogens, and weeds. Although annual pesticide use has increased to approximately three million tons worldwide, crop loss from pests is more than 40% of potential world food production. According to the World Health Organization (WHO), three million pesticide poisonings occurred annually causing 220,000 deaths in 1992 (http://www.wikipedia.org). Besides the environmental hazard, pesticides select for resistant pests, further augmenting the problem.

Sustainable alternatives should be explored to reduce worldwide pesticide use. One such alternative is to use natural enemies for pest control, that is, biological control.

In this chapter, we describe biological control and its advantages over agrochemicals. The basic interactions that can take place between any two species, be it plant–microbe or microbe–microbe, and the mode of action of bacterial biocontrol agents (BCA) is described.

14.2 Biological control

Use of organisms for controlling insect pests, mites, weeds, and plant pathogens is referred to as biological control and abbreviated as biocontrol. These organisms are natural enemies of insect pests, mites, weeds, or plant pathogens. There are different descriptions for the term *biological control*. It has been broadly defined by the U.S. National Research Council as "the use of natural or modified organisms, genes, or gene products to reduce the effects of undesirable organisms and to favor desirable organisms such as crops, beneficial insects, and microorganisms" (Pal and Gardener 2006). According to Cook and Baker, "Biological control is the reduction of the amount of inoculum or disease-producing activity of a pathogen accomplished by or through one or more organisms other than man" (Cook and Baker 1983). This includes the use of hypovirulent or avirulent strains of pathogenic species such as nonpathogenic *Fusarium oxysporum* for biocontrol of pathogenic strains, microbial antagonists such as *Pseudomonas fluorescens* CHA0 and manipulation of host plant for effective defense, which may be done either by genetic engineering or by using a BCA that primes the plant defense system (Alabouvette et al. 2006). The term has also been classified on the basis of the biological field of its intended use (Pal and Gardener 2006). Biocontrol is concerned with two fields of biology, namely, entomology and plant pathology. In the field of entomology, biocontrol refers to the control of pest insects. Similarly, in plant pathology, it refers to the control of pathogen and weed populations. Therefore, the BCA, in terms of entomology, will include predatory insects, entomopathogenic nematodes, and microbes such as entomopathogenic bacteria, fungi, and viruses. While in case of plant pathology, microbial antagonists or host-specific pathogens will be the candidates. Microbial antagonists of plant pathogens include bacteria, fungi, and viruses. However, for the scope of this chapter, we will talk only about bacterial antagonists involved in the biocontrol of plant pathogens. Biocontrol may be achieved by introducing a known BCA or by using native microorganisms that are potentially suppressive to the pathogen.

14.3 BCA versus agrochemicals

Plant growers rely on chemical fertilizers and pesticides besides good agricultural practices for improving crop yield, but these agrochemicals come packed with a bundle of drawbacks. These agrochemicals are usually environmental hazards and have a broad-spectrum action that can harm beneficial microflora of plants, for example, chemical fungicides act on essential fungal functions such as respiration, sterol biosynthesis, or cell division, and have risk of pathogens developing resistance and showing toxicity to plants/crops (Chandrashekara et al. 2012).

BCAs, being natural enemies of the pathogen, offer the following advantages over chemical agents:

- Effective against specific plant diseases
- Not toxic to plants and some show plant growth promoting properties in addition to biocontrol
- The products of microbial origin are biodegradable and hence environmentally safe because no residues are left in the environment

BCAs are not as popular as chemical agents. They suffer from a major disadvantage of short shelf life and are, at the moment, less potent than chemical agents because many factors need to be taken into consideration before using a BCA. An important factor to be taken into consideration is the amount of biocontrol bacteria that is needed to control a pathogen in the field. This is a problem because the amount of pathogen present in the environment cannot be established. In addition, the number of different pathotypes, that is, strain variants of the pathogen in the field are not known. Due to such variations in the environment, the success of BCA in the field does not exactly reflect the success in the laboratory (Erdogan and Benlioglu 2010). However, sustainable agriculture being the need of the hour, developments should be made to overcome the shortcomings of biocontrol.

14.4 Interaction between host–pathogen and BCA

Association of bacteria with plants can result in three different types of interactions, namely, positive, negative, and neutral. Bacteria entering into a positive interaction, that is, beneficial bacteria enhance plant growth and protect it from infection, thus acting as plant growth-promoting bacteria or BCA. Such bacteria helps plants by nitrogen fixation as in case of *Rhizobium* associated with legumes, solubilization of nutrients, for example, iron solubilization and competition with pathogen by siderophore production, production of plant growth regulators, such as indole acetic acid by plant growth-promoting rhizobacteria, or by inducing a primed state of defense in the plant and production of antifungal antibiotics such 2,4-diacetylphloroglucinol (2,4-DAPG) by *Pseudomonas* species (Bhattacharyya and Jha 2012). These bacteria may be mutualists, that is, they also derive benefit from this positive interaction and hence utilize plant exudates such as organic acids, amino acids, and sugars as a source of nutrition, or they may be facultative mutualists as in the case of protocooperation (Pal and Gardener 2006).

Interaction of pathogens with host plant falls into the category of negative interactions because they derive nutrition for their growth and survival at the expense of host plant that is harmed by the interaction. These may be facultative because most saprophytic soil-borne fungi can survive on dead and decaying organic matter besides the living host or obligate parasites such as viruses that require living plant host for their survival.

There is yet another type of bacteria that does not affect plants at all and lives commensally to the plant. Such microorganisms can be considered to have neutral interactions with plants. They may, however, be involved in positive or negative interactions with other microorganisms associated with the plant.

Biological control of a plant disease involves three major parties: BCA, pathogen, and host plant (Figure 14.1). The host is attacked by the pathogen, whose success relies on its virulence as well as the plant defense response. In addition to these two interacting parties, there are some bacteria that can interact with the plant and enhance its defense capability by inducing systemic resistance (ISR) or negatively interact with pathogen (antagonism) and reduce disease severity. Biocontrol is also affected by other microorganisms present in the plant's microenvironment. Some bacteria are regarded as commensals, because they neither harm nor benefit the plant. However, these bacteria may compete with BCA, reducing its efficiency (Berendsen et al. 2012). Nutritional composition of the plant's microenvironment may also affect biocontrol by either favoring or impeding the growth of BCA (Rudrappa et al. 2008). Therefore, biocontrol can be summed up as the net positive result of interactions between host plant and the associated microorganisms. The multiple plant–microbe as well as microbe–microbe interactions involved in the entire process can be simplified as the interaction between host plant, pathogen, and BCA.

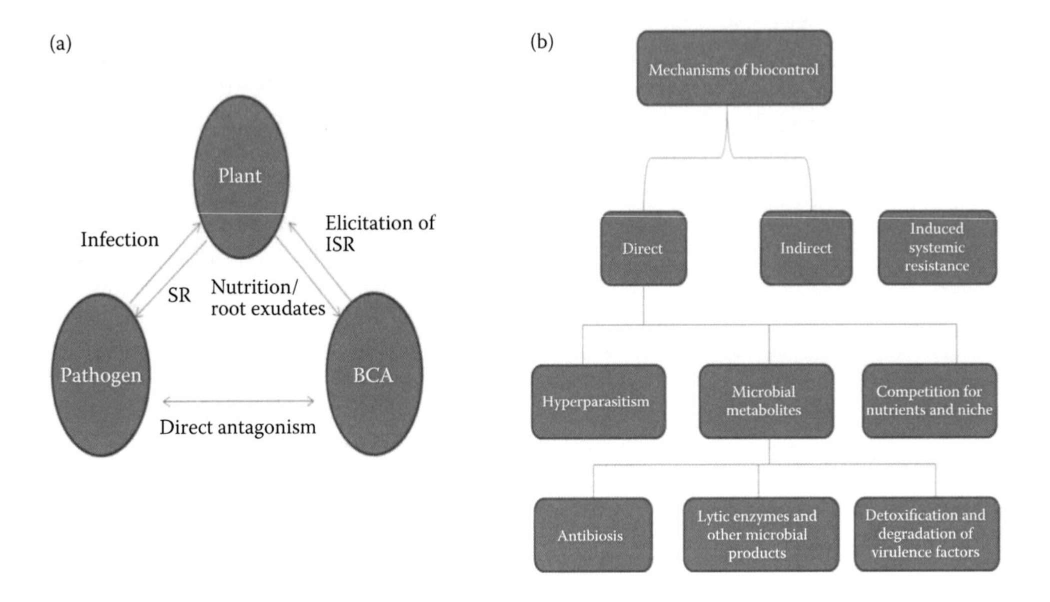

Figure 14.1 (a) Interactions between plant–pathogen and BCA and (b) mechanisms of action of BCA.

14.5 Mechanisms of biocontrol

Biocontrol conclusively is a result of plant-BCA and BCA-pathogen interactions. Several mechanisms have been put forward for the mode by which microbial BCA control plant pathogens (Figure 14.1; Table 14.1). Broadly, these mechanisms can be divided into two main categories, direct antagonism and indirect antagonism.

14.5.1 Direct antagonism

Direct antagonism refers to a situation in which BCA directly interacts with the plant pathogen leading to its suppression. This includes biocontrol caused by hyperparasitism, production of antibiotics and lytic enzymes, and competition for nutrients and niche.

14.5.1.1 Hyperparasitism

Parasitism is a negative interaction in which a parasite grows at the expense of its host. A plant pathogen is the primary parasite of its host plant and derives nutrition by harming its host. However, there are some bacteria (secondary parasites) that parasitize plant pathogens and hence the interaction is known as hyperparasitism. Hyperparasitism ranges in severity from simple attachment to complete lysis and death of the host. For example, attachment of *Enterobacter cloacae* cells to the fungal hyphae of *Pythium ultimum* and complete lysis and degradation of *Pythium debaryanum* hyphae by *Arthrobacter* (Whipps 2001).

Bacterial hyperparasites may be classified as opportunistic, facultative, or obligate parasites. Opportunistic hyperparasites have saprophytic capabilities and act on the fungal cell wall by producing lytic enzymes. Obligate hyperparasites are specific to the host and have more controlled mechanisms and cause less destruction. Facultative hyperparasites are intermediaries that have adapted to colonize host tissue besides their saprophytic capability. For example, *Pasteuria penetrans*, which is a hyperparasite of root-knot nematodes *Meloidogyne* spp. Because an obligate parasite has a high specificity toward its host, its commercialization

Table 14.1 Examples of Biocontrol Bacteria and Their Modes of Action

Biocontrol bacteria	Mode of action	Target pathogen	Reference
E. cloacae	Hyperparasitism	*P. ultimum*	Whipps 2001
Arthrobacter		*P. debaryanum*	Whipps 2001
P. penetrans		*Meloidogyne* spp.	Davies 2009
L. enzymogenes		Fungi, peronosporomyctes, nematodes, and bacteria	Islam 2011
B. laterosporus strain G4 and *Bacillus* sp. B16		*H. glycines*, *T. colubriformis*, and *P. redivius*	Tian et al. 2007
Serratia sp. G3	Lytic enzymes (chitinase and protease), antibiosis (pyrrolnitrin), competition for iron (siderophore)	*B. cinerea*, *Cryphonectria parasitica*, *Rhizoctonia cerealis*, and *Valsa sordida*	Liu et al. 2010
B. subtilis CPA-8	Antibiosis (lipopeptides)	*Monilinia* spp.	Yánez-Mendizábal et al. 2012
Pseudomonas CMR12a	Antibiosis (lipopeptides and phenazine)	*R. solani*	D'aes et al. 2011
Pseudomonas sp. LBUM300	Antibiosis (2,4-DAPG and HCN)	*Clavibacter michiganensis*	Lanteigne et al. 2012
Pseudomonas putida NH-50	Antibiosis (pyoluteorin)	*C. falcatum*	Hassan et al. 2011
P. aeruginosa	Lytic enzymes (chitinase)	*S. sclerotiorum*	Gupta et al. 2006
Bacillus sp. A24	Quorum quenching	*Erwinia carotovora*	Dong et al. 2002
P. fluorescens	Degradation of BS toxin	*B. sorokiniana*	Aggarwal et al. 2011
B. subtilis CAS15	Siderophore (bacillibactin) ISR	*F. oxysporum* Schl. f.sp. *capsici*	Yu et al. 2011
Pseudomonas sp.	Siderophore (pyoverdine and pyochelin) ISR	*B. cinerea*	Verhagen et al. 2010

is prevented by the inability to culture outside the host (Davies 2009). *Lysobacter enzymogenes* as facultative parasites have the ability to produce lytic enzymes as well as antibiotics besides hyperparasitism as a mode of action (Islam 2011). *Brevibacillus laterosporus* strain G4 and *Bacillus* sp. B16 are examples of opportunistic parasites of nematodes such as *Heterodera glycines, Trichostrongylus colubriformis* and *Panagrellus redivius* (Tian et al. 2007).

14.5.1.2 Microbial metabolites

14.5.1.2.1 Antibiotics. Antibiotics are low-molecular weight organic compounds produced by microorganisms to kill and poison other microorganisms. Most of the microorganisms produce one or more antibiotics that enable them to survive and compete under severe environmental conditions. Bacteria produce a diverse range of different types and number of antibiotics (Raaijmakers and Mazzola 2012). Production of several antibiotics by a single BCA may account for the suppression of specific or multiple pathogens. For example, pyrrolnitrin is a broad-spectrum antibiotic produced by *Pseudomonas* and *Burkholderia* sp. that shows antibiotic activity against a wide range of Basidiomycetes, Deuteromycetes, and Ascomycetes, including several economically important pathogens like *Rhizoctonia solani, Botrytis cinerea, Verticillium dahliae,* and *Sclerotinia sclerotiorum* (Raaijmakers et al. 2002). Members of the genus Bacillus and Pseudomonas produce cyclic lipopeptides of three different families, namely, surfactin, iturin, and fengycins (Raaijmakers et al. 2010). Surfactins are heptapeptides interlinked with β-hydroxy fatty acid to form a cyclic lactone ring. These are biosurfactants that interfere with membrane integrity and display antibacterial and antiviral activity but lack antifungal activity. Iturins, on the other hand, display fungicidal activity by affecting membrane permeation. They form ion-conducting pores in the membrane and lead to osmotic imbalance. Fengycins are fungitoxic lipodecapeptides with less activity compared with the other two classes (Ongena and Jacques 2007). Another class of antibiotics produced by the genus Pseudomonas is phenazines, which are heterocyclic nitrogen-containing pigments. For example, phenazine-1-carboxylic acid, 2-hydroxyphenazine-1-carboxylic acid, and 2-hydroxyphenazine, which show antifungal activity. Phenazines have a broad-spectrum activity and are also produced by *Streptomyces, Nocardia, Sorangium, Brevibacterium,* and *Burkholderia* species besides Pseudomonas. Other antibiotics include phloroglucinols, pyoluteorin, polymyxin, circulin, colistin, and zwittermicin A (Compant et al. 2005; Raaijmakers et al. 2002).

14.5.1.2.2 Lytic enzymes and other microbial products. The fungal cell wall is composed of three different polymeric compounds, namely, chitin, proteins, and glucans. The lytic enzymes such as chitinase, proteases, and glucanases act to hydrolyze polymeric compounds of the pathogen such as chitin, proteins, and glucan in the fungal cell wall. These lytic enzymes help in predation or parasitism exhibited by various BCA against fungal plant pathogens.

Chitin is a homopolymer of *N*-aceylglucosamine with residues having β(1-4) linkage and accounts for 10% to 20% of the fungal cell wall (dry weight) and is also found in the exoskeleton of insects. Its microfibrillar structure is formed by interchain hydrogen bonding, which provides tensile strength and integrity to the cell wall. Chitinases belong to the glycosyl hydrolases family 18 or 19 and hydrolyse chitin component of fungal cell wall. Many bacteria with potential biocontrol activity produce chitinases. For example, *Pseudomonas aeruginosa* showing activity against *S. sclerotiorum* and fluorescent *Pseudomonad* strain active against *F. oxysporum* f.sp. *dianthi* and *Alternaria solani* produce chitinases (Gupta et al. 2006; Singh and Shanmugam 2012).

Glucans form the major structural polysaccharide of the cell wall constituting 50% to 60% of its dry weight, itself being composed of glucose monomers linked by β(1-3) linkage, although β(1-4), β(1-6), α(1-3), and α(1-4) are also found. Glucanases are hydrolytic enzymes with lytic activity against glucans. For example, β(1-3) glucanase produced by *Bacillus subtilis* and *Bacillus pumilus* (Essghaier et al. 2009).

All fungal cell walls contain a protein component tightly interwoven within the polysaccharide components, that is, chitin and glucan, accounting for 30% to 40% of the cell wall's dry weight. Traditional cell wall proteins are glycoproteins such as mannoproteins, proteins glycosylated with mannans as in the cell wall of *Saccharomyces cerevisiae* and *Candida albicans,* and proteins glycosylated with galactomannans as in the case of *Aspergillus fumigatus* (Bowman and Free 2006). Thus, extracellular proteolytic enzymes play a role in antagonism by BCA. For example, *Stenotrophomonas maltophilia* inhibits the growth of *P. ultimum* by extracellular proteolytic activity (Dunne et al. 1997).

Besides the hydrolytic enzymes described previously, bacteria also produce enzymes or metabolites that suppress pathogen. Quorum quenching is one such strategy opted by some bacteria, for example, *Bacillus* sp. A24 produces lactonase enzyme, which is capable of degrading *N*-acyl homoserine lactones (AHLs). AHLs are signaling compounds produced by bacteria for quorum sensing that coordinate gene expression in a density-dependent manner. *Variovorax paradoxus* neutralizes AHLs by the production of an aminoacylase that cleaves the fatty acid tail and mineralizes the homoserine lactone ring (Molina et al. 2003; Dong et al. 2002; Leadbetter and Greenberg 2000). Products of BCA are capable of degrading or neutralizing AHLs to suppress plant pathogenic bacteria by blocking their regulation of gene expression and hence affecting their virulence.

14.5.1.3 *Detoxification and degradation of virulence factors*

Virulence factors produced by pathogens are important for causing infection. Some bacteria act as BCA by detoxifying these virulence factors. For example, detoxification of albicidin toxin produced by *Xanthomonas albilineans. Klebsiella oxytoca* and *Alcaligenes denitrificans* produce a protein that reversibly binds to albicidin toxin, whereas *Pantoea dispersa* detoxifies it with the help of an esterase. *Burkholderia cepacia* and *Ralstonia solanacearum* can detoxify fusaric acid, a toxin produced by *F. oxysporum*, which is important for disease development (Walker et al. 1988; Basnayake and Birch 1995; Zhang and Birch 1997; Toyoda and Utsumi 1991; Toyoda et al. 1988). Bacterial strains belonging to the genera *Bacillus, Paenibacillus, Microbacterium, Staphylococcus*, and *Pseudomonas* degrade diffusible signal factor, which is a signal molecule produced by *Xanthomonas campestris* involved in the regulation of virulence (Newman et al. 2008). BS toxin is produced by *Bipolaris sorokiniana*, which causes spot blotch in wheat and barley. *P. fluorescens* reduces the severity of symptoms induced by BS toxin on wheat tissues (Aggarwal et al. 2011). *P. fluorescens* strains have also been shown to detoxify phytotoxin produced by sugarcane red rot pathogen *Colletotrichum falcatum* (Malathi et al. 2002).

14.5.1.4 *Competition for nutrients and niche*

Because microorganisms in the environment are mostly exposed to nutrient-limiting conditions, they compete for whatsoever nutrients available. Plants secrete 40% of their photosynthates as root exudates, making the rhizosphere a nutrient-rich environment favorable for microorganisms compared with the nonrhizosphere (bulk) soil and thus microbial density is higher in the rhizosphere; this is known as the rhizosphere effect (Berendsen et al. 2012). This gives rise to the competition between different species for available nutrients and space. Some bacteria act by competing with the pathogen for its infection sites and nutrients. A BCA with a better ability to colonize plant roots and utilize root exudates will have a competitive advantage over the pathogen.

Iron is an essential element for growth in all microorganisms, being a part of the enzymes involved in electron transport. The availability of iron in soil depends greatly on the pH of the soil, which has low solubility at neutral pH. In a well-aerated soil, iron is oxidized to its ferric form and is insoluble in water (pH 7.4), and the available concentration (10^{-18} M) is quite low as compared with that required (10^{-6} M). Most microorganisms produce low-molecular weight ferric ion chelators called siderophores, which form a complex with ferric ions (Fe^{3+}). Siderophores then bind to iron limitation-dependent receptors at the bacterial cell surface and subsequently release iron into the cytoplasm. Inside the cytoplasm, Fe^{3+} is reduced to its active form, Fe^{2+}. Bacteria producing greater amounts of high-affinity siderophore compared with the pathogen will have a competitive advantage under iron-limiting conditions. Such bacteria will have better ability to colonize the available root niche competing with the pathogen. For example, *Pseudomonas syringae* produces pyoverdin (peptide type), achromobactin (citrate type), and yersiniabactin siderophores, which enhance its colonization and biocontrol ability (Wensing et al. 2010). *B. subtilis* CAS15 is shown to control *Fusarium* wilt of pepper. It produces a catecholic siderophore called bacillibactin (Yu et al. 2011).

14.5.2 *Indirect antagonism*

14.5.2.1 *Induction of host resistance*

When a pathogen attacks its host plant, the plant responds by activating a series of reactions called hypersensitive response (HR) as innate immunity in case of animals. A successful HR is followed by systemic resistance called systemic acquired resistance (SAR) toward a broad spectrum of pathogens. HR response is triggered when a plant recognizes the pathogen, which involves an interaction between the pathogen's virulence gene product and the plant's resistance (R) gene product. This interaction determines the plant's susceptibility toward the pathogen. R genes in plants are highly polymorphic, leading to several different products that help the plant in recognizing the pathogen's virulence products and in developing resistance. After recognition of the pathogen, a cascade reactions starts. This includes an efflux of hydroxide and potassium ions and an influx of calcium and hydrogen ions. Reactive oxygen species (ROS) are generated, which lead to oxidative burst of the cells involved in HR. Oxidative burst causes the death of these cells and the formation of local lesions. Further deposition of lignin and callose around these lesions forms a barrier between the infected and unaffected tissues. However, a virulent pathogen evades the plant defense response and successfully infects the host.

Thus, SAR is developed in response to pathogen attacks and is characterized by the development of necrotic lesions at the site of the attack, accumulation of salicylic acid (SA), increased the expression of pathogenesis-related (PR) genes, and a slowly acquired resistance against a broad spectrum of pathogens. ISR differs from SAR as it is induced by non-pathogenic bacteria. It does not lead to a necrotic lesion and is mediated by Jasmonic acid (JA) or ethylene (ET). However, some bacteria have been shown to induce resistance mediated by SA (De Meyer et al. 1999). Some bacteria activate both pathways, that is, SA and JA dependent. For example, *Bacillus cereus* AR156 simultaneously activates both SA and JA/ET-mediated pathways in *Arabidopsis thaliana* (Niu et al. 2011). The two pathways are antagonistic such that SA leads to the expression of PR genes whereas JA primes the plant for enhanced defense gene expression upon challenge by a broad spectrum of pathogens.

Many bacterial compounds have been illustrated as elicitors of ISR in plants (Table 14.2). For example, in the elicitation of ISR in *Arabidopsis* by *P. fluorescens*, 2,4-DAPG has been found to be a major determinant because 2,4-DAPG minus mutants showed significantly reduced activity (Weller et al. 2012). SA, pyochelin, and pyoverdine (siderophore) have been shown to be important in the elicitation of ISR by *Pseudomonas* sp. against *B. cinerea* in grapevines (Verhagen et al. 2010). Thus, components of direct antagonism may provide additional advantage to biocontrol by elicitation of plant resistance.

Table 14.2 Some Bacterial Elicitors Involved in Induction of Plant Resistance

Elicitor	Bacterial strain	Plant (host)	Reference
Peroxidase, chitinase and β-1,3-glucanase	*Bacillus mycoides* strain Bac J *B. pumilus* 203-6	Sugar beet	Bargabus et al. 2002, 2004
2,3-Butanediol	*B. subtilis* GB03 and IN937a	*Arabidopsis*	Ryu et al. 2004
Antibiotics (DAPG)	*P. fluorescens* CHA0	*Arabidopsis*	Weller et al. 2012
Lipopolysaccharide	*Pseudomonas putida*	*Arabidopsis*	Meziane et al. 2005
Z,3-hexenal	*Pseudomonas putida* BTP1	Bean	Ongena et al. 2004
Siderophore	*Serratia marcescens* 90-166	Cucumber	Press et al. 2001
SA, pyochelin and pyoverdine	*Pseudomonas* sp.	Grapevine	Verhagen et al. 2010
Acetoin	*B. subtilis* FB17	*A. thaliana*	Rudrappa et al. 2010

14.6 BCA with multiple mechanisms for pathogen suppression

Most biocontrol bacteria combine two or more mechanisms, providing efficient suppression of the plant pathogen. For example, *Paenibacillus ehimensis* IB-X-b produces two lytic enzymes, namely, chitinase and β-1,3-glucanase, which show synergistic antifungal activity against several plant pathogenic fungi (Aktuganov et al. 2008). *Lysobacter antibioticus* HS124 shows antifungal activity against *Phytophthora capsici*, which can be attributed to both the production of lytic enzymes such as chitinase, glucanase, and protease as well as 4-hydroxy phenylacetic acid (Ko et al. 2009). Biocontrol bacteria *Pseudomonas protegens* produces two antibiotics, namely, 2,4-DAPG and pyoluteorin (Ramettea et al. 2011).

14.7 Limitations of BCAs

Limitations of BCAs include the following:

- The main limitation of microbial products for use in biocontrol is the lack of consistency. Narrow specificity of the biological control agent limits its efficacy where a diversity of pathotypes is present in the natural population of a pathogen (Schisler et al. 2000).
- To ensure the efficacy of BCA, it is necessary to study the effect of inoculum type, application rate, and time of application. Time and place of application is more important when the mode of action is direct such as antibiosis because these secondary metabolites are produced in small quantities and are not transported a great distance. The size of inoculum must be comparable or larger to that of the pathogen when competition is the mode of action.
- Achieving the optimum number is much more difficult in a natural environment where the pathogen population density is not known.
- In addition to the diversity in pathogen population and variability in climatic conditions, the efficacy of BCA is also subject to its interaction with the natural microflora associated with the plant.

Their use is more common in the United States, Australia, and New Zealand in comparison with other countries (Table 14.3).

Table 14.3 Commercial Biocontrol Products

Microorganisms	Product	Plant, pathogens, or pathosystems	Company
Ageobacterium radiobacter 84	Galtrol	Ornamentals, fruits, nuts	AgBioChem, US
B. subtilis FZB24	FZB24 li, TB, WG, RhizoPlus	Potatoes, vegetables, ornamentals, strawberries, bulbs, turf, and woods	ABiTEP, Germany
Pseudomonas chlororaphis	Cedomon	Leaf stripe, net blotch, *Fusarium* sp., sot blotch, leaf spot, etc., on barley and oats	BioAgri AB, Sweden
Serratia plymuthica HRO-C48	RhizoStar	Strawberries, oilseed rape	Prophytaiologischer Pflanzenschutz, Germany
Streptomyces griseoviridis K61	Mycostop	*Phomopsis* spp., *Botrytis* spp., *Pythium* spp., *Phythophora* spp.	Kemira Agro Oy, Russia

14.8 Conclusion

The role of microbes, bacteria in particular, has been proven beyond a doubt for suppressing various plant diseases and there are many success stories from the laboratory to the market. In the present-day scenario, BCA are not capable of completely replacing agrochemicals. However, some studies have demonstrated a reduction in the percentage of chemical fungicide use, when used in conjunction with BCA. For example, a study on integrated control of fruit rot and powdery mildew of chili has shown a 50% reduction in use of the fungicide Azoxystrobin when used in combination with BCA *P. fluorescens* (Anand et al. 2010). Several areas of research need to be explored for further development in biological control. These include:

1. Study of distribution of pathogens including pathotypes and their antagonists in the environment
2. Effect of BCA on the natural microflora of plants and vice versa
3. Management practices that can aid biocontrol
4. Dynamics of plant defense induction, and
5. Effective combinations of BCA to control multiple pathogens/pathotypes

Thus, biological control agents have the potential to substitute hazardous agrochemicals. However, much research is required to this end to answer several questions that arise about the efficiency of biological control.

References

Aggarwal R., Gupta S., Singh V.B. and Sharma S. 2011. Microbial detoxification of pathotoxin produced by spot blotch pathogen *Bipolaris sorokiniana* infecting wheat. *J. Plant Biochem. Biot.* **20**(1): 66–73.

Aktuganov G., Melentjev A., Galimzianova N., Khalikova E., Korpela T. and Susi P. 2008. Wide-range antifungal antagonism of *Paenibacillus ehimensis* IB-X-b and its dependence on chitinase and beta-1,3-glucanase production. *Can. J. Microbiol.* **54**(7): 577–587.

Alabouvette C., Olivain C. and Steinberg C. 2006. Biological control of plant diseases: The European situation. *Eur. J. Plant Pathol.* **114**: 329–341.

Anand T., Chandrasekaran A., Kuttalam S., Senthilraja G. and Samiyappan R. 2010. Integrated control of fruit rot and powdery mildew of chilli using the biocontrol agent Pseudomonas fluorescens and a chemical fungicide. *Biol. Control* **52**: 1–7.

Bargabus R.L., Zidack N.K., Sherwood J.W. and Jacobsen B.J. 2002. Characterization of systemic resistance in sugar beet elicited by a non-pathogenic, phyllosphere colonizing *Bacillus mycoides*, biological control agent. *Physiol. Mol. Plant Pathol.* **61**: 289–298.

Bargabus R.L., Zidack N.K., Sherwood J.W. and Jacobsen B.J. 2004. Screening for the identification of potential biological control agents that induce systemic acquired resistance in sugar beet. *Biol. Control* **30**: 342–350.

Basnayake W.V.S. and Birch R.G. 1995. A gene from *Alcaligenes denitrificans* that confers albicidin resistance by reversible antibiotic binding. *Microbiology* **141**: 551–560.

Berendsen R.L., Corne Pieterse M.J. and Bakker A.H.M. 2012. The rhizosphere microbiome and plant health. *Trends Plant Sci.* **17**(8): 478–486.

Bhattacharyya P.N. and Jha D.K. 2012. Plant growth-promoting rhizobacteria (PGPR): Emergence in agriculture. *World J. Microb. Biot.* **28**(4): 1327–1350.

Bowman S.M. and Free S.J. 2006. The structure and synthesis of the fungal cell wall. *BioEssays* *28*: 799–808.

Chandrashekara K.N., Manivannan S., Chandrashekara C. and Chakravarthi M. 2012. Biological control of plant diseases, Chapter 10. In: *Ecofriendly Innovative Approaches in Plant Disease Management*. Singh U.K., Singh Y. and Singh A., (eds.) International Book Distributors, Oscar Publication, Delhi, DEL, India, pp. 147.

Compant S., Duffy B., Nowak J., Clement C. and Barka E. 2005. Use of plant growth-promoting bacteria for biocontrol of plant diseases: Principles, mechanisms of action, and future prospects. *Appl. Environ. Microbiol.* **71**(9): 4951.

Cook R.J. and Baker K.F. 1983. *The Nature and Practice of Biological Control of Plant Pathogens*. American Phytopathological Society, St. Paul, MN.

D'aes J., Hua G.K.H., De Maeyer K., Pannecoucque J., Forrez I., Ongena M. and Höfte M. 2011. Biological control of *Rhizoctonia* root rot on bean by phenazine and cyclic lipopeptide-producing *Pseudomonas* CMR12a. *Phytopathology* **101**(8): 996–1004.

Davies K.G. 2009. Understanding the interaction between an obligate hyperparasitic bacterium, Pasteuria penetrans and its obligate plant-parasitic nematode host, *Meloidogyne* spp. *Adv. Parasitol.* **68**: 211–245.

De Meyer G., Audenaert K. and Hofte M. 1999. *Pseudomonas aeruginosa* 7NSK2-induced systemic resistance in tobacco depends on in plant salicylic acid but not associated with PR1a expression. *Eur. J. Plant Pathol.* **105**: 513–517.

Dong Y.H., Gusti A.R., Zhang Q., Xu J.L. and Zhang L.H. 2002. Identification of quorum-quenching N-acyl homoserine lactonases from *Bacillus* species. *Appl. Environ. Microbiol.* **68**: 1754–1759.

Dunne C., Crowley J., Loccoz Y.M., Dowling D.N., de Bruijn F.J. and O'Gara F. 1997. Biological control of *Pythium ultimum* by *Stenotrophomonas maltophilia* W81 is mediated by an extracellular proteolytic activity. *Microbiology* **143**: 3921–3931.

Erdogan O. and Benlioglu K. 2010. Biological control of Verticillium wilt on cotton by the use of fluorescent *Pseudomonas* spp. under field conditions. *Biol. Control.* **53**(1): 39–45.

Essghaier B., Fardeau M.L., Cayol J.L., Hajlaoui M.R., Boudabous A., Jijakli H. and Zouaoui N.S. 2009. Biological control of grey mould in strawberry fruits by halophilic bacteria. *J. Appl. Microbiol.* **106**: 833–846.

Gupta C.P., Kumar B., Dubey R.C. and Maheshwari D.K. 2006. Chitinase-mediated destructive antagonistic potential of *Pseudomonas aeruginosa* GRC1 against *Sclerotinia sclerotiorum* causing stem rot of peanut. *BioControl* **51**: 821–835.

Hassan M.N., Afghan S. and Hafeez F.Y. 2011. Biological control of red rot in sugarcane by native pyoluteorin-producing *Pseudomonas putida* strain NH-50 under field conditions and its potential modes of action. *Pest Manag. Sci.* **67**(9): 1147–1154.

Islam M.T. 2011. Potentials for biological control of plant diseases by *lysobacter* spp., with special reference to strain SB-K88. In: *Bacteria in Agrobiology: Plant Growth Responses*. Maheshwari D.K., (ed). Springer Berlin Heidelberg, pp. 335–363.

Ko H.S., De Jin R., Krishnan H.B., Lee S.B. and Kim K.Y. 2009. Biocontrol ability of *Lysobacter antibioticus* HS124 against *phytophthora* blight is mediated by the production of 4-hydroxyphenylacetic acid and several lytic enzymes. *Curr. Microbiol.* **59**: 608–615.

Lanteigne C., Gadkar V.J., Wallon T., Novinscak A. and Filion M. 2012. Production of DAPG and HCN by *Pseudomonas* sp. LBUM300 contributes to the biological control of bacterial canker of tomato. *Phytopathology* **102**(10): 967–973.

Leadbetter J.R. and Greenberg E.P. 2000. Metabolism of acyl-homoserine lactone quorum-sensing signals by Variovorax paradoxus. *J. Bacteriol.* **182**: 6921–6926.

Liu X., Jia J., Atkinson S., Cámara M., Gao K., Li H. and Cao J. 2010. Biocontrol potential of an endophytic *Serratia* sp. G3 and its mode of action. *World J. Microb. Biot.* **26**(8): 1465–1471.

Malathi P., Viswanathan R., Padmanaban P., Mohanraj D. and Sundar A.R. 2002. Microbial detoxification of *Colletotrichum falcatum* toxin. *Curr. Sci. India* **83**(6): 745–749.

Meziane H., van der Sluis I., van Loon L.C., Hofte M. and Bakker P.A.H.M. 2005. Determinants of *Pseudomonas putida* WCS358 involved in inducing systemic resistance in plants. *Mol. Plant Pathol.* **6**: 177–185.

Molina L., Constantinescu F., Michel L., Reimmann C., Du B. and Défago G. 2003. Degradation of pathogen quorum-sensing molecules by soil bacteria: A preventive and curative biological control mechanism. *FEMS Microbiol. Ecol.* **45**: 71–81.

Newman K.L., Chatterjee S., Ho K.A. and Lindow S.E. 2008. Virulence of plant pathogenic bacteria attenuated by degradation of fatty acid cell-to-cell signaling factors. *Mol. Plant Microbe Interact.* **21**(3): 326–334.

Niu D.D., Liu H.X., Jiang C.H., Wang Y.P., Wang Q.Y., Jin H.L. and Guo J.H. 2011. The plant growth-promoting rhizobacterium *Bacillus cereus* AR156 induces systemic resistance in *Arabidopsis thaliana* by simultaneously activating salicylate and jasmonate/ethylene-dependent signaling pathways. *Mol. Plant Microbe Interact.* **24**(5): 533–542.

Ongena M. and Jacques P. 2007. Bacillus lipopeptides: Versatile weapons for plant disease biocontrol. *Trends Microbiol.* **16**(3): 115–125.

Ongena M., Duby F., Rossignol F., Fouconnier M.L., Dommes J. and Thonart P. 2004. Stimulation of the lipoxygenase pathway is associated with systemic resistance induced in bean by a non-pathogenic *Pseudomonas* strain. *Mol. Plant Microbe Interact.* **17**: 1009–1018.

Pal K.K. and Gardener B.M. 2006. Biological control of plant pathogens. *Plant Health Instruct.* **2**: 1117–1142, doi: 10.1094/PHI-A-2006-1117-02.

Press C.M., Loper J.E. and Kloepper J.W. 2001. Role of iron in rhizobacteria mediated induced systemic resistance of cucumber. *Phytopathology* **91**: 593–598.

Raaijmakers J.M., De Bruijn I., Nybroe O. and Ongena M. 2010. Natural functions of lipopeptides from Bacillus and Pseudomonas: More than surfactants and antibiotics. *FEMS Microbiol. Rev.* **34**(6): 1037–1062.

Raaijmakers J.M. and Mazzola M. 2012. Diversity and natural functions of antibiotics produced by beneficial and plant pathogenic bacteria. *Annu. Rev. Phytopathol.* **50**: 403–424.

Raaijmakers J.M., Vlami M. and de Souza J.T. 2002. Antibiotic production by bacterial biocontrol agents. *Antonie Leeuwenhoek.* **81**: 537–547.

Ramettea A., Frapolli M., Sauxb M.F.L., Gruffazc C., Meyerc J.M., Défago G., Sutrab L. and Loccozd Y.M. 2011. *Pseudomonas protegens* sp. nov., widespread plant-protecting bacteria producing the biocontrol compounds 2,4-diacetylphloroglucinol and pyoluteorin. *Syst. Appl. Microbiol.* **34**: 180–188.

Rudrappa T., Biedrzycki M.L., Kunjeti S.G., Donofrio N.M., Czymmek K.J. and Bais H.P. 2010. The rhizobacterial elicitor acetoin induces systemic resistance in *Arabidopsis thaliana. Commun. Integr. Biol.* **3**(2): 130–138.

Rudrappa T., Czymmek K.J., Paré P.W. and Bais H.P. 2008. Root-secreted malic acid recruits beneficial soil bacteria. *Plant Physiol.* **148**: 1547–1556.

Ryu C.M., Farag M.A., Hu C.H., Reddy M.S., Kloepper J.W. and Pare P.W. 2004. Bacterial volatiles induce systemic resistance in *Arabidopsis. Plant Physiol.* **134**: 1017–1026.

Schisler D.A., Slininger P.J., Hanson L.E. and Loria R. 2000. Potato cultivar, pathogen isolate and antagonist cultivation medium influence the efficacy and ranking of bacterial antagonists of Fusarium dry rot. *Biocontrol Sci. Techn.* **10**: 267–279.

Singh N.A. and Shanmugam V. 2012. Cloning and characterization of a bifunctional glycosyl hydrolase from an antagonistic *Pseudomonas putida* strain P3(4). *J. Basic Microbiol.* **52**: 340–349.

Tian B., Yang J. and Zhang K.Q. 2007. Bacteria used in the biological control of plant-parasitic nematodes: Populations, mechanisms of action, and future prospects. *FEMS Microbiol. Ecol.* **61**: 197–213.

Toyoda H. and Utsumi R. 1991. Method for the prevention of Fusarium diseases and microorganisms used for the same. U.S. Patent 4,988,586.

Toyoda H., Hashimoto H., Utsumi R., Kobayashi H. and Ouchi S. 1988. Detoxification of fusaric acid by a fusaric acid-resistant mutant of *Pseudomonas solanacearum* and its application to biological control of Fusarium wilt of tomato. *Phytopathology* **78**: 1307–1311.

Verhagen B.W., Trotel-Aziz P., Couderchet M., Höfte M. and Aziz A. 2010. *Pseudomonas* spp.-induced systemic resistance to *Botrytis cinerea* is associated with induction and priming of defence responses in grapevine. *J. Exp. Bot.* **61**(1): 249–260.

Walker M.J., Birch R.G. and Pemberton J.M. 1988. Cloning and characterization of an albicidin resistance gene from *Klebsiella oxytoca. Mol. Microbiol.* **2**: 443–454.

Weller D.M., Mavrodi D.V., van Pelt J.A., Pieterse C.M.J., van Loon L.C. and Bakker P.A.H.M. 2012. Induced systemic resistance in *Arabidopsis thaliana* against *Pseudomonas syringae* pv. tomato by 2,4-diacetylphloroglucinol-producing *Pseudomonas fluorescens*. *Phytopathology* **102**: 403–412.

Wensing A., Braun S.D., Büttner P., Völksch B., Ullrich M.S. and Weingart H. 2010. Impact of siderophore production by *Pseudomonas syringae* pv. *syringae* 22d/93 on epiphytic fitness and biocontrol activity against *Pseudomonas syringae* pv. *glycinea* 1a/96. *Appl. Environ. Microbiol.* **76**(9): 2704–2711.

Whipps J.M. 2001. Microbial interactions and biocontrol in the rhizosphere. *J. Exp. Bot.* **52**: 487–511.

Yánez-Mendizábal V., Zeriouh H., Viñas I., Torres R., Usall J., de Vicente A. and Teixidó N. 2012. Biological control of peach brown rot (*Monilinia* spp.) by *Bacillus subtilis* CPA-8 is based on production of fengycin-like lipopeptides. *Eur. J. Plant Pathol.* **132**(4): 609–619.

Yu X., Ai C., Xin L. and Zhou G. 2011. The siderophore-producing bacterium, *Bacillus subtilis* CAS15, has a biocontrol effect on *Fusarium* wilt and promotes the growth of pepper. *Eur. J. Soil Biol.* **47**(2): 138–145.

Zhang L. and Birch R.G. 1997. The gene for albicidin detoxification from *Pantoea dispersa* encodes an esterase and attenuates pathogenicity of *Xanthomonas albilineans* to sugarcane. *Proc. Natl. Acad. Sci.* **94**: 9984–9989.

chapter fifteen

Microbial biodegradation of polycyclic aromatic hydrocarbons

Pankaj Kumar Jain

Contents

15.1 Introduction

Petroleum is a heterogeneous mixture of hydrocarbons, including aliphatic (*n*-alkanes), alicyclic, and aromatic hydrocarbons. It varies in compositional and physical properties according to the reservoir's origin (Van Hamme et al. 2003). The two major groups of aromatic hydrocarbons are the monocyclic aromatic hydrocarbons (MAHs) such as benzene, toluene, ethylbenzene and xylene (BTEX), and the polycyclic aromatic hydrocarbons (PAHs) such as naphthalene, anthracene, phenanthrene, and benzo(*a*)pyrene. PAHs are the most widespread contaminants in the environment. PAHs are a class of organic compounds that consist of two or more fused benzene rings that are arranged in various structural configurations. These hydrocarbons are organic compounds containing carbon and hydrogen, which are highly insoluble in water.

PAHs originate from two main sources: (i) natural (biogenic and geochemical), and (ii) anthropogenic (Mueller et al. 1996). Natural sources of PAHs are volcanic eruptions and forest fires (Blumer 1976). However, the major causes of accumulation in nature are anthropogenic activities such as the combustion of organic materials, for example, fossil fuels, coal, diesel, wood, and vegetation. These are also released in the environment by waste disposal, accidental spills during transportation and storage, and as pesticides, but are also formed during incomplete combustion such as with coal (Freeman and Cattell 1990; Lim et al. 1999).

Interest in the biodegradation of polycyclic aromatic hydrocarbons and compounds (PAHs/PACs) is motivated by their ubiquitous distribution, low bioavailability, high persistence in soils, and potentially deleterious effects to human health.

15.2 *Toxicity of PAHs*

PAHs have a major ecological effect on contaminated marine and terrestrial ecosystems (Santos et al. 2011). Many PAHs are recognized as having acute carcinogenic, mutagenic, and teratogenic properties. Benzo(*a*)pyrene is recognized as a priority pollutant by the U.S. Environmental Protection Agency (U.S. EPA; Renner 1999). This compound is identified as the most potently carcinogenic of all known PAHs and is used as an environmental indicator for PAHs (Table 15.1; Juhasz and Naidu 2000).

PAH concurrent exposure is common. After several experiments, it was proven that exposure to mixtures of PAHs in coke production, roofing, oil refining, and coal gasification are carcinogenic to humans (Integrated Risk Information System [IRIS] 1994; Agency for Toxic Substances and Disease Registry [ATSDR] 1995). Exposures to environmental mixtures, coal combustion effluent, vehicle exhaust, used motor lubricating oil, and tobacco smoke are also thought to be carcinogenic due to the PAH component of the mixtures (IRIS 1994). The effects on health depend on exposure time and the mixture involved. Some studies have suggested that exposure to some of the weakly carcinogenic or noncarcinogenic PAHs such as benzo(*e*)pyrene, benzoic(*g,h,i*)perylene, fluoranthene, pyrene with

Table 15.1 Classification of PAHs by the U.S. EPA, IARC, and DHHS on the Basis of Carcinogenicity

PAHs	U.S. EPA	IARC	DHHS
Acenaphthene			
Acenaphthylene	Not classifiable		
Anthanthrene	Not classifiable	Not classifiable	
Benz(*a*)anthracene	Probably carcinogen	Probably carcinogen	Animal carcinogen
Benzo(*a*)pyrene	Probably carcinogen	Probably carcinogen	Animal carcinogen
Benzo(*b*)fluoranthene	Probably carcinogen	Probably carcinogen	Animal carcinogen
Benzo(*e*)pyrene		Not classifiable	
Benzo(*g,h,i*)perylene	Not classifiable	Not classifiable	
Benzo(*j*)fluoranthene	Not included	Possibly carcinogen	Animal carcinogen
Benzo(*k*)fluoranthene	Possibly carcinogen	Possibly carcinogen	
Chrysene	Probably carcinogen	Not classifiable	
Dibenz(*a,h*)anthracene	Probably carcinogen		Animal carcinogen
Fluoranthene	Not classifiable	Not classifiable	
Fluorene	Not classifiable	Not classifiable	
Ideno(1,2,3-*cd*)pyrene	Probably carcinogen		Animal carcinogen
Phenanthrene	Not classifiable	Possibly carcinogen	
Pyrene	Not classifiable	Not classifiable	

Source: B.-K. Lee and V.T. Vu. Sources, distribution and toxicity of polycyclic aromatic hydrocarbons (PAHs) in particulate matter. In *Air Pollution*, edited by V. Villanyi, 99–122, 2010. Sciyo Janeza Trdine: Rijeka, Croatia.

Note: DHHS, Department of Health and Human Services; IARC, International Agency of Research on Cancer; PAHs, polycyclic aromatic hydrocarbons; U.S. EPA, U.S. Environmental Protection Agency.

benzo(*a*)pyrene, or other carcinogenic PAHs enhances the carcinogenic potential. The noncarcinogenic PAHs induce the initiation and promotion of tumor properties (ATSDR 1995). However, concurrent exposure to other noncarcinogenic PAHs, for example, benzo(*a*)fluoranthene, benzo(*k*)fluoranthene, and chrysene has also been reported to lower the potential carcinogenic effects of benzo(*a*)pyrene (ATSDR 1995). Concurrent exposure to particulate matter and chemicals other than PAHs may also enhance the carcinogenic potential of a mixture (ATSDR 1995). Table 15.2 represents the emission of PAHs in various cities of various countries.

PAHs are recognized as carcinogens but are not genotoxic unless activated by mammalian enzymes, cytochrome P450 monooxygenase, to reactive epoxides and quinones. This enzyme oxidizes the aromatic ring to form epoxide and diol-epoxide reactive intermediates. It has been reported that these intermediates may undergo one of at least four different mechanisms of oxidation and hydrolysis before the intermediates combine with or attack DNA to form covalent adducts with DNA. DNA adducts can lead to mutations of the DNA, resulting in tumors (Harvey 1996). For example, benzo(*a*)pyrene is converted into epoxide by cytochrome P450 monooxygenase, forming different epoxides through the addition of one atom of oxygen across a double bond. The epoxides are short-lived compounds and may rearrange spontaneously to phenols or undergo hydrolysis to dihydrodiols. The dihydrodiols may also act as a substrate for cytochrome P450 once again to form new dihydrodiol epoxides, for example, *trans*-7,8-dihydroxy-7,8-dihydrobenzo(*a*)pyrene-9,10-oxide, which unfortunately are poor substrates for further hydrolysis. These dihydrodiol epoxides may react with DNA and cause mutations and possibly cancer (IARC 1983).

PAHs are present in the environment through anthropogenic or natural activities and exist almost everywhere. These may present in gaseous or particulate phases in ambient air. Due to the persistent nature of PAHs, living beings are always exposed to them and acquire various types of diseases. So, elimination of PAHs from the environment is indispensable.

Table 15.2 Comparison of Total PAHs (ng/m^{-3}) in Various Cities of Various Countries

Country	Area	ΣPAHs	PAHs concentration (ng/m^3)	References
North Chinese Plain	Urban	10	870	Liu et al. 2008
	Rural	10	710	
Flanders, Belgium	Rural	16	114	Ravindra et al. 2006
Seoul, Korea	Urban	16	89	Park et al. 2002
Chicago	Urban	16	13–1865	Li et al. 2008
New Delhi, India	Urban	12	1049–1344	Sharma et al. 2008
	Urban	12	672	Sharma et al. 2007
Chennai, India	Urban	11	326–791	Mohanraj et al. 2011
Agra, India	Urban	16	72.7	Lakhni 2012
Campo Grande, Brazil	Campus	14	8.94	Poppi and Silva 2005
Tai Chung, Taiwan	Urban	21	220	Fang et al. 2004
	Rural	21	831	
	Industry	21	1650	
Brisbane, Australia	Urban	16	0.4–19.73	Lim et al. 2005

Table 15.3 Remediation Technologies and their Benefits and Challenges

Remediation method	Benefits	Challenges	Examples
Bioremediation	Cost-effective, ecological	Possibly a slow process, Possible high monitoring costs	Natural attenuation Biostimulation Bioaugmentation
Phytoremediation	Cost-effective, ecological	Possibly a slow process, Possible high monitoring costs	Phytoextraction Phytostabilization Phytodegradation Phytotransformation Phytovolatilization
Containment	Fast, cheap	Potential leakage, does not reduce contaminant volume or toxicity	Encapsulation Capping
Physical treatment	Comprehensive	Loss of soil characteristics, costs, contaminants have to be discarded or remediated	Heat treatment Soil washing *In situ* flushing Air sparging Soil vapor Extraction Solidification Vitrification
Chemical treatment	Cheap, easy	Loss of soil characteristics, Contaminants have to be discarded or remediated	Liming *In situ* oxidation Permeable reactive barriers

Source: Penn, I. et al., Eco-Industrial Strategies and Environmental Justice: An Agenda for Healthy Communities: Intergrating Brownfields and Eco-Industrial Development. National Center for Eco-Industrial Development, Center for Economic Development, School of Policy, Planning and Development, University of Southern California, 2002.

15.3 *Technologies for removal of PAHs*

Removal of PAHs from the environment is indispensable due to the carcinogenicity and toxicity of these compounds. For the treatment of PAH-contaminated soil, physicochemical and biological remediation is used (Table 15.3).

15.3.1 *Physicochemical treatments for removal of PAHs*

The physicochemical treatments include incineration, thermal desorption, coker, flushing, soil vapor extraction, solidification/stabilization, cement kiln, solvent extraction, pyrolysis-gasification, and landfilling (Figure 15.1).

Currently, physicochemical treatments are dominant over biological treatments; however, these methods lack some desired properties. Incineration is a very effective treatment method, but it is costly and after burning, the soil will lose most of its nutritional value and structure. Landfilling does remove the contaminants, but only relocates the problem (Lageman et al. 2005). Furthermore, in Europe, legislation requires a reduction of the number of landfills. In 2004, the number of landfills with untreated contaminated material in England and Wales was reduced from more than 200 to only 11 (EA 2006). In Finland, the number of landfills has decreased from 232 in 1999 to 80 in 2005 (FEI 2006). As a result, the

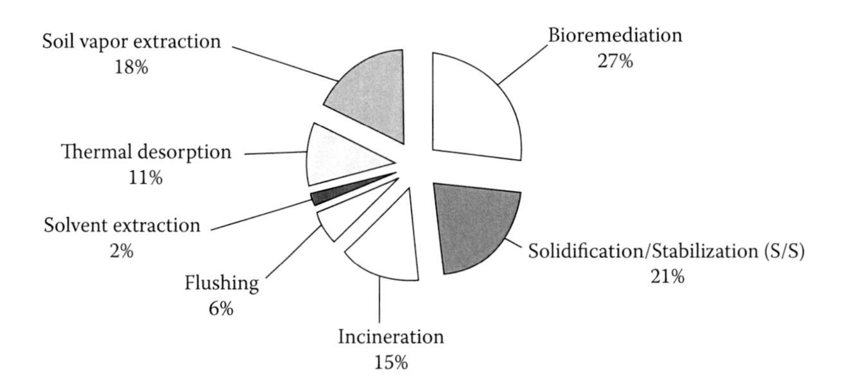

Figure 15.1 Methods used for remediation of PAHs.

Table 15.4 Physicochemical Remediation Technology of Petroleum Sludge Treatment

Technology	Comments
Landfill	Very old but still one of the extensively used technologies, most landfills do not have energy production facilities
Incineration	High-temperature treatment, air pollution risks, expensive control equipment, high capital cost
Thermal desorption	High-temperature oil removal and recovery method from oily solids, high capital and material preparation costs, nonhazardous residues
Pyrolysis gasification	Pyrolysis is the thermal degradation of waste in the absence of air to produce gas, liquid, or solid, which generally takes place between 400°C and 1000°C. Gasification takes place at higher temperatures than pyrolysis (1000°C–1400°C) in a controlled amount of oxygen. Pyrolysis and in particular gasification is obviously very attractive to reduce and avoid corrosion and emissions by retaining alkali and heavy metals
Coker	Complicated sludge preparation for coker feed, some oil recovery, high capital and transportation costs
Cement kiln	Complicated sludge preparation for use of fuel, high material preparation, transportation, and disposal costs
Solvent extraction	Uses solvents and centrifugation or filtration for the separation of oil from sludges, safety hazard with solvent use, high capital cost

cost of dumping contaminated soils into landfills has increased considerably. It is therefore evident that new, innovative methods are needed to treat contaminated soils. Table 15.4 presents the physicochemical treatment technologies used for the removal of PAHs.

15.3.2 Biological treatments for removal of PAHs

Before 1946, Zobell recognized that many microorganisms had the ability to utilize hydrocarbons as the sole source of carbon and energy. He further recognized that the microbial utilization of hydrocarbons were highly dependent on the chemical nature of the components in the petroleum mixture and environmental determinants (Atlas 1981).

Microbial biodegradation of pollutants has intensified in recent years as mankind strives to find sustainable ways to clean-up contaminated environments (Diaz 2008). Biodegradation of hydrocarbons by natural populations of microorganisms represent

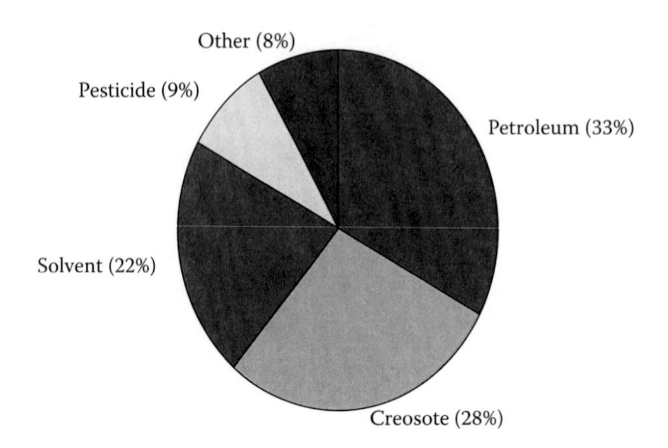

Figure 15.2 Representing bioremediation technology is applicable to all major types of wastage.

one of the primary mechanism by which petroleum and other hydrocarbon pollutants are eliminated from the environment. The effects of environmental parameters on microbial degradation of hydrocarbons, the elucidation of metabolic pathways, genetic basis for hydrocarbon dissimilation by microorganisms, and the effects of hydrocarbon contamination on microbial communities have been areas of intense interest and the subjects of several reviews (Atlas 1981, 1984).

Biodegradation is a process of metabolic ability of microorganisms to transform or mineralize organic contaminants into less harmful or nonhazardous substances. It is a natural process and is influenced by several factors such as nutrients, oxygen, pH value, composition, concentration, bioavailability, chemical, physical characteristics, and the pollution history of the contaminated environment.

Microbial remediation of a hydrocarbon-contaminated site is accomplished with the help of a diverse group of microorganisms, particularly the indigenous bacteria present in soil. These microorganisms can degrade a wide range of target constituents present in oily sludge (Eriksson et al. 1999; Barathi and Vasudevan 2001; Mishra et al. 2001; Figure 15.2). Therefore, we can apply this technology to various contaminants.

Hydrocarbon-degrading bacteria and fungi are widely distributed in marine, freshwater, and soil habitats. Similarly, hydrocarbon-degrading cyanobacteria have been reported, although contrasting reports indicated that the growth of mats built by cyanobacteria in the Saudi coast led to the preservation of oil residues (Barth 2003; Lliros et al. 2003; Chaillan et al. 2004). Typical bacterial groups already known for their capacity to degrade hydrocarbons include *Pseudomonas* spp. (Brito et al. 2006). Molds belonging to the genera *Aspergillus* spp., *Penicillium* spp., *Fusarium* spp., *Amorphoteca* spp., *Neosartorya* spp., *Paecilomyces* spp., *Talaromyces* spp., *Graphium* spp., and the yeasts *Candida* spp., *Yarrowia* sp., and *Pichia* spp. have been implicated in hydrocarbon degradation (Chaillan et al. 2004). Leahy and Colwell (1990) reported the biodegradation of petroleum oil. Biodegradation of oil by the fungi *Rhodotorula*, *Sporobolomyces*, *Aspergillus*, and *Penicillium* has been also studied (Head and Swannell 1999; Table 15.5).

15.3.2.1 Types of bioremediation

Broadly, we can classify bioremediation into two types. Figure 15.3 represents various types of bioremediation methods.

Table 15.5 Genera of Hydrocarbon Degrading Bacteria, Fungi, and Plants which Grow in the Presence of PAHs, BTEX, or Alkanes

		Bacteria	
Achromobacter	*Enterobacter*	*Sinorhizobium*	*Sphingomonas*
Acidovorax	*Flavobacterium*	*Sphingomonas*	*Micrococcus*
Acinetobacter	*Herbaspirillum*	*Staphylococcus*	*Cellulomonas*
Actinomyces	*Methylobacter*	*Stenotrophomonas*	*Gordonia*
Aeromonas	*Microbacterium*	*Streptomyces*	*Dietzia*
Alcaligenes	*Methylococcus*	*Streptococcus*	*Sarcina*
Aquaspirillum	*Micrococcus*	*Spirilum*	*Serratia*
Arthrobacter	*Mycobacterium*	*Variovorax*	*Paenibacillus*
Arthrobacter	*Moraxella*	*Vibrio*	*Rhizobium*
Bacillus	*Nocardia*	*Xanthomonas*	*Rhodomonas*
Boseaa	*Erwinia*	*Yokenella*	*Rhodococcus*
Brevibacterium	*Norcadia*	*Marinobacter*	*Cytophaga*
Burkholderia	*Proteus*	*Alcanivorax*	*Corynebacterium*
Capnocytophag	*Providencia*	*Curtobacterium*	*Corynebacterium*
Cellulomonas	*Pseudomonas*	*Clavibacter*	*Chromobacterium*
Paenibacillus	*Microbulbifer*		
		Fungi	
Aspergillus	*Neosartorya*	*Talaromyces*	*Sporobolomyces*
Penicillium	*Paecilomyces*	*Graphium*	*Amorphoteca*
Fusarium	*Candida*	*Yarrowia*	*Pichia*
Rhodotorula			
		Plants	
Agropyron smithii	*Andropogon gerardi*	*Bouteloua curtipendula*	*Populus deltoides*
Bouteloua gracilis	*Buchloe dactyloides*	*Elymus canadensis*	*Sorghastrum nutans*
Festuca rubra	*Schizchyrium scoparious*		

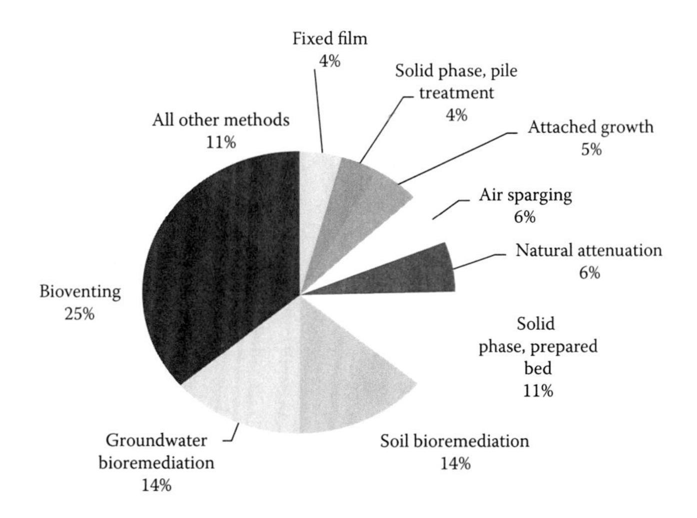

Figure 15.3 Representing various types of bioremediation methods.

i. *In situ bioremediation*—in this method, excavation of soil is not needed. This usually leads to considerable savings in the costs of excavation and transport. According to the calculations of the Federal Remediation Technologies Roundtable (FRTR 2000), the costs of biological *in situ* treatments were approximately 8 to 80 €/m³. Table 15.6 shows types of *in situ* bioremediation.

ii. *Ex situ bioremediation*—this type of bioremediation is generally used only when the site is threatened for some reason, usually by the spill that needs to be cleaned up. *Ex situ* bioremediation is only used when necessary because it is expensive and damaging to the area as the contaminated land is physically removed. Table 15.7 represents the various types of *ex situ* bioremediation.

Selection of the best treatment method depends on the type and quantity of the contaminants, treatment costs, the soil type, and the environmental conditions on the site,

Table 15.6 Types of *In situ* Bioremediation

Intrinsic bioremediation (natural attenuation)	• Intrinsic bioremediation uses microorganisms already present in the environment to biodegrade harmful contaminants • There is no human intervention involved in this type of bioremediation, and because it is the cheapest means of bioremediation available, it is the most commonly used • When intrinsic bioremediation is not feasible, scientists turn next to accelerated bioremediation
Land treatment	• It is a full-scale bioremediation technology in which contaminated soils, sediments, or sludges are periodically turned over (tilled) and allowed to interact with the soil and climate at the site • The advantage of land treatment is that conventional farming equipment is usually the only equipment required
Bioventing	• Contaminated soil is treated by drawing oxygen through it to stimulate microbe growth
Biosparging	• Biosparging involves the injection of air under pressure below the water table to increase groundwater oxygen concentrations and enhance the rate of biological degradation of contaminants by naturally occurring bacteria • It increases the mixing in the saturated zone and thereby increases the contact between soil and groundwater
Bioaugmentation	• The addition of a specific laboratory cultivated microbes or genetically modified organisms to the contaminated sites to increase the rate of biodegradion
Biostimulation	• Use of indigenous microbes to promote the growth of native microbes by providing nutrients • Depends on necessary native microbial and organic material to be present
Electroremediation	• Electrokinetic remediation is traditionally used to remove metals and polar organic compounds from soils, sludges, and sediments • This method uses electrodes with a low-level direct current electric field (usually <10 V/cm or mA/cm²) installed into the contaminated soil • The current mobilizes and transports charged chemicals in the soils liquid phase toward the electrodes. Negatively charged anions and organic compounds will move to the anode, whereas positively charged chemicals such as metals will move toward the cathode

Table 15.7 Types of *Ex situ* Bioremediation

Landfarming	• Landfarming is a simple technique in which contaminated soil is excavated and spread over a prepared bed and periodically tilled until pollutants are degraded • Adaptation of traditional farming techniques (aerating, ploughing) to contaminated areas to increase microbes activity
Composting	• Composting is a technique that involves combining contaminated soil with nonhazardous organic compounds such as agricultural wastes • The presence of these organic materials supports the development of a rich microbial population and elevated temperature characteristic of composting
Biopiles	• This is a hybrid of landfarming and composting • Essentially, engineered cells are constructed as aerated composted piles • Typically used for treatment of surface contamination with petroleum hydrocarbons, this is a refined version of landfarming that tends to control physical losses of the contaminants by leaching and volatilization • Biopiles provide a favorable environment for indigenous aerobic and anaerobic microorganisms
Bioreactors	• Slurry bioreactors or aqueous bioreactors are used for *ex situ* treatment of contaminated soil and water pumped up from a contaminated plume • Bioremediation in reactors involves the processing of contaminated solid material (soil, sediment, sludge) or water through an engineered containment system

among other things (Khan et al. 2004; U.S. Department of Defense 1994). Before the selection of the best available method or methods, it is necessary to conduct a thorough investigation of the properties of the soil. This is especially true for *in situ* remediation because performance of the remediation process is more difficult to monitor and control during treatment than it is in traditional *ex situ* remediation processes (Jeltsch 1990; Morgan and Watkinson 1992).

15.4 Molecular techniques used for identification of PAH degrading microorganisms

Currently, molecular approaches are used to identify microorganisms in PAH contaminated soil and water (Figure 15.4). These techniques are more effective than our classic techniques (Table 15.8). Identifying the diversity of microorganisms that degrade PAHs/PACs can be utilized in the development of bioremediation techniques.

15.5 Role of plasmid in degradation of PAHs

Plasmids are small, circular, or linear extrachromosomal parts of DNA. They contain approximately 2% of the genetic information of a cell and are separate from chromosomes. These can replicate independently from the main chromosome of bacteria. Plasmids are not essential to the life of a bacterium but they determine a cell's resistance to antibiotics (R factor or resistance factor) or degradation to any compounds. These have specific genes that are known as modules. Modules provide the plasmid resistance to a particular antibiotic and degradation of high-molecular weight compounds. The degradative plasmids are known as catabolic plasmids (Table 15.9). These are large (80 → 1100 kb) circular DNA and

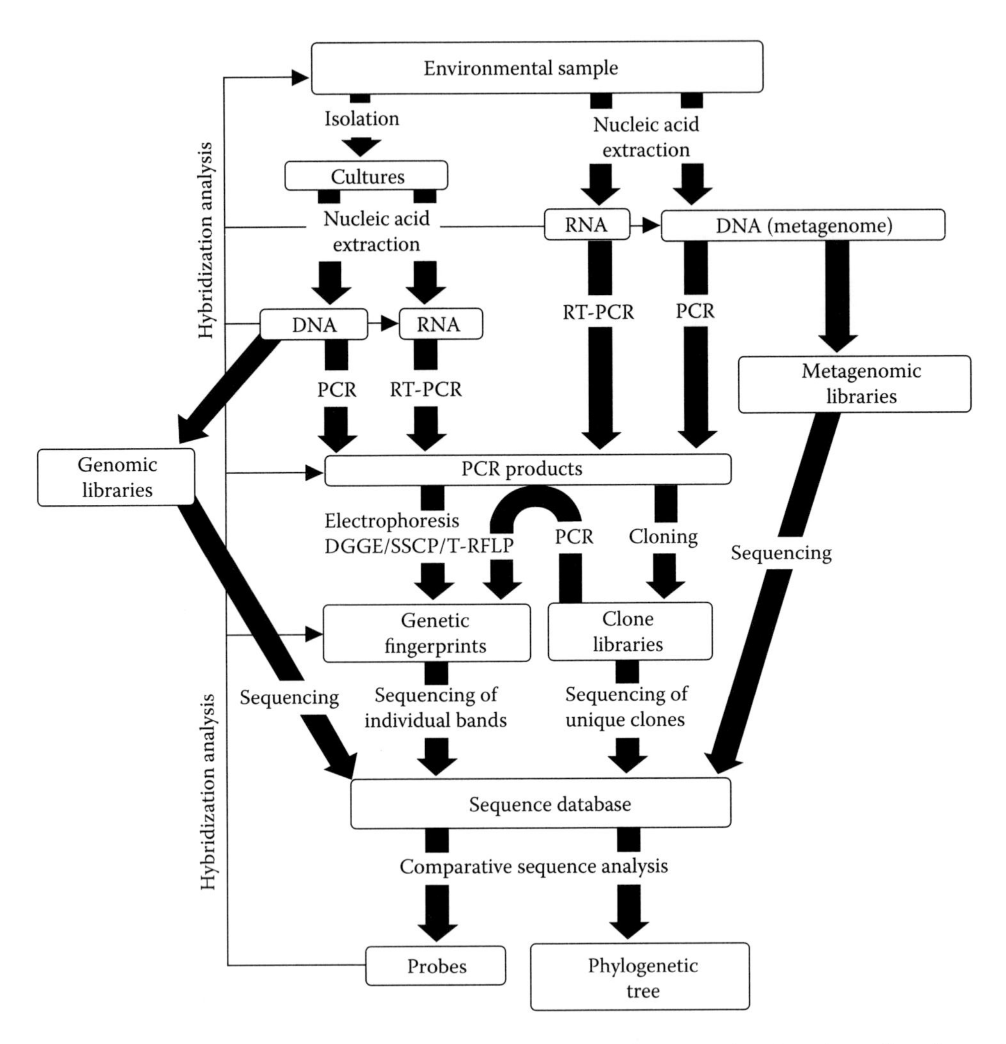

Figure 15.4 Molecular approaches for detection and identification of PAHs degrading bacteria and their catabolic genes. (From Muyzer, G., and Smalla, K., *Antonie Leeuwenhoek* 73: 127–141, 1998; Widada, J. et al., *Applied Microbiology and Biotechnology* 58: 202–209, 2002.)

Table 15.8 Advantages of Molecular Approaches over Classic Approaches

Molecular approaches	Classic approaches
No need for culture microorganisms. Therefore, it can be applied to all types of microorganisms	A culture of microorganisms is essential. So, we cannot apply these techniques on unculturable microorganisms. These unculturable microorganisms may be involved in biodegradation
Molecular methods preserve the *in situ* metabolic status and microbial community composition because samples are frozen immediately after acquisition	Incubation in the laboratory produces artificial changes in the microbial community structure and metabolic activity
Direct extraction of nucleic acids from environmental samples can be used for a very large proportion of microorganisms	We can apply these techniques only for a small population of microorganisms due to handling problems

Table 15.9 Bacteria and their PAHs Degrading Plasmids

Bacterial strain	Plasmid	References
P. putida G7	NAH7	Yen and Serdar 1988; Menn et al. 1993; Simon et al. 1993
P. putida NCIB9816-4	pDTG1	Yen and Serdar 1988; Simon et al. 1993; Resnick et al. 1996
P. putida NCIB9816	pWW601	Yen and Serdar 1988; Yang et al. 1994
Pseudomonas sp. C18	Nah (G7)	Denome et al. 1993
Ralstonia sp. U2	Nag (U2)	Fuenmayor et al. 1998; Kiyohara et al. 1982
A. faecalis AFK2	Phn (AFK2)	Zhou et al. 2001
Pseudomonas fluorescens 5R	pKA1	King et al. 1990
Pseudomonas fluorescens 18H	pUTK21	King et al. 1990
Pseudomonas fluorescens HK44	pUTK21	King et al. 1990
Pseudomonas fluorescens 5RL	pUTK21	King et al. 1990
Sphingomonas aromaticivorans F199	pNL	Romine et al. 1999
Staphylococcus sp. PN/Y	pPNY	Mallick et al. 2007
Terrabacter sp. DBF63	pDBF1	Habe et al. 2005
Sphingomonas sp. HS362	p4	Hwa et al. 2005
Beijerinkia sp.	pKG2	Kiyohara et al. 1983
Pseudomonas fluorescens strain LP6a	pLP6a	Foght and Westlake 1996

can be transferred by cell-to-cell contact to a wide range of bacteria. The role of plasmids in the evolution of bacterial populations in the environment cannot be overemphasized.

The large catabolic gene clusters carried by these plasmids play a central role in the Earth's carbon cycle. Many have a broad host range, enabling their hosts to degrade and recycle complex naturally occurring molecules such as lignin, the second most abundant plant biopolymer on Earth after cellulose. Furthermore, these catabolic gene clusters have evolved the ability to degrade and recycle high-molecular weight petroleum hydrocarbons, which are the most toxic and recalcitrant environmental pollutants, providing a biological solution to environmental pollution. These genes are used for the degradation of PAHs. Many of the genes involved in the degradation of PAHs are often located on plasmids (Johnsen et al. 2005). Thus, a bacterium needs the appropriate catabolic genes to degrade a compound (Table 15.10).

15.5.1 *Bacterial metabolism of naphthalene and phenanthrene*

Naphthalene is the bicyclic aromatic hydrocarbon. It is abundant in soils. Several bacterial species are able to grow on naphthalene as the sole source of carbon and energy (Shamsuzzam and Barnsley 1974; Bosch et al. 1999; Kulakov et al. 2000). It has often been used as a model compound of PAH degradation. The bacterial degradation of naphthalene is well understood. The degradation of naphthalene in soil pseudomonads was first reported by Davies and Evans (1964).

The genes encoding the enzymes required for the catabolism of naphthalene are often carried on plasmids such as the NAH plasmids (Zakharian and Yen 1982). The naphthalene dioxygenase enzyme, encoded by NAH7 plasmid, is known today to be a highly versatile enzyme system, encoding a wide range of reactions (Ensley et al. 1983; Parales et al. 2002). The first evidence of the versatility of the NAH plasmid encoded genes was provided

Table 15.10 Plasmid Encoded PAHs Degrading Gene Clusters of Bacteria

Strain	Substrate	Gene	Function
P. putida	Naphthalene (upper pathway)	nahAa	Reductase
		nahAb	Ferredoxin
		nahAc	Iron sulfur protein large subunit
		nahAd	Iron sulfur protein small subunit
		nahB	cis-Naphthalene dihydrodiol dehydrogenase
		nahC	1,2-Dihydroxynaphthalene oxygenase
		nahD	2-Hydroxychromene-2-carboxylate isomerase
		nahE	2-Hydroxybenzalpyruvate aldolase
		nahF	Salicyaldehyde dehydrogenase
	Salicylate (lower pathway)	nahG	Salicylate hydroxylase
		nahT	Chloroplast-type ferredoxin
		nahH	Catechol oxygenase
		nahI	2-Hydroxymuconic semialdehyde dehydrogenase
		nahN	2-Hydroxymuconic-semialdehyde dehydrogenase
		nahL	2-Oxo-4-pentenoate hydratase
		nahO	4-Hydroxy-2-oxovalerate aldolase
		nahM	Acetaldehyde dehydrogenase
		nahK	4-Oxalocrotonate decarboxylase
		nahJ	2-Hydroxymuconate tautomerase
	Regulator for both operons	nahR	Induced by salicylate
Ralstonia sp. U2	Naphthalene	nagAa	Ferredoxin reductase
		nagAb	Ferredoxin
		nagAc	α Subunit of ISP (iron sulfur protein)
		nagAd	β Subunit of ISP
		nagB	cis-Dihydrodiol dehydrogenase
		nagF	Aldehyde dehydrogenase
A. faecalis AFK2	Phenanthrene	phnAa	Ferredoxin reductase
		phnAc	α Subunit of NDO (naphthalene dioxygenase)
		phnAd	β Subunit of NDO
		phnB	cis-Dihydrodiol dehydrogenase
		phnC	Dihydroxyphenanthrene dioxygenase
		phnD	Isomerase
		phnE	Hydratase-aldolase
		phnF	1-Hydroxy-2-naphtholdehyde dehydrogenase
		phnG	1-Hydroxy-2-naphthoate dioxygenase
		phnH	trans-2-Carboxybenzaldehyde dehydrogenase
		phnI	2-Carboxybenzaldehyde dehydrogenase
		gst	Glutathione-S-transferase

Source: Habe, H., and Omori T., *Bioscience, Biotechnology, and Biochemistry* 67: 225–243, 2003; Van Hamme, J.D. et al., *Microbiology and Molecular Biology Reviews* 67(4): 503–549, 2003.

independently by two research groups (Menn et al. 1993; Sanseverino et al. 1993). This led Sanseverino and coworkers (1993) to conclude that maintaining and monitoring one catabolic bacterial population may be sufficient for the degradation of a significant fraction of PAHs in contaminated soil. However, some organisms, such as *Pseudomonas stutzeri* AN10, carry these genes on their chromosome (Bosch 1999). Transcription of the genes in this operon is regulated by a salicylate-dependent transcription regulator such as NahR for the operon located on the NAH plasmid (Schell 1986; You and Gunsalus 1986).

Pseudomonas putida PpG7 have the capacity to degrade naphthalene because it has plasmid NAH7, which regulates the mineralization of naphthalene. Yen and Serdar (1988) reviewed the biochemistry and genetics of the naphthalene degradation pathway contained on plasmid NAH7. Dunn and Gunsalus (1973) established that the genes involved in naphthalene degradation are located on plasmids and were transferable by conjugation. The size of the *nah* gene is 25 kb and it is located on NAH7 (Grund and Gunsalus 1983; Schell 1983).

It has also been established that the same genes are responsible for partial degradation of phenanthrecene and anthracene. The metabolism of naphthalene has been well studied genetically in *P. putida* strain G7. *P. putida* PMD-1 can utilize naphthalene and salicylic acid as the sole carbon sources due to the presence of plasmid. It has been shown that the plasmid is required for the degradation of salicylate during the conversion of naphthalene (Zuniga et al. 1981).

Three operons are located on this plasmid. The first operon includes genes nahABCDEF. Naphthalene is converted into salicylate by encoding these genes. The second operon (sal) encodes the lower pathway's genes. Salicylate is converted into a TCA cycle intermediate through meta-ring cleavage (oxidation of salicylate via the catechol meta-cleavage pathway) by encoding these genes. This operon includes genes nahGHIJK. The third operon encodes a regulatory enzyme (NahR), which regulates both the upper and lower operons by a transacting positive control. It is located immediately upstream of the nahG gene NahR is needed for the high-level expression of the nah genes and their induction by salicylate (Yen and Gunsalus 1982; Grund et al. 1983; Yen and Gunsalus 1985; Schell 1985, 1986; Schell and Wender 1986).

Phenanthrene is a three condensed rings-fused PAH. These three rings fused in an angular fashion. It is often used as a model substance for microbial metabolism of "bay region" and "K region" containing carcinogenic PAH, such as benzo(*a*)pyrene, benzo(*a*)anthracene and chrysene (Samanta et al. 1999). Recent studies suggest that phenanthrene can be degraded by several Gram-negative and Gram-positive bacterial species (Evans et al. 1965; Kiyohara and Nagao 1978; Kiyohara et al. 1982; Barnsley 1983; Gibson and Subramanian 1984; Houghton and Shanley 1994; Samanta et al. 1999), for example, *Pseudomonas* (Yang et al. 1994; Balashova et al. 1999; Tian et al. 2002), *Mycobacterium* (Kelly and Cerniglia 1991; Churchill et al. 1991; Miller et al. 2004), and *Nocardioides* (Iwabuchi et al. 1998; Saito et al. 2000).

Kiyohara and coworkers (1983) reported that phenanthrene, biphenyl, and other PAH compounds can be degraded by *Beijerinkia* sp. This bacterium has two plasmids, PKG1 and PKG2, with 147 and 20.8 KDa, respectively. The genes for degradation of phenanthrene and biphenyl are localized on PKG2. Also, Okpokwasili et al. (1984) reported a phenanthrene-utilizing *Flavobacterium*. In this bacterium, genes for degradation of phenanthrene are present on a 34 MDa plasmid. Guerin and Jones (1988) reported a phenanthrene degrading *Mycobacterium* strain BG1. In this bacterium, genes for degradation of phenanthrene are present on three plasmids 21, 58, and 77 KDa. Cho and Kim (2001) reported a 500-kb plasmid in *Sphingomonas* strain K514, and it is capable of mineralizing phenanthrene to carbon dioxide. This plasmid is much larger than the NAH plasmid. Hwa et al. (2005) reported the phenanthrene degrading bacterium *Sphingomonas* sp. HS362. It has five plasmids and degradative ability is localized on plasmid P4.

15.5.2 *phn genes for degradation of phenanthrene*

Phenanthrene biodegradation by bacteria has been documented, and various genes are involved in this. *Burkholderia* sp. strain RP007 degrades low-molecular weight PAHs like naphthalene and anthracene. The naphthalene and phenanthrene are degraded through a common upper pathway via salicylate and 1-hydroxy-2-naphthoic acid, respectively. phn operon of *Burkholderia* sp. strain RP007 located on an 11.5-kb *Hind*III fragment, which contained nine ORFs (Table 15.11; Iwabuchi et al. 1998; Samanta et al. 1999; Doddamani and Ninnekar 2000; Prabhu and Phale 2003).

It has been found that the *AaAb* genes, which encode iron sulfur protein α and β subunits of the PAH initial dioxygenase (phn Ac and Ad) are not present on the *phn* locus. Two putative regulatory genes (phnR and phnS) are upstream of the phn catabolic genes. The phnS is a LysR-type transcriptional activator, and phnR is a positive transcriptional regulator. However, the detailed regulatory mechanism of the phn genes has not been described.

Alcaligenes faecalis strain AFK2 can utilize phenanthrene as the sole carbon source through the *o*-phthalate pathway. The phn genes are located in 13 ORF (Table 15.12). However, details have not been reported.

Table 15.11 Plasmid Encoded PAHs Degrading Gene Clusters of Bacteria

S. no.	Protein	Gene name
1	Regulatory protein	phnR
2	Regulatory protein	phnS
3	Aldehyde dehydrogenase	phnF
4	Hydratase aldolase	phnE
5	Extradiol dioxygenase	phnC
6	Isomerase	phnD
7	ISP a subunit of initial dioxygenase	phnAc
8	ISP b subunit of initial dioxygenase	phnAd
9	Dihydrodiol dehydrogenase	phnB

Table 15.12 Plasmid Encoded PAHs Degrading Gene Clusters of Bacteria

S. no.	Protein	Gene name
1	Ferredoxin	phnAb
2	Ferredoxin reductase	phnAa
3	Cis-dihydrodiol dehydrogenase	phnB
4	A subunit of NDO	phnAc
5	B subunit of NDO	phnAd
6	Putative 2-hydroxychromene-2-carboxylate isomerase	phnD
7	Glutathione-S-transferase	Gst
8	Trans-2-carboxybenzalpyruvate hydratase-aldolase	phnH
9	1-Hydroxy-2-naphthoate dioxygenase	phnG
10	2-Carboxybenzaldehyde dehydrogenase	phnI
11	3,4-Dihydroxyphenanthrene dioxygenase	phnC
12	1-Hydroxy-2-naphthoal-dehyde dehydrogenase	phnF
13	Putative-trans-*o*-hydroxybenzylidenepyruvate hydratase-aldolase	phnE

15.6 Conclusion

Contamination of the environment by PAHs has become a significant environmental concern due to their low bioavailability, toxicity, mutagenicity, and carcinogenic properties. PAHs are persistent, mainly due to their high hydrophobicity. The fate of PAHs and other organic contaminants in the environment is related with both abiotic and biotic processes. Abiotic processes include volatilization, photooxidation, chemical oxidation and biotic process including bioaccumulation, and microbial transformation. Microbial action has been recognized as the most influential and significant cause of PAH removal. A variety of microorganisms such as bacteria and fungi can degrade PAHs. Thus, there is a major interest in studying the bacteria present in contaminated environments for bioremediation. Understanding the mechanisms of bacterial populations to adapt in the presence of pollutants, and the identification of key functional genes, will allow the exploitation of microbial PAH-degradative capabilities. This will augment the success of bioremediation strategies.

References

Agency for Toxic Substances and Disease Registry (ATSDR). 1995. Toxicological profile for polyaromatic hydrocarbons (PAHs). Available at http://www.atsdr.cdc.gov/toxprofiles/tp69-c2.html (accessed January 2004).

Atlas, R.M. 1981. Microbial degradation of petroleum hydrocarbons: An environmental perspective. *Microbiol. Rev., 45*: 180–209.

Atlas, R.M. 1984. *Petroleum Microbiology*. Macmillan Publishing Company, New York, pp. 1–618.

Balashova, N.V., Koshelva, I.A., Golovchenko, N.P. and Boronin, A.M. 1999. Phenanthrene metabolism by *Pseudomonas* and *Burkholderia strain. Process Biochem., 35*: 291–296.

Barathi, S. and Vasudevan, N. 2001. Utilization of petroleum hydrocarbons by *Pseudomonas fluorescence* isolated from a petroleum—contaminated soil. *Environ. Int., 26*: 413–416.

Barth, J.H. 2003. The influence of cyanobacteria on oil polluted intertidal soils at the Saudi Arabian gulf shores. *Mar. Pollut. Bull., 46*: 1245–1252.

Blumer, M. 1976. Polycyclic aromatic compounds in nature. *Sci Am., 3*: 35–45.

Bosch, R., Garcia-Valdes, E. and Moore, E.R. 1999. Genetic characterization and evolutionary implications of a chromosomally encoded naphthalene-degradation upper pathway from *Pseudomonas stutzeri* AN10. *Gene, 236 (1)*: 149–157.

Brito, E.M.S., Guyoneaud, R., Goni-Urriza, M., Ranchou-Peyruse, A., Verbaere, A., Crapez, M.A.C., Wasserman, J.C.A. and Duran, R. 2006. Characterization of hydrocarbonoclastic bacterial communities from mangrove sediments in Guanabara Bay, Brazil. *Res. Microbiol., 157*: 752–762.

Chaillan, F., Fleche, A., Bury, E., Phantavong, Y., Grimont, P., Saliot, A. and Oudot, J. 2004. Identification and biodegradation potential of tropical aerobic hydrocarbon degrading microorganisms. *Res. Microb., 155 (7)*: 587–595.

Cho, J.C. and Kim, S.J. 2001. Detection of mega plasmid from polycyclic aromatic hydrocarbon degrading *Sphingomonas* sp. strain KS14. *J. Mol. Microbiol. Biotechnol., 3 (4)*: 503–506.

Churchill, S.A., Harper, J.P. and Churchill, P.F. 1999. Isolation and characterization of a *Mycobacterium* species capable of degrading three and four ring aromatic and aliphatic hydrocarbons. *Appl. Environ. Microbiol., 65*: 549–552.

Davies, J.I. and Evans, W.C. 1964. Oxidative metabolism of naphthalene by soil *Pseudomonads*: The ring fission mechanism. *Biochem. J., 91*: 251–261.

Denome, S.A., Stanley, D.C., Olson, E.S. and Young, K.D. 1993. Metabolism of dibenzothiophene and naphthalene in Pseudomonas strains: Complete DNA sequence of an upper naphthalene catabolic pathway. *J. Bacteriol., 175*: 6890–6901.

Diaz, E. 2008. *Microbial Biodegradation: Genomics and Molecular Biology*, 1st ed. Caister Academic Press.

Doddamani, H.P. and Ninnekar, H.Z. 2000. Biodegradation of phenanthrene by a *Bacillus* species. *Curr. Microbiol., 41*: 11–14.

Ensley, B.D., Ratzkin, B.J., Osslund, T.D., Simon, M.J., Wackett, L.P. and Gibson, D.T. 1983. Expression of naphthalene oxidation genes in *Escherichia coli* results in the biosynthesis of indigo. *Science, 222*: 167–169.

Environment Agency (EA). 2006. Available at http://www.environment-agency.gov.uk/.

Eriksson, M., Dalhammer, G. and Borg-Karlson, A.K. 1999. Aerobic degradation of a hydrocarbon mixture in natural uncontaminated potting soil by indigenous microorganisms at 20°C and 6°C. *Appl. Microbiol. Biotechnol., 51*: 532–535.

Fang, G.C., Chang, K.F., Lu, C. and Bai, H. 2004. Estimation of PAHs dry deposition and BaP toxic equivalency factors (TEFs) study at urban, industry park and rural sampling sites in central Taiwan, Taichung. *Chemosphere, 44*: 787–796.

Federal Remediation Technologies Roundtable (FRTR). 2000. Remediation technologies screening matrix and reference guide, version 4.0. Available at http://www.frtr.gov/matrix2/top_page.html.

Finnish Environment Institute (FEI). 2006. Available at http://www.ymparisto.fi.

Foght, J.M. and Westlake, D.W. 1996. Transposon and spontaneous deletion mutants of plasmid-borne genes encoding polycyclic aromatic degradation by a strain of *Pseudomonas fluorescens*. *Biodegradation, 7 (4)*: 353–366.

Freeman, D.J. and Cattell, F.C.R. 1990. Woodburning as a source of atmospheric polycyclic aromatic hydrocarbons. *Environ. Sci. Technol., 24*: 1581–1585.

Fuenmayor, S.L., Wild, M., Boyes, A.L. and Williams, P. 1998. A gene cluster encoding steps in conversion of naphthalene to gentisate in *Pseudomonas* sp. strain U2. *J. Bacteriol., 180*: 2522–2530.

Grund, E., Denecke, B. and Eichenlaub, R. 1992. Naphthalene degradation via salicylate and genetisate by *Rhodococcus* sp. strain B4. *Appl. Environ. Microbiol., 58*: 1874–1877.

Habe, H. and Omori, T. 2003. Genetics of polycyclic aromatic hydrocarbon degradation by diverse aerobic bcaeria. *Biosci. Biotechnol. Biochem., 67*: 225–243.

Habe, H., Chung, J.S., Ishida, A., Kasuga, K., Ide, K., Takemura, T., Nojiri, H., Yamane, H. and Omori, T. 2005. The fluorene catabolic linear plasmid in *Terrabacter* sp. strain DBF63 carries the β-ketoadipate pathway genes, pcaRHGBDCFIJ, also found in proteobacteria. *Microbiology, 151*: 3713–3722.

Harvey, R.G. 1996. Mechanisms of carcinogenesis of polycyclic aromatic hydrocarbons. *Polycyclic Aromat. Compd., 9*: 1–23.

Head, I.M. and Swannell, R.P.J. 1999. Bioremediation of petroleum hydrocarbon contaminants in habitats. *Curr. Opin. Biotechnol., 10*: 234–239.

Hwa, K.S., Jeong, K.H., Jincchul, A., Hoon, J.J. and Seung-Yeol, S.L. 2005. Characterization of phenanthrene degradation by *Sphingomonas* sp. HS 362. *Korean J. Microbiol., 41*: 201–207.

IARC. 1983. *Monographs on the Evaluation of the Carcinogenic Risk of Chemicals to Humans; Polynuclear Aromatic Compounds, Part 1, Chemical, Environmental and Experimental Data*, Vol. 32. IARC, Lyon, France.

Integrated Risk Information System (IRIS). 1994. U.S. Environmental Protection Agency, Integrated Risk Information Summary. Benzo(a)pyrene (BaP) (CASRN 50-32-8). Available at http://www.epa.gov/subst/0136.htm (accessed January 2004).

Iwabuchi, T., Yamauchi, Y.I., Katsuta, A. and Harayama, S. 1998. Isolation and characterization of marine *Nocardioides* capable of growing and degrading phenanthrene at 42°C. *J. Marine Biotechnol., 6*: 86–90.

Johnsen, A.R., Wick, L.Y. and Harms, H. 2005. Principles of microbial PAH degradation in soil. *Environ. Poll., 133*: 71–84.

Juhasz, A.L. and Naidu, R. 2000. Bioremediation of high molecular weight polycyclic aromatic hydrocarbons: A review of the microbial degradation of benzo(a)pyrene. *Int. Biodeterior. Biodegrad., 45*: 57–88.

Kelly, I. and Cerniglia, C.E. 1991. The metabolism of phenanthrene by a species of *Mycobacterium*. *J. Ind. Microbiol., 7*: 19–26.

Khan, F., Husain, T. and Hejazi, R. 2004. An overview and analysis of site remediation techniques. *J. Environ. Manage., 71*: 95–122.

King, J.M.H., DiGrazia, P.M., Applegate, B., Burlage, R., Sanseverino, J., Dunbar, P., Larimer, F. and Sayler, G.S. 1990. Rapid, sensitive bioluminescent reporter technology for naphthalene exposure and biodegradation. *Science, 249*: 778–781.

Kiyohara, H. and Nagao, K. 1978. The catabolism of phenanthrene and naphthalene by bacteria. *J. Gen. Microbiol., 105*: 69–75.

Kiyohara, H., Nagao, K., Kouno, K. and Yano, K. 1982. Phenanthrene degrading phenotype of *Alcaligenes faecalis* AFK2. *Appl. Environ. Microbiol., 43*: 458–461.

Kiyohara, H., Sugiyama, M., Mondello, F.J., Gibson, D.T. and Yano, K. 1983. Plasmid involvement in the degradation of polycyclic aromatic hydrocarbons by *Beijerinckia* species. *Biochem. Biophys. Res. Commun., 111*: 939–945.

Kulakov, L.A., Allen, C.C., Lipscomb, D.A. and Larkin, M.J. 2000. Cloning and characterization of a novel cis-naphthalene dihydrodiol dehydrogenase gene (narB) from *Rhodococcus* sp. NCIMB12038. *FEMS Microbiol. Lett., 182 (2)*: 327–331.

Lageman, R., Clarke, R. and Pool, W. 2005. Electro-reclamation: A versatile soil remediation solution. *Eng. Geol., 77 (3–4)*: 191–201.

Lakhani, A. 2012. Source apportionment of particle bound polycyclic aromatic hydrocarbons at an industrial location in Agra, India. *Sci. World J., 2012*: 1–10.

Leahy, J.G. and Colwell, R.R. 1990. Microbial degradation of hydrocarbons in the environment. *Microbiol. Rev., 54*: 305–315.

Lee, B.-K. and Vu, V.T. 2010. Sources, distribution and toxicity of polycyclic aromatic hydrocarbons (PAHs) in particulate matter. In: Villanyi, V. (Ed.) *Air Pollution*. Sciyo Janeza Trdine 9, 51000 Rijeka, Croatia, pp. 99–122.

Lim, C.H.M., Ayoko, G.A. and Morawska, L. 2005. Characterization of elemental and polycyclic aromatic hydrocarbon compositions of urban air in Brisbane. *Atmos. Environ., 39*: 463–476.

Lim, L.H., Harrison, R.M. and Harrad, S. 1999. The contribution of traffic to atmospheric concentrations of polycyclic aromatic hydrocarbons. *Environ. Sci. Technol., 33*: 3538–3542.

Liu, W.X., Dou, H., Wei, Z.W., Chang, B., Qui, W.X., Liu, Y. and Shu, T. 2008. Emission characteristics of polycyclic aromatic hydrocarbons from combustion of different residential coals in North China. *Sci. Total Environ., 407*: 1436–1446.

Lliros, M.M., Munil, X., Sole, A., Martinez-Alonso, M., Diestra, E. and Esteve, I. 2003. Analysis of cyanobacteria biodiversity in pristine and polluted microbial mats in microcosmos by confocal laser scanning microscopy CLSM. In: Mendez-Vilas, A. (Ed.) *Science, Technology and Education of Microscopy: An Overview*. Badjz. Formt., pp. 483–490.

Mallick, S., Chatterjee, S. and Dutta, T.K. 2007. A novel degradation pathway in the assimilation of phenanthrene by *Staphylococcus* sp. strain PN/Y via meta-cleavage of 2-hydroxy-1-naphthoic acid: Formation of trans-2,3-dioxo-5-(29-hydroxyphenyl)-pent-4-enoic acid. *Microbiology, 153*: 2104–2115.

Menn, F.M., Applegate, B.M. and Sayler, G.S. 1993. NAH plasmid-mediated catabolism of anthracene and phenanthrene to naphthoic acid. *Appl. Environ. Microbiol., 59*: 1938–1942.

Miller, C.D., Hall, K., Liang, Y.N., Nieman, K., Sorensen, D. and Issa, B. 2004. Isolation and characterization of polycyclic aromatic hydrocarbon degrading *Mycobacterium* isolates from soil. *Microb. Ecol., 48*: 230–238.

Mishra, S., Jyot, J., Kuhad, R.C. and Lal, B. 2001. Evaluation of inoculum addition to stimulate *in-situ* bioremediation of oily-sludge contamination soil. *Appl. Environ. Microbiol., 674*: 1675–1681.

Mohanraj, R., Solaraj, G. and Dhana Kumar, S. 2011. 2.5 and PAH concentrations in urban atmosphere of Tiruchirappalli, India. *Bull. Environ. Contam. Toxicol., 87 (3)*: 330–335.

Morgan, P. and Watkinsons, R.J. 1989. In: Atlas, R.M. (Ed.) *CRC Critical Reviews in Biotechnology*, vol. 5. CRC Press, BOCA Raton FL, pp. 305–333.

Mueller, J.G., Cerniglia, C.E. and Pritchard, P.H. 1996. Bioremediation of environments contaminated by polycyclic aromatic hydrocarbons. In: Crawford, R.L. and Crawford, D.L. (Eds.) *Bioremediation: Principles and Applications*. Cambridge University Press, Idaho, pp. 125–194.

Muyzer, G. and, Smalla, K. 1998. Application of denaturing gradient gel electrophoresis (DGGE) and temperature gradient gel electrophoresis (TGGE) in microbial ecology. *Antonie Leeuwenhoek, 73*: 127–141.

Okpokwasili, G.C., Somerville, C.C., Grimes, D.J. and Colwell, R.R. 1984. Plasmid-associated phenanthrene degradation by Chesapeake Bay sediment bacteria. *Colloq. Inst. Fr. Rech. Explort. Mar., 3*: 601–610.

Parales, R.E., Bruce, N.C., Schmid, A. and Wackett, L.P. 2002. Biodegradation, biotransformation and biocatalysis (B3). *Appl. Environ. Microbiol., 68*: 4699–4709.

Park, S.S., Kim, Y.J. and Kang, C.H. 2002. Atmospheric polycyclic aromatic hydrocarbons. Pickering, R.W. 1999. A toxicological review of polycyclic aromatic hydrocarbons. *J. Toxicol. Cutan. Ocul. Toxicol., 18*: 101–135.

Penn, I., Vos, R., Schlarb, M. and Mitchell, L. 2002. *Eco-Industrial Strategies and Environmental Justice: An Agenda for Healthy Communities: Intergrating Brownfields and Eco-Industrial Development*. National Center for Eco-Industrial Development, Center for Economic Development, School of Policy, Planning and Development, University of Southern California.

Poppi, N.R. and Silva, M.S. 2005. Polycyclic aromatic hydrocarbons and other selected organic compounds in ambient air of Campo Grande City, Brazil. *Atmos. Environ., 39*: 2839–2850.

Prabhu, Y. and Phale, P.S. 2003. Biodegradation of phenanthrene by *Pseudomonas* sp. strain PP2: Novel metabolic pathway, role of biosurfactant and cell surface hydrophobicity in hydrocarbon assimilation. *Appl. Microbiol. Biotechnol., 61*: 342–351.

Ravindra, K., Bencs, L., Wauters, E., de Hoog, J., Deutsch, F., Roekens, E., Bleux, N., Bergmans, P. and Van Grieken, R. 2006. Seasonal and site specifc variation in vapor and aerosol phase PAHs over Flanders (Belgium) and their relation with anthropogenic activities. *Atmos. Environ., 40*: 771–785.

Renner, R. 1999. EPA to strengthen persistent, bioaccumulative and toxic pollutant controls—mercury first to be targeted. *Environ. Sci. Technol., 33*: 62A.

Resnick, S.M., Lee, K. and Gibson, G.T. 1996. Diverse reactions catalyzed by naphthalene dioxygenase from *Pseudomonas* sp. NCIB9816. *J. Ind. Microbiol. Biotechnol., 17*: 438–457.

Romine, M.F., Stillwell, L.C., Wong, K.K., Thurston, S.J., Sisk, E.C., Sensen, C., Gaasterland, T., Fredrickson, J.K. and Saffer, J.D. 1999. Complete sequence of a 184-kilobase catabolic plasmid from *Sphingomonas aromaticivorans* F199. *J. Bacteriol., 181*: 1585–1602.

Saito, A., Iwabuchi, T. and Harayama, S. 2000. A novel phenanthrene dioxygenase from *Nocardioides* sp. strain KP7: Expression in *Escherichia coli*. *J. Bacteriol., 182*: 2134–2141.

Samanta, S.K. and Jain, R.K. 2000. Evidence for plasmid-mediated chemotaxis of *Pseudomonas putida* towards naphthalene and salicylate. *Can. J. Microbiol., 46*: 1–6.

Samanta, S.K., Chakraborti, A.K. and Jain, R.K. 1999. Degradation of phenanthrene by different bacteria: Evidence for novel transformation sequences involving the formation of 1-naphthol. *Appl. Microbiol. Biotechnol., 53*: 98–107.

Sanseverino, J., Applegate, B.M., King, J.M.H. and Sayler, G.S. 1993. Plasmid mediated mineralization of naphthalene, phenanthrene, and anthracene. *Appl. Environ. Microbiol., 59*: 1931–1937.

Santos, H.F., Cury, J.C. and Carmo, F.L. 2011. Mangrove bacterial diversity and the impact of petroleum contamination revealed by pyrosequencing: Bacterial proxies for petroleum pollution. *PLoS One, 6 (3)*: e16943.

Schell, M.A. 1983. Cloning and expression in *Escherichia coli* of the naphthalene degradation genes from plasmid NAH7. *J. Bacteriol., 153 (2)*: 822–829.

Schell, M.A. 1985. Transcriptional control of the nah and sal hydrocarbon-degradation operons by the nahR gene product. *Gene, 36*: 301–309.

Schell, M.A. 1986. Homology between nucleotide sequences of promoter regions of nah and sal operons of NAH7 plasmid of *Pseudomonas putida*. *Proc. Natl. Acad. Sci. USA, 83*: 369–373.

Schell, M.A. and Wender, P.E. 1986. Identification of the nahR gene product and nucleotide sequences required for its activation of the sal operon. *J. Bacteriol., 166*: 9–14.

Shamsuzzaman, K.M. and Barnsley, E.A. 1974. The regulation of naphthalene metabolism in pseudomonads. *Biochem. Biophys. Res. Commun., 60 (2)*: 582–589.

Sharma, H., Jain, V.K. and Khan, Z.H. 2007. Characterization and source identification of polycyclic aromatic hydrocarbons (PAHs) in the urban environment of Delhi. *Chemosphere, 66 (2)*: 302–310.

Sharma, H., Jain, V.K. and Khan, Z.H. 2008. Atmospheric polycyclic aromatic hydrocarbons (PAHs) in the urban air of Delhi during 2003. *Environ. Monit. Assess., 147 (1–3)*: 43–55.

Simon, M.J., Osslund, T.D., Saunders, R., Ensley, B.D., Suggs, S., Harcourt, A., Suen, W., Cruden, D.L., Gibson, D.T. and Zylstra, G.J. 1993. Sequences of genes encoding naphthalene dioxygenase in *Pseudomonas putida* strains G7 and NCIB9816-4. *Gene, 127*: 31–37.

Tian, L., Ma, P. and Zhong, J.J. 2002. Kinetics and key enzyme activities of phenanthrene degradation by *Pseudomonas mendocina*. *Process Biochem., 37*: 1431–1437.

U.S. Department of Defence (U.S. DOD), Environmental Technology Transfer Committee. 1994. Remediation technologies screening matrix.

Van Hamme, J.D., Singh, A. and Ward, O.P. 2003. Recent advances in petroleum microbiology. *Microbiol. Mol. Biol. Rev., 67 (4)*: 503–549.

Widada, J., Nojiri, H., Kasuga, K., Yoshida, T., Habe, H. and Omori, T. 2002. Molecular detection and diversity of polycyclic aromatic hydrocarbon-degrading bacteria isolated from geographically diverse sites. *Appl. Microbiol. Biotechnol., 58*: 202–209.

Yang, Y., Chen, R.F. and Shiaris, M.P. 1994. Metabolism of naphthalene, Fluorene and phenan-threne: Preliminary characterization of a cloned gene cluster from *Pseudomonas putida* NCIB 9816. *J. Bacteriol., 176*: 2158–2164.

Yen, K.M. and Gunsalus, I.C. 1982. Plasmid gene organization: Naphthalene/salicylate oxidation. *Proc. Natl. Acad. Sci., USA, 79 (3)*: 874–878.

Yen, K.M. and Gunsalus, I.C. 1985. Regulation of naphthalene catabolic genes of plasmid NAH7. *J. Bacteriol., 162*: 1008–1013.

Yen, K.M. and Serdar, C.M. 1988. Genetics of naphthalene catabolism in *Pseudomonas. Crit. Rev. Microbiol.,* 15: 247–268.

You, I.S. and Gunsalus, I.C. 1986. Regulation of the nah and sal operons of plasmid NAH7: Evidence for a new function in nahR. *Biochem. Biophys. Res. Commun., 141 (3)*: 986–992.

Zakharian, R.A., Bakunin, K.A., Gasparian, N.S., Kocharian, Sh.M. and Arakelov, G.M. 1980. Plasmid characteristics of naphthalene and salicylate biodegradation in *Pseudomonas putida*. *Mikrobiologiia, 49 (6)*: 931–935.

Zhou, N.Y., Fuenmayor, S.L. and Williams, P.A. 2001. nag genes of Ralstonia (formerly *Pseudomonas*) sp. strain U2 encoding enzymes for gentisate catabolism. *J. Bacteriol., 183*: 700–708.

Zobell, C.E. 1946. Action of microorganisms on hydrocarbons. *Bacteriol. Rev.,* 10: 1–49.

Zuniga, M.C., Durham, D.R. and Welch, R.A. 1981. Plasmid and chromosome mediated dissimilation of naphthalene and salicylate in *Pseudomonas putida* PMD-1. *J. Bacteriol., 10/147 (3)*: 836–843.

Index

Page numbers followed by f and t indicate figures and tables, respectively.

Q

R

S